United States Nuclear Regulatory Commission

Protecting People and the Environment

NUREG-1829
Vol. 2

I0488112

Estimating Loss-of-Coolant Accident (LOCA) Frequencies Through the Elicitation Process

Appendices A through M

Manuscript Completed: March 2008
Date Published: April 2008

Prepared by
R. Tregoning (NRC), L. Abramson (NRC)
P. Scott (Battelle-Columbus)

A. Csontos, NRC Project Manager

Office of Nuclear Regulatory Research

ABSTRACT

The NRC is establishing a risk-informed revision of the design-basis pipe break size requirements in 10 CFR 50.46, Appendix K to Part 50, and GDC 35 which requires estimates of LOCA frequencies as a function of break size. Separate BWR and PWR piping and non-piping passive system LOCA frequency estimates were developed as a function of effective break size and operating time through the end of the plant license-renewal period. The estimates were based on an expert elicitation process which consolidated operating experience and insights from probabilistic fracture mechanics studies with knowledge of plant design, operation, and material performance. The elicitation required each member of an expert panel to qualitatively and quantitatively assess important LOCA contributing factors and quantify their uncertainty. The quantitative responses were combined to develop BWR and PWR total LOCA frequency estimates for each contributing panelist. The distributions for the six LOCA size categories and three time periods evaluated are represented by four parameters (mean, median, 5th and 95th percentiles). Finally, the individual estimates were aggregated to obtain group estimates, along with measures of panel diversity.

There is general qualitative agreement among the panelists about important technical issues and LOCA contributing factors, but the individual quantitative estimates are much more variable. Sensitivity studies were conducted to examine the effects on the estimated parameters of distribution shape, correlation structure, panelist overconfidence, panel diversity measure, and aggregation method. The group estimates are most sensitive to the method used to aggregate the individual estimates. Geometric-mean aggregation produces frequency estimates that approximate the medians of the panelists' estimates and also are generally consistent with both operating experience and prior LOCA frequency estimates, except where increases are supported by specific material aging-related concerns. However, arithmetic-mean and mixture-distribution aggregation are alternative methods that lead to significantly higher mean and 95th percentile group estimates. Because the results are sensitive to the aggregation method, a particular set of LOCA frequency estimates is not generically recommended for all risk-informed applications.

FOREWORD

Estimated frequencies of loss-of-coolant accidents (LOCAs; i.e., pipe ruptures as a function of break size) are used in a variety of regulatory applications, including probabilistic risk assessment (PRA) of nuclear power plants. Currently, the U.S. Nuclear Regulatory Commission (NRC) is using such information to establish a risk-informed alternative to the emergency core cooling system (ECCS) requirements in Title 10, Section 50.46, of the *Code of Federal Regulations* (10 CFR 50.46). Current requirements consider pipe breaks in the reactor coolant pressure boundary, up to and including breaks equivalent in size to the double-ended rupture of the largest pipe in the reactor coolant system. One aspect of this risk-informing activity is to evaluate the technical adequacy of redefining the design-basis break size (the largest pipe break to which 10 CFR 50.46 applies) to a smaller size that is consistent with updated estimates of pipe break frequencies.

To provide the technical basis for a risk-informed definition of the design-basis break size, this study developed LOCA frequency estimates using an expert elicitation process. This process consolidated operating experience and insights from probabilistic fracture mechanics studies with knowledge of plant design, operation, and material performance. Expert elicitation is a well-recognized technique for quantifying phenomenological knowledge when modeling approaches or data are insufficient.

The results from the expert elicitation provide LOCA frequency estimates for piping and non-piping passive systems, as a function of effective break size and operating time through the end of the plant license-renewal period, for both boiling- and pressurized-water reactors (BWRs and PWRs, respectively). The panelists generally agreed on the important technical issues and LOCA-contributing factors. However, as expected, the panelists' estimates exhibit both significant uncertainty and diversity. The uncertainty is reflected in the estimated parameters (mean, median, 5th and 95th percentiles) of the individual LOCA frequency distributions, and the diversity is captured by the confidence bounds on the group estimates. In addition, this study considered the sensitivity of the results to various analysis approaches. The results are most sensitive to the method used to aggregate the individual panelists' estimates to obtain group estimates. In this study, geometric-mean aggregation produces group frequency estimates that approximate the medians of the panelists' estimates and are also generally consistent with both operating experience and prior LOCA frequency estimates except where increases are supported by specific material aging-related concerns. However, arithmetic-mean and mixture-distribution aggregation are alternative methods that lead to significantly higher mean and 95th percentile group estimates.

Because the alternative aggregation methods can lead to significantly different results, a particular set of LOCA frequency estimates is not recommended for all risk-informed applications. The purposes and context of the application must be considered when determining the appropriateness of any set of elicitation results. In particular, during the selection of the BWR and PWR transition break sizes for the proposed 10 CFR50.46a rulemaking, the NRC staff considered the totality of the results from the sensitivity studies, rather than only the summary frequency estimates from this study. The NRC anticipates that a similar approach will be used in selecting appropriate replacement frequencies for the estimates provided in NUREG/CR-5750, "Rates of Initiating Events at U.S. Nuclear Power Plants: 1987 – 1995," and for other applications that require frequencies for break sizes other than those in NUREG/CR-5750.

Jennifer L. Uhle, Director
Division of Engineering
Office of Nuclear Regulatory Research
U.S. Nuclear Regulatory Commission

TABLE OF CONTENTS

LIST OF APPENDICES

LIST OF FIGURES

LIST OF TABLES

EXECUTIVE SUMMARY

The emergency core cooling system (ECCS) requirements are contained in 10 CFR 50.46, Appendix K to Part 50, and GDC 35. Specifically, ECCS design, reliability, and operating requirements exist to ensure that the system can successfully mitigate postulated loss-of-coolant accidents (LOCAs). Consideration of an instantaneous break with a flow rate equivalent to a double-ended guillotine break (DEGB) of the largest primary piping system in the plant generally provides the limiting condition in the required 10 CFR Part 50, Appendix K analysis. However, the DEGB is widely recognized as an extremely unlikely event. Therefore, the staff is establishing a risk-informed revision of the design-basis break size requirements for operating commercial nuclear power plants.

A central consideration in selecting a risk-informed design basis break size is an understanding of the LOCA frequency as a function of break size. The most recent NRC-sponsored study of pipe break failure frequencies is contained in NUREG/CR-5750. Unfortunately, these estimates are not sufficient for design basis break size selection because they do not address all current passive-system degradation concerns and they do not discriminate among breaks having effective diameters greater than 6 inch. There have been two approaches traditionally used to assess LOCA frequencies and their relationship to pipe size: (i) estimates based on statistical analysis of operating experience and (ii) probabilistic fracture mechanics (PFM) analysis of specific postulated failure mechanisms. Neither approach is particularly suited to evaluate LOCA event frequencies due to the rareness of these events and the modeling complexity. In this study, LOCA frequency estimates have been calculated using an expert elicitation process to consolidate operating experience and insights from PFM studies with knowledge of plant design, operation, and material performance. This process is well-recognized for quantifying phenomenological knowledge when data or modeling approaches are insufficient.

The principal objective of this study was to develop separate boiling water reactor (BWR) and pressurized water reactor (PWR) piping and non-piping passive system LOCA frequency estimates as a function of effective break size at three distinct time periods: current-day (25 years fleet average), end-of-plant-license (40 years fleet average), and end-of-plant-license-renewal (60 years fleet average). These estimates are based on the responses from an expert panel and one aim of this study was to obtain estimates that represent a type of group consensus. Additionally, another objective was to reflect both the uncertainty in each panelist's estimates as well as the diversity among the individual estimates.

The elicitation focused on developing generic, or average, estimates for the commercial fleet and the uncertainty bounds on these generic estimates rather than bounding values associated with one or two plants. This approach is consistent with prior studies that did not consider plant-specific differences in developing LOCA frequencies for use in probabilistic risk assessment (PRA) modeling. Consequently, the elicitation panelists were instructed to consider broad differences among plants related to important variables (i.e., plant system, material, geometry, degradation mechanism, loading, mitigation/maintenance) in determining both the generic LOCA frequencies and especially the estimated uncertainty bounds. It is the broad differences in these important variables that contribute most to passive system failure and there is generally sufficient commonality among plants to make such a generic assessment valuable.

The elicitation was solely focused on determining LOCA frequencies that initiate by unisolable primary system side failures that can be exacerbated by material degradation with age. Therefore, active system failures (e.g., stuck open valve, pump seals, interfacing system LOCAs) and consequential primary pressure boundary failures due to either secondary side failures or failures of other plant structures (e.g., crane drops) were not considered. Active system frequency contributions should be combined with the passive system LOCA frequencies to estimate the total system risk. The effects of safety culture on primary side failure (i.e., LOCA) frequencies were also considered.

This study developed LOCA frequency estimates consistent with historical small break (SB), medium break (MB), and large break (LB) flow rate definitions. Additionally, three larger LOCA categories were defined within the classical LB LOCA regime to examine trends with increasing break size, up to and including, a DEGB of the largest piping system in the plant. Contrary to earlier studies, the six LOCA categories are defined in terms of cumulative thresholds rather than break intervals because this definition was more conducive to the elicitation structure. However, the differences between the interval and cumulative estimates are typically much smaller than the estimates' uncertainties. It is therefore recommended that the cumulative threshold estimates from any set of tabular results be used (e.g., Table 1) if interval-defined LOCA frequencies are desired.

Because the LOCA frequency estimates were intended to be both generic and consistent with historical internal-event PRAs, the elicitation primarily considered normal plant operational cycles and loading histories. The loads include representative constant stresses (e.g. pressure, thermal, residual) and expected transient stresses (e.g. thermal striping, heat-up/cool-down, pressure transients) that occur over the license-renewal period. The elicitation implicitly considered all modes of operation based on the loading or operational experience associated with each piping system or non-piping component. Rare event loading from seismic, severe water hammer, and other sources was also not considered in this generic evaluation because of their strong dependency on plant-specific factors. However, a separate research study was conducted to assess the potential impact of seismic loading on the break frequency versus break size relationship. As part of that study, both unflawed and flawed seismic piping contributions were considered. The results of that seismic LOCA analysis are summarized in Section 7.2.

Several important assumptions were made to guide the elicitation process. One such assumption is that plant construction and operation comply with all applicable codes and standards required by the regulations and the technical specifications. The specific impact on the LOCA frequencies of purposefully violating these requirements was not considered. However, it was assumed that regulatory oversight policies and procedures will continue to be used to identify and mitigate risk associated with plants having deficient safety practices. While deviations from these requirements do represent some percentage of the events included in the passive-system failure data, extrapolation of this data implicitly assumes that similar future deviations will continue to occur with similar frequency.

Another important assumption is that current regulatory oversight practices will continue to evaluate aging management and mitigation strategies in order to reasonably assure that future plant operation and maintenance has equivalent or decreased risk. A related assumption inherent in this elicitation is that all future plant operating characteristics will be essentially consistent with past operating practice. The effects of operating profile changes were not considered because of the large uncertainty surrounding possible operational changes and the potentially wide-ranging ramifications of significant changes on the underlying LOCA frequencies.

The expert elicitation process employed in this study is an adaptation of the formal expert judgment processes used to evaluate reactor risk (NUREG-1150), to develop seismic hazard curves (NUREG/CR-6372), and to assess the performance of radioactive waste repositories (NUREG/CR-5411). The process consisted of a number of steps. To begin, the project staff identified many of the issues to be evaluated through a pilot elicitation. The panel members for this pilot elicitation were all NRC staff. A group of twelve panel members was then selected for the formal elicitation. The staff gathered background material and prepared an initial formulation of the technical issues which was provided to the panel. At its initial meeting, the panel discussed the issues and, using the staff formulation as a starting point, developed a final formulation for the elicitation structure. This structure included the decomposition of the complex technical issues which impact LOCA frequencies into fundamental elements so that these important contributing factors could be more readily assessed. Piping and non-piping base cases were also defined for use in anchoring the quantitative elicitation responses. The base cases represent a set of well-defined conditions which could cause a LOCA. A subset of the panel was created to develop quantitative LOCA frequencies estimates associated with the base case conditions. At this initial meeting, the panel was also trained in subjective elicitation of numerical values through exercises and discussion of potential biases.

After this initial meeting, the staff prepared a draft elicitation questionnaire and iterated with the panel to develop the final questionnaire. The panelists quantifying the base case conditions also developed their initial estimates. A second meeting was then held with the entire panel to review the base case results, review the elicitation questions, and finalize the formulation of remaining technical issues. At their home institutions, the individual panel members performed analyses and computations to develop answers to the elicitation questionnaire.

The elicitation questionnaire required panelists to assess the following technical areas: the base case evaluation effort, utility and regulatory safety culture effects on LOCA frequencies, piping system LOCA frequencies, and non-piping system LOCA frequencies. The utility and safety culture questions required the panelists to compare future safety culture with the existing culture and predict the effect on LOCA frequencies. These effects were considered separately from passive system degradation because the panelist judged safety culture and degradation to be independent. The base case evaluation required the panelists to assess the accuracy and uncertainty in the base case analyses, and to also choose a particular base case approach for anchoring their elicitation responses. The piping and non-piping LOCA frequency questions required each panelist to first identify important LOCA contributing factors (i.e., piping systems, materials, degradation mechanisms, etc.) and select appropriate base case conditions for comparison. The panelists were then required to provide the relative ratios between their important contributing factors and the base case conditions based on their knowledge of passive system component failure. Each relative comparison required mid value, upper bound, and lower bound values. The mid value is defined such that, in the panelist's judgment, there is a 50% chance that the unknown true answer lies above the mid value. The upper and lower bounds are defined such that there is a 5% chance that the true answer lies above the upper bound or below the lower bound, respectively. Each panelist was also required to provide their qualitative rationale supporting their quantitative values.

A facilitation team consisting of individuals knowledgeable about the technical issues (substantive members), a staff member with extensive experience conducting expert elicitations (normative member) and two recorders met separately with each panel member in day-long individual elicitation sessions. At these sessions, each panel member provided answers to the elicitation questionnaire along with their supporting technical rationales. The panelists were asked to self-select, based on their expertise, the questions that they addressed. Consequently, several panelists only provided responses for either BWR or PWR plants. After the elicitation sessions, the panel members returned to their home institutions where they refined their responses based on feedback obtain during their session. Upon receipt of the updated responses, the project staff compiled the panel's responses and developed preliminary estimates of the LOCA frequencies. Along with the rationales, these preliminary estimates were presented to the panel at a wrap-up meeting. Panel members were invited to fill in gaps in their questionnaire responses and, if desired, to modify any of their responses based on group discussion of important technical issues considered during the individual elicitations[1]. Final individual estimates of the LOCA frequencies were then calculated and provided to the panel members for final review and quality assurance.

The qualitative insights provided by the panel are reasonably consistent. Panelists identified several advantages and disadvantages with determining the base case LOCA frequencies through operating-experience assessment or PFM analyses. However, there was a general consensus that operating experience provides the best basis for evaluating current-day, small-break LOCA estimates. Hence, many panelists used the operating-experience base cases to anchor their elicitation responses. The panel members also generally expressed the opinions that the future safety culture will not differ dramatically from the current culture, and that the utility and regulatory safety cultures are highly correlated. Many panelists do believe that safety culture can significantly affect LOCA frequencies at specific plants, but there is also an expectation that regulatory actions using existing enforcement measures will diminish both the possibility and impact of deficient safety culture at particular plants. However because it was thought that these plant-specific issues do not affect the generic averages, no specific adjustment to the LOCA frequency estimates was applied to explicitly account for safety culture effects. This decision was endorsed by the elicitation panelists.

There were several technical insights that were consistently identified. Many participants believe that the number of precursor events (e.g., cracks and leaks) is generally a good barometer of the LOCA susceptibility for the associated degradation mechanism. Welds are almost universally recognized as likely failure locations because they can have relatively high residual stress, are preferentially-attacked by many degradation mechanisms, and are most likely to have preexisting fabrication defects. Most panelists also agreed that a complete break of a smaller pipe, or non-piping component, is generally more likely than an equivalent size opening in a larger pipe, or component, because of the increased severity of fabrication or service cracking. Therefore, the biggest frequency contributors for each LOCA size tend to be systems having the smallest pipes, or component, which can lead to that size LOCA. The exception to this general rule is the BWR recirculation system, which is important at all LOCA sizes due to IGSCC. Many panelists thought that aging may have the greatest effect on intermediate diameter (i.e., 6 to 14-inch nominal diameter) piping systems due to the large number of components within this size range and the fact that this piping generally receives less attention than the larger diameter piping and is harder to replace than the more degradation-prone smaller diameter piping.

The participants generally identified thermal fatigue, stress corrosion cracking (SCC), flow accelerated corrosion (FAC), and mechanical fatigue as the degradation mechanisms that most significantly contribute to LOCA frequencies in BWR plants. Generally, the most important BWR degradation mechanism is intergranular SCC (IGSCC), although the panelist's recognize that mitigation has greatly reduced the susceptibility of BWR plants to this mechanism over the past 20 years. With the exception of FAC, similar degradation mechanisms and concerns were also deemed to be important in PWR plants. Specifically, primary water SCC (PWSCC) is a principal concern. Many panelists believe that PWSCC will be mitigated in PWR plants over the next 15 years, but that effective mitigation has yet to be developed and implemented.

The panelists generally agreed on the important technical issues and LOCA-contributing factors. However, the individual quantitative responses are much more uncertain and there are relatively large differences among the panelists' responses. This is to be expected given the underlying scientific uncertainty. The analysis of these responses was structured to account for the uncertainty in the

---

[1] Each panelist's quantitative elicitation responses can be found through the "Electronic Reading Room" link on the NRC's public website (http://www.nrc.gov/) using the Agencywide Documents Access and Management System (ADAMS). The document is found in ADAMS using the following accession number: ML080560005.

individual estimates and the diversity among the panelists. The quantitative responses were analyzed separately for each panel member to develop individual BWR and PWR total LOCA frequency estimates. A unified analysis format was developed to ensure consistency in processing the panelists' inputs. The panelists' mid-value, upper bound and lower bound responses were assumed to represent the median, 95th, and 5th percentiles, respectively, of their subjective uncertainty distributions for each elicitation response. The analysis structure was based on the assumption that all the responses correspond to percentiles of lognormal distributions. These distributions were then combined using a lognormal framework. The final outputs for each panelist are estimates for the means, medians and 5th and 95th percentiles of the total BWR and PWR LOCA frequencies. The panelists' estimates were then aggregated to obtain group LOCA frequency estimates, along with measures of panel diversity.

The individual and group estimates for the means, medians, 5th and 95th percentiles of the LOCA frequency distributions were calculated using the following principal assumptions and choices:

(i)   The mid-value, upper bound, and lower bound supplied by each panelist for each elicitation question were assumed to correspond to the median, 95th percentile and 5th percentile, respectively, of a split lognormal distribution, with the mean calculated assuming that the upper tail is truncated at the 99.9th percentile.

(ii)   Only those panelists whose uncertainty ranges were relatively small were adjusted using an error-factor adjustment scheme to account for possible overconfidence (Section 7.6.2.2).

(iii) Split lognormal distributions were summed by assuming perfect rank correlation among the individual terms.

(iv)   The individual estimates are the total LOCA frequency parameters (i.e., mean, median, 5th percentile, and 95th percentile) determined for each panelist.

(v)   The group estimates of the total LOCA frequency parameters were determined using the geometric means of the individual estimates.

(vi)   Panel diversity was characterized with two-sided 95% confidence intervals based on an assumed lognormal model for the individual estimates.

In the report, these six assumptions and choices define what are termed the *summary* estimates. The report also calculates and discusses *baseline* estimates. These baseline estimates are calculated using all the above assumptions, except no overconfidence adjustment (as described in (ii) above) is applied. Because it is well-established that experts tend to be overconfident, the overconfidence adjustment is deemed to result in improved LOCA frequency estimates

The resultant individual and group summary estimates are consistent with the elicitation objectives and structure and are reasonably representative of the panelists' quantitative judgments. In particular, they are not dominated by extreme results, either on the high or low end, and the geometric means of the individual estimates approximate the medians of these estimates. The median is often used to represent group opinion in elicitations, especially when the individual estimates differ by several orders of magnitude, as they do in this study.

The LOCA frequency summary estimates for the current-day (25 years) and end-of-plant-license (40 years) periods are provided in Table 1 for both BWR and PWR plant types. The aggregated group estimates for the median, mean, 5th and 95th percentiles are presented. Frequency estimates are not expected to change dramatically over the next 15 years for any size LOCA, or even over the next 35 years for BWR LOCA Categories 1- 5 and PWR LOCA Categories 1 - 2 (see results in Appendix L). While order of magnitude increases in BWR LOCA Category 6 and PWR LOCA Categories 3 - 6 over the next 35 years are expected, these increases are largely due to uncertainty about the future and the concern that new degradation mechanisms could arise in the operating fleet. However, while aging will continue, the panelists' consensus is that mitigation procedures are in place, or will be implemented in a timely manner, to alleviate significant increases in future LOCA frequencies for existing degradation mechanisms. Because of the predicted stability in these estimates over the near-term, the current-day (25 year) results can be used to represent the LOCA frequencies over the next 15 years of fleet operation.

The current-day median, mean, and 95th percentile estimates are graphically presented in Figure 1. The 95% confidence intervals calculated for these parameters are also illustrated in this figure. The LOCA frequencies as a function of threshold break diameter were estimated only for the six specified LOCA categories in the elicitation. The plotted points are connected with straight lines in the figure for visual clarity, but this should not be construed as a recommended interpolation scheme. Interpolation of frequencies between category sizes can be done at the user's discretion depending on the conservatism required by the application. Some common interpolation schemes are linear, multi-point nonlinear and cubic spline. A step-wise or stair-step interpolation between two categories where the frequency for the lower category size is used for all flow rates or corresponding break sizes between the two categories provides the most conservative interpolation scheme. Note that any interpolation scheme does not reflect the uncertainty in the interpolated frequencies.

A measure of the individual uncertainties in Table 1 and Figure 1 is given by the differences between the medians and the corresponding 5th or 95th percentile estimates. Panel diversity is reflected in the confidence bounds in Figure 1. The large widths of these confidence bounds (as much as 3 orders of magnitude) reflects the significant diversity of the individual estimates in this study As the LOCA size increased, the panel members generally expressed greater uncertainty in their predictions, and the variability among individual panelists' estimates increased. This is to be expected because of the increased extrapolation required from available passive-system failure data for larger LOCA sizes.

While it is acknowledged that operating experience-based estimates do not necessarily reflect the current state, it is informative to compare such estimates with the elicitation frequencies for the smallest LOCAs (Category 1). This comparison requires the least extrapolation of passive-system failure data since no larger LOCAs have occurred. The BWR and PWR Category 1 LOCA frequencies (including the steam generator tube rupture (SGTR) frequency for PWRs) were estimated up through December 2006. For BWR plants, the average SB LOCA frequency based solely on the number of reported events was estimated to be 5.5E-04 per calendar year. The mean elicitation BWR SB LOCA estimate is 6.5E-04 per calendar year. The BWR elicitation estimate is less than 20 percent higher than the operating-experience estimate which, given the uncertainty of these estimates, is not statistically significant.

For PWR plants, the average SB LOCA frequency estimated analogously is 3.6E-03 per calendar year. The mean elicitation PWR SB LOCA estimate is 7.3E-03 per calendar year. The PWR elicitation estimate is about 100 percent (or a factor of two) higher than the operating-experience-based estimate. Additional insight into this difference can be gained by partitioning the PWR passive-system failure data into frequencies for SGTRs and for all other passive-system SB LOCAs.

Based on reported failures, the mean SGTR LOCA frequency is 3.2E-03 per calendar year. This result is almost identical to the current-day elicitation estimate of 3.7E-03 per calendar year. The frequency of all other Category 1 PWR passive-system failures is 4.0E-04 per calendar year based on operating experience. The corresponding elicitation estimate is 1.9E-03 per calendar year. While this value is 5 times greater than the operating-experience-based estimate, this difference is explained by the elicitation panelists' estimation of the effect of PWSCC on small diameter component failures.

There are several prior studies that also estimated LOCA frequencies. Some care is needed when comparing the elicitation LOCA frequency estimates with these earlier studies because the LOCA categories are defined differently. Specifically, the current-day LOCA Category 1 and 2 estimates in Table 1 are comparable to total system SB, MB, and LB LOCA frequencies, respectively, reported in

NUREG/CR-5750. Additionally, current-day SGTR frequencies (Table 7.18) and PWR LOCA frequencies for all other passive system failures (i.e., frequencies for breaks greater than 100 gpm (380 lpm) in Table 7.19) are comparable to NUREG/CR-5750 SGTR and PWR SB LOCA frequencies when these failure modes are analyzed separately. The NUREG/CR-5750 LB LOCA frequency estimates are best compared to the elicitation LOCA Category 4 frequency estimates because the pipe break sizes are most similar.

After accounting for these differences, the elicitation LOCA frequency estimates are generally much lower than the WASH-1400 estimates and more consistent with the NUREG/CR-5750 estimates. The SB LOCA PWR elicitation estimates after subtracting the SGTR frequencies are approximately 3 times greater than the NUREG/CR-5750 estimates, due to the aforementioned PWSCC concerns. However, the total BWR and PWR SB LOCA frequency estimates are similar once the SGTR frequencies are added to the NUREG/CR-5750 PWR results. The elicitation MB LOCA estimates are higher than the NUREG/CR-5750 estimates by factors of approximately 4 and 20 for BWR and PWR plant types, respectively. These increases are partly due to concerns about PWSCC of piping and non-piping (e.g., CRDM) components as well as general aging concerns with piping in this size range. The NUREG/CR-5750 LB LOCA frequency estimates are slightly higher (less than a factor of 3) than the current elicitation results for both PWR and BWR plants. The generally good agreement between the NUREG/CR-5750 and current elicitation estimates is somewhat surprising given the markedly different methodologies used to arrive at these results.

Table 1 Total BWR and PWR LOCA Frequencies
(After Overconfidence Adjustment using Error-Factor Scheme)

Plant Type	LOCA Size (gpm)	Eff. Break Size (inch)	Current-day Estimate (per cal. yr) (25 yr fleet average operation)				End-of-Plant-License Estimate (per cal. yr) (40 yr fleet average operation)			
			5th Per.	Median	Mean	95th Per.	5th Per.	Median	Mean	95th Per.
BWR	>100	½	3.3E-05	3.0E-04	6.5E-04	2.3E-03	2.8E-05	2.6E-04	6.2E-04	2.2E-03
	>1,500	1 7/8	3.0E-06	5.0E-05	1.3E-04	4.8E-04	2.5E-06	4.5E-05	1.2E-04	4.8E-04
	>5,000	3 ¼	6.0E-07	9.7E-06	2.9E-05	1.1E-04	5.4E-07	9.8E-06	3.2E-05	1.3E-04
	>25K	7	8.6E-08	2.2E-06	7.3E-06	2.9E-05	7.8E-08	2.3E-06	9.4E-06	3.7E-05
	>100K	18	7.7E-09	2.9E-07	1.5E-06	5.9E-06	6.8E-09	3.1E-07	2.1E-06	7.9E-06
	>500K	41	6.3E-12	2.9E-10	6.3E-09	1.8E-08	7.5E-12	4.0E-10	1.0E-08	2.8E-08
PWR	>100	½	6.9E-04	3.9E-03	7.3E-03	2.3E-02	4.0E-04	2.6E-03	5.2E-03	1.8E-02
	>1,500	1 5/8	7.6E-06	1.4E-04	6.4E-04	2.4E-03	8.3E-06	1.6E-04	7.8E-04	2.9E-03
	>5,000	3	2.1E-07	3.4E-06	1.6E-05	6.1E-05	4.8E-07	7.6E-06	3.6E-05	1.4E-04
	>25K	7	1.4E-08	3.1E-07	1.6E-06	6.1E-06	2.8E-08	6.6E-07	3.6E-06	1.4E-05
	>100K	14	4.1E-10	1.2E-08	2.0E-07	5.8E-07	1.0E-09	2.8E-08	4.8E-07	1.4E-06
	>500K	31	3.5E-11	1.2E-09	2.9E-08	8.1E-08	8.7E-11	2.9E-09	7.5E-08	2.1E-07

Sensitivity analyses were also conducted to examine the robustness of the quantitative results to the analysis procedure used to develop the summary estimates. These sensitivity analyses investigated the effect of distribution shape on the means as well as the effects of correlation structure, panelist overconfidence, panel diversity measure, and aggregation method on the estimated parameters. The mean calculation in the analysis procedure used a split lognormal distribution truncated at the 99.9th percentile to obtain reasonably conservative values compared with other possible choices. The correlation structure in the analysis procedure assumed maximal correlation, which is reasonably representative of the elicitation structure. The structure provides conservative 95th percentile estimates. However, based on selected Monte Carlo simulations, an independent correlation structure leads to larger median and 5th percentile estimates. The means are unaffected by the choice of the correlation structure. The analysis procedure also used confidence intervals for the aggregated estimates to measure panel diversity. An alternative approach used quartiles of the individual estimates, leading to comparable, but narrower intervals.

The analysis procedure also adjusted those panelists' responses that had relatively narrow uncertainty bands using an error-factor scheme to account for a known tendency for people, including experts, to be overconfident when making subjective judgments. Sensitivity analyses examined the effects of other overconfidence adjustments of the nominal subjective confidence levels supplied by the panelists. No overconfidence adjustment was also investigated. While blanket overconfidence adjustments can result in large, unsupportable increases in the mean and 95th percentile frequency estimates, no adjustment results in modest decreases in these estimates. Therefore, the error-factor scheme, which adjusts only those panelists who are most overconfident, is deemed to the most appropriate.

Finally, the largest sensitivity is associated with the method used to aggregate the individual panelist estimates to obtain group estimates. The baseline and summary estimates were developed using geometric-mean aggregation. In this study, the geometric-mean aggregation produces frequency estimates that approximate the median of the panelists' estimates and therefore effectively leads to consensus-type results. Therefore, the summary estimates in Table 1 are believed to be a reasonable representation of the expert panel's current state of knowledge regarding LOCA frequencies.

The sensitivity analyses evaluated the effect of using alternative aggregation methods to calculate the group estimates. Specifically, mixture-distribution and arithmetic-mean aggregation were evaluated. For the panelists' responses in this study, these alternative aggregation methods can lead to significantly higher mean and 95th percentile estimates than those obtained using geometric-mean aggregation. Alternative LOCA frequency estimates that are higher than the summary estimates (Table 1) can be derived by using either the summary estimates with 95% confidence bounds (Tables 7.8), the arithmetic-mean aggregated results (Table 7.13), or the mixture-distribution results (Table 7.16). These estimates also incorporate the same overconfidence adjustment as the summary estimates.

Because alternative aggregation methods can lead to significantly different results, a particular set of LOCA frequency estimates is not generically recommended for all risk-informed applications. The purposes and context of the application must be considered when determining the appropriateness of any set of elicitation results. In particular, during the selection of the BWR and PWR transition break sizes for the proposed 10CFR50.46a rule making, the NRC staff considered the totality of the results from the sensitivity studies rather than only the summary estimates from this study The NRC anticipates that a similar approach will be used in selecting appropriate replacement frequencies for NUREG/CR-5750 estimates and for other applications where frequencies for break sizes other than those in NUREG/CR-5750 are required. While the lack of clear application guidance places an additional burden on the users of the study results, those users are in the best position to judge which study results are most appropriate to consider for their particular applications.

BWR: Error Factor Adjustment

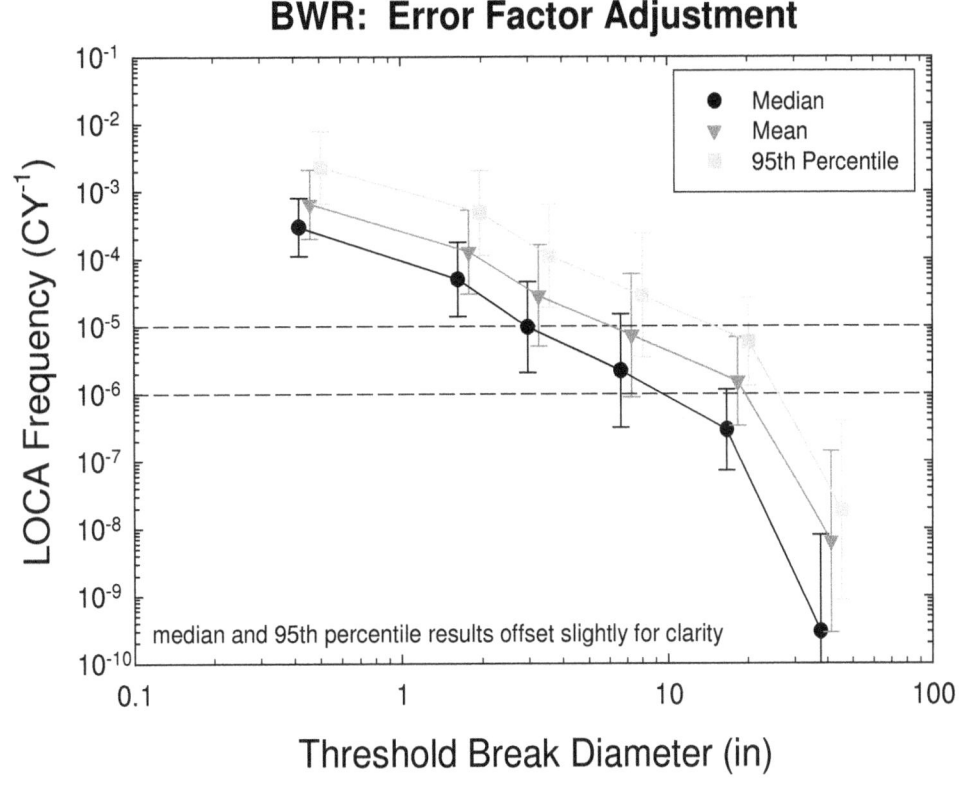

PWR: Error Factor Adjustment

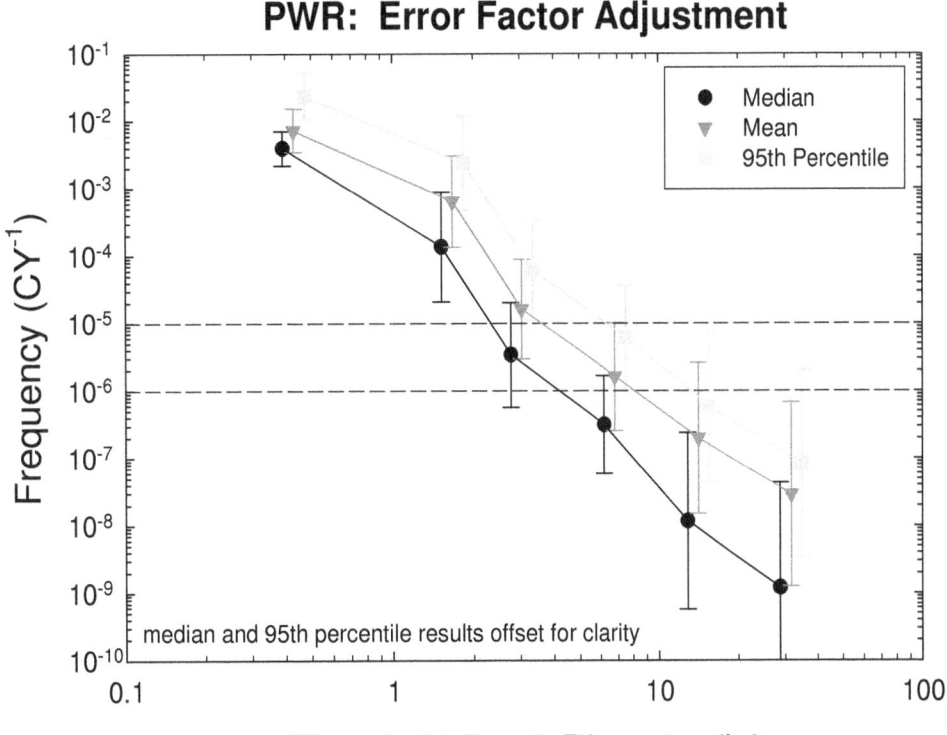

Figure 1 BWR and PWR Error-Factor Adjusted LOCA Frequency Estimates

ACKNOWLEDGMENTS

This work was supported by the United States Nuclear Regulatory Commission (NRC) through the Component Integrity Branch of the Division of Engineering of the Office of Nuclear Regulatory Research under several different contracts. Battelle Memorial Institute support was provided under NRC Contract NRC-04-02-074 (NRC Job Code Y6538). Mr. William Galyean of Idaho National Engineering Environmental Laboratory was supported under NRC Job Code Y6332. Dr. Cory Atwood was supported under NRC Job Code Y6492. Dr. Alan Brothers was supported under NRC Job Code Y6604.

The authors would first like to thank the members of the expert elicitation panel whose insights formed the basis for the results and conclusions reached in this report. The elicitation panel members also provided valuable editorial and technical comments that have been incorporated into this summary document. The panel members included:

- Mr. Bruce Bishop   Westinghouse Electric Co. LLC
- Dr. Vic Chapman   OJV Consultancy Limited
- Mr. Guy DeBoo   Exelon Nuclear
- Mr. William Galyean   Idaho National Engineering Environmental Laboratory
- Dr. Karen Gott   Swedish Nuclear Power Inspectorate
- Dr. David Harris   Engineering Mechanics Technology, Inc.
- Mr. Bengt Lydell   ERIN® Engineering and Research, Inc.
- Dr. Peter Riccardella   Structural Integrity Associates, Inc.
- Mr. Helmut Schulz   Gesellschaft für Reaktorsicherheit (GRS) mbh
- Dr. Sampath Ranganath   Formerly GE Nuclear Energy/Now XGEN Engineering
- Dr. Fredric Simonen   Pacific Northwest National Laboratory
- Dr. Gery Wilkowski   Engineering Mechanics Corporation of Columbus

A special debt of gratitude is expressed to the base case team members (Dr. David Harris, Mr. Bengt Lydell, Mr. William Galyean, and Dr. Peter Riccardella) for the extra effort they provided to conduct the base case and sensitivity analyses discussed in Sections 3 and 4 of this report. In addition, Dr. Riccardella provided base case frequencies for a series of non-piping components and Mssrs. Galyean and Lydell developed the non-piping precursor database. Summaries of the base case team members' analyses are contained in Appendices D – I. Each appendix has been written by the responsible base case team member and these contributions are much appreciated and are vital to the technical basis of this NUREG. Mr. Lydell and Dr. Harris also tirelessly discussed and corrected the authors' characterization of their efforts for accuracy and clarity.

The authors would also like to recognize the facilitation team members who assisted with the development of the technical issues and associated elicitation questions and who also participated in the individual elicitations sessions. Along with the authors, Dr. Ken Jacquay of Casco Services, Inc. was a principal member of the facilitation team

Other members included:
- Ms. Bennett Brady   NRC
- Mr. Frank Cherny   NRC
- Mr. Alan Kuritzky   NRC
- Mr. Arthur Salomon   NRC

The authors are also appreciative of Dr. Corwin Atwood of Statwood Consulting and Dr. Alan Brothers of Pacific Northwest National Laboratory, who provided an external peer review of the elicitation approach and analysis of results. They suggested a number of sensitivity studies to further validate and clarify the conclusions reached in this report. They also provided many helpful comments that have been used to clarify this report. Dr. Atwood, in particular, provided many valuable insights which were used to develop and refine the techniques used to analyze the elicitation results.

Many NRC staff also provided valuable comments and contributions that were used to revise the draft report, including Dr. Arthur Buslik, Mr. Stephen Dinsmore, Mr. Gary Hammer, Mr. Glenn Kelly, and Mr. Arthur Salomon,. In particular, the authors would like to thank Dr. Arthur Buslik for his contribution to Section 5.6.4.4 and Mr. Arthur Salomon for his reviewing and editing assistance. The authors are also grateful to Dr. David Rudland of Engineering Mechanics Corporation of Columbus for assisting with selected Monte Carlo calculations to evaluate the analysis of individual panelist responses.

Finally, the authors would like to thank Dr. Al Csontos of the NRC, who served as the program manager over the final months of this program, and to Ms. Charlotte Matthews and Ms. Patricia Zaluski, both of Battelle-Columbus, for their invaluable assistance in the preparation of this report.

NOMENCLATURE

1. Symbols

a	Flaw depth
a^*	Crack depth having 50% chance of being detected
b	Value of the upper bound supplied by a panelist
b'	Value of the lower bound supplied by a panelist
$b_p(Y)$	pth percentile of distribution Y
C	Parameter in fatigue crack growth relationship
C_1, C_2	Coefficients used in probability of detection curve definition
D	Diameter
DN	Nominal pipe diameter
E	Young's modulus
$E(Y)$	Expected value of distribution Y
$EF(Y)$	Error factor of distribution Y
f	Fraction of welds inspected for cracks or flaws
f	Inspection coverage/scope
g	Median of the lognormal distribution U
g^*	Group estimate of g
J	J-integral fracture parameter
J_D	Deformation J
J_{Ic}	Plane strain J at crack initiation by ASTM813
J_M	Modified J
$J\text{-}R$	J-resistance
k_a	p_ath percentile of the standard lognormal distribution
k_p	pth percentile of the standard lognormal distribution
K	Stress intensity factor
K_I	Mode I stress intensity factor
K_{Ic}	Plane strain stress intensity factor at crack initiation
f_{POD}	Probability of detection function
I	ISI effectiveness factor
L	Length
L_{BC}	Likelihood of a leak due to any degradation mechanism (base case)
L_{PL}	Likelihood of a perceptible leak
L_{TSL}	Likelihood of a technical specification leak
L_{50}	Likelihood of a crack 50% through wall deep
m	Median value
m	Parameter in fatigue crack growth relationship
$m(Y)$	Mean of distribution Y
n	Number of panelists
n_c	Number of cracks or flaws
n_f	Number of failure events
N	Normal operating stress
N	Number of components
N	Number of stress cycles
p	Percentile
P	Pressure
p_a	Value of the percentile in the adjusted distribution corresponding to the original error factor
P_{BC}	Conditional failure probability for the chosen seismic piping base case

P_{FD}	Probability inspected welds will find existing flaw
P_{ND}	Probability of not detecting a crack
P_{PL}	Conditional failure probability of a crack that has just formed a perceptible leak
P_{TSL}	Conditional failure probability of a crack leaking at the technical specification limit
$P_{TSL@SLB}$	Conditional failure probability of a crack leaking at the technical specification limit assuming a Service Level B load
$P_{TSL@SLD}$	Conditional failure probability of a crack leaking at the technical specification limit assuming a Service Level D load
P_{50}	Conditional failure probability of a crack with a maximum depth of 50% of the wall thickness
$p_{L/F}$	Conditional failure probability
p'	Value of the percentile of the panelist's assumed lognormal distribution corresponding to b'
r	Error factor
r_p	Error factor of the panelist's adjusted lognormal distribution
R	Stress ratio
$R_{C/F}$	Number of non-through-wall cracks per leak event
S	Sum of cyclic stress
S	Seismic
S^2	Sample variance
SD(Y)	Standard deviation of distribution Y
t	Wall thickness
T	Thermal
T	Tearing modulus
T	Total time
V(Y)	Variance of distribution Y
Z_1	Anchoring factor
Z_2	Adjustment ratios
ε	Strain
ε	Probability of not detecting a crack regardless of depth
λ	Pipe failure frequency (through-wall crack)
μ	Mean
ν	Parameter controlling slope of P_{ND} curve
φ	Total frequency (cracks and leaks)
σ	Stress
σ	Standard deviation
$σ^2$	Variance
$σ_{DW}$	Dead weight stress
$σ_{NO}$	Normal operating stress
$σ_{te}$	Thermal expansion stress

2.   Acronyms and Initialisms

ACRS	Advisory Committee on Reactor Safeguards
ADAMS	Agencywide Document Access and Management System
AM	Arithmetic mean
ANL	Argonne National Laboratories
ASME	American Society of Mechanical Engineers
ASTM	American Society for Testing and Materials
BINP	Battelle Integrity of Nuclear Piping

BWR	Boiling water reactor
B&W	Babcock and Wilcox
CBP	Conditional break probability
CD	Compact disk
CDF	Cumulative distribution function
CE	Combustion Engineering
CFR	Code of Federal Regulations
CFP	Conditional failure probability
COD	Crack opening displacement
CRD	Control Rod Drive
CRDM	Control Rod Drive Mechanism
CS	Carbon steel
CV	Correct value
CVCS	Chemical Volume and Control System
CY	Calendar year
DEGB	Double ended guillotine break
DVI	Direct Volume Injection
DW	Dead weight
ECCS	Emergency Core Cooling System
ECSCC	External chloride stress corrosion cracking
EDY	Effective degradation years
EF_a	Error factor after overconfidence adjustment
EF_i	Error factor of distribution Z_i
EF_0	Error factor before overconfidence adjustment
$EF(Y)$	Error factor of Y
EFPY	Effective full power years
Emc^2	Engineering Mechanics Corporation of Columbus
EMT	Engineering Mechanics Technology
EPRI	Electric Power Research Institute
EQ	Elicitation question
FAC	Flow accelerated corrosion
FAD	Failure assessment diagram
FDR	Fabrication defect and repair
FS	Flow sensitive
FW	Feed water
GALL	Generic aging lessons learned
GC	General corrosion
GDC	General Design Criterion
GE	General Electric
GL	Generic letter
GM	Geometric mean
GRS	Gesellschaft für Reactorsicherheit
HAZ	Heat affected zone
HPCS	High Pressure Core Spray
HPI/MU	High Pressure Injection/Make-up
HWC	Hydrogen water chemistry
IAEA	International Atomic Energy Agency
IC	Independent correlation
ICI	In-core Instrumentation
ID	Inside diameter
IGSCC	Intergranular stress corrosion cracking

IHSI	Induction heat stress improvements
INEEL	Idaho National Engineering and Environmental Laboratory
IPE	Individual plant evaluation
IPIRG	International Piping Integrity Research Group
IQR	Interquartile range
IRS	Incident Reporting System
ISI	In-service inspection
IS LOCA	Interfacing system loss of coolant accident
LAS	Low alloy steel
LB	Large break
LB	Lower bound
LBB	Leak before break
LC	Localized corrosion
LEF	Lower error factor
LER	Licensee Event Report
LERF	Large early release frequency
LIV	Loop Isolation Valve
LLNL	Lawrence Livermore National Laboratory
LOCA	Loss of coolant accident
LOOP	Loss of offsite power
LPCS	Low Pressure Core Spray
LPHSW	Last pass heat sink welding
LQ	Lower quartile
LTOP	Low temperature over pressurization
MA	Material aging
MB	Medium break
MERIT	Maximizing Enhancements in Risk Informed Technology program
MF	Mechanical fatigue
MITI	Ministry of International Trade and Industry (Japan)
MRP	Materials Reliability Program
MSIP	Mechanical Stress Improvement Process
MSIV	Main Steam Isolation Valve
MV	Mid value
NB	Nickel-based weld
NDE	Non-destructive examination
NG	Nuclear grade
NPP	Nuclear power plant
NPS	Nominal pipe size
NRC	Nuclear Regulatory Commission
NSSS	Nuclear steam supply system
NUPEC	Nuclear Power Engineering Test Center (Japan)
NWC	Normal water chemistry
OBE	Operational basis earthquake
OD	Outside diameter
OECD	Organization for Economic Co-operation and Development
OPDE	OECD Piping Data Exchange
ORNL	Oak Ridge National Laboratories
PFM	Probabilistic fracture mechanics
PIFRAC	Pipe fracture mechanics material property database
PIV	Pressurizer isolation valve
PNNL	Pacific Northwest National Laboratories

POD	Probability of detection
PORV	Power operated relief valve
PRA	Probabilistic risk assessment
PRC	Perfect rank correlation
PSA	Probabilistic safety assessment
PSI	Pre-service inspection
PSL	Pressurizer Spray Line
PTS	Pressurized thermal shock
PVP	Pressure Vessel and Piping
PWHT	Post weld heat treatment
PWR	Pressurized water reactor
PWSCC	Primary water stress corrosion cracking
QA	Quality assurance
QC	Quality control
RCIC	Reactor Core Isolation Cooling
RCP	Reactor Cooling Piping
RCPB	Reactor Coolant Primary Boundary
RCS	Reactor cooling system
RES	Office of Nuclear Regulatory Research
RH	Reactor head
RHR	Residual Heat Removal
RI-ISI	Risk informed in-service inspection
RPV	Reactor Pressure Vessel
RR	Rolls Royce
RS	Residual stress
RV	Random variable
RWCU	Reactor Water Cleanup
SAM	Seismic anchor motion
SB	Small break
SCC	Stress corrosion cracking
SCSS	Sequence Coding and Search System
SG	Steam generator
SGTR	Steam generator tube rupture
SI	Stress improvement
SIS	Safety Injection System
SKI	Swedish Nuclear Inspectorate
SLB	Service Level B
SLC	Standby Liquid Control
SLD	Service Level D
SRM	Staff requirements memorandum
SRV	Safety relief valve
SQUIRT	Seepage Quantification of Upsets in Reactor Tubes
SS	Stainless steel
SSE	Safe shutdown earthquake
TBS	Transition break size
TF	Thermal fatigue
TGM	Trimmed geometric mean
TGSCC	Transgranular stress corrosion cracking
TMI	Three Mile Island
TS	Thermal stratification
TSL	Technical specification leakage

TWC	Through-wall crack
UA	Unanticipated mechanism
UB	Upper bound
UEF	Upper error factor
UQ	Upper quartile
US	United States
USNRC	United States Nuclear Regulatory Commission
VTC	Video Teleconference
WH	Water hammer
WO	Weld overlay
WOR	Weld overlay repair
WOG	Westinghouse Owners Group

APPENDIX A

PANEL MEMBER QUALIFICATIONS

APPENDIX A

PANEL MEMBER QUALIFICATIONS

This appendix contains short resumes/biographical sketches for each of the panel members demonstrating their qualifications and credentials for this expert elicitation.

BRUCE BISHOP
PRINCIPAL ENGINEER
RELIABILITY AND RISK ASSESSMENT GROUP
WESTINGHOUSE ELECTRIC COMPANY'S NUCLEAR SERVICE DIVISION
PITTSBURGH, PENNSYLVANIA

Mr. Bishop has been at Westinghouse for almost 35 years, working primarily in the area of structural reliability analysis, initially on breeder reactor core components and then on light water RPVs, piping and other primary loop components. During this time, Mr. Bishop developed and applied the structural reliability and risk assessment (SRRA) models, methods and software PFM analyses supporting a number of risk informed inspection initiatives. Included are the SRRA applications for most plant piping, reactor internals and reactor coolant pump components, and the irradiated belt line region, head penetration nozzles and nozzle inner radius region of the RPV. Mr. Bishop has been the recipient of six George Westinghouse Signature Awards for Engineering Excellence and two Special Recognition Awards from the ASME Pressure Vessel and Piping Division for conference tutorials on Application of Probabilistic Structural Mechanics. He has participated in the development and presentation of numerous Westinghouse technical reports and more than 35 publications on application of structural reliability to risk informed decisions, including a chapter on pressure vessel and piping applications in the *Probabilistic Structural Mechanics Handbook*. Mr. Bishop, a member of the editorial boards for *Reliability Engineering and System Safety* and *Nuclear Engineering and Design,* is currently a member of the ASME Safety Engineering and Risk Analysis Division and the ASME Codes and Standards Working Groups on Implementation of Risk-Based Inspection and Operating Plant Criteria (for RPV integrity issues). He also actively participates in the PWR Materials Reliability Program (MRP) Issue Task Groups on Reactor Vessel Integrity and Alloy-600 Issues, including the PFM analyses of the RPV during postulated PTS events and the Alloy 82/182 butt welds that are subject to PWSCC.

VIC CHAPMAN
O J V CONSULTANCY, LTD.
DERBY, ENGLAND

Mr. Chapman has a Diploma in Engineering and an Honors Degree in mathematics. He has worked in the pressure system design and maintenance field for over thirty years. Over this period he has gained experience in the design and fracture analysis of pressure vessels, the study of material properties and the statistical interpretation of data. Over the past twenty-five years he has concentrated on the area of PFM, risk based decision-making, risk based inspection and inspection qualification. He was responsible for the introduction of the Risk-Based ISI program into the Royal Naval Nuclear Fleet.

Mr Chapman retired from Rolls Royce (Naval Nuclear Division) four years ago and became an independent consultant, forming his own company, O J V Consultancy Ltd. In addition to his consultancy work, which is funded by many organizations and international bodies, he has played a leading role on several international committees:

He is currently Chairmen of ENIQ-TGR (European Network for Inspection & Qualification-Task Group on Risk).

A member of the European program NURBIM (Nuclear Risk-Based Inspection Methodology).

A member of the ASME Research Committee on Risk Technology.

He chaired the European program EURIS (European Network of RI-ISI).

Was a member of the ASME Research task force on Risk Based Inspection – Developed of Guidelines.

Was a member of the UK Technical Advisory Group on Structural Integrity (TAGSI) sub-committee on defects in welds.

GUY DEBOO
SENIOR STAFF ENGINEER
EXELON NUCLEAR
CHICAGO, ILLINOIS

Mr. Deboo has 28 years experience working in the nuclear power generation field. Mr. DeBoo's recent experience includes fatigue, crack growth and flaw stability analyses necessary to demonstrate operability for most power plant components. These evaluations would include root cause and remaining life determinations. He has extensive experience with IGSCC and other material degradation issues. He also has performed and supervised functionality and operability evaluations of systems and components to address unanticipated operating events or conditions, which do not meet inspection or test requirements.

During his 28 years in nuclear power generation, Mr. DeBoo has worked on three major nuclear projects (six units) including all design, inspection and testing phases leading to commercial operation. This experience included system and component seismic qualification, component fatigue qualification, licensing and design review for fuel load, special analytical assessments of the safety significance of installation discrepancies, and system/equipment/component functionality/operability evaluations resulting from startup and operating test programs.

Mr. DeBoo has extensive experience in the evaluation of fatigue-related problems, material degradation issues and the assessment of remaining life in vessels, piping and supports for nuclear applications. He has performed safety evaluations for unanticipated operating events, and has developed plant-unique acceptance criteria to permit continued operation. Those events include fluid transients, thermally stratified flows, and flow-induced vibration. He supervised the AMSE Class 1 piping fatigue analysis on two BWR units and two PWR units.

Mr. Deboo has a B.S. in Mechanical Engineering from Northwestern University and a M.S. in Mechanical Engineering from the University of Illinois. He is a member of the ASME. Currently he is serving on Section XI of the ASME Boiler and Pressure Vessel Code as a member of the Working Group on Pipe Flaw Evaluations and as the Secretary of the Working Group on Flaw Evaluation.

WILLIAM GALYEAN
SENIOR PRA ANALYST
IDAHO NATIONAL ENGINEERING AND ENVIRONMENTAL LABORATORY
IDAHO FALLS, IDAHO

Mr. Galyean has over 25 years experience in performing PRA on commercial nuclear power plants. After earning his Bachelor of Science degree in Physics at Millersville State College in 1976, he went to the University of New Mexico where he obtained his Master of Science degree in Nuclear Engineering in 1978.

Mr. Galyean is a senior PRA analyst at the Idaho National Engineering and Environmental Laboratory (INEEL), which he joined in 1986. In recent years, Mr. Galyean was the Principal Investigator of two large NRC-sponsored PRA-related programs. The latest entailed performing detailed statistical analyses on nuclear power plant reliability data collected from operating experience. This data was then used to estimate both the historical system reliability for operational missions actually performed and the expected reliability for postulated risk-significant missions. The earlier program was a research program aimed at supporting the resolution of USNRC Generic Issue 105, "Interfacing System Loss-of-Coolant Accidents at Light Water Reactors." This program integrated many disciplines in the area of risk analysis, including: human reliability, thermal-hydraulics, consequences, external events, and stress analysis. In addition a number of innovative techniques were developed for generating human error probabilities and fluid system rupture probabilities. Mr. Galyean has also developed and teaches 1-week courses on PRA modeling techniques and on Level-2 PRA, and developed low-power and shutdown (LP/SD) specific PRAs as part of the Standardized Plant Analysis Risk (SPAR) model program. Other activities include serving on an NRC-sponsored expert panel that was formed to produce updated estimates of loss of coolant accident frequencies, supporting the independent validation of a PFM computer code (FAVOR), participating as the INEEL project manager on a multi-company, multi-lab project funded through the DOE NERI program performing a risk-informed assessment of new reactor design requirements. He has also provided significant contributions to a number of other risk/reliability related programs, including NUCLARR (Nuclear Computerized Library for Assessing Reactor Reliability), a PC-based databank of reliability data for both hardware and human actions; a reliability analysis of the INEEL site power distribution system; and an analysis of the risk significance of possible operator actions for managing severe accidents at a commercial nuclear power plant.

Prior to joining INEEL Mr. Galyean worked for Falcon Research and Development Company, Science Application International Corporation, and the NUS Corporation.

Mr. Galyean is a member of the American Nuclear Society, and has served on an International Atomic Energy Agency (IAEA) review teams (International Peer Review Service – IPERS) that reviewed a Level-1 PRA on the Borssele NPP (Dutch) and a Level-2 PRA on the Krsko NPP (Slovenian).

KAREN GOTT
SWEDISH NUCLEAR POWER INSPECTORATE

Dr. Gott studied metallurgy and materials science at Imperial College, London.

During the more than 20 years she has worked in Studsvik Dr. Gott studied many aspects of the environmental effects on structural materials in nuclear power plants, both through contract research projects and failure analysis. She has held a number of different types of position whilst at Studsvik including project manager, marketing manager and manger of the reactor chemistry group. She was also on periodic loan to a US subsidiary in Richland, WA, to help them establish laboratory support for their decontamination services.

The main areas of her research activities were
- Creep crack formation in stainless steels (mechanical testing, electron and light optical metallography)
- Fracture mechanics (corrosion fatigue, residual stress measurement, non-destructive testing)
- Reactor chemistry (PWR and BWR chemistry, activity build-up including field measurements, decontamination)
- Reactor materials (surveillance testing, failure analysis, metallography of Inconel 182)

In her current position at the Swedish Nuclear Power Inspectorate she has continued to work in the field of environmental degradation of nuclear power plant structural materials. The work covers both the regulatory and the research aspects. On the regulatory side she is involved in the development of regulations, inspection and safety evaluations that form the basis for decisions based on Swedish law and regulations. One of her responsibilities includes the management of the materials and chemistry research area for the Inspectorate. In addition she has built a database covering operationally induced failures and damage to mechanical components in the Swedish nuclear fleet and is responsible for its maintenance and the associated analysis of failure cases. In 2003 she was on a six month job rotation to the Materials Engineering Branch of the NRC's Office of Nuclear Reactor Regulation working amongst other things on PWSCC problems.

She is a member of the international conference committee which arranges the regular water chemistry conferences in the nuclear field, and has also acted on the international committee for the Fontevraud conference in France. She served as chairperson of the steering committees of two large international projects concerning irradiation assisted SCC and the establishment of a pipe failure database.

DAVID HARRIS
PRINCIPAL ENGINEER AND VICE-PRESIDENT
ENGINEERING MECHANICS TECHNOLOGY
SAN JOSE, CALIFORNIA

Dr. Harris is a principal engineer at Engineering Mechanics Technology (EMT), Inc. and has some 30 years of experience in fracture mechanics and solid mechanics analysis and applications. His background is in mechanical engineering, and he has extensive experience in probabilistic structural mechanics, especially as related to fracture mechanics.

Dr. Harris began his career as a mechanical engineer at Lawrence Radiation Laboratory (LRL) in Livermore, California. After several years at LRL, Dr. Harris joined one of the earliest vendors of acoustic emission instrumentation, Dunegan Corporation, as Director of Research. After four years at Dunegan Corporation, Dr. Harris joined Science Application, Inc. (SAI, now known as SAIC) in their Palo Alto office. During his seven years at SAI, Dr. Harris' efforts included performing some of the earliest applications of PFM to nuclear reactor piping. He was the principal developer of the PRAISE code, which was developed for the USNRC. The PRAISE code is based on PFM and is one of the most widely applied tools for evaluation of the reliability of weldments in nuclear reactor piping.

Dr. Harris worked at Failure Analysis Associates for over ten years. During this time he developed and applied fracture mechanics to a wide variety of problems, ranging from railroad wheels to rocket ship engines. These efforts included both deterministic and probabilistic aspects, and involved both computer software development and applications to industrial problems. He was the manager of the Fracture Mechanics section, which included some five engineers involved in fracture mechanics and related finite element stress analysis. He was the principal developer of the NASCRAC code, which is a general purpose code for deterministic analysis of crack growth that was developed for NASA.

Dr. Harris is currently a vice-president and principal engineer at EMT a company that he was involved in founding some seven years ago. EMT is an engineering consulting firm that specializes in fracture mechanics, life prediction and related software – both deterministic and probabilistic. Efforts at EMT include development of the PRAISE code in Windows (WinPRAISE), including enhancements to make the software easier to use in routine applications, and expansion of PRAISE to include crack initiation due to cyclic loading in air and water environments. He was also involved in the development of commercial fracture mechanics software – including linear and nonlinear SmartCrack. BLESS is a code for analysis of reliability of headers and piping in fossil-fired power plants that was developed with support of the Electric Power Research Institute (EPRI). BLESS is a physics-based model that considers both crack initiation and growth due to creep and cyclic loading.

Dr. Harris has been involved in ASME activities related to reliability considerations in design and inspection of nuclear reactor piping. He was an original member of the ASME Research Task Force on Risk-Based Inspection Guidelines, and was the editor of Volume 3 of a series of reports published by this committee. Volume 3 was on applications to fossil fired power plants. He is currently vice chairman of the Risk Technology Committee of the ASME. Dr. Harris is a member of ASTM as well as ASME. He has nearly 100 publications in the open literature, primarily in the areas of acoustic emission and fracture mechanics. He received a B.S. and M.S. in mechanical engineering from the University of Washington and a Ph.D. in applied mechanics from Stanford University.

BENGT LYDELL
SUPERVISOR
ERIN® ENGINEERING AND RESEARCH, INC.
WALNUT CREEK, CALIFORNIA

Mr. Lydell has 30 years of risk and reliability analysis experience. Prior to joining ERIN®, he held positions with the Swedish Nuclear Power Inspectorate (SKI), Pickard, Lowe and Garrick, Inc., and NUS Corporation. Mr. Lydell has extensive, practical experience with applied quantitative risk assessment. In various capacities (systems analyst, human reliability analyst, independent reviewer), he has supported numerous domestic and foreign PSA projects (Level 1 and 2, and internal flooding). As an independent contractor, during the period 1993-99 he performed R&D in piping reliability analysis for the oil and gas and nuclear industries. This work explored field experience data and its role in quantitative piping reliability analysis, including the interfaces between PSA requirements and PFM. The SKI pipe failure database resulted from this work. Under contract to SKI and BKAB (a Swedish utility), during 1998-99 he performed a pilot LOCA-frequency study; a summary report is published as SKI Report 98:30 (May 1999). This particular study was commissioned to address the feasibility of applying BWR pipe operational experience data to the estimation of plant-specific LOCA frequencies. The SKI pipe failure database formed the basis for the OECD Nuclear Energy Agency's "OECD Pipe Failure Data Exchange" Project (OPDE), an international forum for the exchange of pipe failure information. Managed by Mr. Lydell, a clearinghouse is operating the OPDE database and provides the quality assurance function.

SAM RANGANATH
XGEN ENGINEERING
SAN JOSE, CALIFORNIA

Dr. Ranganath has spent 30 years working with BWRs. He spent over 28 years working on BWRs at General Electric (GE) before moving to set up a consulting company - XGEN Engineering that provides fracture mechanics, materials and stress analysis services to the power industry. His last position at GE Nuclear Energy was Engineering Fellow and Manager, Hardware Design. He has also taught graduate courses in structural mechanics and materials at Santa Clara University and San Jose State University for over 10 years

Dr. Ranganath has a Ph. D in Engineering from Brown University and a Masters degree in Business Administration from Santa Clara University. He is a Fellow of the ASME. He has also been an Engineering Fellow at GE Nuclear Energy and was elected to the Engineering Hall of Fame at GE Nuclear Energy.

Dr. Ranganath has been active in the development of the ASME Code for over 20 years. He led the effort on developing flaw acceptance rules for austenitic piping in the ASME Code. He was also played a major role in developing improved rules for seismic design of nuclear power plant piping. He has also been the principal investigator on several materials research programs at the Electric Power Research Institute. He has also been active in the BWR Vessel and Internals Program (BWRVIP) and has been the lead author of several Inspection and Evaluation documents for BWR internal components. His expertise in BWR issues such as IGSCC, corrosion fatigue, fracture mechanics, ASME Section XI and Section III Codes, repair hardware design and BWR design is important in assuring that the LOCA frequency conclusions reflect BWR field experience.

PETE RICCARDELLA
SENIOR ASSOCIATE
STRUCTURAL INTEGRITY ASSOCIATES
GREENWOOD VILLAGE, COLORADO

Pete Riccardella received his PhD. from Carnegie Mellon University in 1973 and is an expert in the area of structural integrity of nuclear power plant components. He co-founded Structural Integrity Associates (SIA) in 1983, and has contributed to the diagnosis and correction of several critical industry problems, including:

- Feedwater nozzle cracking in BWRs
- Stress corrosion cracking in BWR piping and internals
- Irradiation embrittlement of nuclear reactor vessels
- Primary water stress corrosion cracking in PWRs
- Turbine-generator cracking and failures.

Dr. Riccardella has been principal investigator for a number of EPRI projects that led to advancements and cost savings for the industry. These include the **FatiguePro** fatigue monitoring system, the **RRingLife** software for turbine-generator retaining ring evaluation, **RI-ISI** methodology for nuclear power plants, and several **PFM** applications to plant cracking issues. He has led major failure analysis efforts on electric utility equipment ranging from transmission towers to turbine-generator components and has testified as an expert witness in litigation related to such failures.

He has also been a prime mover on the ASME Nuclear ISI Code in the development of evaluation procedures and acceptance standards for flaws detected during inspections. In 2002 he became an honorary member of the ASME Section XI Subcommittee on ISI, after serving for over twenty years as a member of that committee.

In 2003, Dr. Riccardella was elected a Fellow of ASME International.

HELMUT SCHULZ
DEPARTMENT HEAD – COMPONENT INTEGRITY
GESELLSCHAFT FÜR ANLAGEN- UND REAKTORSICHERHEIT (GRS)
KÖLN, GERMANY

Mr. Schulz has over 35 years of experience in nuclear engineering, structural and fracture mechanics, materials, and nuclear safety. At GRS he is Head of the Department of Components Integrity and was/is a member of various national and international advisory bodies regarding nuclear safety, component integrity, and codes and standards. In this role, he is responsible for the safety assessment of nuclear components and structures, as well as related research and verification of fracture mechanics codes for safety applications.

Prior to joining GRS, he worked for Gesellschaft für Reaktorsicherheit where he held various staff positions and project management responsibilities for PWR safety assessment work. Prior to that he was with United Nuclear Corporation where he did work with fuel element design and inspection of reference elements in US Nuclear Power Plants. He was also on staff with AEG Research Center for which he worked in the area of fuel element design, qualification of fabrication processes, testing programs on fuel elements and inspection of reference elements in nuclear pilot plants. Mr. Schulz holds a B.S. /M.S. in mechanical and nuclear engineering and has served on the engineering faculty at Essen.

FRED SIMONEN
LABORATORY FELLOW
PACIFIC NORTHWEST NATIONAL LABORATORY
RICHLAND, WASHINGTON

Since joining the Pacific Northwest National Laboratory (PNNL) in 1976, and before that at the Battelle Columbus Division beginning in 1966, Dr. Simonen has worked in the areas of fracture mechanics and structural integrity. His research has addressed the safety and reliability of nuclear pressure vessels and piping as well as other industrial and aerospace structures and components.

During the 1990's Dr. Simonen was a leader on the behalf of NRC and the ASME in the implementation of risk-informed methods for the inspection of nuclear piping. Dr. Simonen supported NRC staff by writing all chapters on piping reliability that are now part of DG-1063 Regulatory Guide on RI-ISI of Nuclear Power Plant Piping. These chapters provided the first formal set of guidelines to industry for probabilistic structural mechanics calculations for estimating piping failure probabilities. His recommendations impacted the selection of critical piping components that are given high priority for nondestructive examinations. On behalf of NRC and ASME Research, Dr. Simonen led a national effort during 1996-97 to benchmark PFM computer codes. The exercise concluded with PNNL performing the first ever statistically based calculations to quantify the uncertainties in calculated piping failure probabilities.

Since the early 1980's he has led several studies for the USNRC on the effects of PTS on the failure probability of RPVs. This work has advanced the technology of PFM and methods for estimating the number and sizes of flaws in vessel welds. Dr. Simonen's research has corrected longstanding deficiencies in traditional methods used to estimate the number and sizes of the welding flaws that govern the structural reliability of high-energy reactor piping and vessels.

Dr. Simonen was invited during 1995 and 1998 by the IAEA to meetings in Russia and Sweden to participate with a group of experts who evaluated the application of the LBB concept to RBMK reactors. The Central Research Institute of the Electric Power Industry invited Dr. Simonen to Japan during 1998 to present lectures on the reliability of reactor piping and methods to quantify the benefits of ISI programs.

Dr. Simonen has published over 200 papers, articles and reports in the open literature. He is a member/fellow with the ASME and serves on numerous ASME committees and codes and standards bodies, and has been awarded a number of prestigious awards from ASME.

Dr. Simonen holds a PhD. and Masters Degree in Engineering Mechanics from Stanford University and a B.S. in Mechanical Engineering from Michigan Technological University.

GERY WILKOWSKI
PRESIDENT
ENGINEERING MECHANICS CORPORATION OF COLUMBUS
COLUMBUS, OHIO

Dr. Wilkowski is an internationally recognized expert on the fracture behavior of piping in the nuclear as well as oil and gas industries. His areas of expertise include: full-scale pipe and pressure vessel fracture testing, nondestructive examination, J_R-curve testing, high-rate toughness testing, experimental design and instrumentation, elastic-plastic estimation scheme analysis, impact testing, ASME Section XI flaw analyses, LBB analyses, and pipe system fracture behavior under seismic loading.

He was heavily involved in the development and verification of the fracture mechanics analyses for circumferential cracks in nuclear pipe for ASME Section XI. He was also a member of the following review committees:
(1) NRC Pipe Crack Task Group member that developed the NRC LBB procedure,
(2) NRC Peer Review Committee for proposed new seismic design rules for nuclear piping,
(4) NRC CRDM cracking review team member,
(5) NRC Davis-Besse clad integrity review team member,
(6) Consultant to AECB on CANDU pressure tube guillotine break phenomena, and
(7) Member of DOE's Peer Review Groups for: Savannah River plant, New Production Reactor plant, Advanced Neutron Reactor, and uranium hexafluoride storage cylinders.

Dr. Wilkowski is a fellow of ASME. Currently he is a member of the following ASME Boiler and Pressure Vessel Code Section XI groups: Plant Operating Criteria Special Working Group, Flaw Evaluation Working Group, and Secretary of the Pipe Flaw Evaluation Working Group. He is the past chairman of the ASME Materials Fabrication Committee, and past chairman of the Pipe and Support Subcommittee of the ASME Operations, Applications, and Components Committee, all of which are part of the ASME Pressure Vessel and Piping Division. He was a coordinator for the 14th, 16th, and 17th Structural Mechanics in Reactor Technology (SMiRT) Conferences. He is a registered professional engineer in the State of Ohio since 1979.

Dr. Wilkowski has more than 200 technical publications, most on piping fracture. He is currently on the Editorial Board of the *International Journal of Pressure Vessels and Piping*. He is a past Associate Technical Editor of the ASME Journal of Pressure Vessel Technology, and guest editor of the Nuclear Engineering and Design journal. He was editor or co-editor of eleven ASME special technical publications. He was co-editor of four NRC Conference Proceeding Reports on LBB.

Dr. Wilkowski has both a B.S. and M.S. degree in Mechanical Engineering from the University of Michigan and a PhD in Nuclear Engineering from the University of Tokyo.

APPENDIX B

MEETING MINUTES FROM GROUP PANEL MEETINGS

APPENDIX B

MEETING MINUTES FROM GROUP PANEL MEETINGS

In this appendix the meeting minutes from the three group meetings of the expert panel are presented. First the meeting minutes from the kick-off meeting are presented, followed by the meeting minutes from the base case review meeting. Lastly, the meeting minutes from the wrap-up meeting of the elicitation panel are presented.

MEETING NOTES FROM US NRC LOCA ELICITATION KICK-OFF MEETING
DOUBLETREE HOTEL, ROCKVILLE, MD
FEBRUARY 4 – 6, 2003

Day 1 – Tuesday, February 4, 2003

Welcoming Remarks, Agenda Review, and General Information

Rob Tregoning of the USNRC began the meeting with a review of the agenda and general announcements. In addition, the individuals present were asked to introduce themselves with a short background of their experience related to the issue of LOCA frequency estimations.

Mike Mayfield of the USNRC welcomed the group and offered his perspective on the subject. Hossein Hamzehee of the USNRC also stressed the importance of the LOCA frequency determination for the continuing effort to explore risk-informed revision of 10 CFR 50.46, which govern ECCS requirements.

The meeting attendance list was provided to all of the meeting participants.

Presentation: Importance of LOCA Distributions to 50.46

The first presentation on the agenda was made by Alan Kuritzky of the USNRC. Alan laid out the importance of LOCA frequency estimates with respect to the 50.46 revision effort. Some of the key points from his presentation and subsequent discussion are outlined below:

- The NRC staff proposed a plan for risk-informing the technical requirements of 10 CFR Part 50 (Option 3) in SECY 99-264.
- Stakeholder input was considered in the recommendation to focus revision on the ECCS requirements.
- These requirements are covered in three regulations: 10 CFR 50.46, Appendix K to 10 CFR 50.46, and General Design Criterion (GDC) 35.
- Potential changes to 10 CFR 50.46 fall in one of the following areas
  - ECCS reliability (one of the focuses of elicitation – due to the simultaneous Loss of Offsite Power [LOOP] requirement)
  - ECCS acceptance criteria
  - ECCS evaluation model
  - ECCS LOCA size definition (another focus of the elicitation)
- The elicitation results will impact changes to the ECCS reliability areas and the ECCS LOCA size definition.
- ECCS reliability is primarily impacted because of the effort to eliminate the simultaneous LOCA-LOOP requirement. This has two pieces:
  - LOCA initiation frequencies and
  - Conditional probability of LOOP given a LOCA.

The focus of the current expert elicitation effort will be to obtain robust LOCA frequencies for use in the LOCA-LOOP evaluation. Interim LOCA frequencies were developed last spring as part of an internal

NRC elicitation effort and these have been used to demonstrate the technical feasibility of a change in this requirement. The results of this interim NRC effort will not be made available to this panel until after the wrap-up meeting to ensure that the panel results are independent. The objective is to have this project completed by December 2003. More detail on this topic is provided in Rob Tregoning's subsequent presentation.

For the LOCA size definition effort, a computational code is being developed to incorporate LB LOCA contributions from pipe breaks and other component failures. The results of this panel will be used to normalize the analytical code results and also provide distributions for important input variables. The targeted completion date for this technical feasibility study is July 2004. More detail on this topic is also provided subsequently.

Rob Tregoning indicated that the LOCA frequencies would also be used within the 10 CFR 50.61 risk-informed revision effort (PTS Rule) to ensure that current calculations are acceptable.

Presentation: Current LOCA Frequencies and Failure Mechanisms

The next presentation was made by Bill Galyean of INEEL in which he reviewed Appendix J of NUREG/CR-5750. Some of the key points from his presentation and subsequent discussion include:

- There are a number of varieties of LOCA initiating events, including:
  - Traditional pipe break LOCAs
  - Stuck open PORVs and SRVs
  - Steam generator tube ruptures
  - Reactor coolant pump seal failures
  - Interfacing system LOCAs (ISLOCAs) – where primary system coolant is inadvertently introduced into the secondary side piping and a secondary pipe fails creating a leak path of primary coolant outside containment
  - Reactor vessel rupture
- While failure data exists for some of those categories, data for pipe break LOCAs and other similar events simply does not exist because it has never occurred.
- There is methodology for estimating the frequency of an event that has never occurred. A Bayesian update of a non-informative prior can be employed. This assumes that the mean value for the distribution is ½ of a failure over the service life. This can be result in a very conservative estimate because the assumed failure frequency in the prior is so high (pf = 0.5). If the failure rate is not constant over time, one also needs to account for time dependency and this methodology is not equipped for this.
- A primary Appendix J assumption is that you needed a leak before you can get a break. A conditional pipe break probability given a leak was based on the Beliczey-Schulz correlation.
- There was also a presentation of passive LOCA failures that can occur in non-piping systems as well as a list of possible data sources for this information.

Discussion: The elicitation panel discussed the validity of this assumption for degradation mechanisms that result in long surface flaws which are not as likely to leak prior to failure. Also, the expectation is that leaking flaws will be fixed after they are discovered during a plant walkdown or through other leak detection methods.

Discussion: Bruce Bishop indicated the need for very clear definitions of what constitutes a large, medium, small, and very small break LOCA. The concern is that the system response to a DEGB where the flow rates can reach 860,000 gpm (3,250,000 lpm) (according to Westinghouse calculations) is very different from a 5,000 gpm (19,000 lpm) leak which is also often characterized as a large break LOCA. Rob Tregoning indicated that clear definitions will be developed as part of this exercise.

Discussion: Gery Wilkowski relayed information provided by Helmut Schulz that the Beliczey and Schulz correlation of conditional probability of a rupture given a leak was developed for cyclic fatigue crack growth.

Discussion: Rob Tregoning emphasized that Bill Galyean's presentation was provided to recap the last NRC-sponsored work in this area. This NUREG/CR-5750, Appendix J approach is not endorsed for the expert elicitation process; however, it represents one manner in which LOCA frequencies have been developed. Tregoning also emphasized that because substantial LOCAs have not occurred, past operating experience data needs to be augmented by information from other areas. If information was available simply from operating experience, there would be no need for the elicitation.

Discussion: The point was also raised that the panel needs to consider LOCA sources other than traditional pipe LOCAs.

Presentation: LOCA Frequency Determination Using Expert Elicitation

Rob Tregoning then made a presentation on LOCA frequency determination using expert elicitation. The objectives of this presentation are to motivate the expert elicitation effort, discuss the limitations of relying solely on past operating experience; present ongoing NRC-sponsored research in this area; define the objective of the expert elicitation; outline the approach; and discuss the structure of the kick-off meeting. Some of the specific key points from his presentation and subsequent discussion are outlined below:

- The LB LOCA design basis size will be determined by considering all relevant LOCA sources. A PFM-based model is under development for predicting LOCA contributions as a function of break size. This code will also account for LOCAs from non-piping sources, and will include contributions from future unknown failure mechanisms. Expert elicitation input from this panel will be used throughout program development.
- The objectives of the elicitation are to
  - Develop future SB, MB, and LB LOCA frequency estimates extending up through the end of the plant-license-renewal period (approximately 35 years).
  - Develop benchmark problems and standardized inputs for conducting PFM simulations of LB LOCA events in important BWR and PWR systems.
- The elicitation approach will construct base cases. Quantitative LOCA estimates as a function of break size will be developed for these base cases. Then, important variables and issues will be discussed within the relative framework of the base cases.

Discussion: It was stressed that the base cases will just provide a reference point. Adjustments to the base cases will account for the impact of those issues which contribute significantly to the LOCA frequencies. These adjustments must consider their effect on current LOCA frequencies and their time dependence up through the end of the plant license-renewal period ($\approx$ 35 years).

- The programmatic approach for the elicitation was presented. It consists of the following important areas.

- Conduct the kick-off meeting
- Develop elicitation questions
- Allow individual study to develop answers for elicitation questions
- Conduct the individual elicitations
- Analyze results (facilitation team)
- Conduct a wrap-up meeting

Discussion: It was conveyed that when the facilitation team queries each elicitation panel member, best estimate answers will be sought as well as the uncertainty in the estimates. It was also stressed that each panel member need not answer all questions, but only those that they feel that they are qualified and comfortable with answering. However, people will be encouraged to answer all questions, even in areas outside of their specific expertise. Uncertain knowledge should be reflected in the uncertainty estimates. Rob Tregoning indicated that one job of the facilitation team will be to filter out responses which are not well-founded, and exclude them from the final analysis.

- The principal components of the kick-off meeting were reviewed again. There are several major objectives which need to be accomplished during this kick-off meeting:
  - Present the elicitation objectives and define fundamental terms to ensure common understanding.
  - Undergo elicitation training to understand the process and approach.
  - Construct methodology for developing baseline LOCA estimates. Develop a classification scheme and approach for issues which could affect the baseline LOCA estimates.
  - Identify and classify issues for consideration. Discuss issues as necessary for clarification.
  - Agree on significant issues to include in the elicitation.
  - Determine the structure of the elicitation questions.

Presentation: Expert Elicitation Process

Lee Abramson of the US NRC spoke on the expert elicitation process that will be followed in this exercise. Some of the specific key points from his presentation and subsequent discussion are outlined below:

- Key word is "formal" use of expert judgment. Engineers practice informal expert judgment every day.
- It was emphasized that elicitation is a structured process and that the process requires experienced practitioners to conduct the exercise. This is not a "do it yourself" activity.

Discussion: A question was raised if the results of this elicitation or past elicitations could be used as a baseline for future efforts, in much the same way that Bayesian analysis is performed. Lee Abramson indicated that there is no natural means of updating results from prior elicitations based on recent experience or new data. However, it may be appropriate to use the results of a prior elicitation as starting point for future elicitation.

- The need for comprehensive documentation was also stressed to ensure that the process approach, issues, analysis techniques, results and uncertainties are clear. Additionally, follow-on work to refine the results requires comprehensive documentation in order to understand the basis of the initial study.
- The need for an expert panel with a broad range of expertise and experiences was expressed. Also all of the stakeholders (both utilities and regulators) must be represented.

- There are two methods of elicitation: group and individual. The problem with group sessions (versus individual sessions) is that often group dynamics lead to domination of one or two individual opinions. The results then no longer represent everyone's input.
- Elicitation team for this exercise consists of
  - Normative expert – **Lee Abramson**
  - Substantive experts – **Alan Kuritzky, Ken Jaquay, Rob Tregoning, others?**
  - Recorder – **Paul Scott**
  - Documenter – **Paul Scott (could be same as recorder)**
- Panel members need to provide rationale for answers so others can see why certain panelists came up with certain answers. In that way other panelists have the option of changing their answers based on feedback from the group. The panel will largely be provided this feedback at the wrap-up meeting. Panel members can revise answers to any question at any time.

Discussion: It was asked if the response will be weighted in any way to account for expertise in a given area. Lee Abramson replied that the analysis will use unweighted responses so that everyone's response is judged equally. With this size of panel, weighting should not substantially affect the final results. The elicitation will also query the panel member's uncertainty for each answer. If inordinate uncertainty exists, then the response *may be* downgraded. Also, the rationale provided by each panelist will help determine if responses need to be weighted.

- Types of biases present in elicitation processes:
  - Motivational biases (i.e., social pressure or group pressure to make a certain decision). These need to be recognized and avoided at all costs.
  - Cognitive biases --- biases can occur when people have developed an initial answer and more data becomes available which require the initial answer to be modified. Typically people underestimate the impact of the new data. This bias is referred to as anchoring. The elicitation structure will be developed in an attempt to minimize these biases. For instance, initial estimates of the total LOCA frequencies will not be asked.
  - Background biases (i.e., what an individual might see as reasonable, or would expect, based on his background.) For example, an experimentalist might see a high probability of failure of a piping based on the number of experiments he has run in which he saw a failure, but typically the test conditions were such that similar conditions in the field are highly unlikely to ever occur. This bias is natural, but it is important to get each individual to consider all variables which affect the result and break them down into meaningful pieces.
- People are more than likely to underestimate the true uncertainty, by a factor of 1/2.
- People are more likely to anchor on median value, not on the extremes.
- Goal is to make the questions as unambiguous as possible (very precise) and to focus questions on the major issues affecting the LOCA analysis.
- The uncertainty range will be queried during the elicitation by asking for the "number" such that there is 5% chance that the true response is less than this number. A separate number will be provided for the UB such that there is also a 5% change that the true response is higher than this number. This corresponds to the 90% coverage interval of the variable.
- Purpose of elicitation panel members is to come up with individual answers, not a consensus.

Discussion: There was quite a bit of discussion and confusion about the definition of the coverage interval. Lee Abramson said that the uncertainty range (difference between higher and lower response to a given question) should cover the true number for that variable 90% of the time. The true value should fall below the lower response 5% of the time and the true value should land above the higher response 5% of the time. However, Lee cautioned against making the coverage interval inordinately large just to capture uncertainty. If this occurs, the coverage interval contains little useable information.

Elicitation Exercise

Each participant filled out an elicitation questionnaire dealing with age related health issues. The results from this exercise were reviewed with the meeting participants on Wednesday morning. As part of this exercise, Lee Abramson indicated that it is usually easier to determine relative rates versus absolute rates. Various absolute and relative questions were posed in order to demonstrate this concept.

Definition of Terms for Elicitation

Terms used during the elicitation must be commonly understood by the group in order to foster discussion, issue development, and subsequent elicitation. Certain key terms must be defined. Rob Tregoning indicated that, for this exercise, all definitions should be kept generic, not plant specific. The first term to be defined is LOCA. Rob Tregoning presented the NUREG/CR-5750, Appendix J definition as a starting point. This report defines a LOCA as "an unisolable breach of the Reactor Coolant Primary Boundary (RCPB) requiring ECCS initiation."

The group felt that the term "unisolable" was not appropriate because the main point is to limit the scope to Class 1 piping. Also, the merits of the phase ECCS initiation were debated because the ECCS response in some plants requires use of normally operating plant equipment. Therefore, some plants might require a large leak before implementation of standby ECCS systems. There was also a discussion on the merits of using break instead of breach, but the term breach was determine to be more generic than break. The addition of the term "sudden breach" instead of just "breach" was also neglected because of the vagueness of the word sudden.

The group agreed to a definition of a general LOCA as follows. **A LOCA is "a breech of the reactor coolant pressure boundary which results in a leak rate beyond the normal makeup capacity of the plant".**

The next definitions are required to determine the size classifications of LOCAs. Once again, Rob Tregoning presented the definitions used in NUREG/CR-5750, Appendix J as a starting point. These definitions were also used in NUREG-1150 and form the basis of plant PRA event trees. This document defined three LOCA size categories: SB, MB, and LB. The NUREG/CR-5750 definitions are as follows:

- **SB LOCA** - A break that does not depressurize the reactor quickly enough for the low pressure systems to automatically inject and provide sufficient core cooling to prevent core damage. However, low capability systems (i.e., 100 to 1,500 gpm [380 to 5,700 lpm]) are sufficient to make up the inventory completion. For a BWR, this translates to a pipe in the primary system boundary with a break size less than 0.004 ft^2 (370 mm^2), or a 1 inch (25 mm) equivalent inside pipe diameter, for liquid, and less than 0.05 ft^2 (4,600 mm^2), or an approximately 4 inch (100 mm) inside diameter pipe equivalent, for steam. For a PWR, this equates to a pipe break in the primary system boundary with an inside diameter between ½ to 2 inches (13 to 50 mm).

- **MB LOCA** – A break that does not depressurize the reactor quickly enough for the low pressure systems to automatically inject and provide sufficient core cooling to prevent core damage. However, the loss from the break is such that high capability systems (i.e., 1,500 to 5,000 gpm [5,700 to 19,000 lpm]) are needed to makeup the inventory depletion. For a BWR, this translates to a pipe in the primary system boundary with a break size between 0.004 to 0.1 ft^2 (370 to 9,300 mm^2), or an approximately 1 to 5 inches (25 to 125 mm) inside diameter pipe equivalent, for liquid, and between 0.05 to 0.1 ft^2 (4,600 to 9,300 mm^2), or an approximately 4 to 5 inches (100 to 125 mm) inside pipe diameter equivalent, for steam. For a PWR, this equates to a pipe break in the primary system boundary with an inside diameter between 2 to 6 inches (50 to 150 mm).

- **LB LOCA** – A break that depressurizes the reactor to the point where the low pressure system injection automatically provides sufficient core cooling to prevent core damage. For a BWR, this translates to a pipe in the primary system boundary with a break size greater than 0.1 ft^2 (9,300 mm^2), or an approximately 5 inch (125 mm) inside diameter pipe equivalent, for liquid and steam. For a PWR, this equates to a pipe break in the primary system boundary with an inside diameter greater than 6 inches (150 mm).

The elicitation panel questioned the basis of the equivalent pipe diameter relationships to break size provided in the NUREG/CR-5750 Appendix J. Bill Galyean thought that they could be traced back to NUREG 1150 and possibly WASH-1400. It was quickly determined that "break" should be replaced by "breech" everywhere for consistency with the general LOCA definition. Also, the group decided that the formal definitions should be based on leak rate, and not equivalent break area or size.

At this point, the need for additional LOCA size classification was revisited. This request was promulgated by Bruce Bishop based on discussions with the Westinghouse Owner's Group (WOG). The System response and mitigation procedures for a 5,000 gpm (19,000 lpm) LOCA (lower limit LB LOCA within NUREG/CR-5750, Appendix J) and a DEGB of the largest class 1 pipe (flow rate up to 860,000 gpm [3,250,000 lpm] according to WOG) are significantly different. Because, the elicitation results will used in existing PRAs, it was also stressed that the original leak rate classifications in NUREG/CR-5750, Appendix J should also be maintained. While the group also agreed that the leak rate threshold should ideally be based on the equipment needed to mitigate a specific event, this information is highly plant specific and could not be approximated generically.

For the reasons stated in the above paragraph, the group decided to keep the NUREG/CR-5750, Appendix J leak thresholds of 100 gpm (380 lpm), 1,500 gpm (5,700 lpm), and 5,000 gpm (19,000 lpm), but to add several leak rate categories above 5,000 gpm (19,000 lpm). The highest category was set at 500,000 gpm (1,900,000 lpm) to capture the DEGB events of the largest primary system pipes. Additional ranges of 25,000 gpm (95,000 lpm) and 100,000 gpm (380,000 lpm) were chosen to span the range from 5,000 gpm (19,000 lpm) to 500,000 gpm (1,900,000 lpm) in roughly equivalent magnifications. These leak rate categories were also chosen because they tend to group DEGBs by primary system functionality.

The LOCA size classification thresholds adopted by the group are summarized in Table B.1.1[1]. A category 1 LOCA is defined as "a breach of the reactor coolant pressure boundary which results in a leak rate which is greater than 100 gpm (380 lpm). Similarly, a category 6 LOCA is a breach of the RCPB which results in a leak rate which is greater than 500,000 gpm (1,900,000 lpm). It should be stressed that category 1 LOCAs include contributions from all categories. The group preferred the threshold classification of LOCA sizes instead of partitioning the sizes into ranges as in NUREG/CR-5750, Appendix J. Care will be needed during the elicitation to ensure that these definitions are understood.

---

[1] The nomenclature for the table and figure numbers is such that the letter B refers to Appendix B, the first number (1 or 2) refers to a figure associated with either the first or second panel meeting, and the second number refers to the numerical sequence of that particular table or figure in the text for the applicable meeting, i.e., either first or second.

Table B.1.1 LOCA Size Classification Thresholds

Category	Leak Rate Threshold (gpm)
1	> 100
2	> 1,500
3	> 5,000
4	> 25,000
5	> 100,000
6	> 500,000

It was determined by the group that these leak rates should be roughly correlated to breach area, and converted into an equivalent pipe diameter so that the UB leak rates for various piping systems could be determined. There was some concern about the feasibility of developing generic estimates. It was suggested that equivalent pipe sizes could be based on 250 gpm/in^2 (1.47 lpm/mm^2) for liquid PWR lines and 175 gpm/in^2 (1.03 lpm/mm^2) for liquid BWR lines. However, these estimates did not agree with the Westinghouse equivalent pipe diameter estimates.

Presentation: SKI-PIPE Database: Background - Structure - Status - Applications (1994 - 2002)

This presentation by Bengt Lydell discussed the SKI-PIPE database evolution and background; the database structure and content; current database status; and LOCA frequency estimate conducted with the data base. Some of the specific key points from his presentation and subsequent discussion are outlined below:

- Background: The database was motivated to create a tool that would serve both PRA and the PFM/material science practitioners. It's structured to provide information to completely define the piping systems attributes (design characteristics) and the influence functions (operating history) which govern system failure probability. By thoroughly assessing these features it is possible to determine plant specific estimates of piping system reliability.
- Structure and Content: The database covers pipe failures in commercial nuclear power plants from 1970 to the present.
- It should be stressed that SKI-PIPE only includes failures in piping systems, external to the RPV. Non-piping system failures are not included. Also, SKI-PIPE contains only passive piping failures of metallic piping.
- A pipe failure is defined in the database as any degradation that results in piping repair or replacement.
  - Each record in the database is indexed. References to the original data source (e.g., LER report) and supporting information are provided. All the supporting documented is stored electronically.
  - The database is organized by reliability attributes (i.e. design features such as material, dimensions) and influence factors (i.e. unique service conditions, including degradation susceptibility).
  - When the original record is incomplete (such as an LER), a best effort is made to fill in database gaps by directly contacting the plant operators.
  - It is noted in the database when each record consists of multiple flaws at a single component location. However, subsequent data entries are typically associated with only the largest flaw at that location.

- The database includes both surface penetrating flaws and non surface penetrating flaws (i.e., embedded flaws).
- Current Database Status:
  - The database is continually being updated.
  - The current OECD-sponsored OPDE project has participants from 12 nations. The first year of the three year effort is concerned with adding and validating database entries for each of the member countries from 1998 through 2001.
  - Raw data is currently obtained from over 40 different sources
- Applications: Two relevant studies are the determination of LOCA frequencies for the Barsebäck-1 plant and examination of IGSCC in Russian graphite moderated reactors (RBMK).
  - The Barsebäck-1 study employed plant-specific attribute and influence functions which were comprehensively developed for all "known and credible" degradation mechanisms.
  - The Beliczey and Schulz conditional rupture probability was not used in the Barsebäck-1 analysis. Instead a Bayesian update of a Jeffrey's modified non-informative prior was employed.
  - The database results have been compared with PFM predictions for welds in certain systems with some success.
  - The RBMK studied indicated that the experience today with IGSCC in Russia is similar to US BWR IGSCC cracking experience in the late 70's to early 80's, before wide-spread mitigation was adopted

Discussion: The panel asked if they could get copies of the SKI-PIPE database. Karen Gott of SKI indicated that it is possible to distribute a non-proprietary version of the database. This non-proprietary version contains piping failures thru 1998.

Baseline LOCA Determination I

Discussion: Rob Tregoning commented that the panel needed to define baseline LOCA frequencies in order to benchmark relative responses during the elicitation. He also mentioned that the SKI-PIPE database could be used to develop baseline frequencies if the group could develop well-defined "base case(s)". The base case(s) will represent a set of conditions and physical phenomena. In theory, the absolute LOCA frequencies associated with each base case are not important for the elicitation session because all elicitation responses will be judged relative to the base case conditions. The absolute frequencies are only required to reconstruct the final results. However, the panel members decided that their elicitation responses might change depending on the exact LOCA frequencies associated with the base case conditions. That is, if a base case frequency was 10^{-8}/year, the elicitation responses might be quite different than if the frequency was 10^{-2}/year. **The group therefore agreed that they will define rigorous conditions for each base case and also associate absolute LOCA frequencies with these conditions.**

Day 2 – Wednesday, February 5, 2003

Elicitation Exercise Review

Lee Abramson reviewed the elicitation questionnaire results from Tuesday afternoon's session. Overall, the results were good and consistent with expectations. The group tended to perform better on those questions that asked for the ratio of diseases between men in different age ranges (questions 3 and 4 in exercise). The mid value tended to reasonably close to the actual 2000 census values for these questions.

Additionally, the true value was contained with the 50% interquartile region (75% - 25% percentiles) 10 out of 12 times, or 83%, which is quite good.

The average coverage interval for these questions was 71%. The coverage interval should theoretically be 90%, so that group underestimated the uncertainty. Lee Abramson indicated that typically group uncertainty is about ½ of the true value. In other words, people tend to be more confident in their responses than they should be.

The results were not quite as good when the group was asked to provide absolute disease rates for an age category (Question 2). The true value was inside the 50% interquartile range 4/6 times, or 67%. Also, the coverage interval only captured the true value 61% of the time, 10% less than for the relative questions. This performance demonstrates the supposition relative differences between conditions tend to be more accurate than absolute measures for a given condition. This will be a guiding principal in developing the elicitation framework.

LOCA Issue Development

LOCA Issue development required the group to brainstorm important LOCA issues. The group first defined a structure for categorizing issues in the form of a flowchart (Figure B.1.1). It was stressed that the LOCA frequencies in this exercise will consider only passive system failure. Active system failure will not be considered for the following reasons:
1. The panel has no specific expertise in these types of failures.
2. Failure of these components is not as rare and there is adequate data to assess their contribution to the LOCA frequencies.
3. Active components are subject to ongoing maintenance which should diminish the likelihood of future failure rate increases.

However, LOCA frequency contributions from active components will be combined with the passive component contributions to develop final LOCA estimates which can be supplied as PRA input. The estimation of these contributions will occur separately, but will be summarized for the panel at the wrap-up meeting.

The group divided the passive system LOCA sources (Figure B.1.1) into two classes: piping and non-piping. Non-piping contributions include RPVs, steam generators, bolting flange failures, valves, pumps, etc. The distinction between piping and non-piping categorization is useful because piping has unique issues.

Piping LOCA Contributions

The group first defined piping to include vessel penetrations (e.g CRDM housings, instrumentation lines), piping, and safe ends. The boundary between piping and non-piping components (e.g. vessels) was defined as the nozzle (or component) side of the safe-end/piping to nozzle weld. The group then decided that piping should be categorized by the specific plant system. The piping system is important because it defines the functionality and operating history, or influence factors. The plant system is also often associated with specific piping designs and materials, or attribute functions. The relationship between the piping attribute and influence characteristics will determine its failure propensity.

For a given plant system, the variables which affect the LOCA probability fall into one of the following five categories: geometry, materials, loading history, degradation mechanisms, and mitigation or maintenance procedures. The group decided to list all the possible contributors for each variable category and then link the dependencies with a given plant system. Obviously there is a synergistic effect among these variables. The piping system requirements result in geometrical and material selection constraints.

The geometrical and material choices mesh with the system functionality and operating history to determine component loading history. This specific combination dictates the degradation mechanisms that emerge. Mitigation and maintenance procedures are developed to counteract these mechanisms. The effectiveness of these strategies, however, is a function of all the other variables discussed.

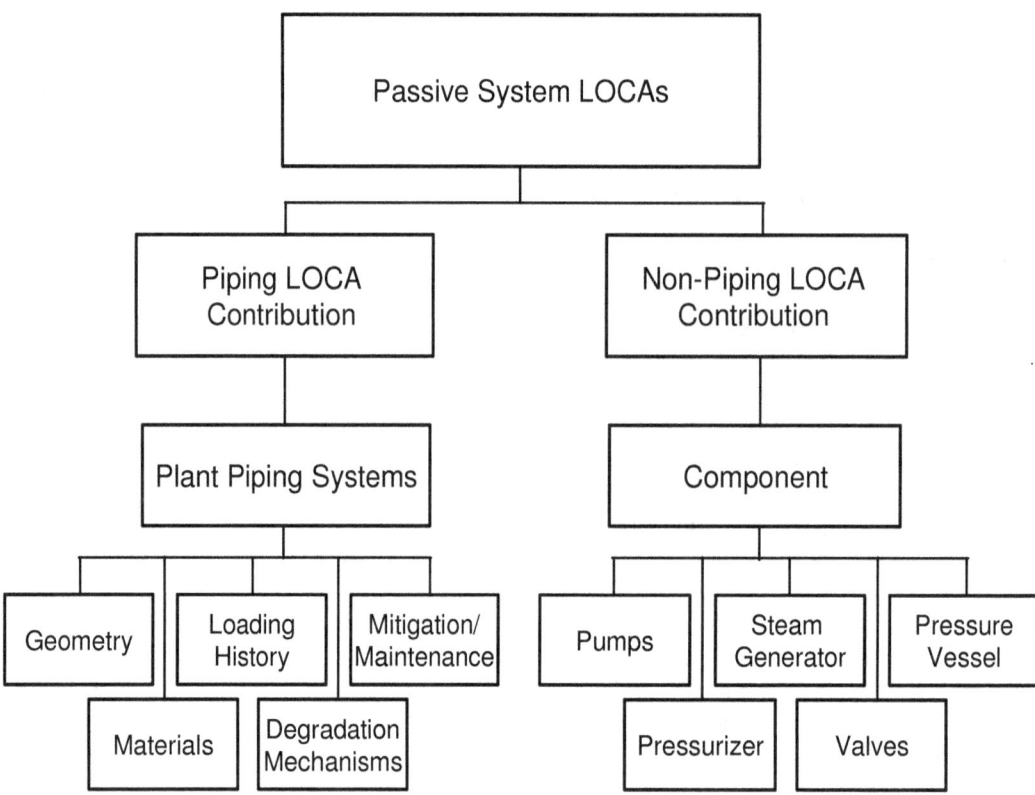

Figure B.1.1 Passive LOCA Contributions

Material Variables: It was quickly determined that the materials category should also include fabrication procedures as well. Material variables are important because they give rise to specific degradation mechanisms. Fabrication variables can lead to variations in defect formation and residual stress which can result in certain locations having a greater propensity for degradation. The important variables which affect the LOCA propensity are summarized in Table B.1.2. Table B.1.2 indicates whether the material is found in the base metal portion of the piping system or the welds. Issues that are strictly related to fabrication are separated in the table.

The circumferential versus axial welded pipe issue was raised to differentiate between seamless and non-seamless pipe. While the bulk of current piping is seamless, there is some remaining seam welded piping in service. The concern is that seam-welded piping is more susceptible to degradation and leaking than is seamless. The shop versus field welded issue was raised due to possible weld quality differences between welds made in controlled conditions during piping fabrication and those made on site during piping system assembly. Finally, it was noted that defects, residual stress irregularities, and poor material properties can be associated with repair welding. Experience has shown that field cracking and leakage is often associated with repair welds.

Table B.1.2 Piping Material and Fabrication Variables

Materials	Base metal	Weld/HAZ
304, 304L, and 304NG stainless steel	X	X
316, 316L, and 316NG stainless steel	X	X
Carbon steel	X	X
Stainless steel clad carbon	X	X
Alloy 600	X	
Alloy 82/182		X
Cast stainless steel	X	
Stainless steel bimetallic		X
Fabrication Issues		
Circumferential vs. axial welded pipe		X
Shop vs. field welds		X
Repair welds		X

Geometric Variables: Table B.1.3 lists the geometric variables which influence the LOCA frequency distributions. These variables affect piping stress, system compliance, the propensity for a given degradation mechanism, and the likelihood of leaking versus catastrophic rupture. Many of the variables are obvious and include general system information such as piping diameter and thickness (NPS and schedule), the number of welds and their location, the types and numbers of specific piping components, and the layout of supports and snubbers. The system configuration is related to the layout, but also specifically considers where active components such as pumps, valves, and flow offices are located. The variable "connections" in this table are meant to distinguish between welded connections which are more typical and flanged connections which can segregate piping from active components.

Table B.1.3 Geometric Piping Variables

NPS and Schedule (diameter and thickness)
Component Type (Elbow, tee, fittings, straight pipe, reducers, sockets, safe ends, end caps, etc.
Field fabricated vs. drawn
Crevice weld (thermal sleeves)
Number of welds & location
Configuration (active components, flow orifices, etc.)
Layout including locations and types of supports and snubbers
Connections (bolted flanges)

The final two variables in this table are a combination of geometric and material/fabrication issues: the existence of crevice welds (e.g. at thermal sleeves), and the difference between drawn piping and piping which is field-fabricated. Drawn piping is cold worked to size and is usually limited to smaller diameter piping (< xx" NPS). Field-fabricated piping may be hot forged or cold work to some extent, but the final size is often achieved by machining. This is more typical for larger diameter pipes (> yy"). Crevice or filet welds are for sleeves and other piping system attachments. Partial penetration welding can result in a greater propensity for flaws and higher residual stresses and constraint than through-wall welding.

Material Degradation Mechanisms: The next variable group describes material degradation mechanisms. As mentioned earlier, the degradation mechanisms are often associated with the piping system material. Each material is susceptible to each of the mechanisms listed, although the degree of susceptibility will greatly vary. The system loading history can also favor certain mechanisms. Table B.1.4 summarizes the mechanisms developed during the group brainstorming session. The mechanisms are segregated by the primary mechanism type. Additional sub-categories are used to identify either specific degradation mechanisms under the appropriate main category or features associated with the main category.

Table B.1.4 Material Degradation Mechanisms

Main Category	Sub-Category 1	Sub-Category 2	Sub-Category 3	Sub-Category 4
Low Cycle Thermal Fatigue	Crack Initiation	Crack Growth		
High Cycle Mechanical Fatigue	Vibration	Pressure	Temperature	
Stress Corrosion Cracking	IGSCC	TGSCC	ECSCC	PWSCC
Localized Corrosion	Pitting	Crevice Corrosion		
General Corrosion	Boric Acid (ID or OD)			
Fretting Wear				
Material Aging	Thermal	Dynamic	Radiation	Creep
Fabrication Defects and Repair				
Hydrogen Embrittlement				
Flow Sensitive	Erosion/ Cavitation	FAC		
Unanticipated (New) Mechanisms				

Fatigue degradation was separated into low cycle fatigue which is primarily driven by thermal loading fluctuations due to plant heat-up and cool-down cycles and high cycle mechanical fatigue which could result from general loading fluctuation on the piping. This loading fluctuation could be induced by vibration, pressure or temperature fluctuation (e.g. striping). High and low cyclic loading are often differentiated with respect to the loading magnitude relative to yield strength or number of cycles. For this elicitation, a rule of thumb differentiation of 1,000 cycles is sufficient to differentiate between low-cycle and high-cycle events. Both crack initiation and crack growth portions of life are important contributors to fatigue life, although crack initiation occupies a greater percentage of life in high cycle fatigue.

Stress corrosion cracking (SCC) is listed as a main category and it includes IGSCC which was prevalent in BWRs in the late 70's, TGSCC which affects casting components, PWSCC which has more recently surfaced in PWRs, and ECSCC. The localized corrosion category includes both general pitting and crevice corrosion which is likely in tight, stagnate areas. While general corrosion is listed as its own category, boric acid corrosion is the principal contributor. Both internal (ID) and external (OD) boric acid corrosion are included in this sub-category. External corrosion can result from leaking fluid from another component in the plant which then impacts the pipe. Fretting wear describes material erosion usually resulting from external contact with other components. This wear usually occurs with vibration loading but should be distinguished from high cycle fatigue which results in crack formation.

Material aging relates to changes in the intrinsic quasi-static material properties at the time when the component is placed into service. These properties include constitutive properties (stress/strain behavior), cyclic, crack growth resistance (da/dN versus ΔK), and resistance to crack initiation and tearing under monotonic loading (J-R behavior). Aging can occur due to thermal or radiation embrittlement; long term, or low temperature creep. Dynamic strain aging refers to the stress/strain and J-R curve resistance changes which result from dynamically applied loading. It is most evident in the ductile to brittle transition temperature shift of ferritic steels. Hydrogen embrittlement is somewhat related to the other material aging mechanisms in that strength and toughness can be affected over time. The distinction is that hydrogen embrittlement only becomes prevalent after a crack has formed. The other aging degradations do not require a preexisting defect although the system impact is certainly enhanced in their presence.

The flow sensitive category is used to capture those mechanisms which are sensitive to the flow characteristics of the piping medium. The erosion/cavitation sub-category refers to material erosion due to cavitation which occurs when vapor bubbles collapse. This is distinct from FAC where downstream turbulence results in erosion/corrosion that is not accompanied by low pressure boiling. The category of unanticipated or new mechanisms covers those mechanisms which could surface in the future which are either unknown at the present time or not deemed to be important at this time. This category is purposely vague to capture the panel's general uncertainty of the completeness of our understanding of future piping degradation mechanisms. For instance, just a few years ago, PWSCC would not likely have been considered to be an important degradation mechanism. Now however, it is of primary concern.

Fabrication defects and the repair of those defects (or lack thereof) are distinct from repair welding mentioned earlier in the material/fabrication issue table. This issue covers the likelihood of repair of fabrication defects and the possibility that these fabrication defects could lead to failure due to one of the other mechanisms listed in Table B.1.4. The repair component of this issue only considers the possibility that these defects are repaired (or not), not any new defects generated by the repair process. Defects generated by repair welds have been captured in the geometric variability section (Table B.1.2).

Loading History Variables: The next variable which contributes to the piping LOCA frequency estimates is the system loading history. The term loading history considers both the magnitude and frequency of the loading applied to the piping system over its service history. The different types of applied loading are summarized in Table B.1.4. Again, as in Table B.1.4, the loading variables are divided into main and sub categories.

The thermal loading category considers loading from differential expansion between dissimilar piping materials. This loading is potentially exacerbated at the ends of the piping if connected to a rigid component (e.g. steam generator or RPV). This loading is categorized as restraint-free expansion in Table B.1.4. Radial thermal gradients are induced in piping due to the temperature difference between the pipe ID and OD. Insulation can diminish this gradient. Thermal stratification can occur under low flow rate conditions when hotter liquid flows on top of cooler water. Boundary layer fluctuations in this interface can induce thermal cyclic loading which is referred to as thermal striping.

Table B.1.5 Loading History Variables

Main Category	Sub-Category 1	Sub-Category 2	Sub-Category 3	Sub-Category 4	Sub-Category 5
Thermal	Differential Expansion	Restraint Free Expansion	Radial Gradient	Stratification	Striping
Water Hammer	Steam Hammer				
Seismic	Inertial	Displacement			
Pressure	Normal	Transients			
Residual Stress	Design	Repair welds	Fabrication	Mitigation-Induced	
Dead Weight Loading					
SRV Loading					
Overload (Ext. and Int.)	Pipe Whip	Jet Impingement	Deflagration		
Support	Snubber malfunction	Hanger Misadjust.			
Vibration	Mechanical	Cavitation			

Water hammer, dead weight, and SRV loading are considered as separate main categories with no corresponding sub-categories. Water hammer is distinguished from other pressure transients because of its potential severity. Safety relief valve loading describes the pressure transient which occurs when SRVs are opened or closed. Dead weight loading is contributed by the weight of the pipe and any unsupported attachments. The pressure loading category includes normal operating pressure and any pressure transients other than those specifically listed in Table B.1.5 (e.g. water hammer, SRV loading, internal overloads, and cavitation).

Residual stress is a prominent loading category and includes contributions from locally-induced welding process stresses related to the as-designed weld (Design sub-category in Table B.1.5) and additional contributions due to weld repair. The Design sub-category also encompasses contributions from the pipe system compliance on the weld restraint during component assembly. System compliance will obviously influence the residual stress distribution which if formed at a particular weld joint. Fabrication residual stresses include cold-springing needed to align piping during plant construction. This residual stress contribution may not be apparent from the system design which is why it is distinguished from the Design sub-category. Finally, a unique sub-category is entitled "Mitigation-Induced" to account for residual stresses which may be applied during plant operation to mitigate certain failure mechanisms. These stresses are induced by processes like weld overlay repairs (used during IGSCC mitigation) and mechanical stress improvement (applied for VC Summer). These stresses are certainly associated with the weld joint being treated, but also affect the residual stress distribution in surrounding welds and piping.

Overloads can result from external failures in other plant systems and internal accidents. These accidents can result in loading which is potentially in excess of the structural design limits of the piping. External overloads include those induced by pipe whip and jet impingement. Both of these categories require failure outside of the reactor pressure boundary as a precursor event. The loading on the reactor pressure boundary piping occurs either by pipe whip of the failed components or through water or steam jet impingement caused by the breech in the other system. A special type of internal overload sub-category is Deflagration. This describes the loading due to hydrogen combustion which occurred at the Hamaoka and Brunsbuettal plants. This category would cover not only direct failure of pressure boundary piping, but also precursor failure of secondary piping that leads to conditional failure (from shrapnel or jet impingement) of pressure boundary piping.

Another loading history category is support structure loading. This includes loading due to a malfunctioning snubber or misadjusted hanger that leads to piping system loading that is beyond the intended design limits. Vibration loading includes classical displacement loading due to nearby active component vibration (Mechanical sub-category) and loading due to cavitation which may exist within the piping system.

The final loading category is seismic which will be treated uniquely from the other loading variables listed. The seismic loading category includes both inertial and displacement loading components. Seismic loading is treated separately within PRA models and the LOCA contribution from seismic loading is also calculated separately. However, many analysts frequently calculate conditional failure probabilities due to seismic loading and this is often the principal transient of interest in most piping systems. The panel therefore seemed generally comfortable with considering the possible effects of seismic loading if the loading magnitude was specified. For these reasons, it was decided to query the panel about seismic effects separate from any other loading history contributions to LOCA. The seismic contributions can then be segregated from the final results and used to examine the effect of conditional seismic events on the LOCA frequency distributions.

Mitigation and Maintenance Issues: The final piping variable which was discussed separately is in the area of mitigation and maintenance. This is an important topic area because these procedures have been developed to ensure piping system integrity and prevent piping rupture. The effectiveness of any particular procedure is often a function of the degradation mechanism, although some issues developed are not specific to any mechanism. These procedures can sometimes result in unintended consequences which actually exacerbate the piping failure likelihood. It should be stressed that the procedures and issues in this table are not just concerned with current practice, but also future application and possible improvements. Each panel member must express his or her expectations about future LOCA performance up to the end of the plant license-renewal period.

The first four mitigation procedures (Table B.1.6) are related to piping system inspection and maintenance. In-service inspection and RI-ISI considers both current application of these programs and future industry trends. Currently, more US plants are adopting RI-ISI. The effectiveness of these techniques to find and determine the extent of degradation is considered in this category. Often a technique's effectiveness is quantified by the POD for a certain degradation mechanism. Leak detection considers the broad array of leak detection methods (including plant walkdowns) and their effectiveness in uncovering piping degradation. Online monitoring considers the effect that current system performance indicators (pressure, temperature, etc.) may have on preventing failures as well as future systems that could be utilized to monitor degradation in real-time.

Planned maintenance accounts for programs which monitor degradation and then replace piping segments once the degradation exceeds allowable limits. This is a popular approach for dealing with FAC in carbon steel piping. Planned maintenance also considers any component cleaning or preparation for inspections and the effect on the failure likelihood. Planned maintenance can be either beneficial or detrimental. For instance, maintenance requires closed systems to be opened which could introduce air if gas blanketing is not sufficient. Maintenance can then lead to more future problems than if the maintenance had not occurred.

Table B.1.6 Mitigation and Maintenance Variables

Mitigation/Maintenance Procedures & Issues	Degradation Mechanism Specific	Plant Specific Variable
ISI/RI-ISI	X	
Leak Detection (Plant Walkdown)	X	
Online Monitoring	X	
Planned Maintenance	X	
Water Chemistry	X	
Decontamination	X	
Internal Linings and Coatings	X	
Weld Overlay	X	
IHSI/MSI	X	
Pipe Replacement (New Materials, New Design & Layout)	X	
Improved Weld Techniques/Materials	X	
Improved Inspection Techniques	X	
Socket Weld Replacement	X	
Plant Operating Conditions	X	
Stratification Mitigation	X	
Utility Safety Culture		X
Regulatory Safety Culture		

The next group of procedures is concerned with changing either the piping medium environment or the metal/medium interface in order to impede degradation. Water chemistry is concerned with additions or changes in the basic water chemistry in order to reduce the degradation rate of a certain mechanism. For instance, hydrogenated and noble metal additions to BWR water have proven effective in impeding the rate of IGSCC. This category also considers fluctuations in water chemistry over the plant's operating cycle and the affect that this may have on failure rates. Decontamination is related to water chemistry and considers the removal of impurities in the water supply and the possible impact that this could have on the degradation rate of certain mechanisms.

The application of internal linings and coatings is used to segregate a susceptible piping material from the environment using a coating or overlay of a more resistant material. The coating or lining performs the same role as stainless steel cladding does in protecting carbon steel piping. Weld overlays, induction heating stress improvement (IHSI), and mechanical stress improvement (MSI) all attempt to change or relieve weld joint residual stress in order to impede crack growth. Normally, they attempt to create compressive residual stresses at the inside surface, or throughout the entire, piping segment. However, as mentioned earlier, they can also affect the residual stress in other sections of the piping.

The next group of mitigation and maintenance procedures is concerned with anticipated future improvements in materials, repair techniques, and inspection methods that could reduce the likelihood of future LOCAs. These improvements are captured by the categories for Improved Weld Techniques/Materials and Improved Inspection Techniques. The Pipe Replacement category considers not just the removal of possibly degraded piping, but also the replacement with new materials that are less susceptible to known, important degradation mechanisms for a certain system. This could also be coupled with new design and layout configurations to reduce residual stresses and improve accessibility for inspections. Socket weld replacement is a specific piping replacement program that is currently being considered. The effectiveness and scope of its implementation still is uncertain which is the reason that this has been included as a separate category.

The next two mitigation procedures are related to current and future plant operating performance. Thermal stratification mitigation is a specific technique employed by some plant operators in order to improve thermal mixing and reduce stratification and striping stresses which can occur in the surge line and other piping systems. General plant operating performance is a category that captures all other similarly related issues. This would include issues such as the effect of possible power upgrades, possible future changes in the heat-up and cool-down cycle, the possibility of increased time periods between outages, etc.

The final two issues (Table B.1.6) are related to utility and regulatory safety culture in general. There are many specific issues that the group lumped into these broad categories. One such issue within the utility safety culture is human error. Human error was defined by the expert panel as the likelihood that incorrect action is taken during mitigation and maintenance. This includes improper application of techniques and procedures, misinterpretation of obvious indications (beyond the POD included in ISI), and omission of a prescribed procedure. Other issues within the broad category of utility safety culture include the adoption and implementation of risk-informed management strategy which requires a detailed understanding of real-time plant risk and the objective to embrace changes that reduce overall plant risk. The impact of economic considerations is important in terms of choosing which mitigation strategies to pursue. All decisions weigh plant risk with economic considerations to hopefully arrive at the optimal mitigation strategy. However, this mitigation strategy might not lead to the lowest possible plant risk. Flow accelerated corrosion monitoring programs illustrate this concept. While the absolute lowest plant risk could be achieved for a system by replacing the pipe with FAC-resistant materials, many plants have chosen to monitor the degradation and replace only when the failure risk becomes unacceptable.

Also part of the general safety culture are the lessons-learned from past problems. This experience may decrease the response time for mitigating future problems. For instance, the industry experience with mitigating IGSCC in the early 1980's may provide some useful strategies for PWSCC issues that are currently surfacing. Response time is generally an important mitigation concept. When degradation mechanisms are identified, the failure likelihood due to these mechanisms may continue to increase with time until effective mitigation strategies are employed which reduce their propensity. Obviously, shorter response times are preferred. The industry required roughly three to four years to fully implement IGSCC cracking mitigation strategies after the issue was fully identified. A final related issue is technology transfer which is the training of and knowledge transfer to the next generation of plant operators and engineers. As the workforce continues to age and is replaced by less experienced workers, it is possible that plant risk may be affected.

The regulatory safety culture also encompasses many of the same issues discussed under utility safety culture. Certainly lessons-learned, regulatory response time, and technology transfer equally apply to the regulatory culture. The regulatory environment is also affected by the agency's interaction with the public and the changing public perception of risk. Management philosophy and the adoption of risk-informed regulations may also influence the regulatory safety culture.

The group next determined if the effectiveness of specific mitigation or maintenance procedures varied as a function of degradation mechanisms and materials being evaluated. This dependency is reflected in the second column of Table B.1.6. Plant and regulatory safety culture were considered to be general issues which do not vary significantly with the degradation mechanism. However, the utility safety culture was considered by the panel to vary from plant to plant, as indicated in Table B.1.6. Regulatory safety culture was not determined to be a plant-specific function.

BWR Piping Systems: The variables just discussed (geometry, materials, loading, degradation mechanism, and mitigation) are important for determining the overall LOCA frequencies. However, these variables are linked to each specific LOCA-sensitive piping system (Figure B.1.1). The next group task was to identify LOCA-sensitive piping and assign only pertinent variables to each system. This task spanned both the Wednesday (2/5) and Thursday (2/6) meeting days, but it is summarized here for continuity. It should also be noted that Sam Ranganath provided most of the initial input for BWRs.

The data presented in Tables B.1.7 (for BWRs) and B.1.8 (for PWRs) summarize information developed by the panelists at several brainstorming sessions. Therefore, they represent panelist expertise rather than a systematic analysis of a database of piping system geometries. In some cases, there are minor discrepancies between the pipe size ranges provided and actual plant survey data. For example, a review of the LBB database developed by Emc^2 shows that there are 16-inch diameter surge lines, such as those at South Texas Units 1 and 2, while Table B.1.8 indicates that the range in surge line diameters is 10 to 14 inches. However, these relatively small discrepancies in the outer ranges of these bound have no significant effect on the results and conclusions which focused on developing generic LOCA frequency estimates. The diameters provided in this table do represent nominal pipe dimensions (e.g., 12-, 16-, 28- inch diameter) for the sake of simplicity, rather than inside pipe dimensions which would have some fractional component. This convention was utilized by consensus among the panelists.

The BWR piping systems that can result in a LOCA were first identified (Table B.1.7). These include the recirculation (RECIRC), feed water, steam line, high pressure (HPCS), and low pressure core spray (LPCS), RHR, RWCU, CRD, standby liquid control (SLC), instrument lines in both the reactor and in other piping systems (INST), drain lines, head spray lines, steam relief valve lines, and the reactor core isolation cooling (RCIC) system. It should be noted that while drain lines are associated with each system, they were segregated into a separate category due to their common functionality. The materials commonly used for the piping within each system are identified (column 2 of Table B.1.7). Similarly, the safe end material (column 4) and the weld material (column 5) are also indicated. The intent of this identification was to be comprehensive and also indicate the most prevalent materials wherever possible. Table abbreviations are provided at the end of the table.

Some additional clarification is required for certain entries in this table. In the recirculation piping system, the safe end materials (stainless steel or Alloy 600) are furnace-sensitized during manufacture. The feedwater safe end is manufactured either by interference fit, butt welded, or by a triple sleeve weld overlay. The HPCS and LPCS contain both creviced and non-creviced welds between the piping and safe end. Also, the bulk of this system's piping material is carbon steel. The CRD system consists of a crevice Alloy 82/182 weld to the RPV head while the stub tube ("safe end" in this system) is stainless steel and alloy 600 which is welded and furnace sensitized. It should be noted that no stainless steel clad carbon steel, cast stainless steel, or bimetallic stainless steel welds were indicated in any of these systems, although they are listed in Table B.1.2.

The next variable listed in Table B.1.7 is the nominal piping size present in the system (column 3 in Table B.1.7). This is the only geometric variable (Figure B.1.1 and Table B.1.3) indicated in Table B.1.7. It was not possible to do an exhaustive listing of the possible geometric variables (as with materials) due to the complexity and plant-specific nature of variables related to layout, configuration, weld location, and component type. Each expert panel member must individually determine variability and influence of these parameters. The rationale for listing the piping size is only to provide the panel with an indication of the piping size for maximum leak rate assessment. Common piping sizes for a system are separated by commas in Table B.1.7. Size ranges are separated by a dash and the maximum piping size is given as < 4 inch for many of the smaller systems. It should be noted that the feedwater system typically consists of 10 or 12 inch diameter pipes. However, a range from 12 to 24 inch is also possible.

Significant degradation mechanisms that could be associated with piping materials in each system are included in the sixth column of Table B.1.7. Unanticipated mechanisms (UA) and fabrication defect and repair (FDR) issues are present in every system. Stress corrosion cracking is listed for all stainless and carbon steel materials, while global corrosion (GC) is associated solely with carbon steel piping. Localized corrosion (LC) is included for carbon steel and stainless steel piping for all systems, except SLC, CRD, and instrumentation lines. Material aging (MA) was considered for the higher temperature lines that see constant use for both stainless and carbon steels. Flow sensitive (FS) degradation is present in all carbon steel piping systems with constant use. Mechanical fatigue was judged to be significant in all of the smaller piping systems (< 4 inch diameter). However, it was also considered important in the SRV and feedwater systems. Thermal fatigue (TF) was judged to be important in the feedwater, RHR, RWCU, HPCS/LPCS, and head spray piping.

Table B.1.7 BWR LOCA-Sensitive Piping Systems

System	Piping Matls.	Piping Size (in)	Safe End Matls.	Welds	Sig. Degrad. Mechs.	Sig. Loads.	Mitigation /Maint.
RECIRC	304 SS, 316 SS, 347 SS	4, 10, 12, 20, 22, 28	304 SS, 316 SS, A600*	SS, NB	UA, FDR, SCC, LC, MA	RS, P, S, T, DW, SUP, SRV, O	ISI w TSL, REM
Feed Water	CS	10, 12 (typ), 12 - 24	304 SS, 316 SS*	CS, NB	UA, FDR, MF, TF, FS, LC, GC, MA	T, TFL, WH,P, S, SRV, RS, DW, O	ISI w TSL, REM
Steam Line	CS – SW	18, 24, 28	CS	CS	UA, FDR, FS, GC, LC, MA	WH, P, S, T, RS, DW, SRV, O	ISI w TSL, REM
HPCS, LPCI	CS (bulk), 304 SS, 316 SS	10, 12	304 SS, 316 SS, A600*	CS, SS, NB	UA, FDR, SCC, TF, LC, GC, MA	RS, T, P, S, DW, TS, WH, SUP, SRV, O	ISI w TSL, REM
RHR	CS, 304 SS, 316 SS	8 - 24	CS, 304 SS, 316 SS	CS, SS, NB	UA, FDR, SCC, TF, FS, LC, GC, MA	RS, T, P, S, DW, TS, O SUP, SRV	ISI w TSL, REM
RWCU	304 SS, 316 SS, CS	8 – 24	CS, 304 SS, 316 SS	CS, SS, NB	UA, FDR, SCC, TF, FS, LC, GC, MA	RS, TS, T, P, S, DW, SUP, SRV, O	ISI w TSL, REM
CRD piping	304 SS, 316 SS (low temp)	< 4	Stub tubes – A600 and SS*	Crevice A182 to head	UA, FDR, MF, SCC	RS, T, P, S, DW, V, O, SRV	ISI w TSL, REM
SLC	304 SS, 316 SS	< 4	304 SS, 316 SS	SS, NB	UA, FDR, MF, SCC	RS, T, P, S, DW, V, O, SRV	ISI w TSL, REM
INST	304 SS, 316 SS	< 4	304 SS, 316 SS	SS, NB	UA, FDR, MF, SCC, MA	RS, T, P, S, DW, V, O, SRV	ISI w TSL, REM
Drain lines	304 SS, 316 SS, CS	< 4	304 SS, 316 SS, CS	SS, NB	UA, FDR, MF, SCC, LC, GC	RS, T, P, S, DW, V, O, SRV	ISI w TSL, REM
Head spray	304 SS, 316 SS, CS	< 4	304 SS, 316 SS, CS	SS, NB	UA, FDR, SCC, TF, LC, GC	RS, P, S, T, DW, SRV, O	ISI w TSL, REM
SRV lines	CS	6, 8, 10, 28	CS	CS	UA, FDR, MF, FS, GC, LC, MA	RS, P, S, T, DW, SRV, O	ISI w TSL, REM
RCIC	304 SS, 316 SS, CS	6, 8	304 SS, 316 SS	SS NB	UA, FDR, SCC, LC, MA	RS, P, S, T, DW, SRV, O	ISI w TSL, REM

* See note in text.

304 SS = 304 series stainless steel
(Table B.1.2)

316 SS = 316 series stainless steel
(Table B.1.2)

A600 = Alloy 600

CS = carbon steel

CS – SW = seam welded carbon steel

NB = Nickel-based weld (Alloy 82/182)

UA = unanticipated mechanisms

MA = material aging
LC = local corrosion
FDR = fabrication defect and repair
SCC = stress corrosion cracking
MF = mechanical fatigue
TF = thermal fatigue
FS = flow sensitive (inc. FAC and erosion/cavitation)
ISI w TSL = Current ISI procedures with technical specification leakage detection requirements considered.
RS = residual stress

P = pressure
S = Seismic
T = Thermal
DW = dead weight
SUP = support loading
SRV = SRV loading
WH = water (and steam) hammer
O = overload
V = vibration
TFL = thermal fatigue loading from striping
TS = thermal stratification
REM = all remaining mitigation strategies possible (eg. not unique to piping system)

Significant loading sources for BWR piping systems are also listed (column 7). All systems undergo residual stress (RS), pressure, thermal, seismic, SRV and dead weight (DW) loading. Water (or steam) hammer (WH) was considered to be important in the feedwater, steam line, and HPCS/LPCS systems. Support loading (SUP) was mainly considered to occur through the snubber support. This is important for the recirculation (RECIRC), RHR, RWCU, and HPCS/LPCS systems. Vibrational loading is listed for the smaller diameter piping systems. This loading is always coupled with mechanical fatigue degradation (column 6). However, vibrational loading is conspicuously absent from the feedwater and SRV lines which both have MF in the list of significant degradation mechanisms.

Overloads (O) are possible for all systems, but they are likely to be external due to pipe whip, jet impingement, or secondary system failure. However, the drain line, CRD, instrument lines, and SLC were deemed to be more likely to be susceptible to internal overloads. The thermal loading was broken down into thermal fatigue loading due to striping (TFL) in the feedwater system, and thermal fatigue loading due to stratification (TS) in the RHR, RWCU, and HPCS/LPCS systems. The head spray line, which is also judged to be TF susceptible, does not have a corresponding thermal fatigue loading source considered.

There were no mitigation and maintenance procedures that were identified by the panel as being unique for any particular BWR piping system. Standard mitigation and maintenance for all systems is ISI with credit given for technical specification leakage (TSL) detection. The technical specification leakage threshold is 1 gallon per minute. The effect of all remaining (REM) mitigation and maintenance procedures and issues (Table B.1.6) on the LOCA likelihood should be considered by the panel.

Table B.1.8 PWR LOCA-Sensitive Piping Systems

System	Piping Matls.	Piping Size (in)	Safe End Matls.	Welds	Sig. Degrad. Mechs.	Sig. Loads.	Mitigation /Maint.
RCP: Hot Leg	304 SS, 316 SS, C-SS, SSC-CS CS – SW	30 - 44	A600, 304 SS, 316 SS, CS	NB, SS, CS	TF, SCC, MA, FDR, UA	P, S, T, RS, DW, O, SUP	ISI w TSL, REM
RCP: Cold Leg/Crossover Leg	304 SS, 316 SS, C- SS, SSC-CS, CS – SW	27 - 34	A600, 304 SS, 316 SS, CS	NB, SS, CS	TF, SCC, MA, FDR, UA	P, S, T, RS, DW, O, SUP	ISI w TSL, REM
Surge line	304 SS, 316 SS, C-SS	10 - 14	A600, 304 SS, 316 SS,	NB, SS	TF, SCC, MA, FDR, UA	P, S, T, RS, DW, O, TFL, TS	TSMIT, ISI w TSL, REM
SIS: ACCUM	304 SS, 316 SS, C-SS	2 - 12	A600, 304 SS, 316 SS,	NB, SS	TF, SCC, MA, FS, FDR, UA (FAC)	P, S, T, RS, DW, O	ISI w TSL, REM
SIS: DVI	304 SS, 316 SS	2 - 6	A600, 304 SS, 316 SS,	NB, SS	TF, SCC, MA, FS, FDR, UA (FAC)	P, S, T, RS, DW, O	ISI w TSL, REM
Drain line	304 SS, 316 SS, CS	< 2"			MF, TF, GC, LC, FDR, UA	P, S, T, RS, DW, O, V, TFL	ISI w TSL, REM
CVCS	304 SS, 316 SS	2 – 8	A600 (B&W and CE)	NB	SCC, TF, MF, FDR, UA	P, S, T, RS, DW, O, V	ISI w TSL, REM
RHR	304 SS, 316 SS	6 – 12			SCC, TF, MA, FDR, UA	P, S, T, RS, DW, O, TFL, TS	ISI w TSL, REM
SRV lines	304 SS, 316 SS	1 – 6			TF, SCC, MF, FDR, UA	P, S, T, RS, DW, O, SRV	ISI w TSL, REM
PSL	304 SS, 316 SS	3 – 6		NB	TF, SCC, MA, FDR, UA	P, S, T, RS, DW, O, WH, TS	ISI w TSL, REM
RH	304 SS, 316 SS	< 2	A600		MF, SCC, TF, FDR, UA	P, S, T, RS, DW, O, V, TS	ISI w TSL, REM
INST	304 SS, 316 SS	< 2	A600		MF, SCC, TF, FDR, UA	P, S, T, RS, DW, O, V	ISI w TSL, REM

C-SS = cast stainless steel
SSC-CS = stainless steel clad carbon steel
FW = fretting wear

TSMIT = thermal stratification mitigation
HREPL = vessel head replacement

PWR Piping Systems: LOCA-sensitive PWR piping systems were also determined (Table B.1.8) along with associated piping safe end, and weld materials; pipe size; significant degradation mechanisms; significant loading sources; and system-dependent mitigation and maintenance procedures. The format of this table is identical to the BWR summary table (Table B.1.7) and the abbreviations have been retained.

Unique abbreviations are defined at the bottom of Table B.1.8. It should be noted that the group did not discuss broad differences between Westinghouse (W), B&W, and Combustion Engineering (CE) designs. In fact the only place it is noted (Table B.1.8) is in the use of A600 safe ends in the CVCS system in CE and B&W plants. However, any plant design distinctions may be important for certain LOCA classes and should be considered by each expert during analysis.

The LOCA-sensitive PWR systems listed include reactor coolant piping (RCP) hot leg, RCP cold leg, and RCP crossover legs, the surge line, SIS accumulator line (ACCUM) and SIS DVI line, drain lines, CVCS, RHR, SRV lines, pressurizer spray lines (PSL), CRDM lines, reactor head (RH), in-core instrumentation (ICI), and instrumentation (INST) lines. The hot leg was segregated from the other RCP components by the group due to its higher operating temperature. The SIS system was divided into the ACCUM and DVI components to account for the piping size, material, and functionality differences. The RH group was intended to capture all the non-CRDM lines that penetrate the upper reactor vessel head. This grouping is distinct from the ICI system. The INST line grouping here considers mainly lines within piping systems and not the reactor. It is worth noting that this grouping is different from the grouping for BWRs where INST lines capture both piping and reactor lines.

The materials utilized in PWR piping systems are similar to those in counterpart BWR systems. One difference is the inclusion of cast stainless steel (C-SS) in RCP, surge line, and ACCUM PWR piping. Also, stainless steel clad carbon steel (SSC-CS) is prominent in certain plant designs within the RCP. There is also less use in general of carbon steel in PWRs. It should be noted that the only material listed in Table B.1.2 which is not explicitly listed in either BWR (Table B.1.7) or PWR (Table B.1.8) piping systems is bimetallic stainless steel welds.

The degradation mechanisms are again tied to the material and functional considerations of the piping system. The FDR and UA categories are included for all systems, as is thermal fatigue. Stress corrosion cracking was affiliated primarily with stainless steel piping, but also for carbon steel. Material aging was listed for several higher-temperature, constant-service piping systems (PSL, RHR, RCP, SIS) and mechanical fatigue was deemed important for smaller diameter piping, including the CRDM. Flow sensitive degradation, specifically FAC, was determined to be important in only the SIS system piping, while fretting wear (FW) is listed only for the ICI system.

Significant loading sources are consistent with the BWR piping sources. Pressure, seismic, thermal, RS, DW, and overload loading histories are sources for all systems. Smaller lines again are again considered to be susceptible to vibration loading and this loading is linked to the MF degradation mechanism. The RCP system is considered to have additional support loading contributions, mainly due to snubber malfunction. Both thermal stratification and thermal fatigue loading due to striping and heat-up/cool-down were listed as significant for the surge line.

Thermal fatigue loading is also important for the RHR and drain lines according to the group, while the reactor head and pressurizer surge lines are influenced by thermal stratification. The PSL also must consider water hammer. Only the SRV lines need to consider SRV transients which is quite different that the BWR classification. All of the major loading variables (Table B.1.5) were considered in either BWR or PWR systems. However, hanger misadjustment and cavitation loading were not specifically mentioned. They would certainly fall under the broader loading categories listed in Table B.1.5, but may need to be considered individually by each expert during the elicitation.

As with BWR piping, ISI with credit for leak detection is existent for all piping systems. All remaining mitigation and maintenance issues should also be considered for their effect on the LOCA frequencies. However, some specific mitigation procedures have been highlighted. This includes thermal stratification mitigation which some operators practice to limit surge line loads. Also, reactor vessel head replacement

(HREPL) is a solution being considered to alleviate CRDM cracking concerns. The group will need to consider the extent and effectiveness (now and in the future) of each of these specific procedures.

Base Case Development

During the elicitation, each panelist will determine the biggest contributors for each LOCA frequency threshold category (Table B.1.1) separately for BWR and PWR plants. The information summarized in Tables B.1.2 – B.1.8, and discussed in the previous sections, is simply intended to identify those issues and variables which contribute to the LOCA frequency distributions. Each panelist will determine the magnitude and likelihood of each variable separately, but more importantly will determine the importance of the interrelationships among the variables.

Each panel member will not be asked to provide absolute LOCA frequencies. All questions will be structured so that relative differences with a specific base case will be queried. The base cases will be associated with absolute frequencies and quantitative LOCA estimates will be derived from these values and the relative relationships provided during the elicitation. The group spent quite a bit of time and effort both understanding the role of the base cases in the elicitation and assessing their importance. Ideally the base cases are chosen to represent significant contributing conditions to the total LOCA frequency estimates in order to minimize the extrapolation required during questioning. Representative, significant base cases should therefore theoretically improve the elicitation accuracy. Once this concept was understood, the group settled on several base case conditions for PWR and BWR systems. Pete Riccardella provided the original suggestions which were largely adopted by the panel after the analysis framework was clarified. The base case discussion evolved over the Wednesday and Thursday meeting days. All of the discussion will be summarized in this section for consistency.

Two base cases were developed for BWR piping systems (Table B.1.9). The first case will examine the recirculation system piping. All the various piping sizes will be considered, and original 304 stainless steel material will be assumed that has not been replaced during plant operation. The safe end is non-creviced Alloy 600 which is connected to the piping and vessel by Alloy 82/182 weld material. Only the IGSCC (subcategory of SCC, Table B.1.2) will be considered as the degradation mechanism. The loading will consist of pressure, residual stress, and dead weight nominal components. Transients to be considered include SRV loads and seismic. The base case will assume that NWC is used in the system.

The next BWR base case will examine the feed water system. A 12 inch diameter carbon steel pipe will be analyzed. The safe end and weld materials were not specified and will need to be defined by the base case analysis team (to be discussed subsequently). The degradation mechanisms for this base are FAC and TF. Loading sources include pressure, thermal, residual stresses, and dead weight nominal components. Thermal fatigue loading from stratification and possibly striping will provide alternating loads, and water flow velocities will also be included to assess fluid loading. Transients for this base case will include water hammer and seismic. Once again NWC will be assumed.

Three base cases were constructed for PWR systems (Table B.1.9). The first will examine the hot leg in the reactor coolant piping system. A 30 inch diameter Type 304 stainless steel pipe will be considered with Alloy 600 safe ends and Alloy 82/182 bimetallic welds. This base case will examine thermal fatigue and PWSCC. Loading will again include pressure, thermal, residual stress, and dead weight nominal loads and thermal fatigue alternating loads. Transients will include seismic loading and a pressure pulse transient, the magnitude and duration of which is still to be determined.

Table B.1.9 Base Case Analyses

Plant Type	System	Piping Size (in)	Piping Material	Safe End Material	Weld Material	Degradation Mechanism	Loading	Mitigation/ Maint.
BWR	RECIRC	12 – 28	Original 304 SS	Non creviced A600	A82	IGSCC	P, S, RS, DW, SRV	NWC, leak detection, ISI
	Feed water	12	CS			FAC, TF	P, S, T, RS, DW, WH, Flow velocities	NWC, leak detection, ISI
PWR	RCP – Hot Leg	30	304 SS	A600	A82	TF, PWSCC	P, S, T, RS, DW, pressure pulse	ISI, leak detection
	Surge Line	10	304 SS		A82 at Pressurizer	TF, PWSCC	P, S, T, RS, DW, pressure pulse	ISI, leak detection
	SIS: DVI HPI/mak eup	4	SS/CS			TF	P, S, T, RS, DW, pressure pulse	ISI, leak detection

The next PWR base case is a 10 inch diameter surge line. The surge line material is Type 304 SS and an Alloy 82/182 bimetallic weld will be included at the pressurizer. No safe end materials will exist. Once again, thermal fatigue and PWSCC will be considered. Loading will include pressure, thermal, residual stress, and dead weight nominal loads and thermal fatigue alternating loads. Transients will include seismic loading and a pressure pulse transient, the magnitude and duration of which is still to be determined.

The final PWR base case was the most ill-defined case because it was added after initial group discussions. It was added to provide a base case for a smaller diameter piping system. This base case will need to be defined more completely prior to analysis. A 4 inch diameter high pressure injection/makeup (HPI/MU) line will be examined for thermal fatigue degradation. The piping material, welding, and safe end materials still need to be specified. Nominal loading is once again provided by pressure, thermal, residual stress, and dead weight loads. Thermal alternating loads will be defined and seismic and pressure pulse transients will be considered.

Absolute LOCA frequencies will be developed for each base case and for each threshold leak rate category defined in Table B.1.1. There are six leak rate categories and five base cases; therefore at least thirty separate calculations will be required to fully define the base case frequencies. The base cases will include analysis of many welds and other piping components. The LOCA frequencies for the system will obviously be the summation of the contributions from all system components. The frequencies will also be determined as a function of time. Three time periods will be evaluated: 25 years after plant startup (current-day), 40 years after start-up (end-of-plant-license), and 60 years after start-up (end-of-plant-license-renewal).

The panel decided that seismic transients would be handled as part of a sensitivity study. As mentioned previously, seismic-induced LOCAs will not be determined as part of this elicitation for several reasons: the PRA models that these estimates will be used for do not consider seismic loading; there has been significant work in developing seismic LOCA estimates; and the group has no specific expertise in seismic analysis. However, many panel members are experienced in conducting analysis when the seismic loading history is provided. Many panel members also seem comfortable with comparing other loading histories to seismic events. Therefore, it was decided that the elicitation would ask for comparisons with conditional seismic loading (probability of occurrence of 1) separately in order to conduct a sensitivity analysis of the effect of seismic loading.

Originally, the group determined that the seismic loading magnitude will be 0.3 g for the sensitivity study. Pete Riccardella suggested changing the 0.3 g criteria for the seismic condition to an ASME Code faulted-stress condition. By doing this soil conditions, damping characteristics, etc. do not have to be considered. While there was some discussion on the merits of this suggestion, the issue was not finalized.

The base cases will also perform a sensitivity analysis to determine the effect of the frequencies both with and without ISI. Standard ISI techniques will be considered in the analysis. Credit will also be given in this analysis for leak detection. The leak detection threshold for the base case analysis will be leak rates which are commensurate with the data defined in the SKI-pipe database. The specific leak rate threshold associated with the database must be defined.

A base case team was established to develop four separate LOCA frequency estimates for each of the five base cases. The base case team will consist of **Vic Chapman, Bengt Lydell, David Harris,** and **Bill Galyean.** The team will model a specific piping system and define plant operating characteristics for each base case. Then, the team will develop input for each of the five LOCA variables (material, geometry, loading, degradation mechanism, and mitigation/maintenance) within the parameter constraints identified in Table B.1.9. The team will share information to ensure that each analysis is considering the same nominal conditions for each base case.

Each base case team member will develop their own LOCA estimates for each system using whatever methodology they choose. The likely general approaches for each team member are summarized in Table B.1.10. Each specific methodology will require additional assumptions. It is incumbent that each team member catalog required assumptions and document the methodology used to arrive at their base case LOCA estimates. This information will then be rigorously, yet concisely, presented to the remaining panel members. This should allow each panel member to completely understand the assumptions, methodology, and results generated by each base case team member. It needs to be stressed that once the general conditions are developed, the base case members should independently develop their estimates without further consultation. This step is necessary to retain realistic sample uncertainty in the calculated results.

Table B.1.10 Base Case Approaches

Base Case Team Member	Analysis Approach
Vic Chapman	PFM using PRODIGAL code
Bill Galyean	Direct analysis of operating experience
David Harris	PFM using PRAISE code
Bengt Lydell	Direct analysis of operating experience

The base case team would collaborate to ensure that the PFM analyses accurately capture that leaking pipe operating experience. This is the one aspect of the exercise that contains plant operating experience data. Initial PFM calculations will be conducted based on best-estimate assumptions and the current leak rate frequency predictions will be compared with the operating experience. At this point, PFM input assumptions may be changed in order to match the operating history. Each PFM model should accurately document the input variables, any model changes, and results both before and after benchmarking. This benchmarking exercise will help the remaining panel members gauge uncertainty in the calculations.

The panel will supply background information to the base case calculation team as required. All requests for background information will be coordinated by Rob Tregoning to ensure proper cataloging and dissemination to the group. Some volunteers already offered certain background information. Bruce Bishop has run his own PFM models for seven Westinghouse plants that could ultimately be used to help verify the base case calculations.

It was also stressed by the panel that it is important to make the PFM modeling conditions as close as possible to the postulated service conditions so that the various base case approaches can be directly compared to assess uncertainty and possible inaccuracies. For instance, many existing PFM models assume that all repairs are perfect (no defects). However, many repairs introduce new defects and most large flaws are associated with repairs. This fact (and other similar issues) is naturally captured within the operating-experience database, and needs to be considered within the PFM modeling if possible.

Confidentiality

A discussion was held on the confidentiality of participant's responses during the exercise. Rob Tregoning indicated that all information provided as part of this exercise will remain confidential and will not be distributed to anyone not specifically involved in the exercise. The kick-off meeting has been videotaped, but this will not be distributed outside of the group. Elicitation sessions will be taped for accuracy, but this information will also not be made public. There will be public reporting of the assumptions, methodology, elicitation results, and calculated LOCA frequencies that stem from this exercise. However, the summary reporting will only identify the names, affiliations, and possibly credentials of the expert elicitation panel and the facilitation team early in the report. No reference to individual opinions will be documented.

Day 3 – Thursday, February 6, 2003

Use of Base Case in Elicitation

The base cases LOCA frequencies will provide absolute LOCA estimates that each panel member will use to anchor the relative likelihood of LOCAs in other (non-base case) piping systems. Each panel member will also determine how well the base cases depict expected current and future LOCA performance in the piping systems that they model. It is therefore *not* important that a panel member agree with the modeling assumptions, approach, and results provided by the base case team. However, it is imperative that the base case development is completely understood by each panel member. Each panel member will be able to correct perceived deficiencies in the calculated base case frequencies during the elicitation. Each panel member will also determine, relative to the base case results, the LOCA contributions of other (non base case) piping systems, and the contributions and uncertainty induced by the primary piping system variables.

Reference Case Development

In order to decompose the elicitation topics further, the group determined that it would be useful to further decompose non base case piping systems and variables. This was accomplished by defining a set of reference conditions for each LOCA-sensitive piping system identified in Tables B.1.7 and B.1.8. The reference conditions are similar to the base cases in that they define a unique set of conditions (materials, geometric variables, mitigation and maintenance procedures, and degradation mechanisms) that can be analyzed. They are different from the base cases in that absolute LOCA frequencies will not be developed for the reference cases by the base case team. The reference cases for various systems will be compared to determine the *relative* LOCA-severity among piping systems. LOCA-severity variability within any specific system can then be gauged with respect to the reference case for that system. It will be up to each panel member to determine the method for developing these relative comparisons.

The BWR reference cases (Table B.1.11) represent specific combinations of the possible BWR piping as previously listed in Table B.1.7. In general, one material and degradation mechanism has been chosen is ideally representative of each piping system. The effect of fabrication defects and repair (discussed earlier) should be considered for its effect on the other degradation mechanism in all cases. Nominal pressure, thermal, RS, and DW loading should be considered for all cases. One loading transient was identified for each system. All the transients should be fairly well identified based on past discussion, but the overload transient for the control rod drive (CRD) piping needs to be better defined. In all cases, the snubber is considered to be functional.

Table B.1.11 BWR Reference Case Conditions

System	Piping Material	Piping Sizes (in)	Safe end	Welds	Degradation Mechanisms	Loading	Mitigation and Maintenance
RECIRC	304 SS	10, 12, 20, 22, 28	304 SS	SS	SCC, FDR	P, T, RS, DW, SRV	NWC, ISI w. TSL, 88-01 (AI), 182
Feed Water	CS	10, 12, 12 - 24	304 SS	CS	FAC, FDR	P, T, RS, DW, WH, TFL	NWC, ISI w. TSL, 88
Steam Line	CS – SW	18, 24, 28	CS	CS	FAC, FDR	P, T, RS, DW, SRV	NWC, ISI w. TSL, 88
HPCS, LPCS	CS	10, 12	304 SS	CS	TF, FDR	P, T, RS, DW, TS, SRV	NWC, ISI w. TSL, 88
RHR	304 SS	8 – 24	304 SS	SS	SCC, FDR	P, T, RS, DW, TS, SRV	NWC, ISI w. TSL, 88
RWCU	304 SS	8 – 12	304 SS	SS	SCC, FDR	P, T, RS, DW, TS, SRV	NWC, ISI w. TSL, 88
CRD piping	304 SS	< 4	A600 and SS	Creviced NB welds	SCC, FDR	P, T, RS, DW, O	NWC, ISI w. TSL, 88
SLC	304 SS	< 4	304 SS	SS	SCC, FDR	P, T, RS, DW, SRV	NWC, ISI w. TSL, 88
INST	304 SS,	< 4	304 SS	SS	MF, FDR	P, T, RS, DW, V, SRV	NWC, ISI w. TSL, 88
Drain lines	304 SS	< 4	304 SS	SS	SCC, FDR	P, T, RS, DW, SRV	NWC, ISI w. TSL, 88
Head spray	304 SS,	< 4	304 SS	SS	TF, FDR	P, T, RS, DW, SRV	NWC, ISI w. TSL, 88
SRV lines	CS	6, 8, 10, 28	CS		MF, FDR	P, T, RS, DW, SRV	NWC, ISI w. TSL, 88
RCIC	304 SS	6, 8	304 SS	SS	SCC, FDR	P, T, RS, DW, SRV	NWC, ISI w. TSL, 88

Specific piping, safe end, and weld materials were not determined by the group. Table B.1.11 represents an initial attempt to select these materials based on the general discussion. However, the group did decide to consider uncreviced Type 304 where proper in the reference cases. The mitigation and maintenance for all systems should assume NWC. Standard ISI with technical specification leakage detection should be considered along with augmented inspection as defined in generic letter 88-01. The mitigation and maintenance also has listed alloy 182 and stress improved, but these concepts need to be better defined and summarized.

The PWR reference case conditions are provided in Table B.1.12. The reference cases were again distilled from the LOCA-sensitive PWR piping systems (Table B.1.8). The philosophy behind this table was consistent with the BWR reference case development (Table B.1.11) with a few notable exceptions.

The effects of fabrication defects and repair on the other listed degradation mechanisms should again be considered in all piping systems. However, several PWR reference cases list multiple other degradation mechanisms which is a departure from the BWR approach (Table B.1.11). The PWR piping reference cases also account for nominal loading supplied by pressure, thermal, residual stress, and dead weight loading to each system. However, very few PWR systems have associated loading transients while all BWR systems do. It may be necessary to add associated PWR transients for consistency. The PWR piping, safe end and weld materials were also not specified. Some initial choices have been made in Table B.1.12, but feedback from the group is required in order to finalize selection. The mitigation and maintenance to be considered for each reference case consists of ISI with TSL. No other special mitigation procedures were identified.

Table B.1.12 PWR Reference Case Conditions

System	Piping Material	Piping Sizes (in)	Safe end	Welds	Degradation Mechanisms	Loading	Mitigation and Maintenance
RCP: Hot Leg	304 SS	30 - 44	A600	NB	TF, SCC,FDR	P, T, RS, DW	ISI w TSL
RCP: Cold/ Crossover Legs	304 SS	22 - 34	A600	NB	TF, FDR	P, T, RS, DW	ISI w TSL
Surge line	304 SS	10 - 14	A600	NB	TF, FDR	P, T, RS, DW, TFL, TS	ISI w TSL
SIS: ACCUM	304 SS	10 - 12	304 SS	SS	TF, FDR	P, T, RS, DW	ISI w TSL
SIS: DVI	304 SS	2 - 6	304 SS	SS	TF, FDR	P, T, RS, DW	ISI w TSL
Drain line	304 SS	< 2"		SS	MF, TF, FDR	P, T, RS, DW, V	ISI w TSL
CVCS	304 SS	2 – 8		SS	TF, MF, FDR	P, T, RS, DW, V	ISI w TSL
RHR	304 SS	6 – 12			TF, FDR	P, T, RS, DW, TS	ISI w TSL
SRV lines	304 SS	1 – 6			TF, FDR	P, T, RS, DW, SRV	ISI w TSL
PSL	304 SS	3 – 6		NB	TF, FDR	P, T, RS, DW, WH	ISI w TSL
RH	304 SS	< 2	A600		TF, FDR	P, T, RS, DW, TS	ISI w TSL
INST	304 SS	< 2			MF, TF, FDR	P, T, RS, DW, V	ISI w TSL

It is important that the baseline and reference case conditions be clearly defined prior to the start of the elicitation so that each panel member understands the general attributes of each of these cases. As mentioned previously, the elicitation questions will be structured to query the variability and uncertainty associated with each piping system with respect to the reference cases. Each panel member will compare the reference and base cases to assess the relative importance of each piping system to the total LOCA frequencies. Every effort will be made to accommodate all requests and information will be shared among the group. Additionally, any areas or issues which are not clear to a panel member should be raised to Rob Tregoning as soon as it arises.

Non-Piping LOCA Contributions

The final portion of the meeting concentrated on developing issues associated with non-piping contributions (passive failures only) to the LOCA frequencies. Active components will be analyzed separately during this program from operating experience. Because active components have maintenance plans, the group in general expects that the failure rate of these components will not increase in the future. Operating experience should therefore adequately represent active component failure rates. This rationale is the basis for considering only passive component failures within this elicitation.

The non-piping contributions will be combined with the piping component to determine the total LOCA frequency (Figure B.1.1). The group decided to break these issues down by component functionality. This is an analogous approach used to tackle the piping contribution which initially segregated piping systems by functionality. Five main components were determined as candidates for passive failures: the pressurizer, the RPV, valves, pumps, and steam generators/steam systems. The valve component category encompasses both pressure isolation valves at Class 1 to Class 2 piping boundaries and also loop-stop valves. The pumps category only considers pumps in the reactor coolant or recirculating water system.

For each component category, the panel developed sub-categories which represent specific possible failure modes (e.g. what portions of the component could fail passively). Each failure mode is governed by the same variables that are important for piping systems (material, geometry, loading, degradation mechanisms, and mitigation/maintenance). Unfortunately the group did not have sufficient time or resources to fully develop comprehensive variable lists in the same manner as for piping systems.

Table B.1.13 illustrates the failure modes developed for pressurizer failures. Please note that all table abbreviations for this and all subsequent tables in this section are as previously defined unless indicated. Bold items in the failure mode sub-category of Table B.1.13 (and all following tables) indicates that operational data exists which captures that component failure. For instance, in Table B.1.13, the group thinks that data is available on heater sleeve failures.

Table B.1.13 Pressurizer Failure Scenarios

Component	Geometry	Material	Degradation Mechanisms	Loading	Mitigation/ Maintenance	Comment
Shell		A600C-LAS, SSC-LAS	GC, SCC, MF, FDR, UA			Boric acid wastage from OD
Manway		NB-LAS, SSC-LAS, LAS, HS-LAS (Bolts)	GC, SCC, MF, SR, FDR, UA			Bolt failures
Heater Sleeves	**Small diam. (3/4 to 1 in)**	**A600, SS**	**TF, MF, SCC, FDR, UA**			**Req. multiple failures**
Bolted relief valves		C-SS	MA, FDR, UA			
Nozzles		SSC-LAS C-SS	CD, TF, SCC, MA, FDR, UA, GC			Same as surge line

NB-LAS = nickel-based clad low alloy steel
SR = Stress Relaxation and loss of preload

The panel identified failures in the pressurizer shell, manway, heater sleeves and nozzles as passive LOCA candidates. Also, the pressurizer bolted relief valves could fail. The group generally did not have information on component geometries and loading and mitigation/maintenance were not discussed by the panel due to lack of time. However, some specific issues were discussed for each of these failure modes. The shell failure envisioned would most likely occur by boric acid wastage from the outer diameter of the shell. Manway failures would result by multiple bolt failures. Heater sleeves fail due to PWSCC, but as a result of their size, multiple failures are required in order to result in a LOCA. Bolted relief valves could fail due to steam cutting or localized bolt corrosion resulting from boric acid leaks.

The RPV failure modes (Table B.1.14) focused on vessel head bolt failure, failure of CRDM connections, nozzle failure, RPV wastage, and RPV corrosion fatigue. Upper head vessel head bolt failure is most likely due to human error during removal at each refueling cycle. Human error could occur as a result of improper installation procedures. Problems, however, could be identified during prestart-up inspection. The lower head bolts are not removed during refueling and they could be susceptible to common cause failure resulting from local bolt corrosion leading to several simultaneous bolt failures. A certain percentage of these bolts are inspected at each outage and the assumption is that inspection would not be effective in identifying the degradation prior to failure. These requirements may uncover the likelihood of common cause errors leading to some latent failure that is not immediately evident and shed light on other possible failure mechanisms. An example of a common cause failure is a torque wrench/tensioner which is out of calibration so that all bolts are improperly installed and then can possibly fail during operation.

CRDM connections far outside of the reactor could be welded, bolted, or threaded and seam welded. The degradation mechanism would be a function of the specific connection. For instance, welded connections would be susceptible to the mechanisms and loading discussed previously for CRDM components. Bolted CRDM connections would be subject to steam cutting, boric acid corrosion, aging and other degradation mechanisms that are unique to bolts. It must be stressed that the CRDM connections in this table refers to the CRDM which connects to the drive mechanism. Inboard connections are considered to be part of the "CRDM piping system" discussed earlier. For bolted connections, this demarcation line is the flange joint. The group identified failure data for CRDM leakage from bolted flanged connections.

Table B.1.14 Reactor Pressure Vessel (RPV) Failure Scenarios

Component	Geometry	Material	Degradation Mechanisms	Loading	Mitigation/ Maintenance	Comment
Vessel Head Bolts		high strength steel	GC, FDR, UA		Human error	Removal leading to human error (common cause failure) during refueling
RPV wastage		SSC-LAS LAS	GC, FDR, UA, MA			LAS = some BWR upper head, Boric acid wastage (upper & lower head, shell)
CRDM connections		**SS**	**FDR, UA**			**welded, bolted, threaded + seal weld**
CRDM	4-6	A600 base nozzle, SS, C-SS, and NB-LAS housing with NB weld	SCC, TF, MF, LC, GC, FDR, UA	P, S, T, RS, DW, O	HREPL, ISI w TSL, REM	Nozzles and piping up to connection
Nozzles		**LAS, SSC-LAS,**	**TF, MF, LC, GC, SCC, FDR, UA**			**LAS = BWR only**
ICI	< 2"	304 SS, 316 SS	MF, SCC, TF, FW, FDR, UA	P, S, T, RS, DW, O, V	ISI w TSL, REM	
RPV Corrosion Fatigue		SSC-LAS LAS	LC, MF, MA FDR, UA			LAS = some BWR upper head, Initiate at cladding cracks (upper & lower head, shell)
BWR penetrations		SS	SCC, LC, FDR, UA			Stub tubes, drain line, SLC, instrumentation, etc.
PWR penetration		SS, A600	SCC, FDR, UA, LC, MF, TF			

NB-LAS = nickel-based clad low alloy steel
SR = Stress Relaxation and loss of preload

There are two RPV degradation mechanisms which were specifically discussed. The first was degradation of the shell, upper head, or lower head due to boric acid corrosion. The second mechanism was corrosion fatigue developed at through-thickness cladding cracks in the shell, upper head, or lower head. The nozzle category is subject to similar degradation mechanisms as in the attached piping. It should be stressed that the nozzle category only considers the flared portion of the nozzle up to the reactor shelf. The nozzle safe end was earlier defined as part of the piping system. The group identified that some data on nozzle issues exists.

Valve failure modes are summarized in Table B.1.15. The cast stainless steel valve bodies are susceptible to an array of potential degradation mechanisms. These include cavitation (CAV), thermal fatigue (TF), and material aging (MA). Casting defects (CD) are another particular concern. Failure due to the other mechanisms listed could initiate at either the defects, or repairs of those defects. The main steam isolation valve (MSIV) body is associated with similar failure modes. Specific failure modes for valve bonnets, and valve bonnet bolts were not discussed. Presumably, bonnet bolt failures would be susceptible to the same failure mechanism of other bolts: aging, boric acid corrosion, steam cutting, etc. The hot leg/cold leg loop isolation valve failure modes were also not discussed. However, failures in these valves could be described in terms of the bonnet, body, or bonnet bolt failure sub-categories listed earlier. It should also be noted that valve sizes are generally consistent with the piping system where they are located.

Table B.1.15 Valve Failure Scenarios

Component	Geometry	Material	Degradation Mechanisms	Loading	Mitigation/ Maintenance	Comment
Valve Body		CS, SS C-SS	FAC, CAV, LC, TF, MA, GC, CD, SCC, FDR, UA			CS, SS = BWR only
Valve Bonnet		CS, SS C-SS	FAC, LC, GC, SCC, MA, CD, FDR, UA			CS, SS = BWR only
Bonnet Bolts		HS-LAS	GC, SCC, FDR, UA SR			
Hot Leg/Cold leg loop isolation valves			FDR, UA			
MSIV Body			CAV, TF, MA, CD			

HS-LAS = High Strength Low Alloy steel (SA540 GrB23, SA193 GrB7)
CAV = Cavitation Damage
SR = Stress Relaxation and loss of preload

Steam generator tube rupture (Table B.1.16) can occur from a variety of different mechanisms including thermal fatigue, mechanical fatigue, SCC, and general corrosion. The tubes can also be degraded by mechanical deformation (MECDEF), or denting, during installation, inspection, or cleaning. Steam generator tubes are too small to lead to a LOCA due to a single tube failure. Therefore, multiple tube rupture needs to also be considered in order to achieve a certain size LOCA. There is data which exists for SGTR.

Steam generator failure can also occur at the manway (specifically bolt failure), the steam generator shell, or the nozzles. These various failure modes were also not sufficiently discussed so little information has been defined in Table B.1.16. However, the nozzle failure issues will likely be similar to the associated piping system, while manway bolt failure would be caused by the same types of mechanisms as for other bolt failures.

The pump failure modes (Table B.1.17) are similar to many of the failure modes already discussed for other components. The cast pump bodies are potentially subject to the same degradation mechanisms (CAV, TF, CD, MA) as other cast components. The recirculation (RECIRC) bonnet bolts and RCP nozzle are also susceptible to mechanisms discussed earlier. The only unique mode considers an incipient failure of a pump flywheel which could initiate collateral damage in other components or in other piping systems. There was no appropriate passive pump failure data that was identified by the group.

Table B.1.16 Steam Generator/Steam System Failure Scenarios

Component	Geometry	Material	Degradation Mechanisms	Loading	Mitigation/ Maintenance	Comment
Tube Rupture	**5/8 to 3/4" diam.**	**A600**	**TF, MF, SCC, GC, LC, FRET, MECHDEF, FDR, UA**			**single and multiple tube rupture**
Manway Bolts		CS, LAS	SCC, GC, LC, SR, FDR, UA			
Shell		CS, LAS,	GC, LC, MF, TF, FDR, UA			
Nozzles to safe end		SSC-LAS CS, LAS SSC-CS	FAC, SCC, FDR, UA			
Tube Sheet Failure		NB-LAS A600	SCC, GC, FRET, MF, FDR, UA			

FRET = fretting or mechanical wear

Table B.1.17 Pump Failure Scenarios

Component	Geometry	Material	Degradation Mechanisms	Loading	Mitigation/ Maintenance	Comment
Pump Body		C-SS, SSC-CS	CAV., TF, CD, MA, SCC, fatigue			
RECIRC Bonnet Bolts		HS-LAS	SCC, GC, SR			
RCP nozzle						
Flywheel failure						initiating collateral damage – secondary pipe failure

HS-LAS = High Strength Low Alloy steel (SA540 GrB23, SA193 GrB7)
SR = Stress Relaxation and loss of preload

It is obvious that the non-piping passive LOCA sources have not been nearly as well-defined as the piping system sources, and they must be better defined prior the elicitations. However, due to the number and complexity of the components, the panel realized that it may not be possible to fully define all the variables listed in the tables above. At a minimum, the group decided that it would need isometric drawings for as many of these components as possible.

The manner for arriving at the LOCA contributions of these other components will be similar to the approach followed for the piping contribution. Reference cases will be developed and absolute LOCA estimates will be assigned to those numbers. However, these reference cases will be based strictly on data. The bolded items in Tables B.1.13 – B.1.16 are component failures that are supported by passive-system failure data. This data will first need to be accumulated and analyzed. Karen Gott and Bill

Galyean are possible sources for some of this data. There is an EPRI database called PM-BASIS which consists of mainly active components, but there may be some data on bonnets and packings. Also, Spence Bush may have some data for these components that might be useful to the group. Finally, the group discussed that there may be data available for feed water nozzles.

Once the data is developed, it will be made available to the group. This data will make up the base case information for the non-piping components. The group will also be asked how representative the base case data is for future (end-of-plant-license-renewal) LOCA estimates. The LOCA propensity (for each leak threshold rate) for the components without data will also be queried relative to these base cases. This approach is identical to the development of the piping LOCA contributions discussed earlier.

MEETING MINUTES FOR SECOND ELICITATION MEETING
FOUR POINTS SHERATON, BETHESDA MD

Day 1 – June 4, 2003 - Base Case Review

Dr. Rob Tregoning (USNRC) welcomed everyone to the Second Elicitation meeting and reviewed the agenda for the two days. Everyone in attendance introduced themselves. A package of the Day 1 presentations was provided to everyone. Rob warned the group that there was a lot of material to cover in each presentation. Furthermore, he indicated that we should not treat what is presented at this meeting as final, but more of a snapshot of where we are presently. Next, Rob reviewed the objectives for the first day of the meeting. The objectives for the first day were:
1. Review base case conditions
2. Understand assumptions, methodologies, and results calculated by each base case member
3. Understand important factors and variables that lead to differences among results
4. Determine what additional calculations are required to complete the base case analysis

Rob also indicated that he was not asking everyone to agree with the results to be presented, but simply to understand what was done. The panel members can state differences of opinion during their individual elicitations.

Next, the second day agenda (Elicitation Coordination) was discussed. The second day agenda has been adjusted slightly to ensure adequate time for the topic of non-piping LOCA frequency determination since it received less attention at the last meeting

Rob then reviewed the meeting objectives for Day 2. The Day 2 objectives were:
1. Finalize elicitation question sets and provide consistent understanding of each question.
2. Determine methodology for evaluating non-piping LOCAs and identification of non-piping base case data.
3. Determine methodology for evaluating conditional seismic loading including determination of seismic loading magnitude.
4. Determine what information panelists will require prior to their elicitations and assign action items for providing information.
5. Develop final schedule and time-frame for upcoming elicitations.

Bruce Bishop (Westinghouse) asked if the panel would get a status report on the new PFM code being developed as part of the USNRC program. Rob indicated that time was not available at this meeting, but that sometime in the fall or winter an initial meeting will be set up where this new code being developed by Battelle and Emc^2 could be presented.

Presentation 1 – Base Case Review and Summary Results
by Rob Tregoning of the USNRC

The purpose of this presentation was to review the conditions analyzed by the base case team and summarize the calculated results to date. This talk served as a prelude for the next four presentations by the base case team members.

The base case results will be used to anchor elicitation responses by the elicitation panel members as part of their individual elicitations. The elicitation members can use one or all of the base case results directly for their anchoring, or provide their own base case analysis if they choose. There were a total of five (5) base cases defined at the first elicitation meeting in February, see Table B.2.1.

Table B.2.1 Base Case Conditions

Base Case Identification	Piping System	Pipe Diameter, inches	Piping Materials	Degradation Mechanisms	Loading	Mitigation
BWR-1	Recirculation	12 to 28	Type 304 stainless (originally), non-creviced A600 safe ends, nickel based (NB) welds	IGSCC	Pressure, RS, DW, safety relief valve transient (SRV)	NWC, leak detection (LD), ISI, augmented inspection per Generic Letter 88-01 (88-01)
BWR-2	Feedwater	12	Carbon steel	FAC, thermal fatigue (TF)	P, RS, DW, thermal (T), water hammer (WH),	NWC, LD, ISI, 88-01
PWR-1	Hot leg	30	Type 304 stainless, A600 safe ends, NB welds	PWSCC, TF	P, RS, DW, T, pressure pulse (PP)	LD, ISI
PWR-2	Surge line	10	Type 304 stainless, A600 safe ends, NB welds at pressurizer	PWSCC, TF	P, RS, DW, T, PP	LD, ISI
PWR-3	High pressure injection makeup nozzle (HPI/MU) (B&W)	4	Stainless and carbon steel	TF	P, RS, DW, T, PP	LD, ISI

As part of the base case effort, the base case members were to evaluate the LOCA frequencies at 25 years (current-day), 40 years (end-of-plant-license), and 60 years (end-of-plant-license-renewal). These results will then be used by the individual elicitation panel members to anchor their respective responses so that they can estimate the various LOCA frequencies at these same time periods.

The goal for the base case members is to calculate results for the set of conditions listed in Table B.2.1. The base case members also shared their results and presentations prior to this meeting so that there was a common format for the presentations. At this time the base case results comparison charts in the handouts should be viewed as works in progress. Furthermore, some results from Vic Chapman (OJV Consultancy) are still forthcoming. Vic and Chris Bell need to provide additional information to the panel members on how they conducted their base case analyses.

The current base case results are summarized in slides 16 through 20 of this presentation. For David Harris's calculations, the frequency results at 60 years were averaged over the 20 year time period from 40 to 60 years while the 25 year estimates were averaged over the first 25 years of operation. It is important that results are consistent (with consistent assumptions and conditions) among each base case team member.

The hot leg results (PWR-1 on page 18 of the handout) indicates large initial uncertainty. Bruce Bishop questioned if the results are for individual welds or the overall system. The response was that the intent was that these results should reflect the frequencies for the overall system. It was pointed out that the hot leg results should reflect the LOCA frequencies for the hot leg only. The results should not consider all the other lines associated with the RCS, i.e., the cold leg, cross over, surge line, etc.

The slides 21 and 22 (Remaining Work and Differences Among Methodologies) of Rob's presentation (Base Case Conditions and Summary Results) were not included in the handout and were to be filled out later by the team members. These updates will be posted on the ftp site once available. It was indicated that the LOCA frequencies for the lower leak rate categories (> 100 gpm [380 lpm]) include all the incidences of LOCAs in the higher leak rate categories, e.g., the 100,000 gpm (380,.000 lpm) bin should include all incidences of LOCAs in the 500,000 gpm (1,900,000 lpm) bin.

Presentation 2 – Report No. 1 by Base Case Team to Expert Panel on LOCA Frequency Distributions
By Bill Galyean, INEEL

Bill employed a "top down" approach in his analysis of operating experience. His database represents approximately 2,600 LWR-years of operating experience. The resultant average age of a plant is 23 years. In that 2,600 years of LWR operating experience there have been no passive system LOCAs with a resultant leak rate greater than 100 gpm (380 lpm) (Category 1 LOCAs).

As part of this analysis, Bill assumed that cracks and leak events are indicators of LOCA frequencies. They indicate system susceptibility. In order to get a LOCA, Bill's analysis assumed that a piping system must first have a leak or a crack. In the first 2,647 years of US LWR experience represented in Bill's database there have been approximately 1,100 crack and leak events, but no LOCAs. Note, at the first elicitation meeting the demarcation between leaks and breaks was set at a 100 gpm (380 lpm).

A comment was made that small pipes are more susceptible to LOCAs than large pipes, i.e., large pipes are less likely to fail catastrophically. A question was raised as to why limit the analysis to US operating experience only. Bill did not categorically know whether there have been any 100 gpm (380 lpm) LOCAs worldwide. Another reason to limit his analysis to US operating experience is that there are some fundamental design differences between US and other overseas plants. Pete Riccardella (Structural Integrity Associates) thought that if Bill had included foreign experience that the number of years of operating experience would have about doubled so Bill's LWR LOCA frequency number of 1.9E-04/year be would reduced by a factor of two to approximately 1.0E-04/year. It is important to understand the basis for this 1.9E-04/year number since everything else is referenced to this number. This number is the total number of LOCAs of all sizes.

It was noted that this is a different approach than followed in NUREG/CR-5750. This analysis was not an attempt to update NUREG/CR-5750. Pete Riccardella asked if this database included all of the small diameter socket weld cracks that occur due to vibration fatigue. Bill indicated that this was the case, even though these small diameter lines could not result in a 100 gpm (380 lpm) leak. It was pointed out that there was no distinction between cracks and leaks in Bill's analysis. Any crack deeper than 10 percent of the wall thickness was included in the analysis.

Gery Wilkowski (Emc^2) asked if the analysis of the feedwater system (BWR-2) included FAC as a failure mechanism. Gery noted that on the secondary side there have been large breaks in some piping systems due to FAC. Bill indicated that he limited his database search to those systems that affected reactor coolant pressure boundary integrity.

Karen Gott was surprised at the low number of incidences for the feedwater system. Most of the problems seen to date with the feedwater systems have been outside the primary portion of system. Furthermore, cracks in nozzles are associated with the RPV and not piping. For the PWR systems there has been a lot of feedwater cracking, but those cracks have been on the secondary side. It was also pointed out to the panel that everyone has access to the SLAP database that Bill used for his analysis. It is now on the ftp site. Rob encouraged everyone to use it as part of their elicitation exercises. The SLAP database is current up to the end of 1998.

A question was raised about the validity of the single IGSCC failure reported in the carbon steel feedwater system. Bill indicated that this was the reported database value. Karen Gott said they've seen such cracking in Sweden as well.

The BWR recirculation system provided a unique problem for the base case analysis. From a materials standpoint, the base case was the old system. Bill segregated the data by old pipe (Type 304 stainless) versus new pipe (Type 316 nuclear grade [NG]). For the old pipe (Type 304 stainless) there were 127 events in 550 years of operating experience versus 3 events in 410 years of operating experience for the new pipe (Type 316NG).

The resultant leak/crack frequency for the old pipe (127 events/550 years = 0.231 events per year) is about a factor of 2 greater than the overall leak/crack frequency for the overall recirculation system history (old plus new), i.e., 130 events/960 years = 0.135 events per year. It was pointed out though that this improvement may be more due to other factors than pipe replacement only. The improvement could also be due to changes in water chemistry, or the installation of weld overlay repairs. Hence, it may be more appropriate to refer to the pipe systems as mitigated (new) or unmitigated (old) pipe systems.

It was stated that the base case is unrealistic in that it is for the old pipe case (Type 304 stainless) and no one uses that material anymore. Also, the base case does not account for the incorporation of water chemistry improvements which all plants have already implemented.

As part of his analysis, Bill made an assumption that the LOCA categories/sizes (e.g., 100, 1500, 5000 gpm, etc. [380, 5,700, 19,000, lpm, etc.]) are related on a logarithmic sense (1, 0.3, 0.1, 0.03, 0.01, 0.003). Half likelihood on logarithmic sense realizing that larger LOCAs are a subset of Category 1 (100 gpm [380 lpm]) LOCAs.

The 40 welds for the PWR-1 case (hot leg) include the cold leg and cross over leg welds. This is inconsistent with the assumption stated above that the PWR-1 case only considers the hot leg (not the cold leg or cross over leg). (Note, there are typically 3 loops in a PWR plant and there can be 5 to 7 welds per loop, but the loading is not the same for all these welds.)

Rob Tregoning indicated that the correlation between pipe size and LOCA size (gpm) that were originally supplied are subject to change.

Bill's aging correction factor is for thermal fatigue and should not be used for the other failure mechanisms.

The non-pipe LOCAs that Bill included are for passive systems only (bolted flanges, etc.). He did not include active system contributions, e.g., stuck open valves, to the non-pipe break frequencies.

Bill's results for the "Current Estimate" are significantly smaller than the LOCA frequencies reported by others (WASH-1400, NUREG-1150, NUREG/CR-5750) due to the larger database (more years of operating experience without a LOCA). The best agreement is with the NUREG/CR-5750 results. Bill employed a Bayesian approach as part of his LOCA frequency analysis by assuming a half of failure for these very low occurrence events. This is a very common data analysis practice. Since Bill's analysis is based on the analysis of past passive-system failure data, the data implicitly include ISI and other mitigation experience implemented by industry.

Each plant has a PRA which includes LOCA frequency estimates for the plant. These LOCA frequency estimates are often based on WASH-1400 or NUREG-1150 and the ranges shown on pages 36 and 37 of Bill's presentation are the ranges for the IPEs. Thus, the IPE range, WASH-1400 and NUREG-1150 frequencies shown by Bill on pages 36 and 37 of his handout are closely related/interlinked. As described in Slides 36 and 37, Bill expressed the uncertainty in his analysis by assuming an error factor of 10. This is somewhat crude and somewhat arbitrary, but the scope of the base case analysis did not ask for uncertainty, just a best estimate.

Bill indicated that he has no strong technical basis for extending his "Current Estimate" analysis out to 40 or 60 years. With Bill's approach, the LOCA frequencies will go down as more years of operating experience are accumulated, unless a LOCA event occurs in the future. Bill presented information predicting 8 years to double the frequencies for thermal-fatigue events. This estimate conservatively assumes no industry-wide mitigation programs. Rob Tregoning indicated that he did not include results for Bill in the summary table for 40 and 60 years of operation because they are most appropriate for current estimates.

Bruce Bishop was very uncomfortable predicting the future out to 40 or 60 years. He felt that the uncertainties are going to increase dramatically. Lee Abramson (USNRC) responded that this is a natural experience and likely shared by others on the panel. However, it's incumbent that each member to attempt these predictions. Rob and Lee stressed again that the panel will not be forced to answer any questions that they are very uncomfortable with. Rob further stressed that the March 2003 SRM specified that the NRC staff needs to revisit the LOCA frequency estimates every 10 years and the effort will be most concerned with the next 10 years. However, it is still important to gain longer-term insights.

A question was raised as to whether or not to have Bill extend his analysis out to 40 and 60 years? Gery Wilkowski said, yes, he wanted to see Bill's assumptions. Fred Simonen (PNNL) would like to see more partitioning by pipe size as part of Bill's analysis. Dave Harris (Engineering Mechanics Technology) concurred. Sam Ranganath (formerly of General Electric) specifically indicated that it would be helpful for the BWR recirculation system because they replaced the smaller diameter pipes, but not the 28-inch diameter pipes. Bengt Lydell is to check on the statistics in his database to see if any of the 28-inch diameter BWR recirculation lines had leaks and provide this information in his final base case report. Pete Riccardella and Sam Ranganath were unaware of any General Electric large diameter recirculation pipes that leaked.

Report No. 2 by Base Case Team to the Expert Panel on LOCA Frequency Distributions by Bengt Lydell (Erin Engineering and Research)

Bengt, like Bill Galyean, used passive-system failure data in his analysis, but in a much different manner. Whereas Bill Galyean followed a "Top Down" approach, Bengt followed a "Bottoms Up" approach. Bengt's presentation assumed that all BWR welds were category D & E welds. For BWRs, Bengt indicated that Category D & E welds specify inspection criteria based on Generic Letter 88-01. Category D welds have been subjected to weld overlay repairs and Category E welds are subject to IGSCC.

Bengt's used a different database than Bill Galyean did is his analysis. Bengt's database is proprietary (PIPEex) and the panel will not have access to this during the elicitation. The SLAP database that Bill used is available to the panel members on the website. The cut-off date for PIPEex events is the end of 2002 while the cut-off date for the SLAP database is the end of 1998. PIPEex has about twice the number of data entries as does SLAP and includes international experience.

Bengt only looked at welds in his analysis. He did not consider base metals. His database did not show any occurrences of base metal indications. However, he invoked a wide definition for what was encompassed by the term "weld". He included the heat-affected-zone (HAZ) and counter bore region into his definition of what a weld was. Bengt also did not consider degradation of non-piping passive components in his analysis.

For PWR systems, he assumed that the V.C. Summer and Ringhals cracks were circumferentially oriented cracks, not axially oriented. Whereas Bill did a "top down" analysis, Bengt did a "Bottoms up" analysis. The failure rates were derived for individual welds, and then an integration system level model was formed by combining the contributions from each individual weld failure to an overall pipe system failure frequency.

The term "prior" has very specific meaning in this analysis. It means "before _mitigation_/remedial action in response to a significant pipe failure". Hence, failure rates that are input to LOCA frequency calculations explicitly account for reliability improvements (mitigation methods) made in response to past pipe degradation histories. Failure in Bengt's analysis is defined as a "through-wall flaw resulting in leakage". Bengt did not include surface cracks found during ISI in his data reduction process.

Karen Gott has a report in Swedish that may be valuable in this effort. This report gives the number of leaks and the number of ISI detected surface cracks. Gery Wilkowski thought that was important information since leaking through-wall cracks are readily detected, but the surface cracks that would grow to be long in length are more of a LOCA threat. The number of records shown on slide 10 from Bengt's presentation includes both leaks and cracks.

There was much discussion among the group in an effort to understand slides 11 through 14. In slide 13, Bengt did not eliminate welds if mitigation was performed prior to 15 years of operation in developing his prior distribution. There were no leaks in BWRs after 15 years. From years 10 to 15, it is possible that there may have been some plants that used mitigation, but those mitigated plant weld numbers were still included in the weld failure rate analysis, i.e., that may account for why the weld failure rate was not accelerating as the number of years increase. Bengt used a Monte Carlo simulation to create this plot, where he needed to estimate the number of welds for the number of plants that were at a certain age. Results in slide 14 include results from US, Spanish, Swedish, and Japanese plants. The results were adjusted by the number of welds and type of welds.

These preliminary results are used to determine the "prior" LOCA frequencies (weld failure rate). The next step is to determine the "posterior" frequencies based on "prior" distributions. Bengt accounts for uncertainty in the knowledge base, which is fundamental difference between his analysis and Bill's. Dave Harris thought the "posterior" frequencies should be equal to the Prior LOCA frequencies times the Likelihood Function. Lee Abramson explained that the likelihood function was built in.

It is noted in slide 19 that the Bayesian update strategies are different for each base case. It was also noted again that the weld failures represent leaks only and not cracks. Weld failure was defined as a through-wall flaw with leakage less than or equal to the tech spec limit for undefined leakage.

Gery Wilkowski noted that in January 2003, PWSCC was found in a surge line bimetallic weld at the pressurizer in a Belgium plant. It was noted that some transient event is needed to cause a tech spec limit flaw to propagate to a higher category (well beyond 100 gpm [380 lpm]) LOCA. Slide 23 shows that the conditional probability of failure ($P_{L/F}$) is a function of the nominal pipe diameter (DN), i.e., $P_{L/F} = a \times DN^b$, much like what is in NUREG/CR-5750 (Beliczey and Schulz, i.e., $P_{R/TWC} = 2.5/DN$). While the exact formulations are different, the conditional failure probability in all cases is an inverse function of pipe size.

The form of the conditional leak probability given a failure is inconsistent with the original development of the Beliczey and Schulz correlation which was developed to relate leaks to breaks as a function of pipe size. This use (see slide 24) may not be physically realistic because it assumes that larger diameter pipes are less likely to result in a category 0 leak. The implication is that larger diameter pipes are less likely to reach a Category 0 leak, and then progress to higher leak rates. Slide 26 presents information on the aspect ratios of IGSCC cracks. It was asked how the deep, full circumference cracks (a/t = 0.5, 2/B = 1.0) formed. The expectation is that these are likely crevice cracks. It was suggested that it would be nice to break down this data by pipe size in order to assess the relevance.

Bengt assumes that a through-wall crack can only propagate into a large leak if there is a large transient event. Slide 27 documents the loading categories assumed to drive the crack among various LOCA categories. Category 0 to Category 1 LOCA progression can occur assuming moderate loading, while to go from a Category 0 or Category 1 LOCA to a Category 6 LOCA, would require an extreme loading transient. The general consensus of the panel members was that this was a very subjective approach. It was noted that there were about 400 water hammer events reported, but Bruce Bishop and Guy DeBoo said that if there were this many water hammers, then the plant piping system was probably redesigned.

The extrapolation of results from the "Current Estimate" to 40 and 60 years is based on posterior analysis of prior results assuming no additional failures. If one assumes no additional failures, the failure rates will go down with time. Bengt will examine possibly extrapolating his base case results out to 60 years using another assumption.

Report No. 3 by Base Case Team to the Expert Panel on LOCA Frequency Distributions by Dave Harris (Engineering Mechanics Technology)

Dave prefaced his comments with the thought that he thinks that the base case results are surprisingly close considering the differences in the approaches. Dave indicated that an important input to any PFM analysis is the stress history. This requirement is contrary to the operating-experience approaches where the stress history is indirectly reflected in the incidence of cracking events. Dave's analysis is performed on individual pipe locations which are then integrated to determine the overall system frequency.

Some key points from the crack initiation and crack growth portion of Dave's presentation are that piping failures occur due to the initiation and growth of cracks. Cracks initiate due to stress corrosion or fatigue. Growth is controlled by fracture mechanics (other than early SCC). The question was asked as to what assumption Dave used as to the size of the crack once it initiates. Dave's PRAISE code assumes that fatigue cracks are 0.3 inch (7.6 mm) deep (per a criteria proposed by Argonne National Laboratories [ANL]). This is based on an assumed 25 percent load drop definition of crack initiation from an S-N specimen test. In addition, in PRAISE the SCC rules differentiate between early SCC growth and fracture mechanics growth since the early growth is faster than calculated by fracture mechanics analysis. The SCC rules in PRAISE assume a 0.001 inch (0.025 mm) deep surface crack with some distribution function on length. The latest version of PRAISE (2002) includes updates to the S-N curves that incorporate environmental effects. Pete Riccardella noted that Art Deardorff of SAI was doing an update of some of the environmental S-N results from EPRI. For crack growth, the focus of PRAISE is semi-elliptical part-through surface cracks. PRAISE considers crack growth in both the depth and length directions (K at both the maximum depth and at the ends of crack.)

Dave pointed out that fatigue failure of welds is dominated by growth from pre-existing crack-like fabrication defects. Thus, the flaw distribution of initial fabrication defects is an important parameter to define. PRAISE stipulates that the final failure is controlled by tearing instability. PRAISE treats the stresses at maximum load as load-controlled stresses from stability analysis perspective.

The leak rate in PRAISE is computed based on the length of the leaking through-wall crack on the inside pipe surface using the SQUIRT leak-rate code. The SQUIRT code has recently been updated with numerous technical enhancements as part of the USNRC Large Break LOCA program. PRAISE includes some of the mechanistic dependent crack morphology parameters from some of the earlier versions of SQUIRT, but not the new COD-dependent roughness, number of turns, and flow path length to thickness ratio parameters. In addition SQUIRT has been made more user friendly by incorporating a graphical user interface. Note, WinPRAISE is PC-PRAISE with a Windows pre-processor for entering input parameters.

Dave used a stratified sampling technique that allows for the evaluation of extremely small probabilities events such as the case of fatigue crack growth from pre-existing defects (10E-17 frequencies). The approach also assumes that all cracks with leak rates greater than 5 gpm (19 lpm) are discovered and subsequently removed from service which implies that getting a higher Category LOCA (e.g., a 1,500 gpm [5,700 lpm] LOCA) would most likely result from the growth of a long surface crack that pops through the wall thickness and immediately becomes a TWC with a length equal to length of the surface crack on inside pipe surface. (The only other means of achieving a higher Category LOCA would be through some sort of transient event.)

Dave postulates that inaccuracies in leak-rate calculations will not significantly impact final LOCA frequencies. Even assuming the leakage detection capability equals zero should not have a large effect. Sam Ranganath felt that a surface crack grows 3 or 4 times faster in the length direction than it does in depth. He wasn't sure if that was due to multiple initiation sites, or the surface growth rate being higher. The analysis in PRAISE considers crack growth in both the depth and length directions (K at both depth and ends of the crack) with an RMS value of K in each direction.

The detection probabilities shown in slide 14 are based on depth only, not length. It is likely that current technology has better performance. It was also noted the fatigue crack growth parameter used could be improved based on newer results. Dave Harris indicated that sensitivity studies show that using improved crack growth parameters in PRAISE (e.g., including environmental effects) will result in changes in LOCA frequencies on the order of a factor of 2. Slide 16 presents an example S-N initiation curve for a low alloy steel. Vic Chapman raised the concern that there may not be a plateau or fatigue limit with the higher number of cycles. Gery Wilkowski commented that the default flow stress values (slides 17 and 18) are very tight from a standard deviation perspective, especially in light of what was seen in NUREG/CR-6004 from an analysis of PIFRAC data.

The NUREG-6674 stresses have been downgraded from the design basis stresses by Jack Ware of INEEL to make them more realistic with fewer transients. The stresses may have been elastically calculated values, which can go well above yield and still be allowed for secondary stresses by the Code. PRAISE uses the most realistic values available. The final results (frequencies) from Dave's analyses are very dependent on stress input values. Dave commented that one only needs stresses at the high stress locations. These high stress locations dominate the final frequency answers.

Slide 21 shows the surge line stresses. These stresses are probably for the flank of the elbow, not the weld. These values can't be used for the girth weld location (i.e., they are too high). The stresses shown are stress amplitudes (the stress ranges will be twice these values). In addition to stresses and number of occurrences, one also needs some input as to the spatial distribution for these stresses. One of the short comings of PRAISE is that PRAISE doesn't have a model for FAC for the feedwater lines. The fatigue initiation models are only in latest versions of PC-PRAISE, i.e., from NUREG/CR-6674 published in June 2000.

An ad hoc procedure was used with pc-PRAISE in order to obtain results for larger leak rates (stratified sampling for fatigue crack growth is not available for fatigue crack initiation). One of the handouts provided shows this ad hoc procedure.

A question arose with slide 26 concerns the fact that Dave's analysis shows that the cumulative failure probabilities continue to increase after 20 years after a weld overlay repair is applied at 20 years. Past experience at Battelle as part of the Degraded Piping Program showed that weld overlays are very effective. They have much higher strength than the base metal. Another aspect of their application is that they apply a very high compressive stress at the crack plane. These high compressive stresses should restrict any further crack growth of the surface crack. In addition, it was noted that these high compressive stresses may preclude the environment from getting to the crack tip. Dave noted that he put in a linear approximation of the stresses through the thickness, the increased thickness of the overlay, and the crack growth equations for Type 316 NG into his analysis. WIN-PRAISE uses an adjusted weld residual stress pattern (linear gradient) for the case of post-weld overlay residual stresses (see slide 38). Dave will also present the failure probabilities without the weld overlay so that an assessment of its effect can made.

Bruce Bishop asked how much was Dave's results affected by inspection. Dave didn't think that the final frequencies would be affected that much. Dave's results showed that there was a minimal change in LOCA frequencies for the hot leg as a result of the application of a 5SSE earthquake. This was not surprising to Dave since he has found similar behavior in a previous study. Lee Abramson pointed out that one means of seeing the effect of the earthquake is to compare conditions for equal probabilities. For example, for the 40 year time period analysis, no earthquake results in 1.3E-18 LOCA frequency for the no leak case while for the same 40 year time period analysis, a 5SSE earthquake results in 1.3E-18 frequency for a DEGB.

Dave, generally found that the LOCA frequencies were not highly dependent on J_{Ic} and dJ/da. Gery Wilkowski thought that if the toughness was low enough that one was operating in the EPFM regime then the LOCA frequencies may be more dependent on toughness. Gery indicated that he thought that the toughness values used in Dave's base cases were too high for weld crack locations, or aged cast stainless steel pipe and fittings.

PRAISE can't account for time dependent material properties. Thus, to account for aging, one would need to input aged properties in at time equal to zero.

The very low LOCA frequencies for the hot leg in slide 31 may be an artifact of the failure mechanism (fatigue) chosen for analysis. Higher frequencies may be seen for some other mechanism, such as PWSCC. This is a case that we may want to analyze in future analyses.

Report No. 4 by Base Case Team to the Expert Panel on LOCA Frequency Distributions by Vic Chapman and Chris Bell

The first part of the presentation was presented by Chris Bell (Rolls Royce) with the final few slides presented by Vic Chapman. Chris presented the general assumptions and methodology of the PRODIGAL code. Note that PRODIGAL actually has several modules. One of them is to determine weld defect size from welding information. Another is to determine failure probabilities for navy nuclear power plants. All results presented are for a per weld basis.

In PRODIGAL, surface imperfections with a depth of 0.004 inch (0.1 mm) are assumed, but there are no SCC initiation/growth models in PRODIGAL. Past study has shown that there is little sensitivity to the assumed depth and sizes much less than 0.004 inch (0.1 mm) have little effect. For surge line case, there is sensitivity to the existence of 0.04 inch (1 mm) defects, see slide 19.

Based on some work of Ritchie, Pete Riccardella felt that the 0.004 inch (0.1 mm) initial defect size was near the LB of the region where fatigue crack growth (da/dN) models were valid. It was noted that Omesh Chopra from Argonne thinks that this lower limit is closer to 0.02 inch (0.5 mm). It was suggested that this limit is dependent on the grain size of the material. Note, cast stainless steels can have very large grain sizes.

Vic and Chris felt that they could not do a generic analysis for seismic considerations since the effect of seismic has been found to be highly dependent on plant layout. Therefore they didn't consider seismic stresses in their analysis. In addition, they only considered the three PWR base cases since they had little experience with BWRs.

They used the same stress data as Dave Harris did for consistency purposes. As Dave did, they used a second order distribution of stresses thru the thickness per NUREG/CR-5505 criteria that was done by PNNL in 1998.

The failure criterion for their instability analysis is based on the FAD approach in R6 using K_{IC}, i.e., crack initiation would equal failure. The crack initiation used for stainless steel was closer to wrought base metal rather than weld metal which would be an order of magnitude lower, or aged cast stainless steel which could be another factor of 3 lower than the weld metal toughness.

Bruce Bishop asked what the mean temperature in the analysis is used for. It is used for material property considerations, such as flow stress, but not for subcritical crack growth. Slide 16 from Vic and Chris' presentation shows the cumulative probability of a TWC (probability of a leak occurring). The implication from this slide is that if you are going to have a leak, it will occur in the first 25 years of operations. It was noted that slide 16 is conditional on having a crack (probability of having a crack is 1).

Rob Tregoning indicated that he wants each of the panel members in the next week to make list of what they want to see (e.g. data they used in doing their analysis) from either individual participants or from the group as a whole.

It was noted that the dominant hot leg cycles are those due to heat up and cool down. There are only on the order of 5 of these cycles per year.

Pete Riccardella thought that there were a lot of cycles on the PWR HPI/Make up nozzle each year. Dave Harris countered that he thought that there were only about 40 of these cycles in 40 years of operations. Bengt Lydell agreed with Pete and thought there were a lot of thermal cycles. Pete thought that something was missing here. Bengt said there was a nice ASME paper on the cyclic stress history of these nozzles that we could use in the analysis.

Chris indicated that although they typically keep the aspect ratio constant in their PRODIGAL runs, they have the ability to grow cracks in length, and often get very irregular crack shapes.

In slide 24, the "separation" referred to is the crack-opening displacement. In this slide, at the surge line elbow, Vic speculates that the crack starts to act as a hinge so that crack opening becomes very large for the longer crack lengths. This assertion is the basis for slide 25 illustration of the crack frequency versus angle distribution that was assumed in the analysis. This relationship was created by Vic with guidance from metallurgist to rectify predicted and experimental crack lengths.

It was commented that for the same input, we are seeing similar results from PRAISE (Dave Harris' results) and PRODIGAL (Vic Chapman and Chris Bell's results). It was noted that as far as this exercise is concerned, PRODIGAL's strength is the defect distribution analysis. PRODIGAL only looks at thermal fatigue while PRAISE can look at other failure mechanisms. PRAISE can also look at crack initiation while PRODIGAL cannot. However, neither accounts for FAC or PWSCC at this time. PRAISE can also look at crack initiation whereas PRODIGAL assumes crack growth from weld defects or surface imperfections.

Vic pointed out that embedded cracks can straddle the compressive zone of the residual stress field through the thickness so that once they break thru to inside surface they are already through the compressive zone. This is in contrast to case where a crack is growing through the wall thickness from the inside pipe surface and the crack gets trapped in the compressive zone near mid thickness. There was also the question of whether embedded cracks are affected by the environment.

The next topic for discussion was to plan the next step for the base case calculations.

Bruce Bishop would like to bench mark the PFM results against the 25 operating-experience estimates developed by Bill Galyean and Bengt Lydell, and then use the PFM models (PRAISE and PRODIGAL) to predict the 40 and 60 year results. Rob Tregoning thought that this was an excellent idea. Rob also indicated that we could do this for some cases, but not all cases, e.g., FAC in the feedwater system or PWSCC.

Sam Ranganath would like to know when we compare results where do we get good agreement and where not. Rob indicated that we haven't made comparisons on a consistent basis as of this date but that this would be rectified.

The next issue focused on additional stresses to consider in the PFM results. The surge line and HPI/MU cases were mentioned. Dave Harris has already done some additional analysis for the surge line, based on refined stresses developed by Art Deardorff. Gery Wilkowski noted that he had surge line stresses from a Westinghouse Owner's Group report used for LBB analyses. Pete agreed to verify that the previous stresses provided by Art are appropriate. Pete also agreed to provide more accurate HPI/MU stresses.

Sam Ranganath would like to lower the stresses to 10 ksi (70 MPa) for the recirculation line (BWR-1) and to lower the feedwater line stresses (BWR-2) by 20 percent.

The next area where the panel thought we may want to focus is some sensitivity analysis using different material properties. Gery Wilkowski volunteered to supply some distributions of material properties (mainly toughness values for welds and aged cast stainless steels) developed as part of NUREG/CR-6004.

Gery Wilkowski also wanted to see the ratio of surface crack to through-wall cracks removed from service, and the distribution of the lengths and depths of those service removed surface cracks. He then wanted to see a comparison of the PFM probability of leaks for IGSCCs from PRAISE and the distribution of surface cracks that might exist up to the time that piping might have been replaced/repaired (15 service years?). Karen suggested looking at the more recent results on a yearly basis because ISI wasn't sensitive enough to find surface defects in early years. Cracks were only discovered once they became a leaking crack. Also, in Sweden, even if whole pipe sections were removed, all the welds were inspected, whereas in the US if the pipe system was replaced they did not spend the effort to inspect welds that were being removed from service. Hence, the database of cracks removed from service should separate Swedish and US plants. The important aspect of this comparison is to see the population of the ISI remove surface crack lengths compared to the surface flaw sizes calculated by PFM. If the service removed crack lengths from ISI are much longer than calculated by the PFM analyses, then the PFM analysis should underestimate the future failure probabilities. Gery noted several times that failure probabilities should be controlled by development of long surface cracks, not the growth of leaking through-wall cracks. Bengt Lydell has some papers relating ISI-detected surface-crack geometries to through-wall crack leaks that he can provide on the ftp site and can provide to Rob Tregoning.

It was suggested to include PWSCC in the hot leg base case analysis. Dave Harris has initially done this using IGSCC relationships for preexisting flaws. Initiation is not accounted for. Gery Wilkowski noted that the crack growth rate through the weld metal (along the dendritic grains) is much faster than IGSCC. Gery will work with Karen Gott and Bill Cullen to see how Dave can adjust the PRAISE model to get initiation and growth for PWSCC. Gery will provide the IGSCC initiation and growth equations that PRAISE uses to Bill Cullen so that appropriate constants can be provided for PWSCC in PRAISE for base case calculations. Gery and Karen will provide information to Dave Harris on PWSCC crack initiation and growth models that he can use to evaluate the impact of PWSCC on the appropriate base case calculations.

The panel thought it was important to address Dave Harris' strange results for weld overlay repairs. What is leading to high growth rates after the overlay is applied. Can a comparison be shown with what would happen if no weld overlay had been applied? Dave Harris is to address this concern of the unexpectedly high growth rates after the weld overlay repair is applied.

When the summary comparison tables are completed, the PFM subgroup needs to clearly identify where the PFM conditions do not agree with operating experience. Rob Tregoning indicated that the base case subgroup needs to finish up the base case calculations by the end of the month. We cannot delay individual elicitations any longer.

Day 2 (June 5) – Elicitation Coordination

Rob Tregoning started the morning by reviewing the agenda for the second day. There were six (6) basic items to cover. These were:
- Reviewing the elicitation questions
- Reviewing the leak rate versus pipe break size evaluation
- Addressing the non-piping LOCA evaluations
- Reviewing the conditional seismic evaluation
- Addressing the additional information required prior to the individual elicitations for piping and non-piping evaluations
- Elicitation scheduling

The specific objectives for the second day included:
- Finalize the elicitation question sets and to provide a consistent understanding of each question.
- Determine the methodology for evaluating non-piping LOCAs and the identification of non-piping base case data.
- Determine the methodology for evaluating conditional seismic loading including determination of the seismic loading magnitude.
- Determine what information the panelists will require prior to their elicitations and assign action items for providing information.
- Develop the final schedule and time-frame for the upcoming elicitations.

**Presentation 1 (Day 2) – Elicitation Questions: Structure and Review
by Rob Tregoning (USNRC)**

Rob stressed that prior to their individual elicitations, each panel member needs to do their homework and answer as many of the elicitation questions as possible. Panel members can change their answers at any time during this exercise, including during the actual elicitation and afterwards.

It was indicated that the term "LOCA frequencies" should really be "LOCA probabilities" in this presentation during the analysis of the conditional emergency loading. Rob agreed to make this change to the presentation.

Pete Riccardella questioned whether we should expand the conditional seismic loads to conditional emergency and faulted loads which include transients like water hammer as well as seismic. Fred Simonen thought that we needed to talk with the PRA people. Alan Kuritsky (USNRC) indicated that these other potential LOCA causing events were currently not included in the PRAs. Rob suggested tabling this discussion until later in the agenda. Bruce Bishop thought that if we continued with the traditional seismic approach then we need to consider the fact that the snubbers may not work probably. Bruce indicated that there is a significant probability that the snubbers may not work as advertised.

Rob next addressed the top down elicitation structure. Bill Galyean used a top down approach where he assigned an overall piping LOCA contribution, and then looked at the breakdown in the contribution due to piping system, geometry, load history, mitigation, materials, and degradation mechanisms. Vic Chapman questioned what the top down approach gave us (we start with the final answer that we are looking for). Lee Abramson indicated that at the end of the day we will get numbers in each of the blocks on slide 3 so that we can decompose the problem into the small pieces. The panel members can initially choose either a top down or bottoms up approach.

Bruce Bishop asked how soon the panel members are going to have the final base case results prior to the first elicitation. Rob indicated that he would be working with the base case members the week of June 9 to finalize their answers. He hoped to have the final results by the end of June. However, the base case conditions are well-known

The panel members need to supply their answers (on a pre-established form) and the facilitation team will work with the individual panel members individually. The important point of the pre-elicitation exercise is to quantify the median value results, provide some qualitative uncertainty and also rationale. During the elicitations the panel members can change answers as they interact with the facilitation team and points are clarified.

Bruce Bishop felt that the "utility safety cultural" should be "utility operations cultural" in that safety and economic drivers both feed into the operations cultural. Karen Gott indicated that the IAEA definition of safety cultural (and how we defined safety cultural at the kick off meeting) includes both safety and economic aspects. With regards to the ratios on safety cultural issues, ratios greater than 1.0 indicate that things are getting worse; less than 1.0 means things are getting better.

Slide 6 shows a flow chart in which the question is which variables are independent. Variables in this context are geometry, load history, mitigation, materials, and degradation mechanisms. For these variables, each panelist may need to estimate the future impact of that variable, e.g., what new materials, or what new degradation mechanisms, should be expected in the future. We may want to look at the past to estimate what might happen in the future (e.g., Gery Wilkowski's plot of new failure mechanisms with time which shows a new mechanism approximately every 7 years).

There ensued a long discussion on the comparisons of <u>reference</u> cases (defined at the kick off meeting) with <u>baseline</u> cases. We defined a reference case for each piping system (e.g., diameter, material, degradation mechanism, etc) whereas we only defined a few baseline cases to which the reference bases are to be compared (i.e., anchored). For example, the hot leg (base case) may be a natural comparison for the reference case for the cold leg.

For Questions 3A.1 and 3A.2 (slide 7) and all related questions, Rob will change the "surge line" to "cold leg" example so questions don't refer to a system that is both a base case and a reference case, realizing that we will end up asking same question for a surge line as well. For those systems which have both a base case and a reference case, these comparisons may be more natural and easier than inter-system comparisons.

It was decided to eliminate the assessment of which variables are independent or dependent. All variables will be considered to be dependent as originally defined during the kick-off meeting. Thus, the original Question 3A.2 will be eliminated. Other References to correlated or independent variables in other questions should be eliminated in the final presentation version. For original Question 3A.3, the requirement to list at least 80 percent gets the most significant contributions, but not all of them.

There was considerable discussion on what the panel members would need to provide for Question 3A.3 and 3A.4. We need to know what variables have a major impact on LOCA frequencies to quantitative that impact as best as possible. Lee Abramson tried to make the point that things would become clearer once individual panel members got into their elicitations and tried to put numbers to the answers to the questions.

A question was asked if any attempt is going to be made to look at plants and determine how many plants have a certain combination of variables (V1, V2, V3, etc.), and how many plants have another set of variables (V2, V4, V5, etc.), and how many have another set. It was noted that some variables will be important at certain plants and other variables will be important at other plants. Rob indicated that it would be nice to have such information, but it was not practical to get such information in the time frame we have. This could possibly be a follow-on effort. However, it should be stressed that the plant design information will not likely result in a significant change in the analysis. If certain designs do not contribute to the LOCA frequencies then they are not significant contributors and the panelists can focus on designs that do as long as they exist in several plants. If the population of the significant contributors is in error by 2 to 3, it will likely not matter.

The only issue to avoid during quantification is if you believe that only a few (1 – 2) plants of a certain design, operating experience, etc. significantly contribute to the generic LOCA frequencies. These plants should not be explicitly considered in these generic estimates. However, but possibly applicability of these generic results to those design conditions should be discussed during the elicitation.

Elicitation Question (EQ) 3A.1 compares a base case to a reference case, then EQ 3A.4 will use the impact of the important variables to compare other similar piping systems to the reference cases. Reference cases are the link back to the base case for which we will have actual LOCA frequency estimates. The panel members don't have to do the mapping back to the base cases, the facilitation team will do that. Then the facilitation team will filter the results up (bottoms up approach) to get an overall LOCA frequency.

There was some discussion about how the facilitation team would integrate these results and how the panelists could account for these individual contributions.

Again, it was emphasized that if panel members are not comfortable in answering specific questions, then they need to say so. If the panel members need to make a crude assumption, then do so, but indicate that during the individual elicitations so the facilitation team can help estimate the level of uncertainty. Dave Harris thought that it won't be clear to him how this all fits together until he starts the process. Then he is sure that he will have a lot of questions. Rob Tregoning and Lee Abramson told him, and the rest of the panel members, to call them for any clarification of questions. Rob also indicated that he will be contacting each panelist prior to their elicitation to discuss issues.

The flow chart on slide 9 is for a "top down" approach, much in the motif of what Bill Galyean discussed on Day 1. As part of this approach, one only needs to tie one system (of those identified as being important contributing systems to LOCA frequencies) to the base case since previously we had identified individual piping system contributions. One can make this connection through a reference case if not a base case system, or can tie directly to the base case if the system is a base case system. This approach may be more straightforward for people who need to integrate variables in mind, which might lead to more uncertainty.

Panel members can chose which approach to follow (top down or bottoms up), and they can switch back and forth depending on different systems. As part of their homework prior to their elicitation, they only need to do one approach, but the facilitation team may ask about each approach during the elicitation. There will be different tables to fill out with each approach. The bottoms up approach may be more rigorous with less subjectivity, but the top down approach may be easier to understand. One can get top down answers from the bottoms up approach, but can't do reverse. By doing both approaches for certain systems, panelists can search for consistency in their analysis.

Rob discussed the elicitation questions related to non-piping components starting at slide 12. At the kick-off meeting in February we didn't spend as much effort developing the base case and reference cases for non-piping components as we did for the piping systems. Thus, these gaps needed to be filled during the remainder of this meeting.

Slide 12 illustrates the flow chart for the "bottoms up" approach for the non-piping components. At the kick-off meeting, we had identified five (5) non-piping components to consider (pressurizer, valves, pumps, RPVs, and steam generators). In order to estimate the frequencies for these non-piping components, the panel members need to pick either a piping or non-piping base case for comparison. The facilitation team will integrate the results in a manner similar to the bottoms up approach for piping systems.

There was a question about the nature of the non-piping base case conditions, especially in light of the thorough discussion about the piping base cases during the previous day. We had originally planned to have precursor data for the non-piping base cases for comparison. However, we do not have data identified yet. We may have to drop the idea of using non-piping base cases and only have piping base cases to compare to. We will come back to this issue later in the afternoon.

Bruce Bishop felt that tying non-piping components back to piping base cases would be difficult. He foresaw lots of dissimilarities between non-piping and piping in failure mechanisms, etc. He suggested that we try our best to come up with some non-piping base cases. Even if we can't come up with base cases for all 5 of the components, if we could come up with base cases for a few, that would be better than nothing.

Pete Riccardella asked if we have defined the failure mode for these non-piping components. Rob felt that we don't have a clear definition at this time. Rob felt that we had to take more of a mechanistic viewpoint. There was also general confusion about the definition of "failure mode" for the non-piping issues. Rob indicated that this means the failure mechanism. It was agreed that "failure mechanism" is a preferable term and Rob will change the phrasing for the elicitation questions from "failure mode" to "failure mechanism" to avoid any confusion.

It was again emphasized that the panel members will not be asked to provide absolute LOCA frequencies. However, if a panel member prefers to think in terms of frequencies, they should feel free to do so. Elicitation questions will ask for relative comparisons with the base cases and other conditions. The facilitation team will then make the calculations to get the absolute LOCA frequencies. The panel members should make the best comparisons possible and are not compelled to answer questions in areas where they have no expertise.

Rob showed an example of a table that he may provide the panel members for them to fill out for EQs 3B.1 and 3B.2 for the top down approach, see Table B.2.2.

The complete set of tables to be filled out for the elicitation will be provided electronically by Rob. There will be a space for comments in each table row to initiate discussion during the elicitation process. The tables will be provided in Excel format. If a panel member wants to change the Excel spreadsheet format they should feel free to do so as long as the cell references for each answer remains unchanged. The final calculations will be done in Excel. Therefore, the elicitation results should be provided to Rob in the excel spreadsheets if at all possible. Hardcopies or MS Word versions of the tables can provide upon request.

Rob will provide the panel members with a copy of the spreadsheet that he will use to calculate LOCA frequencies sometime during the elicitation process. Rob will attempt to complete this spreadsheet to the individual elicitations so that the panel members can see how their responses reflect their calculated LOCA frequencies. However, this will be a lower priority than coordinating the information exchange among the expert panel and finishing the base case calculations.

Table B.2.2 Elicitation Questions 3B.1 & 3B.2

BWR Piping Systems: Important System Contributions to LOCAs

LOCA Cat.	25 Years of Plant Operation				40 Years of Plant Operation				60 Years of Plant Operation			
	Systems	System Cont.	5% LB	5% UB	Systems	System Cont.	5% LB	5% UB	Systems	System Cont.	5% LB	5% UB
1												
	Total				Total				Total			
2												
	Total				Total				Total			
3												
	Total				Total				Total			
4												
	Total				Total				Total			
5												
	Total				Total				Total			
6												
	Total				Total				Total			

Presentation on Non-Piping LOCA Evaluation: Base Case Data and Remaining Issues by Rob Tregoning (USNRC)

Prior to this presentation, Rob noted that this presentation was not included in the handout, but will be put on the ftp site. Rob also asked that the base case team provide him in an electronic format with any references that they used so that the references can be put on the ftp site.

It was first noted that we have not been successful in locating additional failure data for several of the components where we were lacking data. Fred Simonen had spoken with Spencer Bush and was not able to locate additional data. Rob and others were also somewhat unsuccessful.

Based on the leak rate versus opening area from one of the prior presentations, Pete Riccardella questioned if we needed multiple failures for the heater sleeves as shown in slide 3 in Rob's presentation. Bruce Bishop indicated that thermal fatigue needed to be added to the degradation mechanisms for vessel head bolts (slide 4). Rob indicated that these tables are not filled in completely at the present time. These tables are neither comprehensive nor as complete as the piping component table. We focused more on the piping issues at the kick off meeting than we did on the non-piping issues.

For all of the major groups, we initially listed at least one component (**bolded item** in the original non-piping tables in presentation and kick-off meeting notes document) where the panel thought that failure (leaks or cracks) data is available. It was thought that these **bolded items** represented potential non-piping base cases for anchoring. If these non-piping base cases can't be develop, then we will have to use a piping base case for anchoring. As Bruce Bishop indicated earlier that would be an unnatural comparison, making it somewhat difficult. The question was asked how we develop these non-piping base cases. It was thought that areas where we could come up with passive-system failure data, e.g., steam generator tubes, would be logical first choices. It was thought that data on steam generator tube failures should be easy developed.

Bruce Bishop indicated that there was an INEEL NUREG report by Vic Shah that has piping and non-piping failure data that could possibly be used for base cases. Obviously, this would be a good place to start. It was noted though that this INEEL report is for PWRs only, and the steam generator tube failure data will be included in this report. Bill Galyean and Rob Tregoning are to locate and distribute copies of this report to the panel members.

Pete Riccardella volunteered to run a base case analysis for feedwater nozzles and the belt line region for the RPV due to LTOP. Pete thought he could have some analysis results by the end of the month. These would be for BWRs and will be done using predetermined flaw distributions. Note, to date there have only been cracks, there have been no leaks to date.

Another potential non-piping base case for anchoring is the PTS study for the belt line region of the RPV. This would be for PWRs. Rob Tregoning will extract this data. Gery Wilkowski noted that we should use caution in using the PTS results and should only use the contribution from non-pipe break transients to ensure that the comparison is consistent. In a related action, Bruce Bishop volunteered to provide a Westinghouse nozzle study for PWRs and will also provide PWR vessel failure probability for areas outside the beltline region covered by the PTS study.

Pete Riccardella suggested classifying the alloy 600 penetrations as non-piping failures for consistency with other nozzles. A suggestion was made to move the in-cores and CRDMs to the non-piping category. Fred Simonen suggested putting them all under a separate category called vessel penetrations, with a separate bin for RPVs and pressurizers. Karen Gott and Pete Riccardella are to create a base case for penetrations using the CRDM data based on a prior MRP study. Rob Tregoning will move all the vessel penetrations from piping to the vessel bin and supply updated tables.

Bengt Lydell has a non-piping data base that he will query by end of the month. He will also query the IRS data base (Incident Reporting System by INEA) by end of the month. Note the IRS database only includes data countries chose to include. It was also noted that MITI and NUPEC (both in Japan) have a data base for non-piping components. The NRC supposedly has a copy of this database. Rob will check into relevancy and availability. Bill Galyean volunteered to examine his database to see any relevant non-piping events, i.e., stream generator tube rupture statistics, incidents of bolting connections, etc., by the end of the month. Rob will be the conduit for getting the results from the various individuals searching the databases out to the rest of the group. Results will be posted to the ftp site and more important items will be bulk emailed to the panel.

The question was asked if the panel members could take home the modified tables for the pressurizer, RPV, pumps, valves, and steam generators and fill them out and return them back to Rob within two weeks. Rob will modify these non-piping tables (Tables B.1.7 and B.1.8 from the First Elicitation Minutes) with the additions made during the second meeting discussions and send them out to the panel members. Each panel member is to modify these non-piping tables and get them back to Rob Tregoning.

Fred Simonen has an electronic version of the GALL report (Generic Aging Lessons Learned) which identifies the key degradation mechanisms that could be used to help fill out these tables. Fred will extract the relevant tables to be used in identifying the key degradation mechanisms.

Presentation on Conditional Seismic Evaluation
By Rob Tregoning (USNRC)

Each base and reference case includes at least one transient since transients are needed to initiate a LOCA event. At this time the transients are poorly defined.

Fred Simonen has seen reports that show seismic stresses, but he has no idea of the magnitude of the non-seismic transients (e.g., water hammer, safety relief valve transients). He and Gery Wilkowski would like help in establishing a rough order of magnitude for these types of transients. This information could be best expressed as a percentage of the Service Level stresses. Pete Riccardella indicated that SRV transients could be on the order of a small earthquake, just with a higher frequency. Bruce Bishop agreed to provide some water-hammer transient stresses for the pressurizer. Gery Wilkowski volunteered to provide some summary information from past probabilistic LBB analyses

Bengt Lydell indicated that the water hammer frequency is about 5E-3. There are more water hammer events on the secondary side, but there are design basis events that can cause water hammer on the primary side. It as noted again that without a transient a large LOCA is highly unlikely. The cracks will just leak until they are detected, and then will be repaired. Long surface cracks that don't leak are drivers for large LOCAs.

Guy Deboo (Consolidated Edison) volunteered to provide stresses and frequencies for transients (e.g., feedwater line water hammer, SRV, and seismic from the LaSalle plant). Gery Wilkowski to provide some tables from NUREG/CR-6004 showing the N+SSE stresses for about 30 piping systems. This data was originally developed for the ASME Section XI Working Group on Pipe Flaw Evaluation. Additionally, Gery Wilkowski and Guy Deboo will provide stresses (not frequency) for some large faulted loads that are not really expected to occur over the life of plant. Gery is to examine the N+SSE stresses from the USNRC LBB submittal database. Guy noted that a 1SSE amplitude earthquake, based on seismic hazard curves, is expected to occur once over 40 years (design basis). The frequency (not amplitudes) of the seismic hazard curves are generally considered to be conservative. Gery noted that the design basis apparently is conservative since he is sure that an SSE event has ever occurred at any US (or other) plant. Hence, the seismic event frequency could perhaps be down graded to 0.5/2,600 events/year rather than 1/40 events per year.

For the smaller transients, Dave Harris will extract stresses from a NUREG report by Fred Simonen. Guy Deboo, Pete Riccardella, and Sam Ranganath will provide some data on normal operating transients. Pete Riccardella will get some data showing comparison of design versus actual transients based on some thermal fatigue analysis from a Sandia report.

Bruce Bishop has some plant specific ISI data that he could provide which provides transient information, but he won't be able to provide it expeditiously due to other commitments. In fact, it is unlikely he will be able to provide this during the timeframe of this effort. Pete Riccardella concluded that the actual transients were not as severe as the design basis transients, but there are typically more of them.

It was also agreed that the group should have isometric drawings for the LOCA-sensitive piping. Bengt Lydell will inventory his electronic drawing database. The purpose is to look at generic systems to get an idea of how many welds are involved, pipe sizes, etc. Guy DeBoo will also evaluate the ISO drawings in his archive. Bengt will coordinate with Guy Deboo on this effort. For the time being it was decided that we would not seek additional isometric drawings until we determine if we are missing any of the major piping systems. Guy Deboo will then help obtain drawings (e.g. ISI drawings) for missing systems that at least indicate the number of welds. In general, multiple isometrics of similar systems are useful since each plant design is somewhat unique.

Dave Harris thought we only needed census data, i.e., number of welds as a function of pipe sizes, etc. Dave will provide a census that he has developed of number of welds for the base case piping systems. Dave and Bengt Lydell will coordinate on this action. Gery Wilkowski indicated that there is a MRP report with locations and numbers of bimetal welds that may be useful to review. Gery will provide a table of Inconel weld locations in different piping systems from the MRP-44 report.

Rob Tregoning asked if the panel should consider redefining the base and reference cases. The biggest concern is that the loadings and the mitigation/maintenance may need to more accurately reflect the operating experience. The panel could redefine these variables in a way that is more consistent with the quantitative analyses that was presented for the base cases earlier. The approach is to define these base and reference cases load and mitigation variables to reflect historical plant operating experience for the first 25 years of plant life (e.g. the current LOCA estimate). The consensus opinion agreed with this proposal. Guy Deboo added that he would like to see the base cases run out to 40 and 60 years, with a seismic event included. It was again noted that the objective of the "current estimate" analyses (i.e., out to 25 years of plant life) was to provide a benchmark against historical data.

Rob also asked if the panel wanted to consider more than one degradation mechanism for each reference case since the operating experience contains contributions from all applicable degradation mechanisms. Rob Tregoning wants to make sure each panel member is making the same relative comparisons with the reference cases and that these relative comparisons are natural. However, the group consensus was that it is easier to use the reference cases for anchoring when only considering one degradation mechanism and that no other changes in the reference cases should be adopted.

Vic Chapman argued that if we run the probabilistic fracture models for the first 25 years for benchmarking purposes, and we see a failure, then we should exclude that result since in reality we have not seen any failures to date. Lee Abramson agreed with this thought. The operating experience-based estimates developed by Bill Galyean and Bengt Lydell are inherently benchmarked in this manner. The probabilistic fracture models (David Harris and Vic Chapman) still need to benchmark their data.

Bruce Bishop would like Karen Gott, Bill Galyean, and Bengt Lydell to extract the failure mechanisms as a function of piping system from their databases. Bill will have one of his colleagues do this. Bengt and Karen will query their databases to get a list of degradation mechanisms as a function of piping system. Bruce Bishop would like someone to publish a list of plants by design type. Rob Tregoning indicated that he would provide this information.

Several people wanted Dave and Vic to run their models using refined stress histories. Pete Riccardella agreed to redefine the loads for the HPI/MU nozzle. Gery Wilkowski volunteered to get some more realistic surge line stresses for the surge line elbow case. It should actually be for a crack in the girth weld at the elbow, not a crack in the body of the elbow.

Bill Galyean and Bengt Lydell will determine the frequency of IGSCC leaks and surface cracks as a function of time and pipe size using their respective databases. The surface crack data should be characterized by the length and depth of the flaws if possible. The Swedish data should be separately characterized from the US plant data, since the US plants may not have characterized the flaw shapes by ISI if the piping was being removed from service. In Sweden, flaws from removed pipe systems have been characterized. A further requirement of this query to examine the recirculation system piping studied in the base case would also be helpful. Bengt will provide all this information in an Excel spreadsheet.

Presentation on Leak Rate versus Pipe Break Size Evaluation
By Rob Tregoning (USNRC)

Rob started by reviewing the history behind this analysis. Rob was not able to find a reference for the basis of PWR correlations developed in NUREG-1150. The BWR correlations were extracted from GE NEDO studies performed in the early 1980's. New correlations have been based on several closed-form solutions appropriate for BWR steam and liquid lines and PWRs.

Rob still needs to add a column correlating the pipe diameter to the leak rate/area in slides 8 of 9 of this presentation. It was agreed that the pipe diameter should assume a single ended guillotine break. Rob also agreed to change the word "axial" to "transverse" on slide 9 of the original presentation which relates the pipe fracture area to transverse piping displacement. Gery Wilkowski noted that once a full DEGB occurs that the two ends of the pipe are jets that move away from each other so that he would be hesitant to use the analysis on slide 9 which assumes a transverse displacement. The base case team members will update their results using the new pipe break size to leak rate correlations developed for this presentation.

The final topic on the agenda was a discussion on the schedule for the elicitations.

Rob Tregoning and Lee Abramson want to do 2 elicitations early, possibly the week of July 14th. Then take a month off and restart the elicitations in mid August with the goal of completing all of the elicitations by the end of September. That would leave about a month to analyze the results. Rob is looking for volunteers to be the first two individuals to go through the elicitation process. All panel members should give their schedule to Rob Tregoning so that the individual elicitations can be scheduled.

Once the elicitations are complete and the results analyzed, a wrap-up meeting will be held in the October/November timeframe. As part of this meeting the calculated LOCA frequencies will be presented and any interesting and surprising results from individual questions will also be presented. The panel will also be solicited for feedback on the process. We will try to identify strengths and weaknesses with the process. Any necessary follow-on work will be defined during this meeting. We are also open to suggestions for conducting a reanalysis of these results in ten years. It is anticipated that the wrap-up meeting will take two days.

Again, it was noted that the panel members can change the results after their individual elicitations and after the wrap-up meeting. However, the final report must be submitted by the end of December. It is too premature to speculate on the form of the final report. However, confidentiality of the individual elicitation opinions will be maintained.

MEETING NOTES FROM US NRC LOCA ELICITATION WRAP-UP MEETING
DOUBLETREE HOTEL, ROCKVILLE, MD
FEBRUARY 10-12, 2004

Day 1 – February 10, 2004

Rob Tregoning from the USNRC welcomed everyone and explained logistics for the meeting. Rob had everyone introduce themselves. Next, Rob reviewed the agenda for the three-day meeting. Day 1 would focus mainly on piping; Day 2 on non-piping, and Day 3 on emergency and faulted loads plus soliciting feedback on the process. There are no results to present on the topic of emergency and faulted loads. Only the basic approach will be shown.

Presentation #1 – Elicitation Project Plan, Schedule, and Milestones
By Rob Tregoning

The NRC has initiated some ongoing work looking at active mechanisms, e.g., stuck open valves. Bill Galyean is doing this.

There is a SECY paper due to Commissioners on March 31, 2004 with LOCA frequencies for normal operating loads.

Rob will distribute a draft NUREG documenting expert elicitation results so the panel can provide feedback on the NUREG. Rob expects that the panelists will only a have short time (~2 weeks) to provide feedback.

For the April-June public meetings, Bruce Bishop from Westinghouse suggested meeting with the WOG risk-based group.

Most critical future milestone is finalizing individual expert responses for normal operating loading frequencies by February 25th.

Presentation #2 – Elicitation Results (Box and Whisker Plots
By Rob Tregoning

Rob made a presentation on the details of the box and whisker plots that will be shown over the next 3 days. Many different methods of calculating percentiles; we used Standard method; fundamental message is that doesn't make much difference in final analysis.

Presentation #3 – Safety Culture
By Rob Tregoning

Bruce commented on VG4 with respect to poor US safety culture that didn't see a problem until starting seeing circumferential cracks; Pete commented that it was an economic issue since US utilities charge 7 to 10 cents per KW-hr while overseas may charge 40 cents per KW-hr

Karen Gott pointed out that her experience was that the first time a plant experiences a problem it is a big problem so she would agree with second major bullet on VG#4; on second sub-bullet she thought a better word is experience instead of sensitive.

On VG#5 it was pointed out that while the NRC may have only one vote on code changes, it has the ultimate veto vote.

Bruce felt day of single plant utility is numbered. Helmut Schulz and Karen disagreed with the last sub-bullet on VG6; utilities are willing to invest in older plants since they are already paid for (less capital investment); may be a difference in international experience and US practice.

Helmut commented with regards to VG9 on decommissioning that they have not seen any increase in LER events in the last few years before decommissioning for those plants that were decommissioned.

With regard to VG10 and negative bullets related to risk-informed regulation. Bill Galyean commented that utilities have limited resources and risk informed process helps prioritize; Vic Chapman cautioned against turning crank and getting an answer without thinking of why.

Bruce commented that end of plant license renewal may be 80 years not 60; based on comment made at NRC recent meeting.

Helmut pointed out that boric acid corrosion of manway bolts of 15 to 20 years ago was more serious from a LOCA perspective that Davis Besse head problem of today.

Bottom line is no effect of safety culture on LOCA frequencies. There will be no adjustments to frequencies; no major discussion on part of panel with regards to this bottom line conclusion.

Presentation #4 – Piping Base Case Evaluation I
By Bengt Lydell

Bengt used a bottoms up evaluation based on operating experience.

Markov is standard approach common to any advanced reliability approach; this was the technical basis used by Bengt to develop time dependency.

Bengt's model allows for imperfect repairs or inspections.

Can go from S (unflawed condition) to R (rupture condition) if have some extraordinary event such as gas accumulation at Hamaoka in Japan.

VG13 results are per "weld year"; some of earlier plots are "per reactor year".

Presentation #5 – Piping Base Case Evalution II
By Dave Harris

Vugraph #3 is a summary of revised results since July 2003.

Could use VC Summer Hot Leg/RPV nozzle weld crack as benchmark, but need to be careful to consider all aspects such as differences in weld residual stresses due to repairs.

VG12 shows an order magnitude difference in LOCA frequency for a difference of 2 ksi (14 MPa) in normal operating stress; this was thought to be pretty sensitive result.

Dave assumed a linear residual stress field through the thickness due to the weld overlay repair.

The solid line in VG13 is PRAISE result while prior and post symbols come from Bengt's results.

Presentation #6 – Elicitation Question I: Base Case Evaluation
By Rob Tregoning

There was quite a bit of disagreement on first bullet on with regards to the perceived disadvantages of the various approaches.

Bruce argued that PFM approaches have been benchmarked against operating experience and shown to agree well.

Helmut and Bruce stressed that we are not in a position to review various approaches and we should not provide such a review in the NUREG report.

Need to stress in NUREG that general comments are not a group consensus, but individual responses.

Presentation #7 – Elicitation Calculation Framework
By Lee Abramson

Split distributions necessary if UB and LB are not symmetric with respect to mid value.

Used log normal distribution since results provided by participants fit log-normal distribution; also log-normal distribution easily to manipulate; also tradition is that log-normal is used in risk based approaches; bottom line is that due to variability in responses should not make that much difference as to what distribution chosen.

Analysis will yield medians of mid values and bounds as well.

Presentation #8 – PWR Piping
By Rob Tregoning and Paul Scott

The second main bullet on VG7 should be decrease with "decreasing" piping size.

Dave Harris disagreed with comment that PFM models had problems modeling mitigation.

There was a problem with VG16 with interpreting results for Panelist L.

A number of the panelists were surprised with VG17 that surge line results for Cat 5 are comparable to that of cold leg.

There is a discrepancy in maximums for Category 6 LOCAs at 25 years between VG19 and 20. VG20 shows participant L having the maximum value while VG19 shows participant B as having the maximum value.

Participant J showed the most impact of age on the LOCA frequencies; really obvious in VG21.

Bruce pointed out that for higher Category LOCAs that there was less uncertainty; may be an artifact that there are less systems than can contribute (only large diameter); also these larger systems are better inspected, i.e., better controlled.

Every plot shown is for 25 years unless specifically stated.

Presentation #9 - BWR Piping
By Rob Tregoning and Paul Scott

Similar degradation mechanisms as with PWR mechanisms: thermal fatigue, mechanical fatigue,

Sam commented with regards to Slide #4 that analysis of BWR feedwater says it should crack, but don't find these cracks in service.

For Slide 8 it was suggested to change "risk" to "high" in last bullet.

Maximum diameter of RWCU system is 6 inch; not 24 inch as shown in VG17.

Day 2: February 11, 2003

Rob started the second day at 0810 by reviewing agenda for the next two days.

Presentation #10 – Non-Piping Database Development
By Bill Galyean

LER database at ORNL will no longer be available after 2/29/04; the database will be moving to INEEL but in a different format.

Failures defined as leaks or cracks.

Presentation #11 – Non-Piping Base Case Development: CRDM and LTOP LOCAs
By Pete Ricaradella

Pete used the VIPER program to predict beltline failure frequencies (per vessel year) for typical BWRs.

VG3 and 4 are frequency plots and not probability plots.

For VG7 for large Category LOCAs, see big impact with time between 40 and 60 years; attributed to effect of radiation embrittlement; for smaller LOCAs don't see much effect of time.

EDY stands for Effective Degradation Years; used to normalizes degradation to a reference of a 600°F operating temperature.

For CRDM nozzle ejection probability the assumption in VG11 that immediately have circumferential TWC of 30 degrees is highly conservative according to Bruce in that most are axially oriented.

Of 30 plants, there were 11 nozzles that had circumferential cracks; all of these plants were at about 20 EDY so Pete could take time factor out; total number of nozzles in 30 plants was 881

POD curve for NDE; cracks were EDM notches that were compressed to make them tight; eventually will get some real cracks from the North Anna head that can be used for calibration/validation

VG15 shows the probability of leak in one of 98 nozzles in this plant per vessel year; shows effect of NDE on probability of leakage

VG15 and 16 show effectiveness of NDE and how PFM models can account for inspection in their analyses.

VG17 shows decreasing frequency with time which reflects benefit of inspection.

To get 5,000 gpm (19,000 lpm) leakage, need ejection of 2 nozzles; most likely scenario is for collateral damage as one ejects and causes damage to adjacent nozzles.

Presentation #12 – Steam Generator Tube Rupture Frequencies
By Rob Tregoning

Almost everyone agreed that for PWRs the dominant failure scenario for Category 1 LOCAs was steam generator tube failures.

Used non-piping database which was augmented back to 1987 to capture 2 major events in '87 and '89. There have been 15 leaks since 1990 with 4 events over 100 gpm (380 lpm) since 1987.

On VG2 the reference to Nine Mile Point should be to Indian Point.

Fred Simonen stated that usual scenario assumed for higher category LOCAs is common cause failures such as losing pressure on secondary side causing pressure differential across tube and failure of multiple already degraded tubes.

Rob's analysis of independence (ignoring common cause) was viewed with a great deal of skepticism; Bill Galyean commented that if look at LERs always see multiple tube degradation but only see single tube rupture in the LERs.

This analysis is for 25 years, panel members left to their own devices for later years.

Presentation #13 – Overview of PTS Re-Evaluation Project
By Rob Tregoning

For this analysis all of the crack growth is from a PTS event; not fatigue.

For plants with multiple pass cladding there is a very low probability of flaw penetrating multiple passes; exception was Oconne that was single pass cladding.

Big driver were flaws between plate and axial flaw region.

For those that want to anchor against PTS, Rob will use updated results base on some average values for Oconee, Beaver Valley, and Palasdies

Presentation #14 – Elicitation Question VI: PWR Non-Piping
By Rob Tregoning

Analyses tend to get easier as go up on LOCA sizes in that less systems to worry about.

As a group the panel was uncomfortable with comments on Effect of Operating Time for Category 4 LOCAs.

Bill Galyean suggested that for presentation to public, may want to consider some other means of reporting extremely low frequencies (~1e-15); Rob wasn't sure he could define a cut off value; also when look at median values, these low numbers don't impact final answer; Helmut supported this approach.

Helmut suggested that we explicitly state that a major assumption was that everything was fabricated in accordance with Code standards; no counterfeit bolts, etc.

Participant J only has fatigue in his analysis, once he includes other mechanisms, then his frequencies will go up by 2 or 3 orders of magnitude, but will still be low and probably will not alter the final median results.

Presentation #15 – Elicitation Question VI: BWR Non-Piping
By Rob Tregoning

Two of the six respondents who responded to this question only provided frequencies out to Category 5 LOCAs even though BWR non-piping could cause a Category 6 (Vessels) LOCA; thus Rob only provides results for Category 5 LOCAs here.

Concern with thermal aging of cast stainless steel is when it is present in concert with some other degradation mechanism.

CRDM refers to stud tube housing on bottom head of BWRs.

Panelist F didn't consider RPV for Category 1 LOCAs; question was whether they didn't think important or did they not have a means of making an evaluation.

There is much less spread in results for RPV than valves and pumps; panelists spent more time and more work in past on RPV than pumps and valves.

Panelist C sees a decrease in freq with time but may be an artifact of the fact that he anchored against BWR recirculation line that shows a decrease in LOCA frequency with time; he provides no rationale for why he would expect non-piping LOCA frequency to decrease with time.

Presentation #16 Piping and Non-piping combined Results
By Rob Tregoning

Ratios of non-piping to piping for various category LOCAs are for 25 years only.

Pete made the point that most of the plots that Rob has shown are for 25 years, while he thought 40 and 60 years more important since 25 years is in past and the associated problems have been addressed while 40 and 60 years are for future; Rob responded that not that much difference between 25 and 40 years with some effect for certain category LOCAs at 60 years.

Sam commented that he was somewhat surprised that non-piping contribution less than piping contribution for BWRs in that piping has some active mechanisms that have been successfully mitigated in past; Karen responded that non-piping components more robust.

Bill Galyean warned about combining group distributions and panel distributions that may introduce a bias in that various members of group defined boundaries of system differently.

Much discussion on whether to chose group median or individual medians; Rob and Lee haven't done panel distributions yet.

Comparison of BWR and PWR – effect of mitigation encompassed in results for BWRs, but not PWRs – BWRs have been doing mitigation for 15 to 20 years whereas PWRs are just starting with mitigation for PWSCC.

Inclusion of S/G tube rupture for Category 1 LOCAs will be problematic for some people in that PRA people aren't used to accounting S/G tube failures in with rest of data; typically S/G tube rupture data is presented separately; in future Rob will present data both with S/G tube rupture data and without.

Categories 1, 2, and 3 are historically same as small, medium, and large break LOCAs respectively.

For PWR MB LOCAs, major contributor is CRDM, not S/G tube ruptures.

Discussion of whether MB LOCA was Category 2 or Category 3; some thought that MB LOCA was more in line with Category 3 LOCA.

Bengt felt we are comparing apples and oranges as we try to compare our results with historical results; NUREG/CR-5750 didn't look at non-piping per se whereas we did, although Bill Galyean indicated that if there had been indications of TWC in non-piping components then he would have included that data in his analysis in 5750; some thought that due to apples and oranges nature of our approach with 5750 that we shouldn't present these comparisons but Rob argued that if we don't present these comparisons then others will; Some argued that we should present frequencies for multiple LOCA categories when we compare with small, medium, and large break LOCAs for 5750

Lee reviewed the feedback questionnaire.

Day 3: February 12, 2004

Presentation 17: Emergency and Faulted Loading: Elicitation Approach and Responses

Water hammer type loadings should be in normal operating loading history.

What we are asking panel members to estimate is only the conditional failure probability given a stress with magnitude i ($P_{L/Si}$)

Ken commented that the seismic anchor motion (SAM) stress which is a secondary stress may be a bigger contributor than some primary stresses such as inertial stresses.

On VG entitled Elicitation Requirements, we are asking panel to do first bullet, we will do 2nd bullet, and plants would do 3rd and 4th bullets.

Asking them for a given system and degradation mechanism for their estimate of L50, P50, Ppl, L_{pl}, L_{tsl}, and P_{tsl} and then we will interpolate to get entire curve.

Some people argued during their elicitations that non-piping and large piping are non contributors to LOCA frequency due to seismic; we will look at results from piping and then decide what and if we will do anything for non-piping considerations.

Bruce argued that can get some very high loads due to malfunctions of snubbers.

P_{bc} and L_{bc} are probabilities for base case and likelihood of base case

Presentation 18: Remaining Work
By Rob Tregoning

Rob discouraged panel to make changes to bring their results more in line with others, would encourage panel to make changes if they heard something technically that made them rethink their answers.

Everyone will be involved in reviewing and critiquing NUREG reports; everyone wanted to be involved with the process.

The question of how: possibly another meeting, VTC, circulate vugraphs for review and feedback (electronically); possibly couple with some other meeting (ASME, PVP, etc).

Karen and Dave would want to meet before the NUREG was finalized; others seemed to agree with this.

Ideally we would circulate draft, we would then get comments back, we would then synthesize comments and then feed them back to the group and then meet; all before finalizing NUREG.

Helmut and Bruce supported idea of VTC (maybe limit to a few sites).

Other option is provide slides; review slides on computer and then have a conference call to review; limit to a few hours at a time (bite off small chunks).

Presentation #19 – Remaining work on Active Systems
By Bill Galyean

PORV stands for pilot operated relief valve.

Difficult to correlate stuck open valves categories with leak rate sizes/categories; size of valves will vary between plants.

Presentation #20 – Emergency and Faulted Loading Base Case Development
By Gery Wilkowski

No uncertainties applied to loads in Gery's analysis.

Base case assumes idealized TWC geometry.

Did not do any subcritical crack growth; thus did not consider residual solutions.

LBB.ENG2 is in form of closed form solutions.

For predicting large crack growth in a pipe tests it is better to use J-M than J-D.

Duane Arnold crack would have failed if subjected to Level B type loading.

Global secondary stresses act as primary stresses if crack large enough such that failure stress is below yield strength.

Presentation #21 – Lee reviewed feedback

It would have helped to have had Gery's presentation earlier, before the panel tried to answer seismic question.

Would have been nice to have a video of plants showing various systems as one tours plants with video camera.

Amount of information available was overwhelming; try to do a division of labor so one or two people review something and provide a tutorial to others so everyone is working from same basis; otherwise everyone is inventing the wheel themselves; maybe have a meeting to review these tutorials.

Need a roadmap of where information can be found.

Periodic/weekly update of changes made to ftp site; alternatively an alert message when something added to site; maybe a readme file when something added and what was added and when.

NRC management must make sure that staff are available to panel members during the process; Rob getting pulled off for Davis Besse was a problem; delayed things and then panel members only had a few weeks to respond at the end.

Bruce would like time at meetings to do actual work on elicitations because once they get back home they will get pulled off on to other things and won't be able to get back to answering questions for a long time.

APPENDIX C

ELICITATION TRAINING EXERCISE RESULTS

APPENDIX C

ELICITATION TRAINING EXERCISE RESULTS

As part of the panel member kick-off meeting in February 2003, elicitation training was provided for the elicitation panel. The training involved the panel members answering a series of almanac-type questions for which numerical answers were available. The panel members provided both their best estimate of the answer as well as relative ratios with respect to other quantitative responses. In this way the panel members got an appreciation of the benefits of the anchoring process used throughout the elicitation process.

C.1 Training Questions

The following questions were used in the training exercise.

Q1. According to the 2000 census, how many men 65 or over were there in the U.S.?

Q2. In 1995, how many American men age 65 or older suffered from the chronic conditions listed?

Q3. What is the <u>ratio</u> of the rate for men 45- 64 years old to the rate for men 65 and older for each of the conditions listed?

Q4. What is the <u>ratio</u> of the rate for men under 45 years old to the rate for men 45 - 64 years old for each of the conditions listed?

The answer to Q1 is 14.4 million. The chronic conditions referred to in Q2, Q3, and Q4 and the corresponding answers are listed in Table C.1.

Table C.1   Correct Value (CV) Results to Elicitation Training Questions

	Q2	Q3	Q4
Condition	Rate per 1000	(Age 45-64 rate) / (Age 65+ rate)	(Under 45 rate) / (Age 45-64 rate)
Arthritis	404.7	0.44	0.13
Cataracts	125.1	0.13	0.11
Diabetes	123.6	0.50	0.10
Hearing Loss	366.8	0.56	0.20
Heart Disease	362.4	0.40	0.17
Prostate Disease	118.0	0.30	0.054

C.2 Elicitation Training Responses

As described in Section 3.3.2, the panelists were asked to supply three numbers for each question: a MV, a LB, and a UB. The MV has a nominal 50/50 chance of falling above or below the correct value. The interval (LB, UB) has a nominal 90% chance of covering the correct value.

The following tables summarize the responses made in the training exercise. There were between 15 and 17 sets of responses to each question. (Although there were only 12 panelists on the panel, members of the facilitation team were also invited to participate.) The number of respondents is indicated following each question. The table columns summarize the responses relative to the CV. The first column indicates the number of respondents where CV < LB, i.e., where the coverage interval fell above the CV; the third column indicates the number of respondents where CV > UB, i.e., where the coverage interval fell below the CV. Thus, the total of the first and third columns is the number of respondents whose coverage intervals did not cover the CV. The second column lists three numbers that summarize the set of MVs provided for each row of the table. These are the lower quartile (LQ), median and upper quartile (UQ), respectively. About one quarter of the MVs are less than the LQ and about one quarter of the MVs are greater than the UQ. Hence the interquartile interval (LQ, UQ), denoted by IQI, contains about one half of the MVs. (These three summary statistics are used to construct box and whisker plots, as described in Appendix L.) For ease of reference, the rounded correct values are listed following the conditions for the Q2 - Q4 tables.

Q1. According to the 2000 census, how many men 65 or over were there in the U.S.? (N = 17)
(CV = 14.4 million)

Table C.2 Summary of Respondent Results for Question Q1

CV < Coverage Interval	LQ, Median, UQ	CV > Coverage Interval
N = 3	16, 20, 28	N = 0

Respondents tended to overestimate the CV. Since LQ = 16, about three quarters of the MVs were larger than the CV. However, percent coverage at 82% was near the nominal 90%, with 3 (18%) lying above the CV and none lying below.

Q2. How many American men age 65 or older suffered from the following chronic conditions in 1995? (N = 15)

Table C.3 Summary of Respondent Results for Question Q2

Rate per 1000

Condition	CV < Coverage Int.	LQ, Median, UQ	CV > Coverage Int.
Arthritis (405)	N = 1	135, 200, 400	N = 9
Cataracts (125)	N = 2	50, 150, 200	N = 2
Diabetes (124)	N = 0	90, 150, 250	N = 3
Hearing Loss (367)	N = 1	200, 300, 500	N = 5
Heart Disease (362)	N = 1	150, 200, 375	N = 6
Prostate Disease (118)	N = 3	125, 200, 375	N = 2

Four of the six IQIs covered the CV, and the two which did not almost did. Three of the medians were above the CV and three were below. Thus, the MVs for the six conditions as a whole exhibited no systematic bias in estimating the CVs. However, the coverage intervals tended to underestimate the CVs. Of the 90 coverage intervals, 27 (30%) lay below the CV and 8 (9%) lay above. The average percent coverage of all 90 intervals was 61%. Over the six conditions, the percent coverage ranged from a low of 33% to a high of 80%.

Q3. What is the ratio of the rate for men 45- 64 years old to the rate for men 65 and older for each of the conditions listed? (N = 16)

Table C.4 Summary of Respondent Results for Question Q3

(Rate for ages 45-64) / (Rate for age 65+)

Condition	CV< Coverage Int.	LQ, Median, UQ	CV > Coverage Int.
Arthritis (0.44)	N = 1	0.20, 0.30, 0.50	N = 5
Cataracts (0.13)	N = 2	0.10, 0.20, 0.30	N = 3
Diabetes (0.50)	N = 0	0.25, 0.40, 0.50	N = 3
Hearing Loss (0.56)	N = 0	0.25, 0.30, 0.30	N = 6
Heart Disease (0.40)	N = 0	0.30, 0.30, 0.50	N = 2
Prostate Disease (0.30)	N = 2	0.20, 0.20, 0.40	N = 3

Five of the six IQIs covered the CV, but respondents tended to underestimate the CV. Five of the six medians were below the CV. Of the 96 coverage intervals, 22 (23%) lay below the CV and 5 (5%) lay above. The average percent coverage of all 96 intervals was 72%. Over the six conditions, the percent coverage ranged between 62% and 88%.

Q4. What is the <u>ratio</u> of the rate for men under 45 years old to the rate for men 45 - 64 years old for each of the conditions listed? (N = 16)

Table C.5 Summary of Respondent Results for Question Q4

(Rate for under 45) / (Rate for ages 45-64)

Condition	CV < Coverage Int.	LQ, Median, UQ	CV > Coverage Int.
Arthritis (0.13)	N = 2	0.10, 0.20, 0.30	N = 2
Cataracts (0.11)	N = 1	0.05, 0.10, 0.20	N = 1
Diabetes (0.10)	N = 8	0.20, 0.30, 0.50	N = 0
Hearing Loss (0.20)	N = 1	0.10,.0.20, 0.30	N = 2
Heart Disease (0.17)	N = 2	0.12, 0.20, 0.30	N = 2
Prostate Disease (0.054)	N = 6	0.05, 0.10, 0.20	N = 1

Five of the six IQIs covered the CV, but respondents tended to overestimate the CV. Four of the six medians were above the CV, and two were equal or almost equal to the CV. Of the 96 coverage intervals, 20 (21%) lay above the CV and 8 (8%) lay below. The average percent coverage of all 96 intervals was 71%. Over the six conditions, the percent coverage ranged between 50% and 88%.

C.3 Discussion

The results of the training exercise were consistent with several of the basic premises underlying the elicitation structure and methodology. First, apart from Q1, the responses to the other three questions as a whole did not exhibit any systematic over- or under- estimation bias. Q2 had no systematic bias, Q3 tended to underestimate, and Q4 tended to overestimate the CVs. This result is consistent with the basic premise of the elicitation process, which is that the panel responses as a whole have no systematic bias (see Section 3.3).

Second, the percent coverage of the (LB, UB) intervals were less than the nominal 90% for all four questions. Q1 had the highest percent coverage at 82%, perhaps because the question dealt with demographic data with which the respondents were relatively more familiar. Q3 and Q4 had the next highest percent coverage at about 71% each and Q2 had the lowest percent coverage at 61%. This result is consistent with the rationale for the overconfidence adjustments made to the panelists' uncertainty intervals (see Section 5.6.2).

Third, the two questions (Q3 and Q4) that asked about ratios of rates had higher percent coverage than the question (Q2) that asked about absolute rates. This result is consistent with the rationale for the basic structure of the elicitation questions, which ask about relative rather than absolute LOCA frequencies (see Section 3.8).

APPENDIX D

PIPING BASE CASE RESULTS OF

BENGT LYDELL

APPENDIX D

PIPING BASE CASE RESULTS OF BENGT LYDELL

An Application of the Parametric Attribute-
Influence Methodology to Determine Loss of
Coolant Accident (LOCA) Frequency Distributions

Report No. 2 to the NRC Expert Panel on
LOCA Frequency Distributions

Prepared for

U.S. Nuclear Regulatory Commission
Washington (DC)

June 2004

ACKNOWLEDGEMENTS

The work documented in this report was performed for the U.S. Nuclear Regulatory Commission under Subcontract No. 177115 (Battelle Memorial Institute).

Mr. Karl N. Fleming (Technology Insights, Inc., San Diego, CA) provided constructive review comments on a draft of this report. Mr. Fleming also provided the Markov model solutions supporting the calculation of time-dependent LOCA frequencies.

TABLE OF CONTENTS

LIST OF TABLES

LIST OF FIGURES

ABBREVIATIONS

ASME	American Society of Mechanical Engineers	UT	Ultrasonic Testing
ASTM	American Society for Testing and Materials	**Notation**	
CL	Cold Leg (of a PWR RCS)	A	Attribute
CV	Chemical and Volume Control	a/t	Ratio of crack depth to pipe wall thickness
DN	Nominal Pipe Size, metric [mm]		
DPD	Discrete Probability Distribution	C	Conditional failure probability given a flawed weld and an unusual or severe loading condition
EPRI	Electric Power Research Institute		
FW	Feedwater		
HAZ	Heat Affected Zone (of weld)		
HL	Hot Leg (of a PWR RCS)	F	Failure (= through-wall flaw)
HPI	High Pressure Injection	L	Large leak
HWC	Hydrogen Water Chemistry	P	Probability
IGSCC	Intergranular Stress Corrosion Cracking	S	Susceptibility (to degradation)
		W	Weld count
IHSI	Induction Heat Stress Improvement	$\varnothing$	Pipe diameter
ISI	Inservice Inspection	ϕ	Crack occurrence rate
LOCA	Loss of Coolant Accident	λ	Failure rate (frequency of PBF resulting in leak rate $\leq$ TS limit for unidentified leakage)
MSI	Mechanical Stress Improvement		
MV	Mid Value (50% percentile)		
NDE	Nondestructive Examination	ρ	Rate of large leak event
NMU	Normal Makeup	ν	Leak/spill rate (gpm)
NWC	Normal Water Chemistry	σ_{NO}	Normal operating weld stress (ksi)
NPS	Nominal Pipe Size, US [inch]		
PBF	Pressure Boundary Failure	ω	Repair rate
PWSCC	Primary Water Stress Corrosion Cracking		
RCPB	Reactor Coolant Pressure Boundary		
RCS	Reactor Coolant System		
RR	Reactor Recirculation		
SC	Sensitivity Case		
SI	Safety Injection		
SS	Stainless Steel		
TS	Technical Specifications[1]		

---

[1] For TS leak rate limits see for example NUREG-1431 (Vol 2, Rev. 2, June 2001): Standard Technical Specifications Westinghouse Plants – Bases, Section B 3.4.13, RCS Operational Leakage. For unidentified leakage the Limiting Condition for Operation (LCO) is 3.8 lpm (1 gpm), and for identified leakage the LCO is 38 lpm (10 gpm). See also NUREG-1433 (Vol. 2, Rev. 2, June 2001): Standard Technical Specifications General Electric Plants, BWR/4 – Bases, Section B 3.4.4, RCS Operational Leakage. For unidentified leakage the LCO is 19 lpm (5 gpm). Further, if an unidentified (BWR) leakage has been identified and quantified, it may be reclassified and considered as identified leakage; however, the total leakage limit would remain unchanged.

D.1 Background

Limited to consideration of Code Class 1 piping failures, Base Case Report Number 2 documents an assessment of BWR and PWR loss of coolant accident (LOCA) frequency distributions. The assessment is a demonstration of the role of statistical analysis of service experience data and Markov modeling in a "bottom-up" approach to piping system reliability analysis.

D.1.1 Objectives

Using primary coolant piping design information for three reference plants (one BWR plant and two PWR plants), the overall objective is to determine LOCA frequency distributions that are representative of currently operating U.S. nuclear power plants, including current in-service inspection (ISI) practices and degradation mitigation strategies. This determination is done analytically using a parametric model of piping reliability. The LOCA frequency distributions are determined for three time periods. To address today's piping reliability state-of-knowledge the LOCA frequency is determined at $T = 25$ years. Next the LOCA frequency is extrapolated to $T = 40$ years to represent the primary system piping reliability status at the end of a 40-year operating license. Finally an extrapolation is made to $T = 60$ years to account for a possible license renewal. Analytically, this extrapolation is concerned with the potential impact on the structural integrity of the piping by material aging as well as by reliability improvement efforts.

As implied by the report title, the objective is to develop LOCA frequency distributions. The report addresses two aspects of LOCA frequency distributions. It develops LOCA frequencies associated with a distribution of flow rate threshold values ranging from 380 lpm (100 gpm) at the low end to beyond 380,000 lpm (100,000 gpm) at the high end. Additionally the study develops statistical uncertainty distributions for each set of LOCA frequencies to account for the uncertainty in the input parameters to this piping reliability analysis.

D.1.2 Base Case Definition

During a meeting in Rockville (MD) in February 2003 [D.1], the Expert Elicitation Panel members defined five Base Cases that are denoted as BWR-1, BWR-2, PWR-1, PWR-2 and PWR-3, respectively. The five Base Cases are:

BWR Base Case (Plant 'B')
- BWR-1; Reactor Recirculation (RR) System. This reference case includes one-of-two RR System loops. Each loop consists of one NPS28 recirculation pump loop with a NPS22 manifold with five NPS12 risers; NPS is nominal pipe size in inch. The reference case excludes any small-diameter piping or tubing attached to the main RR piping. With a few exceptions, the selected piping system layout is representative of a BWR/4 reference plant as described in NUREG/CR-6224 [D.2]. The Base Case RR System does not include the NPS4 bypass line, however. The RR piping is fabricated from austenitic Cr-Ni stainless steel of Type A-304 ($\geq 0.035\%$ carbon).

- BWR-2; Feedwater (FW) System. As defined by isometric drawings, this reference case includes Loop B of the Class 1 portion of the FW System (i.e., the part of the FW System that is located in the drywell containment structure). This system of two loops includes NPS12, NPS14 and NPS20 piping. The FW piping is fabricated from carbon steel of Type A-333 Gr. 6.

- Section D.1.3 includes additional information on the BWR Base Case system definitions.

PWR Base Case (Plant 'A.a/b')
- PWR-1; Reactor Coolant (RC) System. As defined by an isometric drawing, this reference case includes one of the NPS30 hot leg (HL) in the RCS.

- PWR-2; Pressurizer Surge Line. As defined by an isometric drawing, this reference case includes the NPS14 piping, which connects the pressurizer to the cold leg (CL).

- PWR-3; High Pressure Injection/Normal Makeup (HPI/NMU) System. As defined by an isometric drawing, this reference case includes the 2-½ inch schedule 160 line between the containment isolation valve and the RCS cold leg (CL).

- The PWR base cases associated with the RC hot leg and pressurizer surge line are typical of a 3-loop Westinghouse PWR (Plant A.a). The PWR base case associated with the HPI/NMU line is typical of a Babcock & Wilcox PWR (Plant A.b).

- Section D.1.4 includes additional information on the PWR Base Case system definitions.

D.1.3 BWR Base Case System Descriptions

Plant B is a BWR/4 assumed to have been in commercial operation for at least 25 years. Similar to many other operating BWR/4 plants in the USA, Plant B is also assumed to be operating with a combination of IGSCC Category D and E welds, according to the nomenclature of U.S. NRC Generic Letter 88-01 [D.3, D.4]. In other words, the plant has experienced some IGSCC and the affected welds have been reinforced by weld overlays. It is further assumed that none of the IGSCC susceptible welds have been subjected to any stress improvement (SI) process such as induction heat stress improvement (IHSI) or mechanical stress improvement process (MSIP). It is also assumed that the weld overlay repairs (WOR) were all performed in the 1982-1988 timeframe. Finally, Plant B is assumed to have been operating with normal water chemistry (NWC) at all time.

The system descriptions in this section are extracted from design information supplied by members of the Expert Elicitation Panel. The BWR-specific system information is included in the following documents and drawings:

- Document No. EPRI-156-310: Degradation Mechanisms Evaluation for Class 1 Piping Welds at Plant B [D.5].

- Excel-file entitled "PlantBWelds." This Excel-file includes weld lists with locations for the RR and FW ASME Section XI Code Class 1 piping. The lists are organized by weld identification numbers (as they appear on the isometric drawings identified below) nominal pipe size and pipe schedule. The Excel file forms the basis for the LOCA frequency model used to derive the LOCA frequency distributions.

- Isometric drawing numbers 6M721-5358-5 (RR System Loop B Ring Header), 6M721-5359-5 (RR Loop B Suction & Discharge Piping), 6M721-2336-1 (FW System Inside Drywell), and 6M721-3537-5 (FW System Inside Drywell).

D.1.3.1 Reactor Recirculation (RR) System - The RR System evaluated in this study consists of two recirculation pump loops external to the reactor pressure vessel (RPV). These loops provide the piping path for the driving flow of water to the RPV jet pumps. Each loop contains a variable speed recirculation pump and two motor operated isolation valves (one on each side of each pump). The recirculation loops are part of the nuclear system process barrier and are located inside the drywell containment structure. The pipe segments that are subject to evaluation in this study consist of:

Loop A: The Class 1 portion starts at the RPV nozzle N1A and is reconnected to the RPV at nozzles N2F, N2G, N2H, N2J, and N2K. Class 1 lines for the Residual Heat Removal (RHR) and Reactor Water Cleanup (RWCU) Systems are connected to this loop. These particular Class 1 lines are excluded from the study scope, however. Loop A is excluded from the BWR Base Case.

Loop B: The Class 1 portion starts at RPV nozzle N1B and is reconnected to the RPV at nozzles N2A, N2B, N2C, N2D, and N2E. Part the original design, a NPS4 bypass line at valve F031B has been removed from the system. Class 1 lines for the RHR and RWCU Systems are connected to this loop. These particular Class 1 lines are excluded from the study scope, however.

D.1.3.2 Feedwater (FW) System - The FW System provides feedwater to maintain a pre-established water level in the RPV during normal plant operation. The Condensate and the FW Systems take water from the main condenser and deliver it to the RPV after passing it through the feedwater heaters and demineralizer system. The Class 1 portion of the FW System consists of two loops:

Loop A: Loop A starts at valve F076A and a connection to the High Pressure Coolant Injection (HPCI) discharge line (at valve F006), and connects to the RPV at nozzles N4A, N4B, and N4C. The HPCI discharge line is excluded from the study scope. Loop A is excluded from the BWR Base Case.

Loop B: Loop B starts at valve F076B, connection to the Reactor Core Isolation Cooling (RCIC) discharge line at valve F013, and a discharge from the RWCU System (at valve F220), and connects to the RPV at nozzles N4D, N4E, and N4F. The RCIC and RWCU discharge lines are excluded from the study scope.

D.1.4 PWR Base Case System Descriptions

The system descriptions in this section are extracted from design information supplied by members of the Expert Elicitation Panel. The PWR-specific system information is included in the following documents and drawings:

- Document No. EPRI-156-330: Degradation Mechanism Evaluation for Class 1 Piping Welds at Plant A.a [D.6]. This document summarizes the degradation mechanisms applicable to a Westinghouse 3-loop PWR.

- References [D.7-D.9] include design information as well as degradation mechanism information applicable to the HPI/NMU system of Plant A.b.

- Excel file entitled "PlantAWelds." This Excel file includes weld lists for the RC system of Plant A.a. The lists are organized by weld identification numbers (as they appear on the isometric drawings identified below), nominal pipe size and pipe schedule. This Excel-file forms one of the bases for the PWR LOCA frequency model used to derive the LOCA frequency distributions.

- Isometric drawing numbers 1MS-22-2262 and CGE-1-4100A (RC Hot Leg), C-314-601 and CGE-1-4500A (pressurizer surge line), and 17-MU-23 (HPI/NMU piping).

D.1.4.1 Reactor Coolant (RC) System (Plant A.a) - The RC System evaluated in this study consists of three similar heat transfer loops connected in parallel to the RPV. Each loop contains a reactor coolant pump (RCP), steam generator, and associated piping and valves. In addition, the system includes a pressurizer, a pressurizer relief tank, interconnecting piping, and instrumentation necessary for operational control. The analysis in this report is concerned with a portion of one of the three RC loops; the portion from the RCP to the RPV (this is one of the hot legs). The pressurizer surge line connects the pressurizer to the RC cold leg Loop A. In summary, the piping sections that are subject to evaluation in this study consist of.

RC-HL: The analysis is concerned with 1-of-3 hot legs. The Loop A HL starts at the RPV, includes an RCP and connects to the 'A' steam generator (S/G). The HL piping is fabricated from stainless steel piping. The section of the HL from the RPV to the RCP is of 31 inch inside diameter, while the section from the RCP to S/G is of 27.5 inch inside diameter piping.

Surge Line: The single surge line is fabricated form NPS14 stainless steel piping and connects to the 29-inch RC cold leg.

D.1.4.2 High Pressure Injection (HPI)/Normal Makeup (NMU) Line (Plant A.b) - In Plant A.b, each of the four RCS cold legs is equipped with high-pressure injection piping. Two of these 2 ½ inch (ID, or approximately NPS3-¾) stainless steel piping lines provide the normal makeup flow to the RCS and they connect to the cold leg via nozzle assemblies. Each of the nozzles is comprised of a base nozzle and a safe-end. To prevent thermal cycling of the base metal each nozzle is equipped with a 1.5-inch thermal sleeve. The analysis is concerned with one of the two HPI/NMU lines.

D.1.5 Summary of Scope Limitations

As outlined above, this LOCA frequency assessment is limited to specific portions of BWR and PWR Code Class 1 piping. The BWR base cases include contributions from potential pipe breaks in Loop B of the respective RR and FW System. Pipe break frequency contributions from normally pressurized sections of HPCI, RCIS, RHR or RWCU piping are not considered in this study. Piping system design information beyond that itemized above is not accounted for in this study. The PWR base cases include contributions from 3-of-3 RC hot legs and 2-of-2 HPI/NMU lines, respectively.

Excluded from the analysis are LOCA frequency contributions due to degradation and failure of cast stainless steel components such as valve bodies. While there is some documented evidence of degradation of such components, (e.g., [D.10]) the frequency of a through-wall defect in valve bodies and pump casings is viewed as being considerably lower than for welds in Class 1 systems.

D.1.6 Technical Approach to LOCA Frequency Estimation

Existing service experience with piping systems shows a strong correlation between failures and presence of an active degradation mechanism in combination with service conditions and transient loading conditions. It is therefore possible to estimate piping reliability parameters through statistical analysis of service experience data. Such analysis includes data processing whereby the appropriate reliability attributes are correlated with influence factors as described in SKI Report 97:26 [D.11].

In this Base Case Report the technical approach to LOCA frequency estimation builds on statistical analysis of service data associated with ASME XI Class 1 piping in the BWR and PWR operating environments. The study accounts for two kinds of uncertainties in piping reliability analysis, namely data uncertainty and state-of-knowledge uncertainty. The pipe failure database on which this study is based is called PIPExp [D.12], which is the extended version of the OPDE pipe failure database [D.13]. A description of PIPExp is included in Appendix A. The uncertainty analysis is performed by using a Monte Carlo merge technique to develop the LOCA frequency distributions. A commercial software package called Crystal Ball (Version 2000.2.2), which is an add-on for Microsoft Excel, is used to perform this Monte Carlo merge operation. Time-dependent LOCA frequencies are developed using a Markov modeling approach [D.14].

The BWR Base Case analysis is based on the degradation mechanism analysis as documented in Reference [D.5], and it builds on insights from an earlier BWR LOCA frequency pilot study [D.15-D.16]. The PWR Base Case analysis is based on the degradation mechanism analysis as documented in References [D.6, D.8], and builds on insights and results from an earlier sensitivity analysis performed in support of a risk informed inservice inspection (RI-ISI) evaluation [D.12]. That sensitivity analysis addressed the impact of using a different pipe failure database on the RI-ISI weld selection.

D.1.7 Study Conventions

Throughout this report, pipe sizes are referenced by nominal pipe size (NPS), which indicates standard pipe size without an inch symbol. The smallest pipe size considered in this study is NPS3-¾ (Plant A.b). All references to specific material types are made according to designations by the American Society for Testing and Materials (ASTM). The term "weld failure" is used to indicate a rejectable (non-through-wall or through-wall) flaw.

During the NRC LOCA Elicitation Kick-off Meeting [D.1], a LOCA was defined as "a breach of the reactor coolant pressure boundary which results in a leak rate greater than 380 lpm (100 gpm)." Instead of using the traditional (or historical) LOCA size classes (small – medium – large) that are based on break size, this study uses LOCA sizes that are based on leak rate threshold values as indicated in Table D.1 (adapted from [D.1]) and Table D.2.

Table D.1 LOCA Size Classification Threshold Values

LOCA Category	Flow Rate (v) Thresholds gpm (lpm)	Comment
0	$v > 10$ (38)	Cat0 corresponds to a pressure boundary failure (breach) resulting in a leakage exceeding the T.S. limit for identified leakage.
1	$v > 100$ (380)	Breach in piping of up to 1.8-inch diameter (BWR), and 1.7-inch diameter (PWR); see Table D.3.
2	$v > 1,500$ (5,700)	Breach in piping of up to 3.3-inch diameter (BWR), and 3-inch diameter (PWR)
3	$v > 5,000$ (19,000)	Breach in piping of up to 7.3-inch diameter (BWR), and 6.8-inch diameter (PWR)
4	$v > 25,000$ (95,000)	Breach in piping of up to 18.4-inch diameter (BWR), and 14-inch diameter (PWR)
5	$v > 100,000$ (380,000)	Breach in NPS28 RR piping (BWR) yields on the order of 230,000 gpm. Breach in RCS hot leg piping of up to 31-inch diameter.
6	$v > 500,000$ (1,900,000)	Applies to PWR RCS-HL base case only, and only for a relatively short time following a postulated DEGB

Table D.2 Estimated Flow Rates from Restrained Double-Ended Guillotine Break (DEGB)[2]

Pipe Size [NPS]	Restrained DEGB (Plant A – PWR)			Restrained DEGB (Plant B – BWR)		
	Break Size [sq. in.]	Press. [psig]	Max. Flow Rate [gpm]	Break Size [sq. in.]	Press. [psig]	Max. Flow Rate [gpm]
1	.41	2250	540	.41	1250	467
2	1.65	2250	2158	1.65	1250	1869
4	6.60	2250	8633	6.60	1250	7476
6	14.84	2250	19424	14.84	1250	16823
8	26.39	2250	32280	26.39	1250	29908
12	59.37	2250	72495	59.37	1250	42411
14	80.81	2250	98624	80.81	1250	57698
22	199.54	2250	243542	199.54	1250	142478
28	323.22	2250	394497	323.22	1250	230790
30	371.05	2250	452867	N/A	--	--

---

[2] Technical basis for leak rate calculation is documented in an attachment to Minutes of Meeting (2nd Elicitation Meeting), Bethesda (MD), June 4-5, 2003.

The estimation of weld failure rates uses Bayesian reliability analysis methodology, and involves the development of prior and posterior failure rate distribution. In this study the term 'prior' refers to piping reliability characteristics before the implementation of industry programs to mitigate or eliminate susceptibilities to certain degradation mechanisms. The term 'posterior' refers to observed or expected reliability characteristics after reliability improvement actions have been implemented.

D.1.8 Report Organization

This report consists of eight sections and four appendices. Section D.2 is an overview of the analysis steps. Section D.3 summarizes the service experience applicable to the BWR and PWR Base Cases, respectively. Using the PIPExp database, Section D.4 includes a summary of the data interpretation and data processing steps necessary to derive piping reliability parameters that apply to the base case definitions. Section D.5 documents the results of the pipe failure rate estimation while Section D.6 is a documentation of the models used for estimating LOCA frequency, while Section D.7 is a summary of results. Section D.8 is a list of references. Note that the Base Case results used in Table E.1 in the main body can be obtained from Tables 16, 17, and 20 in this report.

Appendix A summarizes the PIPExp database structure. Appendix B includes the Excel spreadsheets that are used as the basis for the LOCA frequency models, and Appendix C includes the Excel spreadsheets for the calculation of time-dependent LOCA frequencies. Finally, Appendix D is a summary of selected, significant Code Class 1 and 2 pipe failures in commercial nuclear power plants worldwide.

D.2 Technical Approach

Base Case Report 2 develops BWR and PWR LOCA frequency distributions using a 'bottom-up approach.' Statistical analysis of relevant service experience data is used to quantify the weld failure rate and rupture frequency of individual welds. Next the failure rate and rupture frequency (= LOCA frequency) for an entire system is calculated by concatenating the individual weld failure rates and rupture frequencies. Markov model theory is used to evaluate the influence of alternate strategies for in-service inspection and leak detection on the frequency of leaks and ruptures.

D.2.1 Overview of Analysis Steps

Different approaches have been applied to estimating pipe failure rates and rupture frequencies; from probabilistic fracture mechanics, via direct statistical estimation to expert judgment. The most straightforward approach is to obtain statistical estimates of piping component failure rates based on data collected from field experience. A variation of this approach is to augment statistical estimates of pipe failure parameters with simple correlations that express the problem in terms of a failure rate and a conditional probability for each failure mode of interest such as the approach used in NUREG/CR-5750 [D.17].

A limitation of the statistical analysis approach is that attempts to segregate the service data to isolate the impact of key design parameters and properties of various degradation and damage mechanisms often leads to subdividing a database into very sparse data sets. If not optimized properly, this approach may introduce large uncertainties in the failure rate estimates. In addition, historical data may reflect the influence of no longer relevant inspection programs. If changes to these programs have been implemented, such changes may render the failure rate estimates no longer relevant. In risk-informed applications, the failure data and analysis methods need to provide future predictions of piping system reliability that can account for changes in the inspection strategy or improvement in the NDE technology.

An objective of the work documented in this report is to demonstrate the utility of a pipe failure data collection. Time-dependent LOCA frequencies are calculated by making full use of the PIPExp database in combination with Markov model theory [D.14]. The LOCA frequency calculation in this report is structured to support the Expert Elicitation and consists of four steps; each step is addressed in a separate report section:

- Section D.3. The service experience that is applicable to the five bases cases is summarized in this section. The data summaries correspond to queries in the PIPExp database.

- Section D.4. The approach to calculating time-dependent LOCA frequencies is presented. A Bayesian update process is used to derive failure parameters that reflect the attributes of respective base case definition. The results of this analysis step are in the form of generic weld failure rate distributions. These distributions represent the industry-wide service experience prior to the implementation of the specific pipe failure mitigation programs that are currently in place.

- Section D.5. In this section current state-of-knowledge (or base case specific) weld failure rate distributions are develop. The chosen estimation approach includes a formal uncertainty analysis that accounts for uncertainty in the failure data and exposure data. Engineering judgment and insights from the review of service data are used to address the conditional probability of pipe failure given presence of through-wall flaws.

- Section D.6. An Excel spreadsheet format is used to develop LOCA frequency models corresponding to each of the five base cases. These models generate LOCA frequency distributions at $T = 25$ years. A Markov model is used to investigate the time-dependency of LOCA frequencies. The output of this model consists of LOCA frequencies at $T = 40$ years and $T = 60$ years.

D.2.2 Sensitivity Analyses

Two types of sensitivity analysis are included in this report. The first type addresses the impact on results by an assumed incompleteness of the failure data collection. The second type relates to the sensitivity of the time-dependent LOCA frequencies to different assumptions about leak detection and in-service inspection. The sensitivity analysis results are included in Section D.6.

D.3 Service Experience Data Application to the Base Case Study

The PIPExp database documents service experience with Code Class 1, 2 and 3 and non-safety related (or Class 4) piping in commercial nuclear power plants worldwide. For the time period 1970-2002, this database was queried for service experience data specific to the Base Case piping systems. The results of the database queries are summarized here, and they form the input to the data processing and failure parameter estimation in Sections D.4 and D.5.

D.3.1 PIPExp Database, Revision 2003.1

The pipe failure database utilized in the Base Case Study is called PIPExp. It is an ACCESS database and an extension of the OPDE database [D.12-D.13]. Since the conclusion of the original work in 1998 [D.11, D.17], the pipe failure database has been significantly expanded both in terms of the absolute number of event records and the depth of the database structure (Appendix A provides additional details). Lessons learned through database applications have been used to enhance the structure. In this study of HPI//NMU-, FW-, RC- and RR-piping reliability the statistical analysis is based on service data as recorded in PIPExp and with cutoff date of December 31, 2002. The analysis is inclusive of applicable worldwide BWR- and PWR-specific service experience with Code Class 1 piping. As of 12-31-2002 the database accounted for approximately 1,992 and 3,621 critical reactor-years of operating experience with commercial BWR and PWR plants, respectively.

The database is actively maintained and periodically updated. The effort involved in populating the database while at the same time assuring data quality is not trivial. As an example, changing regulatory reporting thresholds imply that an ever increasing volume of raw data reside in restricted and proprietary database systems rather than in the public domain. For an event to be considered for inclusion in the database it undergoes screening for eligibility. For example:

- The equipment failure must be positively identified as a piping component failure external to the reactor pressure vessel (RPV). A failure involves a pressure boundary degradation, which can be non-through-wall (crack with a/t-ratio $\geq$ 10%, where a = crack depth and t = wall thickness) or a through-wall leak.

- There must exist documented evidence in the form of a hard copy (e.g., USNRC Inspection Report, Licensee Event Report, ISI Summary Report, Problem Identification Form, Condition Report, ASME Code Repair Relief Request, etc.) from which a sufficiently detailed case history is developed. The documented evidence of pipe degradation/failure must contain information on its location within a piping system (e.g., with reference to an isometric drawing and/or P&ID), metallurgy, operating conditions, impact on operation, method of discovery, failure history, etc. so that a data classification may be independently verified.

- Where the documented evidence is deemed incomplete, additional information is solicited through direct contact with plant personnel or by accessing supplemental data.

- There must be sufficient technical information available to fully address the complex relationships between piping reliability attributes (or design parameters) and influence factors (e.g., fabrication/welding techniques, environmental conditions such as water chemistry, flow conditions) on the one hand and degradation/failure mechanisms on the other.

- Differentiation between UT indications versus confirmed crack indications. Only the latter are included in the database given an a/t-ratio $\geq$ 10%.

Following on the initial data screening, each event selected for inclusion in the database is subjected to a classification so that the unique reliability attributes and influence factors are identified. Including memo fields, text fields, numerical fields and data filters, up to 114 database fields describe each record of the database.

D.3.2 Review of BWR-Specific Piping Service Experience
Limited to the BWR Base Case systems, this section summarizes the worldwide service experience with Code Class 1 piping. The results of this review are input to the pipe failure rate estimation.

D.3.2.1 RR Piping Service Experience - The original piping material in BWR plants commissioned prior to mid-1980 is austenitic stainless steels that contain more than 0.03% carbon. During welding these steels are susceptible to sensitization that results in a loss of corrosion resistance. Intergranular stress corrosion cracking (IGSCC) occurs when the sensitized steel is subjected to stresses and corrosive environment. Sensitization can be avoided by controlling the carbon content to below 0.03%. Another approach to controlling sensitization is to add strong carbide formers such as titanium or niobium to the steel. Stainless steels with additions of titanium or niobium are called "stabilized." It is noted that low-carbon content unstabilized stainless steel or stabilized stainless steels are not completely immune to IGSCC, however [D.18].

For Plant B, IGSCC is the predominant degradation mechanism acting on the RR piping welds, including heat-affected zones. During early plant life some weld reinforcements were performed where the inservice inspection revealed presence of surface penetrating, and subsurface cracking due to IGSCC. Since the analysis of LOCA frequency distributions is based on a degradation mechanism evaluation, the PIPExp database is queried for service data including IGSCC. The database queries are summarized in a set of charts and tables below. The database currently includes a total of about 1000 records on IGSCC in BWR piping. Figure D. 1 shows the number of weld failures due to IGSCC by calendar year and Figure D. 2 shows the number of weld failures by year of operation. Here a weld heat affected zone with a/t ≥ 10% is characterized as a "weld failure."

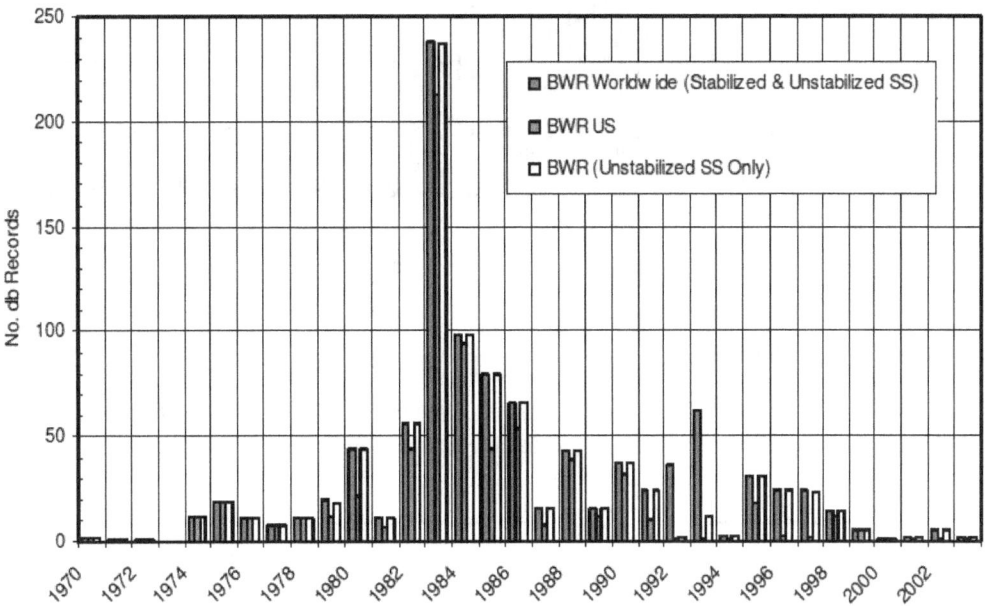

Figure D.1 Weld Failures Due to IGSCC in Code Class 1 & 2 Piping (1970-2002)

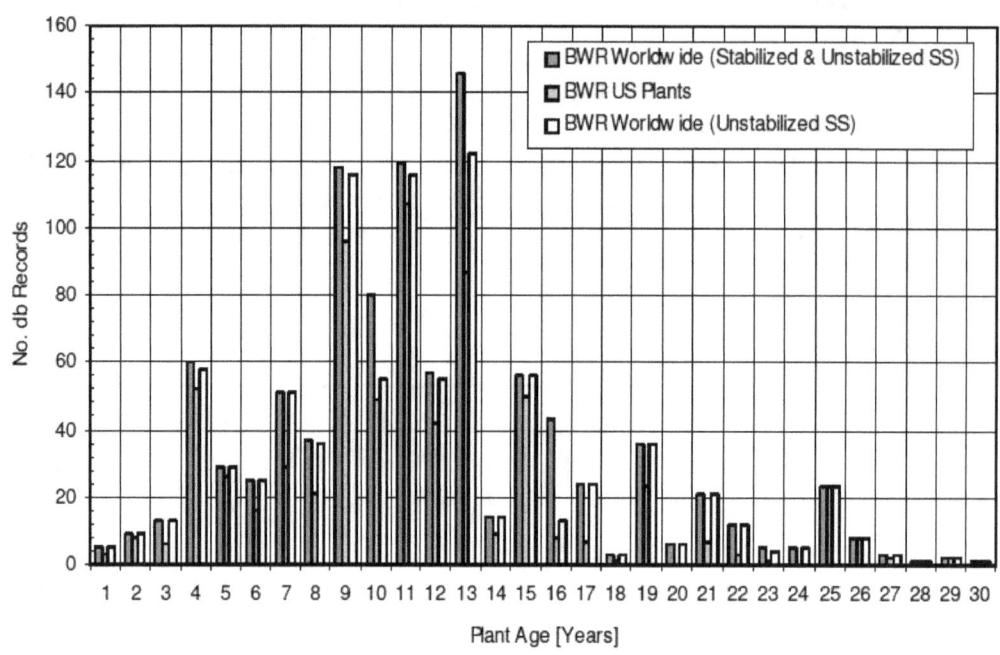

Figure D.2 IGSCC Experience by Year(s) of Operation

In Figure D. 3 the IGSCC data is organized by mode of failure (crack – pinhole leak – leak) and pipe size. Figure D. 4 shows the IGSCC data by size and material type.

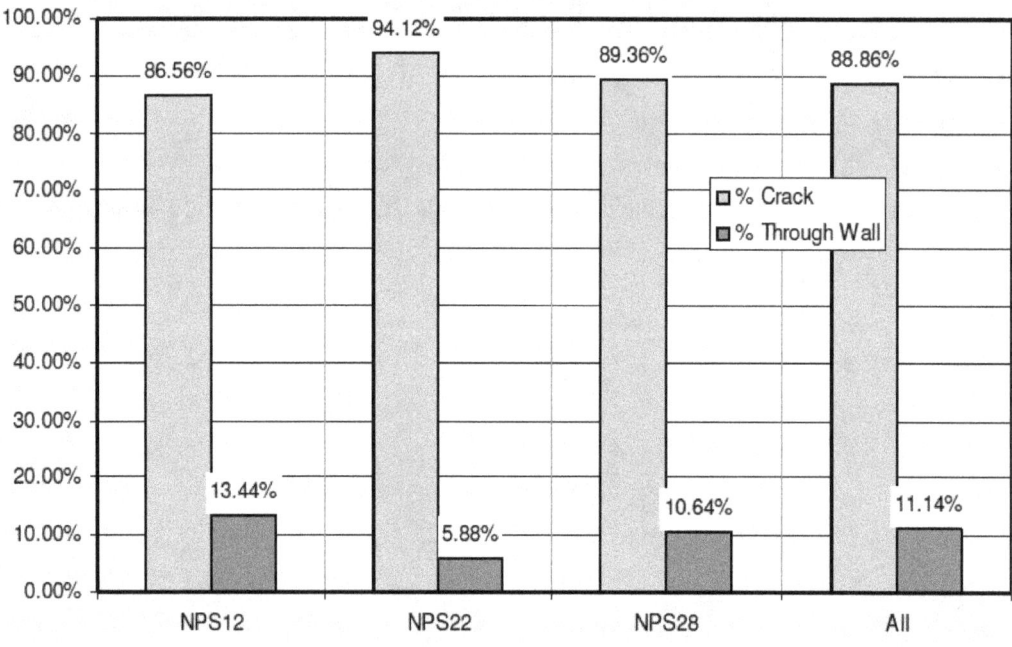

Figure D.3 IGSCC Data by Failure Mode

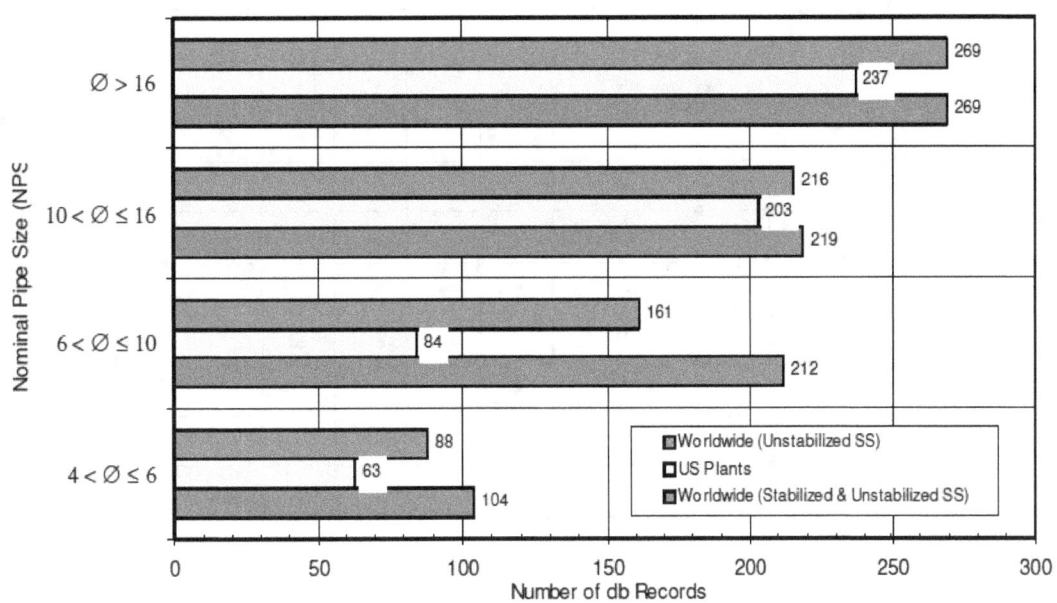

Figure D.4 IGSCC Experience by Pipe Size and Material Type

Figures D. 5 through D.7 include plots of crack depth versus crack length (L) in the circumferential (C) direction. An 'a/t-ratio' of 100% indicates a crack, which has penetrated the outside pipe wall. An 'L/C-ratio' of 100% indicates a crack, which spans the entire inside pipe circumference. Limited to part through-wall cracks, Figure D.8 summarizes the data by a/t-ratio.

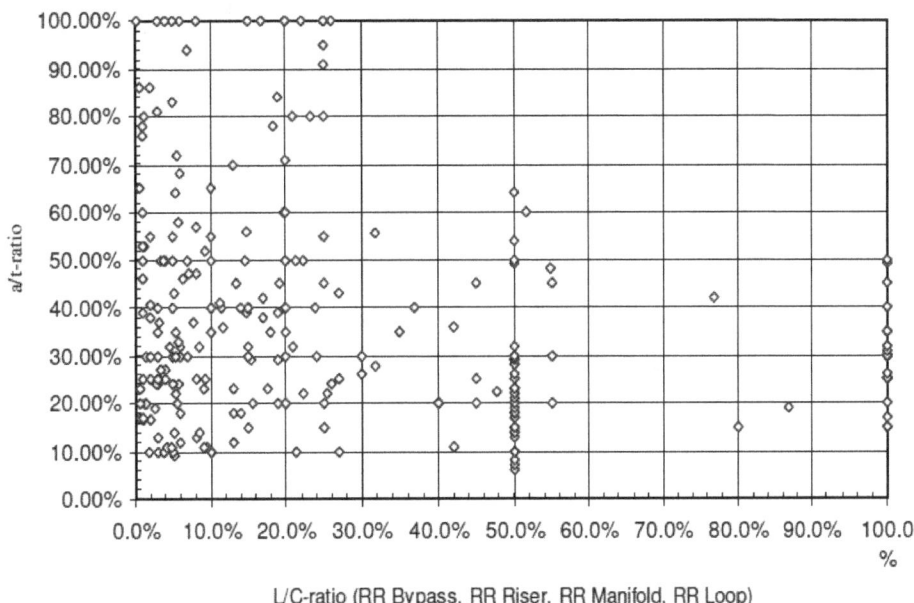

Figure D.5 Crack Depth Versus Crack Length in Austenitic Stainless Steel

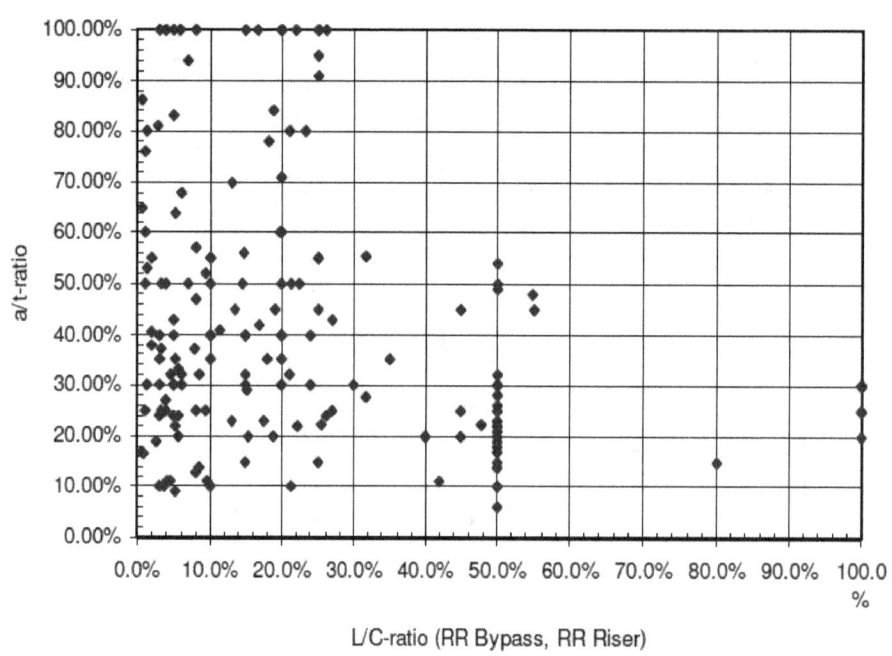

Figure D.6 Crack Depth Versus Crack Length in Austenitic Stainless Steel (NPS4, NPS12)

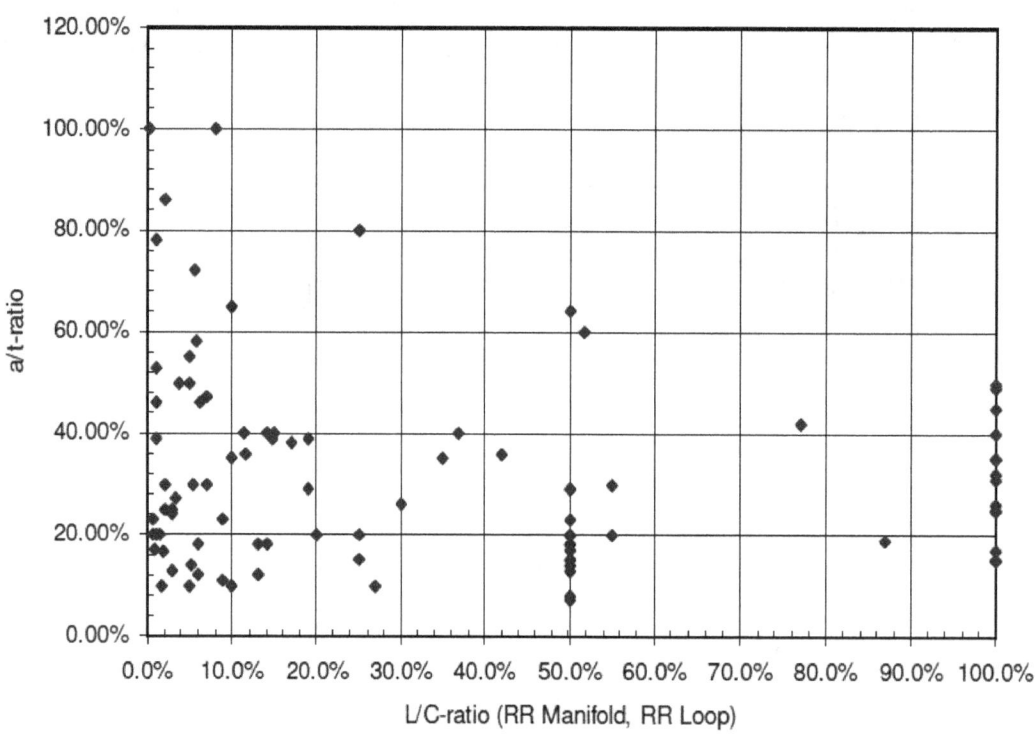

Figure D.7 Crack Depth Versus Crack Length in Austenitic Stainless Steel (NPS22, NPS28)

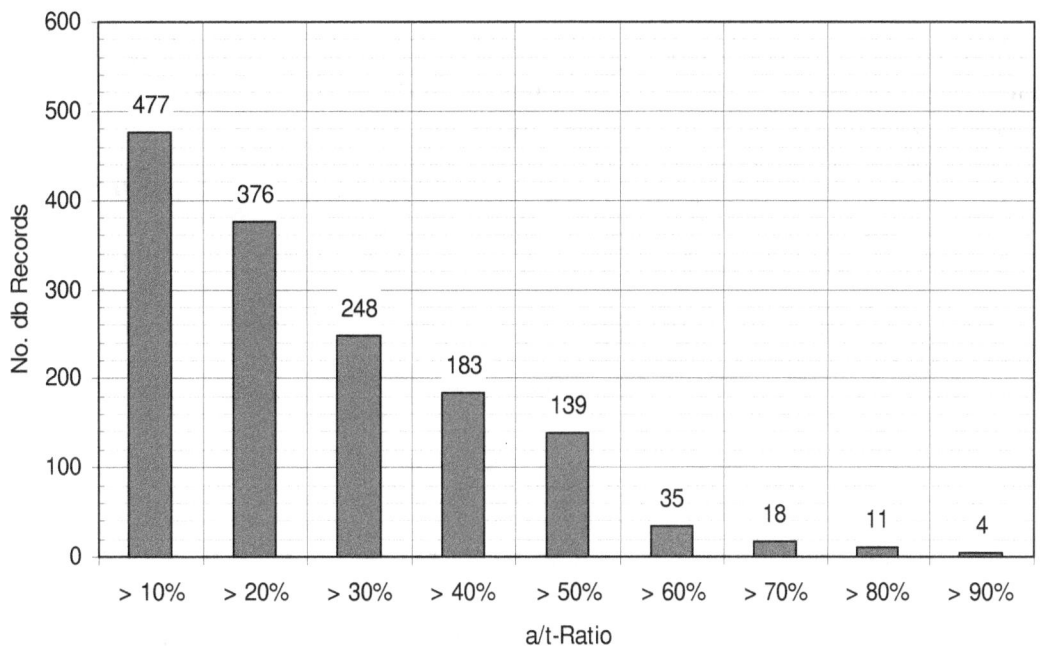

Figure D.8 Summary of IGSCC Non-Through Wall Flaws by 'a/t-Ratio'

In Figure D.5, the combination of a/t = 100% and L/C = 100% would indicate a case of DEGB where the pipe ends are separated from each other. As a rule-of-thumb, a through-wall crack (a/t = 100%) with L/C ≥ 40% is unstable and may exhibit unstable crack growth if it were to be left in place.[3]

As seen from the above, there have been a limited number of cases of leaks in large-diameter Reactor Recirculation piping. Only a small fraction of the total number of through-wall flaws have been active leaks; i.e., leaks that have developed during routine power operation. The majority of the through-wall flaws have been "non-active leaks." That is, leaks that have developed while shutting down for drywell inspection, during performance of weld crown grinding in preparation for ultrasonic examination ("ISI-leaks"), or during the performance of induction heat stress improvement (IHSI – "IHSI-leaks"). There are also some cases where leaks have been discovered during hydrostatic pressure testing to verify the integrity of weld repairs.

Like Figure D.5, Figure D. 8 includes data on all IGSCC-susceptible, Code Class 1 and 2 piping systems in BWR plants. While Figure D.5 includes approximately 300 data points, Figure D.8 includes on the order of 500 data points. This difference in the number of reports represented in respective chart is due to the fact that not all reports on IGSCC include complete details on the crack morphology (dimensions, orientation).

Where through-wall flaws have been observed leak rates have been small. In terms of leak rate and operational impact, so far the two most significant instances of IGSCC occurred at Duane Arnold in 1978 and at the Spanish plant Santa Maria de Garona in 1980. In the former case the leak rate was about 11 lpm (3 gpm) with L/C = 22%. In the latter case the observed leak rate was about 3.0 lpm (0.8 gpm) with L/C = 4.5%.

---

[3] See for example the report EPRI NP-2472 (The Growth and Stability of Stress Corrosion Cracks in Large-Diameter BWR Piping, July 1982).

D.3.2.2 FW Piping Service Experience - Figures D.9 and D.10 summarize the service experience with FW piping. With respect to plant designed by General Electric, the Code Class 1 portion of BWR carbon steel feedwater piping has performed well in the field. There are no reported leaks in medium-or large-diameter RCPB piping. Foreign plants have experienced (and in some cases, continue to experience) thermal fatigue damage due to thermal mixing and stratification. In fact, 80% of the degradation of the RCPB portions of FW piping has occurred in foreign plants with a piping system design that differs from that of U.S. BWR plants.

The U.S. service experience includes a few instances of non-through wall cracking of FW nozzle-to-safe-end (bimetallic) welds. The root cause of the cracking is attributed to weld defects from original construction. As documented in Information Notice 92-35 [D.19], Susquehanna Unit 1 has experienced flow-accelerated corrosion damage about 250 mm (10 inches) from a weld connecting NPS12 piping to a 20-inch by 12-inch reducing tee. There have been no reported flaws in any U.S. plant beyond T = 15 years of operation.

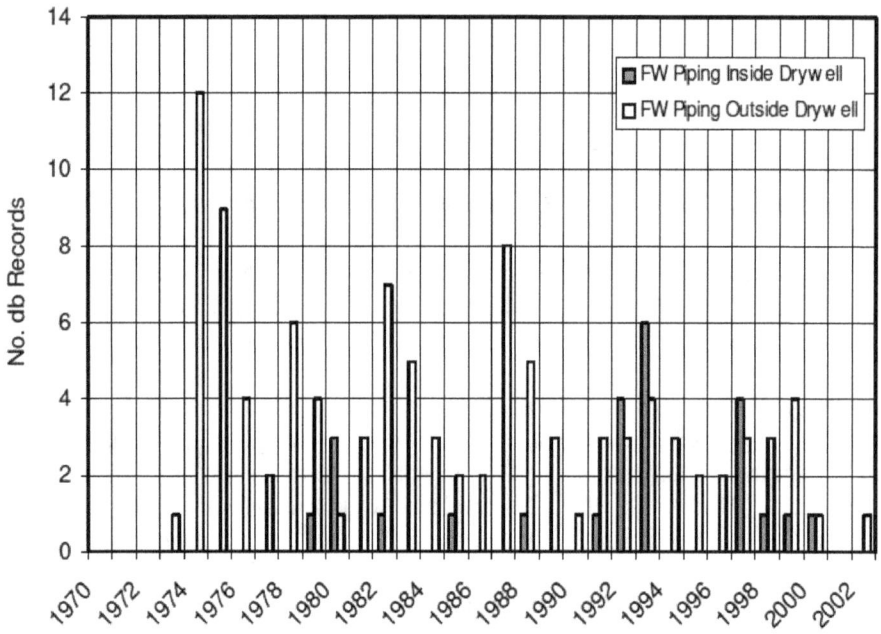

Figure D.9 Service Experience with FW Piping (i)

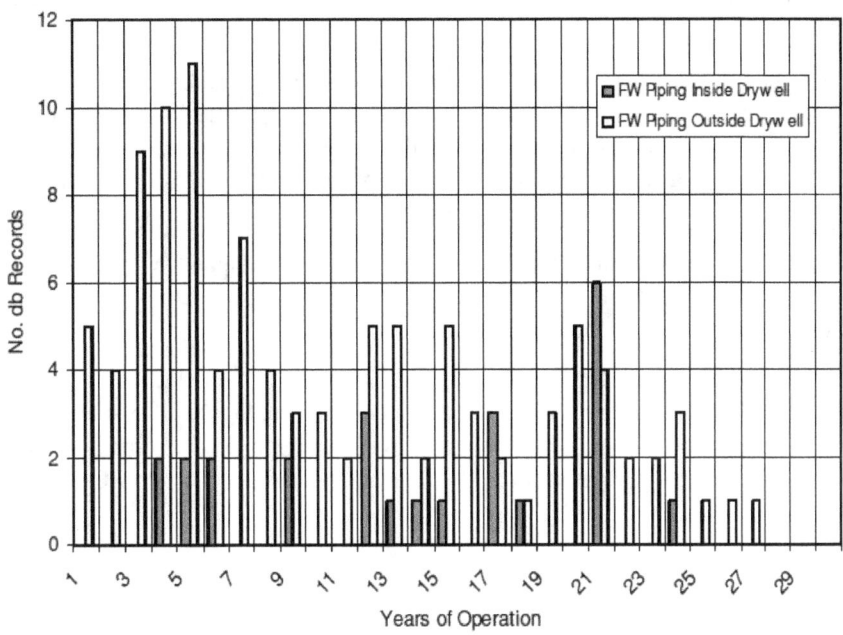

Figure D.10 Service Experience with FW Piping (ii)

D.3.3 Review of PWR-Specific Piping Service Experience

Limited to the PWR Base Case systems, this section summarizes the service experience with Code Class 1 piping. The results of this review are input to the pipe failure rate estimation.

D.3.3.1 RC & HPI/NMU Piping Service Experience - There have only been a limited number of events involving through-wall cracks in the large-diameter RC piping and the Class 1 portion of SI/CV piping. Evidence of axial primary water stress corrosion cracking (PWSCC) in the bimetallic safe-end to RPV nozzle welds of the RC-HL piping has been reported at Ringhals [D.20] and V.C. Summer [D.21].

During an eight-year period, the now decommissioned Trojan nuclear power plant experienced pressurizer surge line movement, which was attributed to thermal stratification [D.22]. In response, the NRC issued Bulletin 88-11 in December of 1988 [D.23] requesting that licensees perform visual inspections of the pressurizer surge line at the first available cold shutdown. Purpose of the inspections was to determine presence of any "gross discernible distress or structural damage in the entire pressurizer surge line, including piping, pipe supports, pipe whip restraints, and anchor bolts."

The current version (June 2004) of the PIPExp database includes four records associated with degradation of pressurizer surge lines:

- Record # 19849; during the Three Mile Island-1 2003 Refueling Outage (18-Oct-2003 to 3-Dec-2003), a UT examination found an axial flaw about 13 mm (0.5-inch) deep in the surge line nozzle-to-safe end interface in dissimilar metal weld No. SR0010BM. This weld connects a 10-inch Schedule 140, carbon steel nozzle to stainless steel safe end.

- Record # 19736; in November 2002 during UT examination of RC piping in the Belgian plant Tihange-2 (a 900 MWe series plant designed by Framatome), code rejectable indications were discovered in the 14-inch Inconel safe-end to nozzle weld. The flaw is believed to be an original construction defect.

- Record # 1119; while in hot shutdown condition, a non-isolable weld leak developed in a 1-inch drain line off the pressurizer surge line of Oconee-1 (LER 50-269/1998-002-01). The through-wall crack had initiated by TGSCC and propagated through-wall by vibratory fatigue. Small-diameter piping connecting to a pressurizer surge line is not part of the PWR-2 Base Case definition.

- Record # 420; during the 1988 annual refueling outage a pinhole leak was discovered in a 10-inch pressurizer surge line bi-metallic weld of Loviisa-1 (a Soviet designed WWER-440/213 plant located in Finland). The weld degradation was attributed to poor weld penetration and high residual stresses. This event was screened out from the data analysis.

Figures D.11 and D.12 summarize relevant service experience with medium- and large-diameter RC and safety injection (SI) and normal makeup (CV) piping. For comparison, Figure D.13 shows the service experience with small-diameter RC and SI/CV piping (≤ NPS2). Figure D.14 is a summary of the worldwide, PWR-specific data pipe failures that are attributed to thermal fatigue. In addition to RC-, SI- and CV-piping this figure includes failures in FW- and RHR-piping.

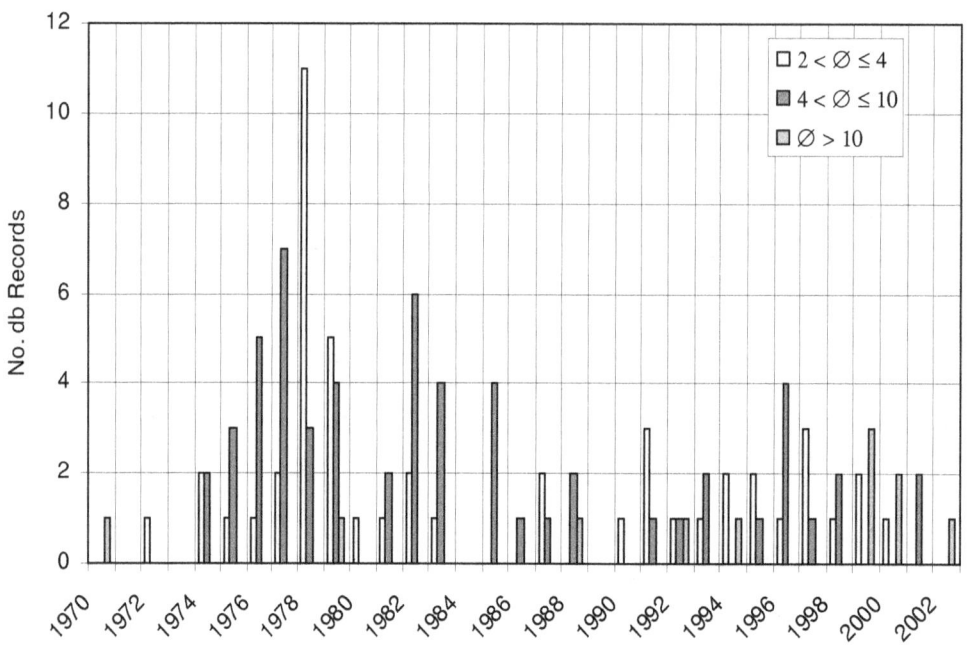

Figure D.11 Weld Failures in PWR RC-, CV- and SI-Piping (1970-2002)

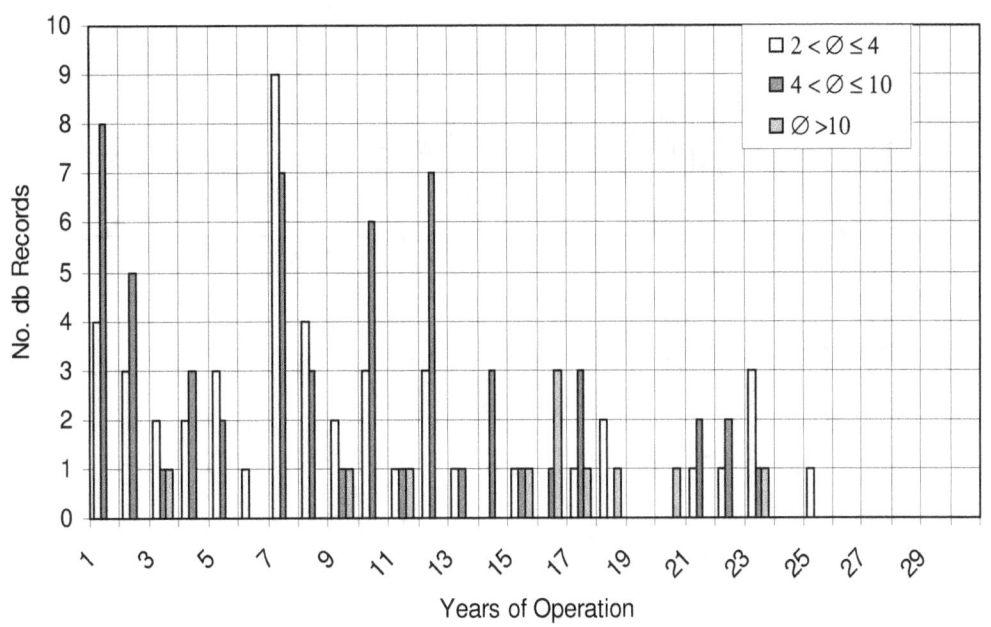

Figure D.12 Weld Failures in PWR RC-, CV- and SI-Piping (1970-2002)

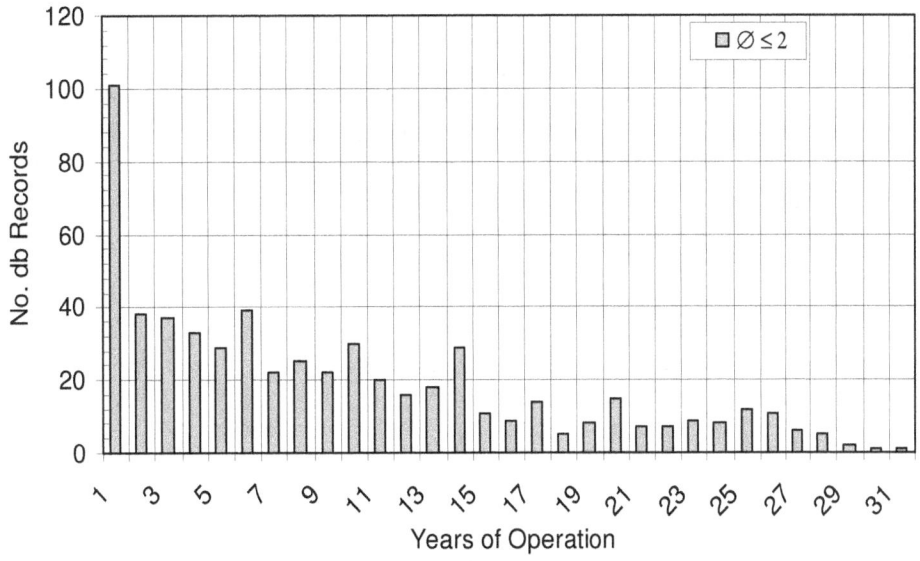

Figure D.13 Weld Failures in Small-Diameter PWR Piping

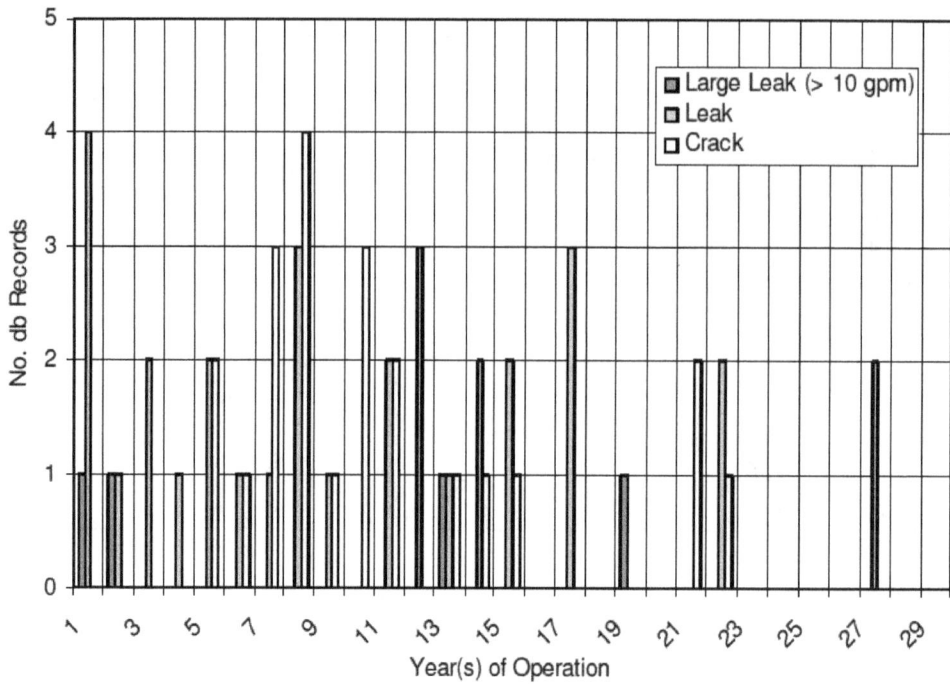

Figure D.14 Pipe Failures Attributed to Thermal Fatigue in PWRs Worldwide

Figure D.14 includes four significant ($v > 38$ lpm [10 gpm]) events, three in foreign plants (Civaux-1 in France, Tsuruga-2 in Japan and Biblis-B in Germany) and one in a domestic plant (Oconee-2). The latter event involved a failure of a weld between the HPI/NMU and the RCS cold leg (= PWR Base Case Plant A.b). The plant operators correctly diagnosed the leak and brought the plant to safe shutdown. Subsequent to the weld failure in Oconee-2, limited to small-diameter piping the Electric Power Research Institute issued the "Interim Thermal Fatigue Guideline" [D.9] for evaluating and inspecting regions where there might be high potential for thermal fatigue cracking. Additional perspectives on thermal fatigue mitigation are included in an OECD-NEA report [D.24]. The Babcock &Wilcox-designed plants now include a new design thermal sleeve to mitigate or prevent thermal fatigue cracking of welds.

Prior to these 'four significant events', thermal fatigue damage occurred at Farley-2 and Tihange-1 (a Belgian plant) during 1988. At these plants, thermal fatigue initiated from cold water leaking through closed check or globe valves in safety injection lines. At Farley-2, the damage occurred in piping connected to the RCS cold leg, and at Tihange-1 in piping connected to the RCS hot leg. In these events the leak rates were 2.6 lpm (0.7 gpm) and 23 lpm (6 gpm), respectively. The U.S. NRC issued Bulletin 88-08 in response to these events.

D.4 Data Processing and Data Reduction

The objective of data processing is to extract from a pipe failure data collection relevant case histories that reflect specific combinations of reliability attributes and influence factors. Next, the data reduction prepares the input to the statistical parameter estimation in the form of event counts and exposure terms to develop Bayesian prior and posterior distributions.

D.4.1 Strategy for Data Processing and Data Reduction

Shown in Figure D.15 is a general four-state Markov model of piping reliability. All failure processes of this model can be evaluated using service data, assuming that such a data collection is of sufficient technical detail and completeness. This model is used in Section D.6 to develop time-dependent LOCA frequencies.

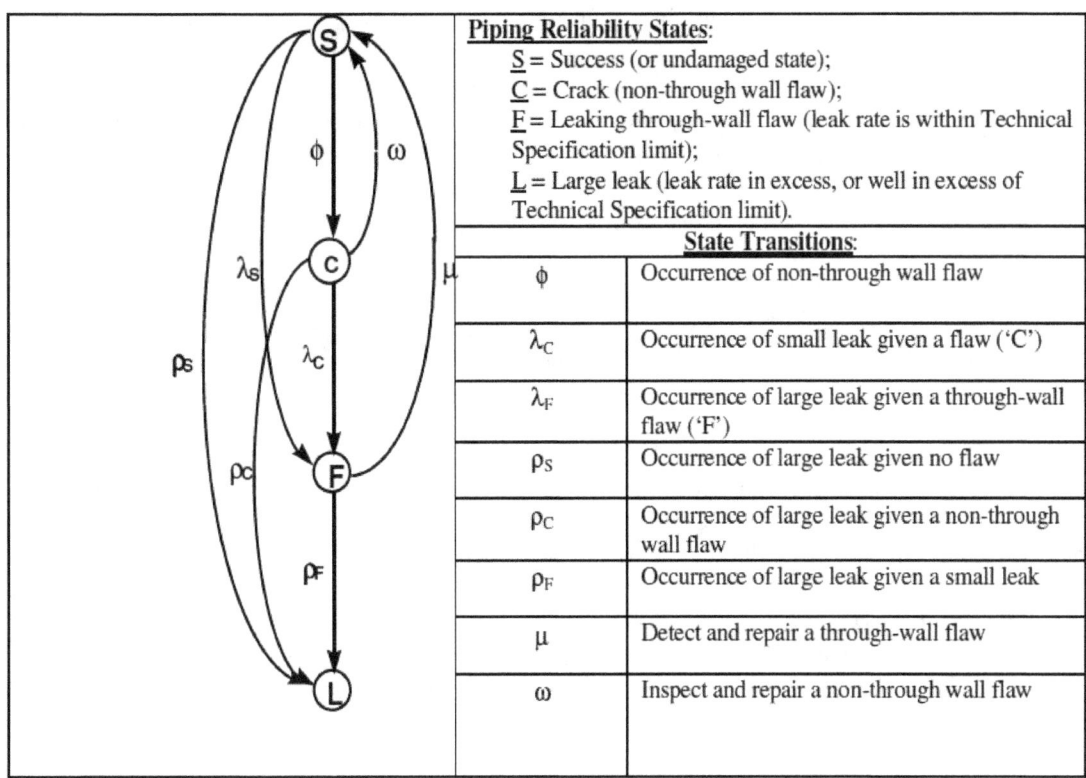

Piping Reliability States:	
$\underline{S}$ = Success (or undamaged state);	
$\underline{C}$ = Crack (non-through wall flaw);	
$\underline{F}$ = Leaking through-wall flaw (leak rate is within Technical Specification limit);	
$\underline{L}$ = Large leak (leak rate in excess, or well in excess of Technical Specification limit).	
State Transitions:	
ϕ	Occurrence of non-through wall flaw
λ_C	Occurrence of small leak given a flaw ('C')
λ_F	Occurrence of large leak given a through-wall flaw ('F')
ρ_S	Occurrence of large leak given no flaw
ρ_C	Occurrence of large leak given a non-through wall flaw
ρ_F	Occurrence of large leak given a small leak
μ	Detect and repair a through-wall flaw
ω	Inspect and repair a non-through wall flaw

Figure D. 15 Four-State Markov Model of Piping Reliability[4]

In this and subsequent report sections, a pipe failure (F) is defined as a through-wall defect resulting in a non-active leak or small, active leak. The frequency of a large leak (L) in excess of Technical Specification limits is estimated using the following simple model:

$$F_L = \lambda \times p_{L|F} \tag{D.1}$$

---

[4] The figure is reproduced courtesy of K.N. Fleming (Technology Insights, Inc., San Diego, California).

$$\lambda = [\textit{No. Failures}] \, / \, [\textit{Exposure} \, (= T \times \textit{No. Components})] \qquad\qquad\text{(D.2)}$$

Where:

F_L	=	Frequency of a large leak [1/Reactor-year].
λ	=	Failure frequency [1/Reactor-year.Extension]; where 'extension' refers to the piping component boundary definition. Depending on the intended application and type(s) of degradation mechanism, the extension could be formed by counts of bends, pipes, tees, welds or length of piping. In Equation (4.2), the exposure term reflects the total component population in the data survey.
T	=	Exposure time (or reactor operating years)
$p_{L|F}$	=	Conditional probability of a large leak given a through-wall defect. Section D.5 includes a technical basis for estimating conditional failure probabilities.

The parameter estimation uses a Bayes' update process that begins with the development of prior distributions for each of the terms in equation (D.1). These prior distributions are shaped by our knowledge about the susceptibility of different piping systems to degradation. The input to the Bayes' update process comes from a small subset of PIPExp after it has been subjected to screening for pipe failures that meet certain selection criteria. A software tool (Bayesian Analysis Reliability Tool -BART™) is used to perform the updates.[5]

Piping reliability is a function of pipe size (diameter and wall thickness), and metallurgy, process medium, environment and design requirements; or attributes and influences, respectively. The purpose of data processing and data reduction is to extract from the total PIPExp database those subsets of service data that correspond to the attributes and influences of the Base Case definitions.

D.4.1.1 Informative versus Noninformative Prior Distributions - The type, extent and quality of applicable service data will determine the actual implementation of the Bayesian update process. Where sufficient service data is available an empirical Bayes approach is used. In this case classical estimation techniques are used to fit a prior distribution to the available data. When no or sparse service data is available a non-informative prior is defined. Relative to the five Base Cases the following approaches are used to determine the prior failure rates distributions:

- BWR Base Case – RR Loop B. There is ample service data on IGSCC. Our prior state-of-knowledge consists of service data before implementation of IGSCC mitigation strategies (mid-1980s). A prior failure rate is derived through classical statistical estimation.

- BWR Base Case – FW Loop B. Given the scarce service data, a lognormal distribution with a mean value of 1.0E-06 per weld-year and range factor (RF) equal to 100 is used. This is a noninfomative prior distribution.

- PWR Base Case – RC Hot Leg. The only available service data involves axial cracks in RPV nozzle-to-safe-end welds at three PWR units. A point estimate for the failure rates is calculated for

---

[5] Details on Bayesian reliability analysis is found in text books on statistical analysis of reliability data; e.g., Martz and Waller (1991): Bayesian Reliability Analysis, Krieger Publishing Company, Malabar (FL), ISBN 0-89464-395-9. For conjugate functions like the gamma and beta distributions a Bayesian point estimator for the failure rate is the mean of respective posterior probability density function, or:

$\lambda = (\delta + r)/(\rho + T)$ – gamma
$\lambda = (\delta + r)/(\delta + \rho + n)$ – beta

Where, (δ, ρ) are the parameters of respective distribution and (r, T, n) correspond to new evidence (i.e., 'r' failures in 'T' hours, or 'r' failures in 'n' tests).

the period 1970 through 2000. This point estimate is approximated by a lognormal distribution with range factor of 100; i.e., essentially a noninformative prior.

- <u>PWR Base Case – RC Surge Line</u>. For the pressurizer surge line there is no service data including non-through wall or through-wall cracking. Again, a lognormal distribution with a mean value of 1.0E-06 per weld-year and RF = 100 is used.

- <u>PWR Base Case – HPI/NMU Line</u>. Service data exists, which is directly applicable to this base case. To account for design changes that have been implemented post-1997, a non-informative prior is combined with B&W-specific failure data and exposure data through end of calendar year 1997. The resulting failure rate represents a prior distribution, which is applicable to this Base Case.

D.4.2 BWR-Specific Apriori Pipe Failure Rates

The failure rate development consists of determining an industry generic pipe failure rates for RR and FW piping, respectively. Next a Bayesian update is performed to generate failure rates that best represent the design and operating conditions of Plant B. Rather than taking an apriori failure from some published source, the approach in this study is to derive apriori failure information from the PIPExp database. The information that is summarized in Section D.3 provides some insights into the time-dependency of failure rates. These insights are explored further below.

D.4.2.1 RR Pipe Failure Data - Programs to mitigate the effects of certain degradation mechanisms strongly influence the achieved piping reliability. As an example, all BWR plants commissioned prior to the early to mid-1980s have experienced IGSCC. Industry initiatives to mitigate or eliminate the influence by IGSCC were implemented by the mid-1980s, and thereafter the rate of IGSCC has dropped sharply. The trend in the IGSCC rate is established by normalizing the data displayed in Figure D.2 (IGSCC by Years of Operation). Calculating the rate of IGSCC per weld-year for a given system and pipe size performs the normalization. Before performing this normalization the database is subjected to additional processing to exclude from further consideration any IGSCC data not directly applicable to the RR System that is representative of the Base Case. Similarly the part of the database including plant population and weld population data must be processed in such a way that an appropriate exposure term is developed commensurate with the failure data. In developing the RR-specific exposure term the following exclusion criteria were applied:

Plant Population Exclusion Criteria Applicable to RR Piping
- BWR plants without external recirculation loops;
- BWR plants in which the RR piping is fabricated from IGSCC resistant material e.g., Nuclear Grade stainless steel.

Table D.3 includes selected weld counts used to derive an exposure term according to Equation (D.2). Organized by pipe size and years of operation, Table D.4 is a summary of weld failures in RR piping. Noteworthy is the observation that there have been no reported through-wall defects in any plant beyond T = 15 years of operation. Using the information in Tables D.3 and D.4, Figure D.16 shows the calculated rate of RR pipe failure per weld-year.

The failure rates in Figure D.16 assume that all RR welds of a certain size to be equally susceptible to IGSCC. As was shown in SKI Report 98:30 [D.15], a correlation exists between weld failure rate and weld configuration. This correlation is assumed to be attributed to the piping layout, complexity of welding operation, and the associated weld residual stresses. The chart in Figure D.17 shows the weld configuration versus fraction of weld failures.

Table D.3 Selected Weld Counts in RR Piping

Plant ID (NSSS Type)	Weld Count by Pipe Size (NPS)								
	4	6	8	12	14	16	22	24	28
1 (AA/3)[6]	30	42	23	--	--	--	--	54	--
2 (BWR/4)	8	--	--	58	--	--	12	--	33
3 (BWR/4)	--	--	--	34	--	2	--	4	24
4 (BWR/5)	36	--	--	51	10	16	--	50	--
5 (BWR/5)	39	--	--	63	--	16	--	45	--
6 (BWR/5)	--	--	--	130	--	--	24	--	97
7 (BWR/5)	--	--	--	138	--	--	16	--	97
8 (BWR/4)	--	--	--	24	--	--	4	--	28
9 (BWR/4)	--	--	--	25	--	--	4	--	38
10 (BWR/4)	12	--	--	59	--	--	10	--	36
11 (BWR/4)	12	--	--	62	--	--	12	--	36
Plant B (BWR/4)	--	--	--	50	--	--	16	--	56
Mean:	23	--	--	63	--	--	12	--	49

Table D.4 Number of Through-Wall Flaws in RR Piping Attributed to IGSCC[7]

Pipe Diameter (∅) [NPS]	Years of Operation															
	Total	1	2	3	4	5	6	7	8	9	10	11	12	13	14	15
$3 < \emptyset \leq 6$	15	0	1	1	2	4	1	0	2	2	0	2	0	0	0	0
$6 < \emptyset \leq 12$	9	0	0	0	0	1	0	1	0	1	1	2	1	1	0	1
$12 < \emptyset \leq 22$	4	0	0	0	0	0	0	1	0	0	0	2	1	0	0	0
$\emptyset > 22$	16	0	0	0	0	0	0	0	0	5	0	6	4	0	1	0
	44	0	1	1	2	5	1	2	2	8	1	12	6	1	1	1

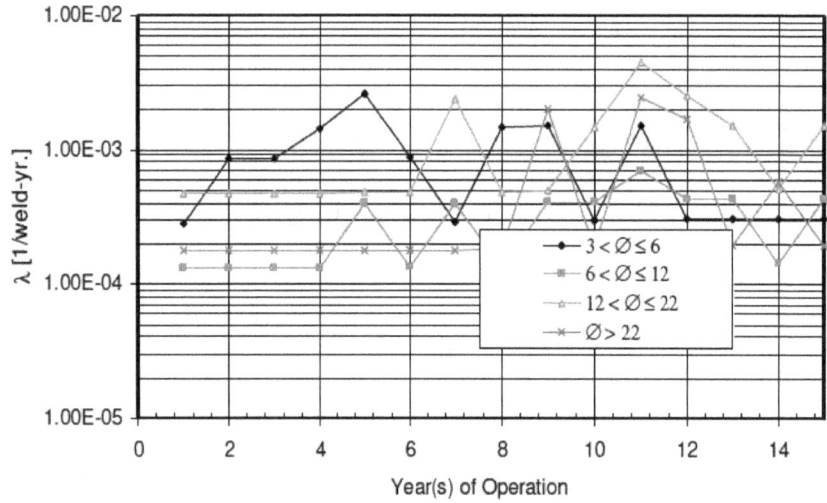

Figure D.16 Rate of IGSCC-Induced RR Pipe Failure ('Prior State-of-Knowledge')

---

[6] See SKI 98:30 [D.15] for details.

[7] This table includes active leaks (= leaks detected during routine power operation) and 'non-active' leaks (= leaks discovered during change of plant mode of operation), but it excludes 'ISI-leaks.' Appendix A, Table A-5 includes details on the through-wall cracks in NPS12, NPS22 and NPS28 Reactor Recirculation piping as included in Table D.4 above.

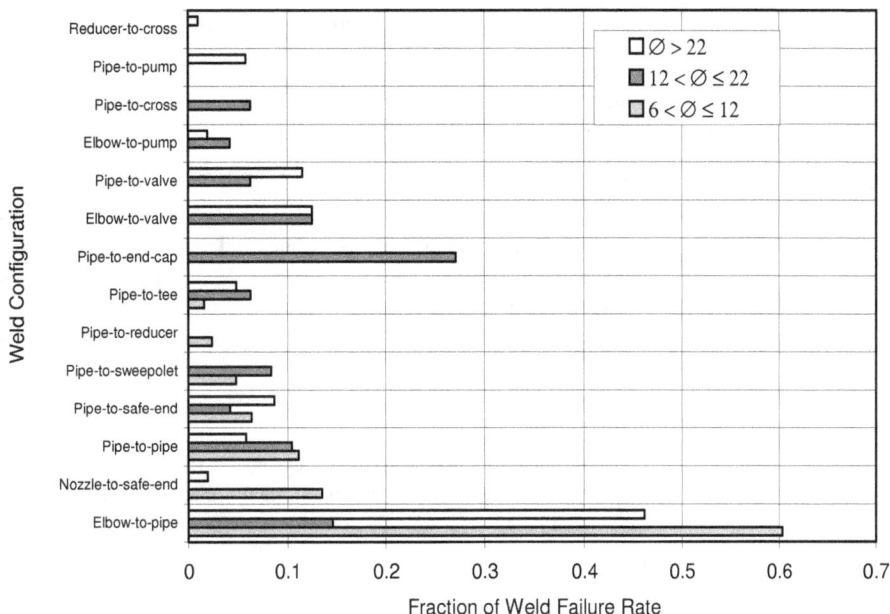

Figure D.17 RR Weld Failures as a Function Weld Configuration

In addition to weld configuration, the likelihood of failure is also a function of pipe size. For a weld of type "i" and size "j" the failure rate is expressed as follows:

$$\lambda_{ij} = F_{ij}/(W_{ij} \times T) \tag{D.3}$$

and with

$$S_{ij} = F_{ij} / F_{j} \tag{D.4}$$
$$A_{ij} = W_{j}/W_{ij} \tag{D.5}$$

the failure rate of weld of type "i" and size "j" is expressed as

$$\lambda_{ij} = (F \times S_{ij})/(W_{ij} \times T) \tag{D.6}$$
$$\lambda_{ij} = S_{ij} \times A_{ij} \times \lambda_{j} \tag{D.7}$$

Where:

λ_{ij}	=	Failure rate of an IGSCC-susceptible weld of type "i", size "j"
λ_{j}	=	Failure rate of an IGSCC susceptible weld of size 'j'
F_{j}	=	Number of size "j" weld failures
F_{ij}	=	Number of type "i" and size "j" weld failures
W_{j}	=	Size "j" weld count
W_{ij}	=	Type "i" and size "j" weld count
Susceptibility (S_{ij})	=	The service experience shows the failure susceptibility to be correlated with the location of a weld relative to pipe fittings and other in-line components (flanges, pump casings, valve bodies). For a given pipe size and system, the susceptibility is expressed as the fraction of welds of type "ij" that failed due to a certain degradation mechanism). This fraction is established from the PIPExp database; Table D.5.
Attribute (A_{ij})	=	In the above expressions the attribute (A) is defined as the ratio of the total number of welds of size "j" to the number of welds of type "i". In expression (4.7), A_{ij} is a correction factor and accounts for the fact that piping system design & layout constraints impose limits on the number of welds of a certain type. For example, in a given system there tends to be more elbow-to-pipe welds than, say, pipe-to-tee welds.

Combining the global (or averaged) failure rates in Figure D.16 with the information summarized in Figure D.17 and Table D.5 provides the apriori failure rates that are input to Equation (D.1). The results are summarized in Section D.4.3.

Table D.5 IGSCC Susceptibility by Weld configuration – Selected Parameter Values

RR System [NPS]	Weld Configuration	Configuration Dependent Parameters	
		Susceptibility (S_{ij})	Attribute (A_{ij})
12	Elbow-to-pipe	6.03E-01	2.8
	Nozzle-to-safe-end	1.35E-01	5.0
	Pipe-to-reducer	2.38E-02	25.0
22	Pipe-to-end-cap	2.71E-01	4.0
	Pipe-to-sweepolet	8.33E-02	2.0
	Pipe-to-cross	6.25E-02	4.0
28	Elbow-to-pipe	4.62E-02	5.6
	Pipe-to-pipe	5.77E-02	3.1
	Cross-to-reducer	9.60E-03	28.8

D.4.2.2 FW Pipe Failure Data - The estimation of failure rates for FW piping uses a non-informative prior distribution together with the weld population data in Table D.6. This approach is selected based on the available, limited service experience with ASME XI Class 1 FW piping; Table D.7. In developing the data summary in Table D.7 the following FW exclusion criteria were used to develop a point estimate of the failure rate:

Failure Data Exclusion Criteria Applicable to FW Piping
- Piping external to the drywell containment structure;
- Non-US data.

Table D.6 Selected Weld Counts in ASME XI Class 1 FW Piping

Plant ID (NSSS Type)	Weld Count by Pipe Size (NPS)								
	8	10	12	14	16	18	20	22	24
1 (AA/3)	1	61	--	14	--	--	--	--	--
2 (BWR/4)	--	--	28	--	--	25	--	--	--
3 (BWR/4)	--	--	28	--	--	26	--	--	--
4 (BWR/5)	--	--	40	12	--	6	--	--	38
5 (BWR/5)	--	--	41	6		6	--	--	42
6 (BWR/5)	13	--	66	--	8	--	7	--	22
7 (BWR/5)	9	--	68	--	8	--	7	6	19
9 (BWR/4)	--	--	50	--	--	--	6	--	45
10 (BWR/4)	--	--	32	1	--	30	--	--	--
11 (BWR/4)	--	--	30	1	--	29	--	--	--
Plant B (BWR/4)	--	--	63	5	--	--	53	--	--
Mean:	--	--	42	--	--	--	41[8]	--	--

Table D.7 Summary of FW Pipe Failure Data

Pipe Size (∅) [NPS]	Location		Data Origin		Failure Mode			
	Drywell	Ex-drywell	US	Non-US	Crack / Wall Thinning	P/H-leak	Leak	Rupture
$3 < \varnothing \le 6$	YES		YES		0	0	0	0
$6 < \varnothing \le 12$	YES		YES		4	0	0	0
$\varnothing > 12$	YES		YES		1	0	0	0
$3 < \varnothing \le 6$	YES			YES	5	0	0	0
$6 < \varnothing \le 12$	YES			YES	9	0	0	0
$\varnothing > 12$	YES			YES	5	0	0	0
$3 < \varnothing \le 6$		YES	YES		1	2	19	3
$6 < \varnothing \le 12$		YES	YES		1	1	7	0
$\varnothing > 12$		YES	YES		2	0	3	0
$3 < \varnothing \le 6$		YES		YES	3	0	1	0
$6 < \varnothing \le 12$		YES		YES	2	0	2	0
$\varnothing > 12$		YES		YES	0	1	1	0

D.4.3 PWR-Specific Apriori Pipe Failure Rates

As summarized in Section D.3.3, there have been only a few through-wall defects in Class 1 PWR piping. For the RC-HL, the rate of PWSCC per weld-year is established using the normalization process discussed in Section D.4.2.1 and with the following specializations:

RC-HL Apriori Failure Rate

---

[8] The mean of weld count in NPS20-, 22- and 24-piping.

- The apriori failure rate is derived using the PIPExp for the time-period 1970 through 2000 to include the consideration of the through-wall defect at V.C. Summer.
- Table D.8 includes the weld population data used to calculate $\lambda_{NPS30} = 8.12E\text{-}05$ per weld-year.
- Failure rate "post-processing" to account for different weld configuration susceptibilities to PWSCC is done consistent with Section D.4.2 and with the S_{ij} and A_{ij} assumed values shown in Table D.9.

Table D.8 Selected Weld Counts in Code Class 1 PWR Piping

Plant ID (NSSS Type)	Weld Count by Pipe Size [NPS]				
	3-¾	10	12	14	30[9]
1 (WEST/4)	--	24	18	1	84
2 (WEST/4	--	24	18	1	52
3 (WEST/4)	--	24	18	--	92
4 (WEST/4)	--	24	14	--	68
Plant A.a (WEST/3)	--	--	5	14	50
Plant A.b (B&W; HPI/NMU system only)	9	--	--	--	--

Table D.9 Degradation Susceptibility by Weld Configuration

RC System (NPS)	Weld Configuration	Configuration Dependent Parameters	
		Susceptibility (S_{ij})	Attribute (A_{ij})
30 (RC Hot Leg)	Nozzle-to-safe-end	8.00E-01	12.5
	Elbow-to-safe-end	8.00E-02	12.5
	Elbow-to-pump	5.00E-02	12.5
	Pipe-to-pump	4.00E-02	12.5
	Elbow-to-pipe	3.00E-02	1.5
14 (Surge Line)	Nozzle-to-safe-end	5.00E-01	14.0
	Pipe-to-safe-end	2.50E-01	14.0
	Branch-to-pipe	5.00E-02	14.0
	Branch-to-HL	1.50E-01	14.0
	Elbow-to-pipe-	5.00E-02	1.40
3-¾ (HPI/NMU)	Elbow-to-nozzle	8.50E-01	9.0
	Elbow-to-pipe	1.00E-01	2.25
	Elbow-to-valve	4.50E-02	3.0
	Pipe-to-pipe	5.00E-03	9.0

In contrast to the BWR weld susceptibility factors in Table D.5, the weld susceptibility factors in Table D.10 are assumed values that reflect the applicable service experience. As an example, for the RC Hot Leg the nozzle-to-safe-end weld is assigned the highest value in view of the available service experience; i.e., the Ringhals and V.C. Summer hot leg cracking as described in Section D.3.3.1. As another example, for the RC Surge Line, relatively high weld susceptibility factors are assigned the safe-end welds and Hot Leg branch connection. In view of the recent experience at TMI-1, the nozzle-to-safe-end weld is given a greater weight than other weld configurations. The uncertainty in the PWR weld susceptibility factors is not evaluated

---

[9] NPS30 is used to characterize the CL- and HL-piping.

further in this study, however. EPRI TR-111880[10] is used for characterizing the prior knowledge about pipe failure due to thermal fatigue.

D.4.4 Prior Distributions for Bayesian Updating

Included for comparison, Figure D.18 shows the calculated flaw rates (non-through wall). Listed in Table D.10 are the RR and FW weld failure rates that represent the state-of-knowledge at T = 15 years of operation. Listed in Table D.11 are the RC and HPI/NMU prior weld failure rates. The failure rates represent the frequency per weld-year of a through-wall flaw resulting in a leakage of less than or equal to the Technical Specification limit for unidentified leakage. In summary, the derivation of prior weld failure rates includes the following steps:

- Determine the number of through-wall leaks from PIPExp database. Includes performing a trend analysis.
- From the PIPExp database, determine the appropriate exposure terms.
- Establish the susceptibility of different weld locations to degradation.
- Combine the output from the previous steps to determine prior failure rates applicable to welds in the RR and FW systems.

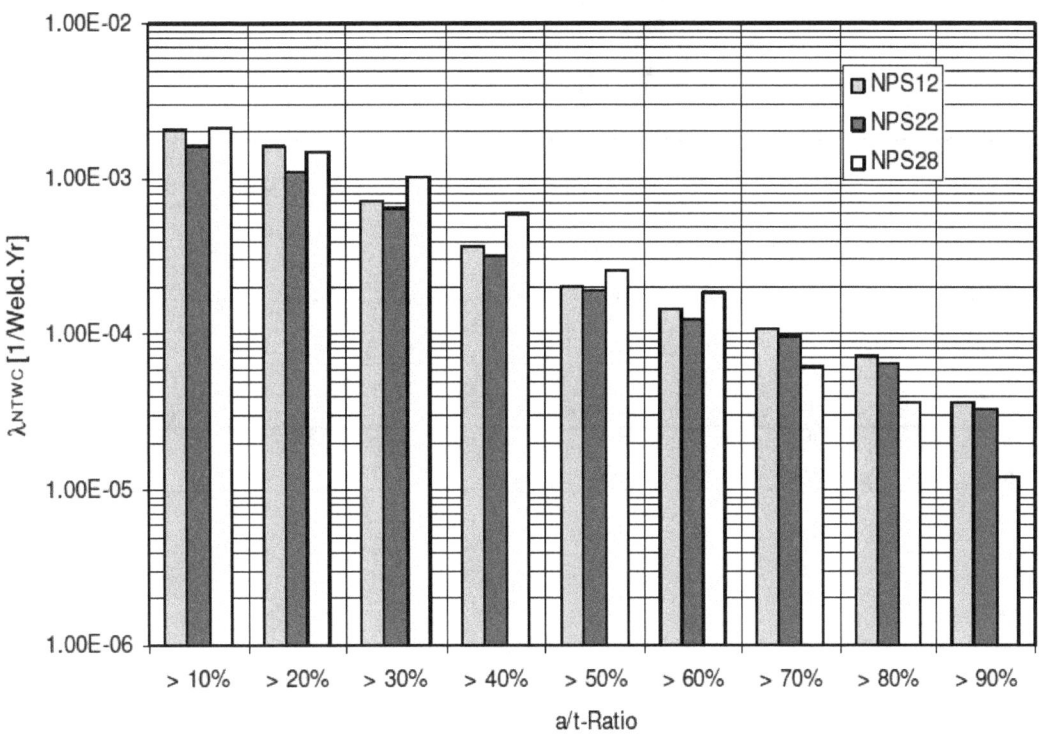

Figure D.18 Prior Frequency of Non-Through Wall IGSCC in RR Piping

---

[10] Table 2-3, page 2-10; λ_j = 1.34E-05 (RF = 100). TR-111880: Piping System Failure Rates and Rupture Frequencies for Use in Risk Informed In-Service Inspection (September 1999).

Table D.10 Prior RR and FW Weld Failure Rates

System	Pipe Size [NPS]	Weld Configuration	Lognormal Distribution Parameters	
			Mean [1/Weld-yr]	Range Factor (RF)
RR	12	Elbow-to-pipe	3.95E-04	10
		Nozzle-to-safe-end	1.59E-04	10
		Pipe-to-safe-end	7.49E-05	15
		Pipe-to-sweepolet	7.02E-05	15
		Pipe-to-reducer	1.40E-04	15
RR	22	Pipe-to-end-cap	7.67E-04	10
RR	22	Pipe-to-sweepolet	1.18E-04	15
		Pipe-to-cross	1.77E-04	15
RR	28	Pipe-to-elbow	7.19E-04	10
		Nozzle-to-safe-end	1.50E-04	10
		Pipe-to-safe-end	6.74E-04	10
		Pipe-to-valve	2.25E-04	10
		Pipe-to-pump	2.25E-04	10
		Pipe-to-tee	1.25E-04	10
		Pipe-to-pipe	4.99E-05	20
		Pipe-to-cross	7.49E-05	20
		Pipe-to-reducer	7.49E-05	20
FW	12	Nozzle-to-safe-end	6.20E-06	100
		Elbow-to-pipe	3.32E-07	100
		Pipe-to-pipe	5.17E-07	100
		Pipe-to-reducer	1.55E-06	100
		Pipe-to-reducing-tee	4.65E-07	100
FW	20	Elbow-to-pipe	4.00E-06	100
		Elbow-to-valve	6.00E-07	100
		Pipe-to-reducing-tee	4.80E-07	100
		Pipe-to-reducer	7.20E-07	100

No failure rates derived for welds in NPS14 piping; it is assumed that the data on welds in NPS12 piping is also representative of welds in NPS14 piping.

Table D.11 Prior RC and HPI/NMU Weld Failure Rates

System	Pipe Size [NPS]	Weld Configuration	Lognormal Distribution Parameters	
			Mean [1/Weld-yr]	RF
RC (Hot Leg)	30	Nozzle-to-safe-end	8.12E-04	100
		Elbow-to-pump	5.07E-05	100
		Pipe-to-pump	4.06E-05	100
		Elbow-to-pipe	3.65E-06	100
RC (Surge Line)	14	Branch-to-HL	2.10E-06	100
		Nozzle-to-safe-end	7.00E-06	100
		Pipe-to-safe-end	3.50E-06	100
		Branch-to-pipe	7.00E-07	100
		Elbow-to-pipe	7.00E-08	100
HPI/NMU	3-¾	Elbow-to-nozzle	9.86E-04	16
		Pipe-to-pipe	6.48E-06	100
		Elbow-to-pipe	1.90E-06	100
		Elbow-to-valve	1.31E-06	100

Note to Data on Pressurizer Surge Line:
• Susceptibility factors from Table D.9 applied directly to prior knowledge.

Notes to Data on HPI/NMU Line:
• Susceptibility factors from Table D.9 applied to prior knowledge.
• To develop a B&W-specific weld failure rate, the prior knowledge (data from TR-111880; see footnote #18) was combined with B&W-specific service data and exposure data from PIPExp through end of calendar year 1997. The failure rate for weld type 'elbow-to-nozzle' accounts for the weld failure at Oconee-2 in 1997.

D.5 Data for LOCA Frequency Estimation

Using the information in Section D.4, this section documents the input data to the LOCA frequency model. A Bayesian update is performed to develop posterior weld failure rates. The frequency of leaks exceeding Technical Specification limits are developed through estimates of the conditional probability of a large leak given a small through-wall flaw.

D.5.1 Posterior Weld Failure Rates
The failure rate calculation involves two factors, the number of applicable failures and the exposure data. To account for uncertainty in the exposure data, which could influence the failure rate calculation the following process is used. First, a best estimate update is performed using the appropriate number of failure events and the number of welds of exposure. To account for plant-to-plant variability in the weld exposure term, a second update is performed using the same failure data but an exposure estimate that is 50% higher, and a third update using an exposure estimate that is 50% lower. Each of the three updates is combined in a posterior weighting process using the following weights: 50% for the best estimate, 25% for the high exposure case and 25% for the low exposure case. The result is an uncertainty distribution for each failure rate, which reflects greater uncertainty than the best estimate data would imply alone. The results are given in Tables D.12 and D.13; Attachment B includes the input to the failure rate calculation. Figure D.19 displays posterior IGSCC flaw frequencies. Figures D.20-D.23 compare the prior and posterior non-through wall crack frequencies.

Table D.12 Posterior RR and FW Weld Failure Rate Distributions – BWR Base Cases

System	Pipe Size [NPS]	Weld Configuration	Failure Rate Uncertainty Distribution Parameters [(≤TS Leak)/Weld-yr]			
			Mean	5%-tile	50%-tile	95%-tile
RR	12	Elbow-to-pipe	4.32E-05	8.48E-06	3.17E-05	1.16E-04
		Nozzle-to-safe-end	4.38E-05	5.52E-06	2.72E-05	1.36E-04
		Pipe-to-safe-end	2.99E-05	2.98E-06	1.70E-05	9.64E-05
		Pipe-to-sweepolet	3.14E-05	2.80E-06	1.71E-05	1.06E-04
		Pipe-to-reducer	7.82E-05	5.71E-06	3.97E-05	2.77E-04
RR	22	Pipe-to-end-cap	1.54E-04	2.28E-05	1.01E-04	4.52E-04
		Pipe-to-cross	4.24E-05	4.38E-06	2.47E-05	1.37E-04
		Pipe-to-sweepolet	7.37E-05	7.02E-06	4.09E-05	2.40E-04
RR	28	Pipe-to-elbow	8.52E-05	1.59E-05	6.07E-05	2.33E-04
		Nozzle-to-safe-end	6.55E-05	5.95E-06	3.61E-05	2.15E-04
		Pipe-to-safe-end	1.44E-04	2.11E-05	9.36E-05	4.28E-04
		Pipe-to-valve	5.96E-05	7.68E-06	3.75E-05	1.84E-04
		Pipe-to-pump	8.36E-05	8.68E-06	4.85E-05	2.71E-04
		Pipe-to-tee	5.78E-05	5.06E-06	3.13E-05	1.96E-04
		Pipe-to-pipe	1.29E-05	5.74E-07	5.25E-06	4.78E-05
		Pipe-to-cross	3.86E-05	7.89E-07	1.08E-05	1.50E-04
		Reducer-to-cross	3.86E-05	7.89E-07	1.08E-05	1.50E-04
FW	12	Nozzle-to-safe-end	2.29E-06	8.61E-10	6.88E-08	5.29E-06
		Elbow-to-pipe	1.75E-07	4.61E-11	4.28E-09	3.74E-07
		Pipe-to-pipe	2.78E-07	7.39E-11	6.89E-09	6.33E-07
		Pipe-to-safe-end	2.43E-07	6.20E-11	5.97E-09	5.50E-07
		Pipe-to-reducer	9.73E-07	2.38E-10	2.24E-08	2.12E-06
FW	12	Elbow-to-reducing-tee	3.33E-07	7.46E-11	7.11E-09	6.94E-07
FW	20	Pipe-to-elbow	1.62E-06	5.57E-10	4.61E-08	3.71E-06
		Pipe-to-pipe	4.10E-07	9.57E-11	8.77E-09	8.00E-07
		Pipe-to-valve	3.38E-07	7.09E-11	6.78E-09	6.53E-07
		Elbow-to-valve	3.54E-07	9.22E-11	8.80E-09	8.39E-07
		Pipe-to-tee	4.50E-07	7.33E-11	7.20E-09	7.25E-07
		Pipe-to-reducer	5.68E-07	1.17E-10	1.14E-08	1.14E-06
		Tee-to-valve	2.18E-07	4.06E-11	4.12E-09	4.06E-07

Table D.13 Posterior RC and HPI/NMU Weld Failure Rate Distributions – PWR Base Cases

System	Pipe Size [NPS]	Weld Configuration	Failure Rate Uncertainty Distribution Parameters [(≤TS Leak)/Weld-yr]			
			Mean	5%-tile	50%-tile	95%-tile
RC (Hot Leg)	30	Nozzle-to-safe-end	7.64E-05	2.12E-07	7.34E-06	2.61E-04
		Elbow-to-pump	1.96E-05	7.59E-09	5.36E-07	4.02E-05
		Pipe-to-pump	1.24E-05	5.78E-09	4.47E-07	3.17E-05
		Elbow-to-pipe	1.05E-06	5.38E-10	3.94E-08	3.04E-06
RC (Surge Line)	14	Branch-to-CL	1.14E-06	2.90E-10	2.69E-08	2.42E-06
		Nozzle-to-safe-end	2.95E-06	9.94E-10	7.99E-08	6.49E-06
		Pipe-to-safe-end	1.75E-06	4.72E-10	4.00E-08	3.64E-06
		Branch-to-pipe	4.76E-07	1.04E-10	1.04E-08	9.87E-07
		Elbow-to-pipe	4.60E-08	1.06E-11	1.04E-09	9.66E-08
HPI/ NMU	3-¾	Pipe-to-safe-end	6.56E-04	1.45E-05	1.99E-04	2.53E-03
		Elbow-to-pipe	1.58E-06	3.30E-10	3.43E-08	3.39E-06
		Pipe-to-valve	1.96E-06	2.36E-10	2.39E-08	2.35E-06
		Pipe-to-pipe	4.55E-06	1.13E-09	1.13E-07	1.11E-05

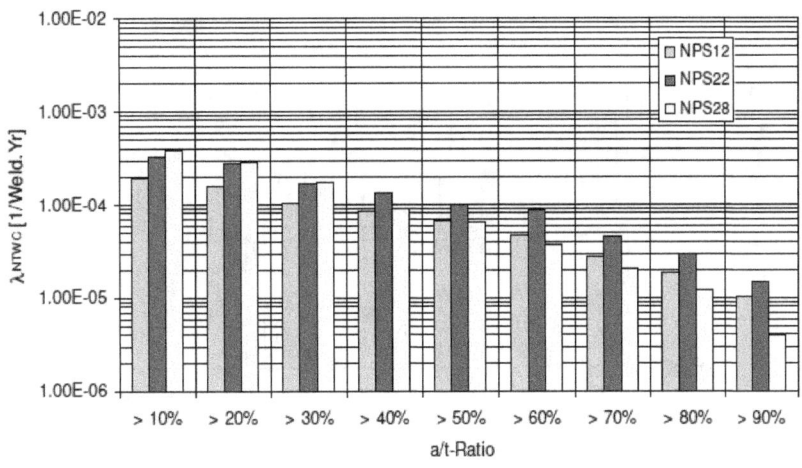

Figure D.19 Posterior IGSCC Frequency (Non-Through Wall)

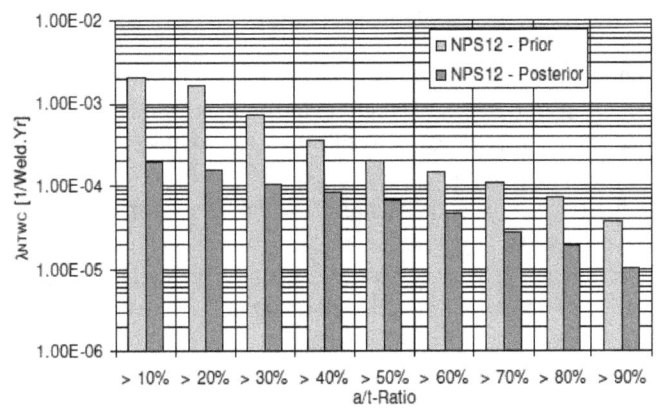

Figure D.20 Prior and Posterior IGSCC Frequency (Non-Through Wall) for NPS12 Welds

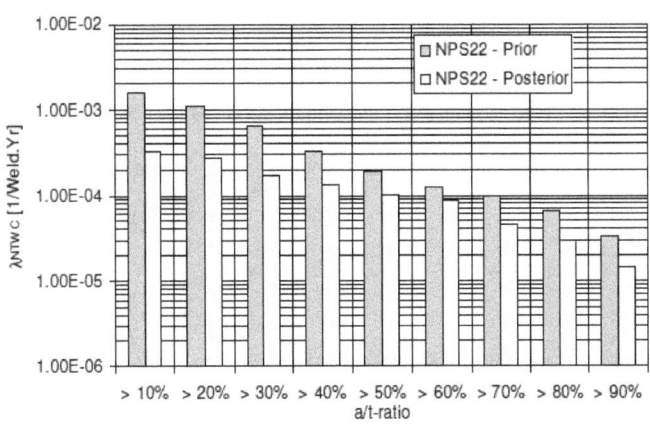

Figure D.21 Prior and Posterior IGSCC Frequency (Non-Through Wall) for NPS22 Welds

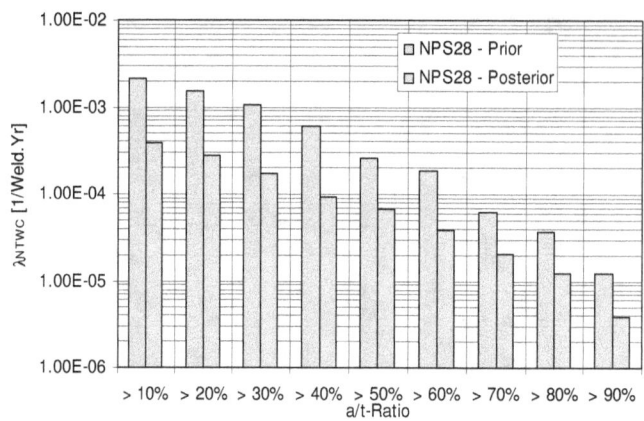

Figure D.22 Prior and Posterior IGSCC Frequency (Non-Through Wall) for NPS28 Welds

D.5.2   Conditional Weld Failure Probability

This section develops a basis for calculating conditional weld failure probabilities. In the model of piping reliability, the conditional weld failure probability, $p_{L/F}$, represents the likelihood of a weld flaw propagating to a significant structural failure. However, for Code Class 1 piping there is no service experience data available to support a direct estimation of $p_{L/F}$. Therefore, the options for calculating the conditional failure probability when the service experience data consists of zero events include applications of (1) probabilistic fracture mechanics modeling and (2) Bayesian modeling. Since the objective of this Base Case was to directly utilize insights from service experience data reviews, the latter approach was selected. Before defining the input to the Bayesian modeling, it is useful to organize the available service experience according to piping classification and severity of observed failures. Figure D.23 shows the conditional pipe failure probability as a function of observed through-wall flow rate for reactor coolant pressure boundary (RCPB) piping (Code Class 1), Code Class 2 and 3 piping, and ASME B31.1 (non-Code) piping.

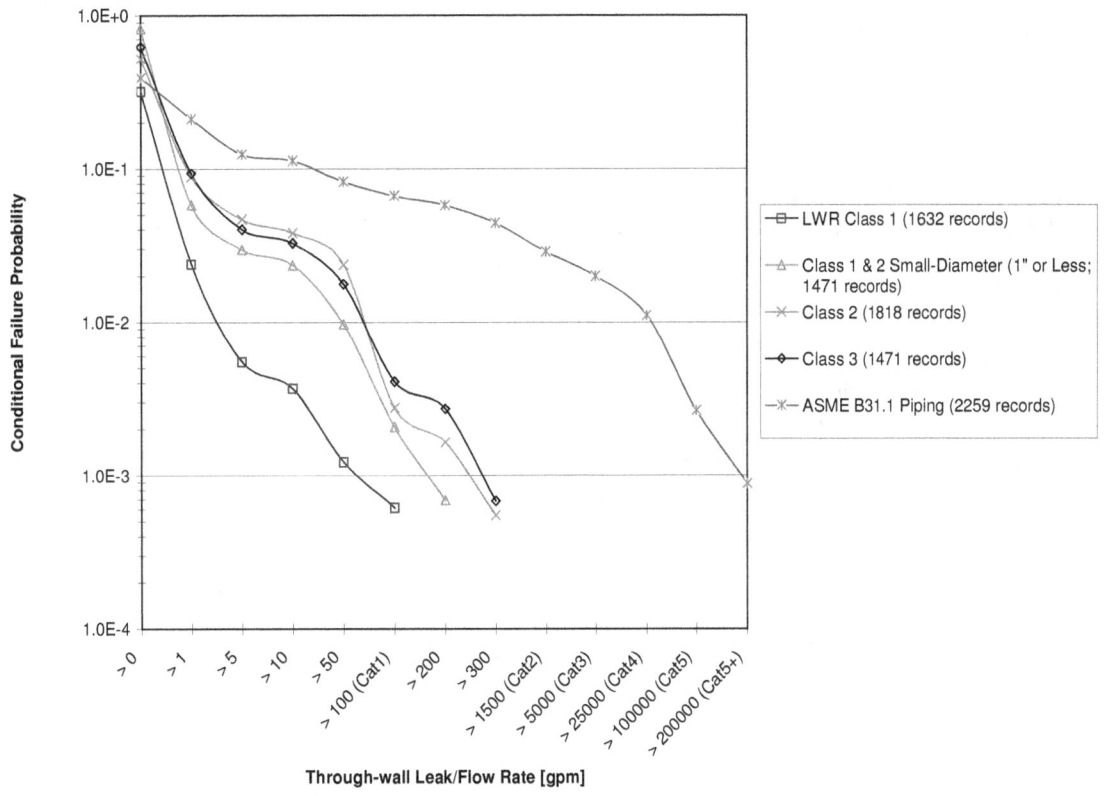

**Figure D.23 Likelihood of Structural Failure According to Service Experience with
Light Water Reactor Piping System**

The empirical data that is used to construct the chart in Figure D.23 represents approximately 7,200 recorded pipe failure events representing almost 9,000 reactor-years of commercial nuclear power plant operation. It is important to note that the chart covers a wide variety of piping systems, from Reactor Coolant Pressure Boundary (RCPB) piping, Safety Injection and Recirculation piping, and Auxiliary Cooling piping to ASME B31.1 piping (or non-Code piping). Figure D.23 represents our current state-of-knowledge with respect to the probability of pipe failure and it provides an upper bound for estimates of conditional pipe failure probability. Noteworthy is the fact that only for ASME B31.1 (non-Code) piping systems have significant structural failures been observed (that is, failures equivalent to double-ended guillotine breaks). For the other classes of piping it is necessary to do data extrapolations beyond the Category 1 through-wall flow rates. In this Base Case study, a Bayesian approach is used for this extrapolation and by acknowledging service data insights. For the Code Class 1 piping, it is also important to recognize that according to the available service experience data, through-wall flow rates greater than 38 lpm (10 gpm) have been primarily attributed to failure of small bore piping. It is unlikely that the structural integrity of the Class 1 piping is equal to or less that that of Code Class 2 and 3 piping for which Category 1 LOCA events have been observed.

The Beta Distribution has some convenient and useful properties for use in Bayes' updating. The analysis starts by defining a prior distribution that represents the analyst's understanding of piping performance given the presence of some sort of degraded condition. The prior distribution is defined by selecting an appropriate set of initial values for parameters A(α) and B(β), denoted as A_{Prior} and B_{Prior}. Then, when looking at the relevant service experience data, if there are "N" structural failures of a certain magnitude and "M" successes

---

(or degraded conditions that were repaired before progressing to a structural failure), the Bayes' updated, or posterior distribution is also a Beta Distribution with the following parameters:

$$A_{Posterior} = A_{Prior} + N$$
$$B_{Posterior} = B_{Prior} + M$$

The above explains how the Beta Distribution can be used to estimate conditional weld failure probabilities. The challenge is to justify the selected parameters when the evidence is zero structural failures. Certainly, it can be argued the ASME B31.1 service experience data represents a very conservative upper bound for the conditional weld failure probability.

Selecting a well justified set of "A" and "B" parameters is not a trivial task. One basic ground rule should be for the "weight" of the field experience data to determine the shape of the posterior Beta Distribution. However, many different parameter combinations will produce the same predicted mean value. Where very little evidence is available about the parameters, constrained non-informative priors may be selected. For such a case, one can say that the "A" parameter has to be a small number.

In this Base Case study, the prior "A" and "B" parameters are defined by first deriving a constraint for the prior mean value of the conditional failure probability and then fixing the "A" parameter at 1.0 for stress corrosion cracking and 2.0 for thermal fatigue to account for the fact that according to available service experience data, thermal fatigue cracks propagate in the through-wall direction considerably faster than flaws caused by stress corrosion cracking. The process for developing conditional failure probabilities starts by deriving a point-estimate of p_{LIF} for small-diameter piping given susceptibility to stress corrosion cracking (SCC). This point estimate is based on Jeffrey's non-informative prior and service experience data. There have been 42 through-wall flaws and zero large leaks in small-bore BWR piping. This gives a point estimate of 1.2×10^{-2}, which is used as a "fix point" for determining conditional weld failure probabilities for other pipe sizes. The relationship between pipe size (diameter and wall thickness) and the conditional failure probability is assumed to follow a power law of the form:

$$p_{LIF} = a \times DN^b \tag{D.8}$$

Where, "DN" is the nominal pipe size in [mm]. Decreasing trends correspond to negative values of b. Parameters a and b are determined for $p_{LIF} = 1.2 \times 10^{-2}$ and DN = 25. Point estimates of p_{LIF} for other pipe sizes are derived using the power law for conditional failure probabilities and assuming that the general shape of the curve is similar to that of piping susceptible to vibratory fatigue for which $p_{LIF} = 2.5/DN$. Next the predicted conditional weld failure probability using the power law approach is used to determine the Beta Distribution parameter "B." For Class 1 piping, engineering judgment, as portrayed by Figure D.23 is used to assign values to the prior Beta Distribution parameters. The proposed Beta Distribution posterior parameters for this Base Case study are summarized in Table D.14.

Table D.14 Proposed Beta Parameters for Code Class 1 Piping

Degradation Mechanism	Pipe Size		Parameter B in Beta Posterior ("Large Leak")
	DN	NPS	
SCC ($A_{Prior} = 1$)	300	12	1,262 ($A_{Post} = 1; M = 0$)
	550	22	1,496 ($A_{Post} = 1; M = 0$)
	700	28	1,700 ($A_{Post} = 1; M = 0$)
Thermal Fatigue ($A_{Prior} = 2$)	90	3-¾	227 ($A_{Post} = 2; M = 0$)
	350	14	592 ($A_{Post} = 2; M = 0$)

D.5.3 Conditional Failure Probability and Flow Rate

The conditional failure probabilities derived in the previous section are assumed applicable to Cat0 LOCA. It is furthermore assumed that for a *significant* primary piping breach to occur there has to be a through-wall flaw coinciding with a plant operational mode change or an unusual or severe loading condition such that the leakage exceeds a Cat0 LOCA. The service data collection (e.g., PIPExp) includes numerous examples where pressure pulses or spikes caused by changing flow conditions following a plant operational mode change have resulted in non-active leaks[11] becoming active leaks. The physics of such transitions from non-active to active leaks are complex and location-dependent (e.g., function of flaw size and pipe stresses). Some published work exists on the correlation between crack propagation and plant transient history [D.25]. Using available empirical data, the uncertainties in such crack growth assessments are considerable, however.

In this analysis a simple parametric approach is applied to the estimation of weighted conditional failure probabilities (C_L) of a pressure boundary breach that exceeds a Cat0 flow rate threshold value. This approach is described through the event tree in Figure D.24. An undetected, or detected but monitored through-wall is exposed to a pressure pulse or unusual loading condition before a decision to perform manual, controlled reactor shutdown. The pressure pulse or unusual loading condition is characterized as a subjectively defined probability distribution.

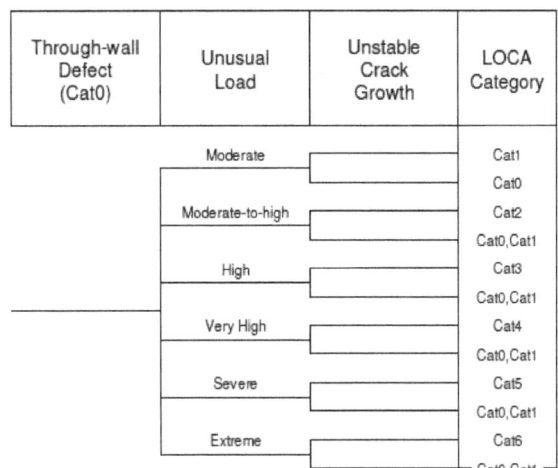

Figure D.24 Event Tree for Definition of LOCA Categories

In the cases of "moderate-to-high" to "extreme", the term "unusual" implies a loading condition beyond that resulting from anticipated transients including manual and automatic reactor/turbine trips. The conditional probability of an unusual or severe loading condition is described by five sets of subjective 3-bin discrete and overlapping probability distributions as summarized in Table D.15. These DPDs are combined with the weld failure rate distributions and conditional weld failure probability distributions by using a Monte Carlo merge technique.

---

[11] The term 'non-active leak' is taken to mean a through-wall flaw without visible leakage or with a small, detectable leakage that stays relatively constant over time.

Table D.15 Probability of LOCA Given Severe Overloading

Category	Flow Rate (ν) Intervals [gpm]	DPD for Severe Loading					
		$C_{L\text{-High}}$	$C_{L\text{-Med}}$	$C_{L\text{-Low}}$	p_{High}	p_{Med}	p_{Low}
0	$10 < \nu \le 100$	N/A	N/A	N/A	N/A	N/A	N/A
1	$100 < \nu \le 1500$	.80	.50	.20	.2	.6	.2
2	$1500 < \nu \le 5000$	.32	.20	.08	.2	.6	.2
3	$5000 < \nu \le 25{,}000$	.13	.08	.03	.2	.6	.2
4	$25{,}000 < \nu \le 100{,}000$	.05	.03	.01	.2	.6	.2
5	$100{,}000 < \nu \le 500{,}000$	.02	.01	.005	.2	.6	.2
6	$\nu > 500{,}000$	.01	.005	.002	.2	.6	.2

Service data on water hammer events provides a justification for the chosen DPDs. From PIPExp, a point estimate for $C_{L\text{-WH-Cat6}}$ is approximately 4.9E-03, which is based on two events involving severe overloading (including plastic deformation) of a pipe section in 411 recorded water hammer events. This is taken as a best estimate C_L-value for calculating a Cat6 LOCA. Figure D.24 includes the rules for how the DPDs are applied to the LOCA frequency calculation. The Cat0 and Cat1 LOCAs include contributions from each loading condition associated with Cat2 or larger pressure boundary breach. In other words, the calculation accounts for the possibility that an 'unusual' loading condition may not result in a global or catastrophic pressure boundary breach. Given a through-wall flaw and severe overload, Figure D.25 shows the conditional failure probability as a function of pipe size.

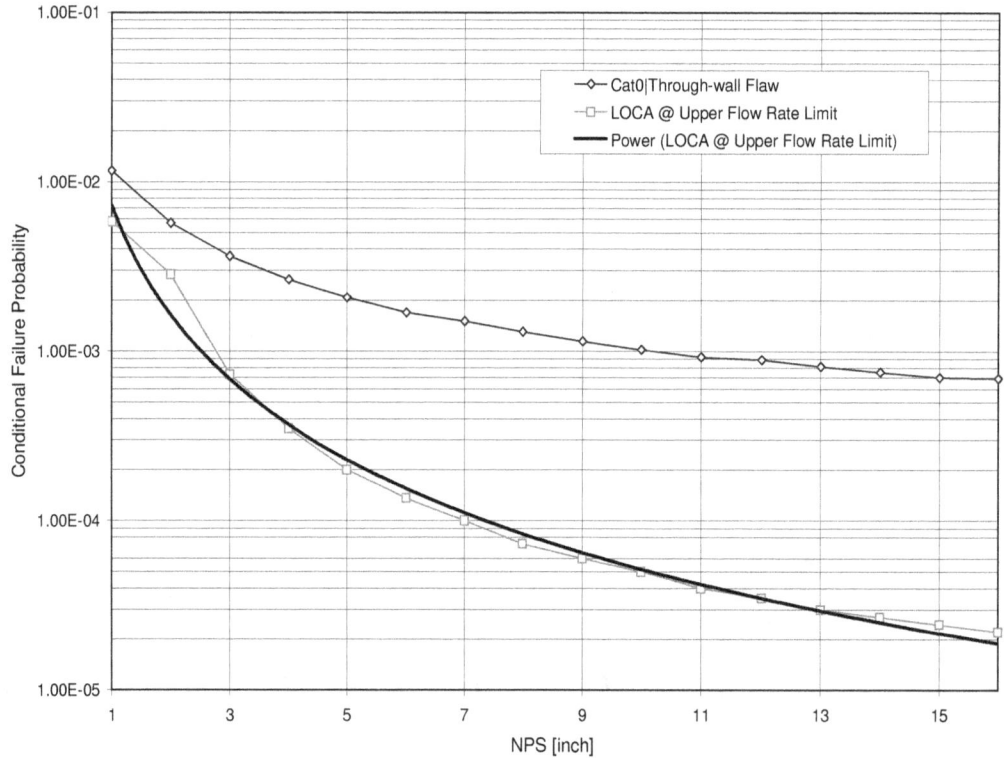

Figure D.25 Conditional Probability of Weld Failure Given Through-Wall Flaw and Severe Overloading

D.6 BWR and PWR Base Case LOCA Frequencies

The Base Case LOCA frequency models are based on three Excel files entitled "PlantBWelds" (BWR Base Case), "PlantA.aWelds" (PWR RC-HL and Surge Line), and "PlantA.bWelds" (PWR HPI/NMU). These Excel files are part of the plant design information supplied by members of the Expert Elicitation Panel. Adding information on weld type by location and weld failure parameters to respective Excel files provides the basis for calculating LOCA frequencies including uncertainty distributions.

D.6.1 LOCA Frequency Models
For the BWR Base Case, the original "PlantBWelds" includes two spreadsheets, one for the feedwater system and one for the reactor recirculation system. In a first step to create a LOCA frequency model, each spreadsheet was split into two, one for Loop A and one for Loop B of respective system. Next a new column was added to each new spreadsheet to include information on weld type by location. A review of isometric drawings provided the input to the new columns.

The statistical information that is summarized in Section D.5 of this report is included in five separate spreadsheets of the modified Excel file. The posterior weld failure rates are included on Tab "Parameters," each parameter assigned a unique variable name. The calculation of LOCA frequency, including Monte Carlo merge operations are performed on Tabs "Intermediate FW_A", "Intermediate FW_B", "Intermediate RR_A", and "Intermediate RR_B." These intermediate calculation sheets are linked to the design information for FW Loop A, FW Loop B, RR Loop A and RR Loop B, respectively. Using the variable names as defined on Tab "Parameters," each weld is assigned an appropriate failure rate (including uncertainty distribution as defined using a Crystal Ball "assumption"). Finally, an integrated calculation of LOCA frequency by leak threshold category (0 through 5) is performed on a separate spreadsheet, which is linked to the intermediate calculations. Each integrated LOCA frequency calculation is defined as a Crystal Ball "forecast."

For the PWR Base Case, the original "PlantAWelds" includes a single spreadsheet with the ASME XI Class 1 Category B-F/B-J welds. In a first step to create a LOCA frequency model, each spreadsheet was split into two, one for the RC-HL and one for the pressurizer surge line. A third spreadsheet with the HPI/NMU weld listed was added to form a new Excel-file corresponding to the PWR Base Case LOCA frequency model.

D.6.2 BWR Base Case LOCA Frequencies at T = 25
Summarized in Table D.16 are the LOCA frequency uncertainty distributions that are derived for BWR Base Case Cat0 through Cat6 LOCA. The results are representative of Plant B after 25 years of operation (T = 25).

Table D.16 Plant B LOCA Frequencies at T = 25 Years

System	LOCA Category	Uncertainty Distribution			
		Mean	5%-tile	50%-tile	95%-tile
RR Loop A NPS12	Cat0	2.06E-06	9.72E-08	1.38E-06	6.28E-06
	Cat1	1.33E-06	6.29E-08	8.95E-07	4.09E-06
	Cat2	1.44E-07	6.04E-09	8.66E-08	4.73E-07
	Cat3	5.81E-08	2.34E-09	3.51E-08	1.91E-07
	Cat4	2.31E-08	9.08E-10	1.39E-08	7.54E-08
	Cat5	N/A	N/A	N/A	N/A
RR Loop B NPS12	Cat0	2.07E-06	1.00E-07	1.36E-06	6.41E-06
	Cat1	1.34E-06	6.45E-08	8.81E-07	4.15E-06
	Cat2	1.44E-07	5.85E-09	8.65E-08	4.72E-07
	Cat3	5.84E-08	2.26E-09	3.47E-08	1.95E-07
	Cat4	2.31E-08	9.15E-10	1.40E-08	7.48E-08
	Cat5	N/A	N/A	N/A	N/A
RR Loop A NPS22	Cat0	1.23E-06	4.88E-08	7.30E-07	4.07E-06
	Cat1	8.58E-07	3.37E-08	5.12E-07	2.89E-06
	Cat2	7.36E-07	2.42E-09	3.94E-08	2.57E-07
	Cat3	2.95E-08	1.01E-09	1.60E-08	1.01E-07
	Cat4	1.16E-08	3.94E-10	6.33E-09	4.04E-08
	Cat5	4.72E-09	1.57E-10	2.52E-09	1.63E-08
RR Loop B NPS22	Cat0	1.24E-06	4.97E-08	7.26E-07	4.06E-06
	Cat1	8.66E-07	3.39E-08	5.10E-07	2.88E-06
	Cat2	7.37E-08	2.48E-09	3.95E-08	2.59E-07
	Cat3	2.91E-08	9.77E-10	1.59E-08	1.01E-07
	Cat4	1.18E-08	3.86E-10	6.36E-09	3.97E-08
	Cat5	4.73E-09	1.56E-10	2.55E-09	1.67E-08
RR Loop A NPS28	Cat0	2.73E-06	1.31E-07	1.80E-06	8.41E-06
	Cat1	1.91E-06	9.05E-08	1.24E-06	5.91E-06
	Cat2	1.64E-07	6.27E-09	9.69E-08	5.38E-07
	Cat3	6.54E-08	2.49E-09	3.92E-08	2.16E-07
	Cat4	2.61E-08	1.03E-09	1.55E-08	8.65E-08
	Cat5	1.05E-08	4.24E-10	6.20E-09	3.50E-08
RR Loop B NPS28	Cat0	2.76E-06	1.35E-07	1.82E-06	8.44E-06
	Cat1	1.93E-06	9.31E-08	1.27E-06	5.92E-06
	Cat2	1.65E-07	6.32E-09	9.73E-08	5.42E-07
	Cat3	6.62E-08	2.57E-09	3.93E-08	2.18E-07
	Cat4	2.64E-08	1.10E-09	1.56E-08	8.74E-08
	Cat5	1.06E-08	4.40E-10	6.33E-09	3.40E-08
FW Loop A NPS12	Cat0	6.69E-07	2.78E-08	4.05E-07	2.20E-06
	Cat1	4.34E-07	1.80E-08	2.61E-07	1.42E-06
	Cat2	4.75E-08	1.61E-09	2.58E-08	1.61E-07
	Cat3	1.89E-08	6.41E-10	1.03E-08	6.49E-08
	Cat4	7.53E-09	2.62E-10	4.07E-09	2.53E-08
	Cat5	N/A	N/A	N/A	N/A
FW Loop B NPS12	Cat0	6.83E-07	2.89E-08	4.16E-07	2.24E-06
	Cat1	4.44E-07	1.90E-08	2.68E-07	1.45E-06
	Cat2	4.86E-08	1.70E-09	2.64E-08	1.67E-07
	Cat3	1.93E-08	6.58E-10	1.06E-08	6.71E-08
	Cat4	7.75E-09	2.80E-10	4.20E-09	2.62E-08
	Cat5	N/A	N/A	N/A	N/A
FW Loop B NPS14	Cat0	6.02E-08	1.52E-09	2.61E-08	2.26E-07
	Cat1	4.19E-08	1.06E-09	1.84E-08	1.55E-07
	Cat2	3.59E-09	7.64E-11	1.42E-09	1.37E-08
	Cat3	1.45E-09	3.12E-11	5.69E-10	5.51E-09

Table D.16 Plant B LOCA Frequencies at T = 25 Years

System	LOCA Category	Uncertainty Distribution			
		Mean	5%-tile	50%-tile	95%-tile
	Cat4	5.80E-10	1.19E-11	2.30E-10	2.21E-09
	Cat5	2.32E-10	4.74E-12	9.11E-11	8.83E-10
FW Loop A NPS20	Cat0	8.43E-07	3.47E-08	5.08E-07	2.74E-06
	Cat1	5.92E-07	2.37E-08	3.55E-07	1.95E-06
	Cat2	5.08E-08	1.75E-09	2.75E-08	1.74E-07
	Cat3	2.03E-08	7.15E-10	1.10E-08	6.83E-08
	Cat4	8.10E-09	2.77E-10	4.33E-09	2.81E-08
	Cat5	3.26E-09	1.12E-10	1.74E-09	1.13E-08
FW Loop B NPS20	Cat0	1.00E-06	4.40E-08	6.13E-07	3.24E-06
	Cat1	7.014-07	3.01E-08	4.28E-07	2.12E-06
	Cat2	6.03E-08	1.03E-09	3.32E-08	2.04E-07
	Cat3	2.41E-08	8.45E-10	1.33E-08	8.00E-08
	Cat4	9.64E-09	3.44E-10	5.27E-09	3.32E-09
	Cat5	3.84E-09	1.38E-10	2.14E-09	1.31E-08
RR Total Loops A & B	Cat0	1.21E-05	3.02E-06	1.03E-05	2.70E-05
	Cat1	8.24E-06	2.02E-06	6.98E-06	1.86E-05
	Cat2	7.64E-07	1.40E-07	6.07E-07	1.92E-06
	Cat3	3.07E-07	5.43E-08	2.44E-07	7.79E-07
	Cat4	1.22E-07	2.22E-08	9.73E-08	3.06E-07
	Cat5	3.05E-08	3.59E-09	2.19E-08	8.52E-08
FW Total Loops A & B	Cat0	3.26E-06	5.55E-07	2.60E-06	8.13E-06
	Cat1	2.21E-06	2.72E-07	1.75E-06	5.56E-06
	Cat2	2.10E-07	2.70E-08	1.54E-07	5.85E-07
	Cat3	8.40E-08	1.10E-08	6.18E-08	2.30E-07
	Cat4	3.36E-08	4.40E-09	2.47E-08	9.35E-08
	Cat5	7.33E-09	3.95E-10	4.33E-09	2.44E-08
RR + FW Total	Cat0	1.53E-05	5.24E-06	1.37E-05	3.10E-05
	Cat1	1.05E-05	3.51E-06	9.30E-06	2.14E-05
	Cat2	9.75E-07	2.24E-07	8.24E-07	2.26E-06
	Cat3	3.90E-07	9.00E-08	3.32E-07	9.05E-07
	Cat4	1.56E-07	3.64E-08	1.31E-07	3.57E-07
	Cat5	3.78E-08	6.43E-09	2.93E-08	9.73E-08

D.6.3 PWR Base Case LOCA Frequencies at T = 25

The PWR Base Case Cat0 through Cat6 LOCA frequencies including uncertainty distributions, are summarized in Table D.17. These results are representative of Plant A.a/A.b after 25 years of operation (T = 25).

Table D.17 Plant A.a/A.b LOCA Frequencies at T = 25 Years

System	LOCA Category	Uncertainty Distribution			
		Mean	5%-tile	50%-tile	95%-tile
RC Hot Leg	Cat0	8.94E-07	4.84E-09	1.27E-07	2.88E-06
(3-of-3); Plant A.a	Cat1	6.65E-07	3.55E-09	9.39E-08	2.14E-06
	Cat2	4.87E-08	2.10E-10	6.15E-09	1.49E-07
	Cat3	1.83E-08	8.33E-11	2.42E-09	5.95E-08
	Cat4	6.99E-09	3.03E-11	8.93E-10	2.21E-08
	Cat5	2.55E-09	1.16E-11	3.29E-10	8.29E-09
	Cat6	1.26E-09	5.44E-12	1.58E-10	4.04E-09
RC Surge Line	Cat0	1.44E-07	2.65E-09	2.98E-08	5.02E-07
Plant A.a	Cat1	1.14E-07	2.13E-09	2.36E-08	3.94E-07
	Cat2	9.60E-09	1.48E-10	1.88E-09	3.46E-08
	Cat3	3.84E-09	5.78E-11	3.50E-10	1.35E-08
	Cat4	1.44E-09	2.01E-11	2.77E-10	5.06E-09
	Cat5	5.31E-10	8.23E-12	1.01E-10	1.87E-09
	Cat6	N/A	N/A	N/A	N/A
HPI/NMU (2-of-2)	Cat0	2.72E-05	4.64E-07	6.90E-06	1.07E-04
Plant A.b	Cat1	1.60E-05	2.62E-07	3.93E-06	6.09E-05
	Cat2	2.33E-06	3.30E-08	5.40E-07	9.02E-06
	Cat3	9.22E-07	1.28E-08	2.14E-07	3.59E-06
	Cat4	N/A	N/A	N/A	N/A
	Cat5	N/A	N/A	N/A	N/A
	Cat6	N/A	N/A	N/A	N/A

D.6.4   Time-Dependency of LOCA Frequency Results

For respective Base Case Plant, the LOCA frequencies are determined for three time periods: T= = 25 years after plant startup (corresponding to today's state-of-knowledge), T = 40 years after plant startup (corresponding to original design life), and T = 60 years after plant startup (corresponding to end-of-life extension). The time-dependent analysis is performed in two different ways. First a 'prospective analysis' is performed based on a Markov model of piping reliability (Figure D.15). Second, a 'retrospective analysis' is performed by using Bayesian statistics.

D.6.4.1 Use of Markov Model to Determine Time-Dependency - According to the Markov model diagram in Figure D.15, a piping component can be in four mutually exclusive states: S (= Success), C (= Cracked), F (= Leaking, non-active leakage, or active leakage with leak rate within Technical Specification Limit) or L (= Leaking, with leak rate in excess of Technical Specification Limit). The time-dependent probability that a piping component is in each state S, C, F, or L is described by a differential equation. Under the assumption that all the state transition rates are constant the Markov model equations will consist of a set of coupled linear differential equations with constant coefficients. The reliability term needed to represent LOCA frequency is the system failure rate or hazard rate $h\{t\}$, which is time-dependent. The hazard rate is defined as:

$$h\{t\} = (1/(1 - L\{t\})) \times dL\{t\}/dt \tag{D.9}$$

Where:

$$1 - L\{t\} = S\{t\} + C\{t\} + F\{t\} \tag{D.10}$$

The hazard rate is a function of time and the parameters of the Markov model; $h\{t\}$ is the time-dependent frequency of pipe rupture. Reference [D.14] provides solutions to the Markov model and derives an expression for $h\{t\}$ as a function of the six parameters associated with the 4-state Markov model: An occurrence rate for detectable flaws (ϕ), a failure rate for leaks given the existence of a flaw (λ_F), two rupture frequencies including one from the initial state of a flaw (ρ_F) and one from the initial state of a leak (ρ_L), a repair rate for detectable flaws (ω), and a repair rate for leaks (μ). The latter two parameters dealing with repair are further developed by the following simple models.

$$\omega = \frac{P_{FI} P_{FD}}{(T_{FI} + T_R)} \qquad (D.11)$$

Where:

P_{FI} = probability that a piping element with a flaw will be inspected per inspection interval. This parameter has a value of 0 if it is not in the inspection program and 1 if it is in the inspection program. For the inspected elements, a value of 1 is used for any ISI inspection case and 0 for the case of no ISI. The element may be selected for inspection directly by being included in the sections sampled for ISI inspection, or indirectly by having a rule such that if degradation is detected anywhere in the system, the search will be expanded to include examination of that element.

P_{FD} = probability that a flaw will be detected given this element is inspected. This is the reliability of the inspection program and is equivalent to the term used by NDE experts, "Probability of detection (POD)." This probability is conditioned on the occurrence of one or more detectable flaws in the segment according to the assumptions of the model. Also note that

T_{FI} = mean time between inspections for flaws, (inspection interval).

T_R = mean time to repair once detected. Depending on the location of the weld to be repaired, the actual weld repair could take on the order of several days to much more than a week. Accounting for time to prepare for repair, NDE, root cause evaluation, etc., the total outage time attributed to the repair of a Class 1 weld is on the order of 1 month or more. However, since this term is always combined with T_{FI}, and T_{FI} could be 10 years, in practice the results are insensitive to assumptions regarding T_R

Similarly, estimates of the repair rate for leaks can be estimated according to:

$$\mu = \frac{P_{LD}}{(T_{LI} + T_R)} \qquad (D.12)$$

Where:

P_{LD} = probability that the leak in the element will be detected per leak inspection or detection period

T_{LI} = mean time between inspections for leaks. For RCPB piping the time interval between leaks can be essentially instantaneous if the leak is picked up by radiation alarms, to as long as the time period between leak tests performed on the system.

T_R = as defined above but for full power applications, this time should be the minimum of the actual repair time and the time associated with cooldown to enable repair and any waiting time for replacement piping.

A summary of the root input parameters of the Markov model and the general strategy for estimation of each parameter is presented in Table D.18.

Table D.18 Four-State Markov Model Root Input Parameters

Parameter	Assumed or Estimated Value	Basis
ω	2.1 x 10^{-2}/year $\{=(.25) \times (.90)/(10+(200/8760))\}$	Element assumed to have a 25% chance of being inspected for flaws every 10 years with a 90% detection probability. In the given example detected flaws will be repaired in 200 hours
μ	7.92 x 10^{-1}/ year $\{=(.90) \times (.90)/(1+(200/8760))\}$	Element is assumed to have a 90% chance of being inspected for leaks once a year with a 90% leak detection probability
ρ_C	Table D.13, D.14 and D.15	The basis is developed in Sections D.4 and D.5.
λ_C	Table D.13 and D.14	The basis is developed in Sections D.4 and D.5.
ρ_F	2.0 x 10^{-2}/year	If the element is already leaking, the conditional frequency of ruptures is assumed to be determined by the frequency of severe overloading events; the given value is equal to the frequency of severe water hammer (from PIPExp database).
ϕ	Variable (for IGSCC $\phi = 7.58 \times (\lambda_C + \rho_C)$)	The occurrence rate of a flaw is estimated from service data. As an example, IGSCC in the BWR operating environment will create ca. 7.58 flaws for every through-wall leak that is observed.
P_{FI}	1 or 0	Probability per inspection interval that the pipe element will be included in the inspection program.
P_{FD}	Variable (see text above for details)	Probability per inspection interval that an existing flaw will be detected. A chosen estimate is based on NDE reliability performance demonstration results and difficulty and accessibility of inspection for particular weld.
P_{LD}	Variable (0 – no leak detection to 0.9 for leak detection using current methods/technology)	Probability per detection interval that an existing leak will be detected. Estimate based on system, presence and type of leak detection system, and locations and accessibility.
T_{FI}	10 years (per ASME XI)	Flaw inspection interval, mean time between in-service inspections.
T_{LD}	Variable (1.5 – once per refueling outage / 1.92E-2 – weekly / 9.13E-4 – each shift)	Leak detection interval, mean time between leak detections. Estimate based on method of leak detection; ranges from immediate/ continuous to frequency of routine inspections for leaks (incl. hydrostatic pressure testing).
T_R	Variable (see text above for details)	Mean time to repair the affected piping element given detection of a critical flaw or leak. Estimate of time to tag out, isolate, prepare, repair, leak test and tag into service.

In addition to generating a time-dependent LOCA frequency, the Markov model provides a basis for investigating the sensitivity of LOCA frequency to different in-service inspection and leak detection strategies. The Markov model determines the inspection effectiveness factor, I, which is the ratio of the LOCA frequency with credit for inspections to that given no credit for inspections:

$$I_j = \frac{h_{25}\{inspprog'\,j'\}}{h_{25}\{noinsp\}} \tag{D.13}$$

Where:

$h_{25}\{inspprog\,'j'\}$ = hazard rate at T = 25 given inspection strategy 'j.'
$h_{25}\{noinsp\}$ = hazard rate given no inspections.

The solutions to the Markov model for time dependent hazard rates are developed in terms of closed form analytical solutions using an Excel spreadsheet. In this study the time-dependent LOCA frequencies are determined for twelve cases that are defined by varying the following parameters (Table D.19):

• Whether or not the piping segment is subjected to any ISI program;
• The extent of the ISI program ('Caused-Based' vs. 'Extensive', all encompassing);
• The inspection interval of the ISI program;
• Type and frequency of leak detection. The different leak detection methods include primary system mass balance calculations, visual observation (through video monitor), (PWR) containment sump level and flow rate monitors, airborne particulate radioactivity and gaseous radioactivity monitors, and different main control room monitors for primary system temperature, pressure, etc.

Table D.19 Inspection Cases Evaluated for Selected Pipe Segments

Leak Inspection Strategy	In-Service Inspection Strategy		
	None	Cause-Based $[P_{FD} = 0.50]$	Comprehensive $[P_{FD} = 0.90]$
None	Case 1	Case 5	Case 9
Refueling Cycle (Hydro Test)	Case 2	Case 6	Case 10
Weekly	Case 3	Case 7	Case 11
8 Hour Shift	Case 4	Case 8	Case 12

The time-dependent LOCA frequencies associated with the five Base Cases are summarized in Figures D.26 through D.40. Figure D.26 is assumed to be representative of Base Case 1; the LOCA frequency at T = 25 years is equal to the calculated point estimate of 8.24E-06 per reactor-year under an assumption of "caused-based" ISI with POD = 0.5 and leak detection (e.g., hydrostatic pressure testing prior to exiting a refueling outage). This assumption is applied to the other base cases as well (Figures D.29, D.32, D.35, and D.38).

It is noted that the service data input to the calculation is associated with piping that has been subjected to different inspection strategies. In some cases flaws have been detected fortuitously and in other cases the flaw detection has resulted from augmented IGSCC inspection programs. The results in Figure D.26 are based on an assumed 'cause-based' inspection strategy whereby the inspection sample is determined by an initial discovery of a flaw. If a flaw is found, the inspection is immediately expanded to cover other similar locations. The combination of inspection sample and rules for expanded search for flaws are sufficient to result in an average probability of detection (POD) of 0.50. The analysis also considers what in this study is termed "comprehensive" ISI, which implies 100% ISI coverage using state-of-the-art NDE technology. Such a program is assumed to result in an average probability of detection of 0.90.

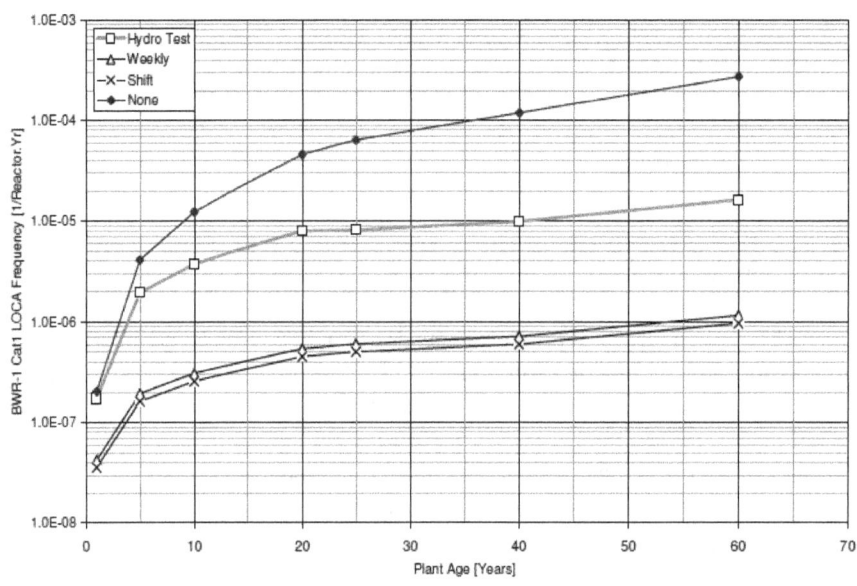

Figure D.26 Time-Dependent BWR-1 Cat 1 LOCA Frequency Given 'Cause-Based' ISI

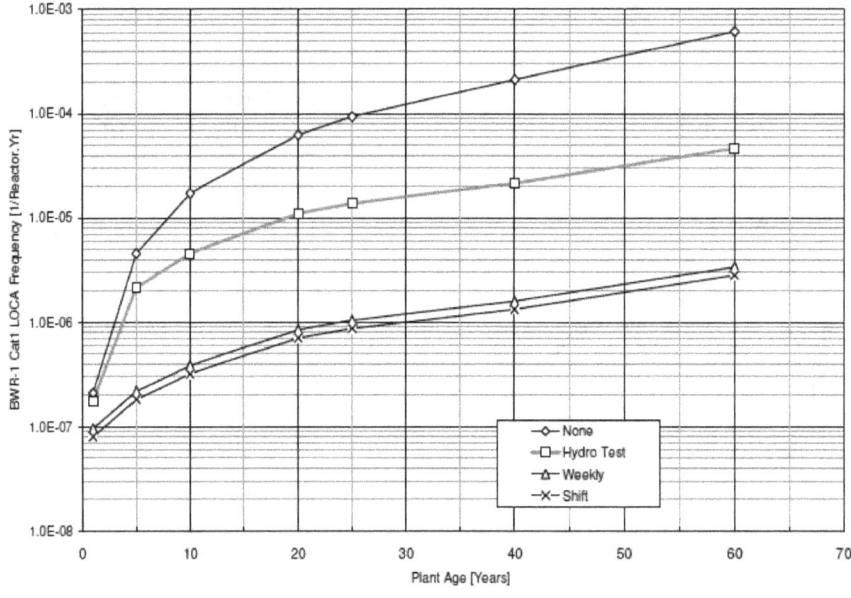

Figure D.27 Time-Dependent BWR-1 Cat 1 LOCA Frequency Assuming no ISI

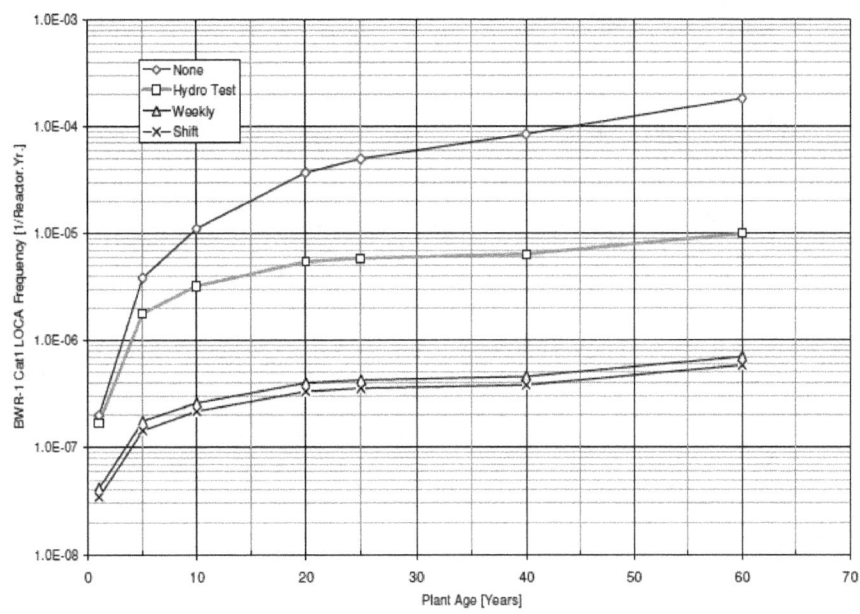

Figure D.28 Time-Dependent BWR-1 Cat 1 LOCA Frequency Given 'Comprehensive ISI'

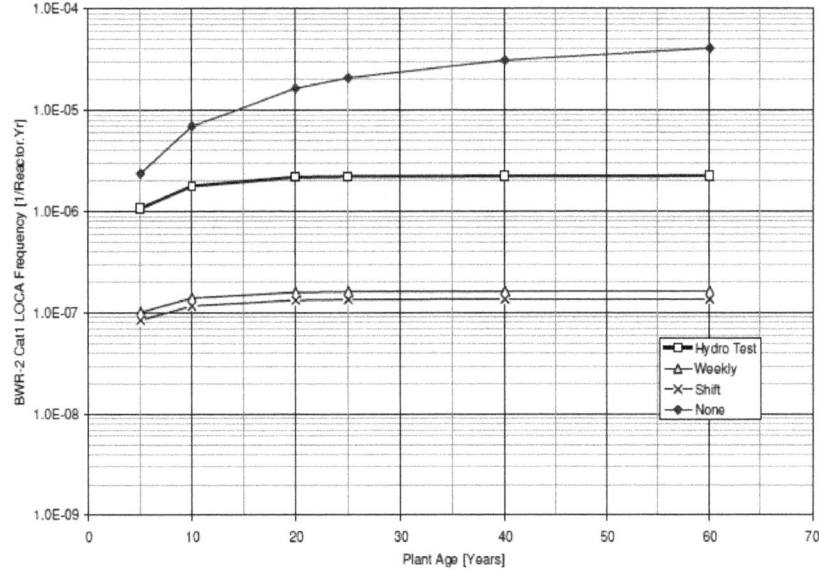

Figure D.29 Time-Dependent BWR-2 Cat 1 LOCA Frequency Given 'Cause-Based' ISI

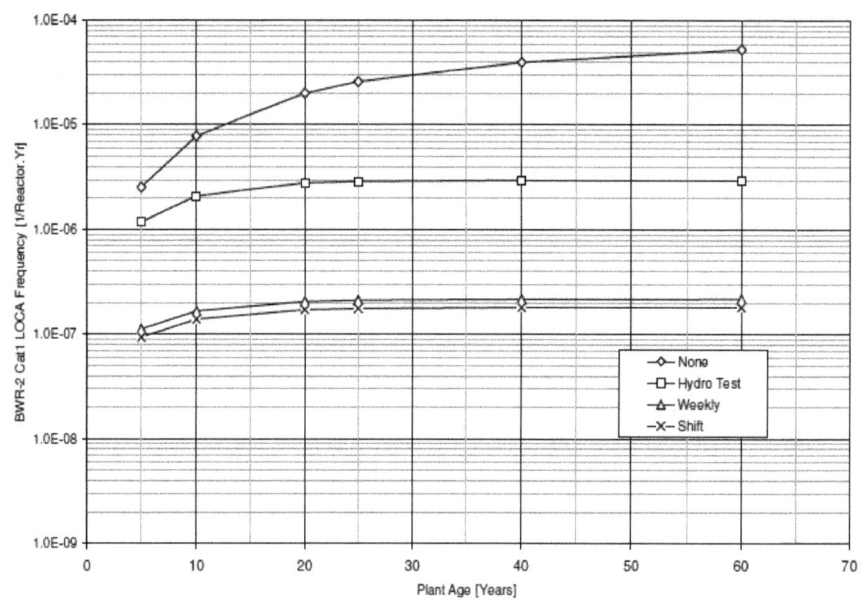

Figure D.30 Time-Dependent BWR-2 Cat 1 LOCA Frequency Assuming no ISI

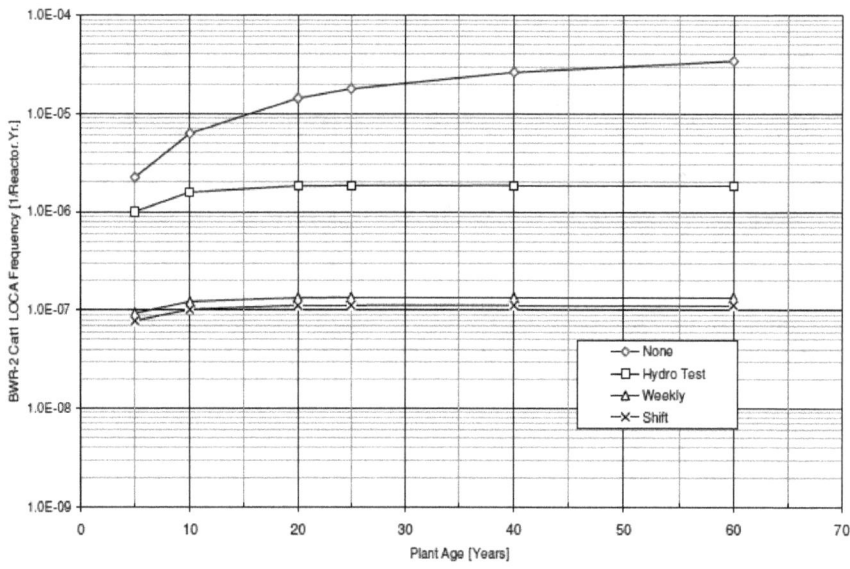

Figure D.31 Time-Dependent BWR-2 Cat 1 LOCA Frequency Given 'Comprehensive ISI'

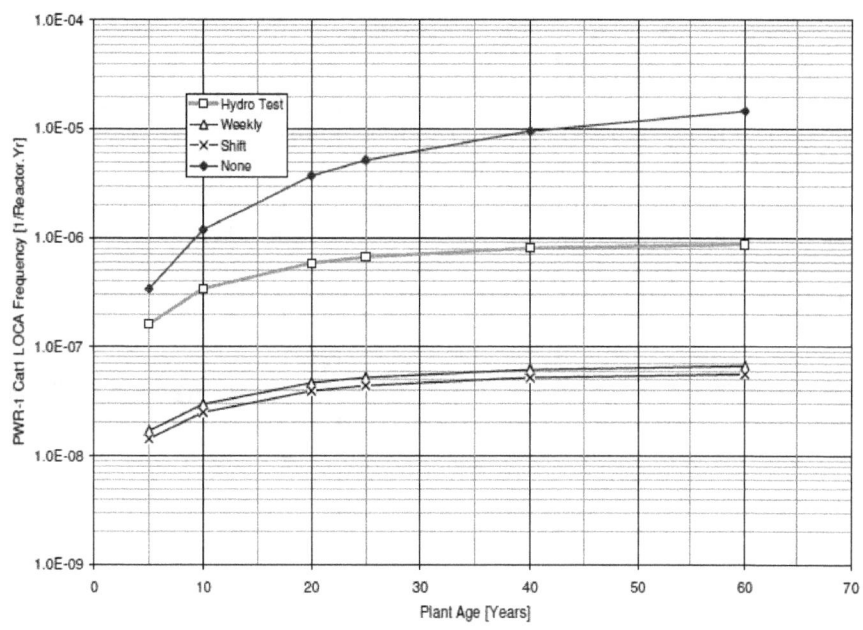

Figure D.32 Time-Dependent PWR-1 Cat 1 LOCA Frequency Given 'Cause-Based' ISI

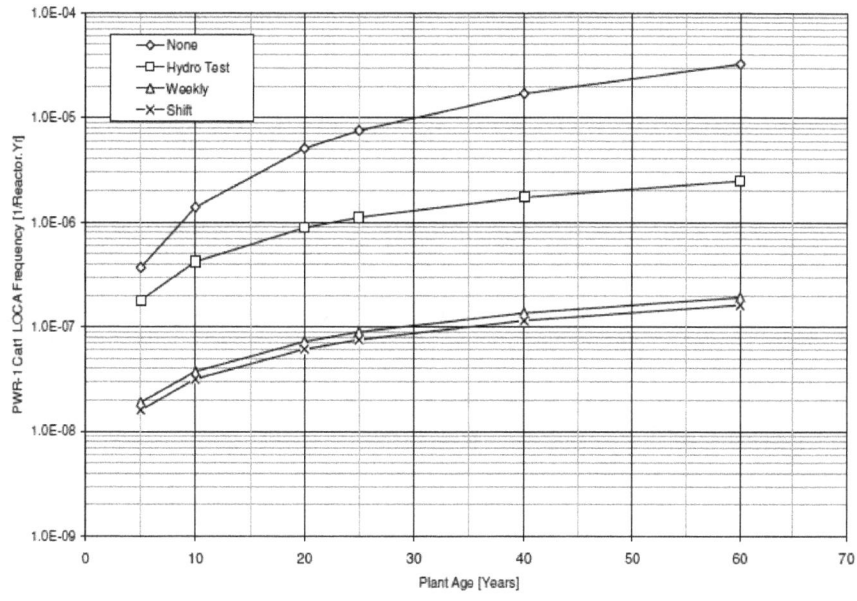

Figure D.33 Time-Dependent PWR-1 Cat 1 LOCA Frequency Assuming no ISI

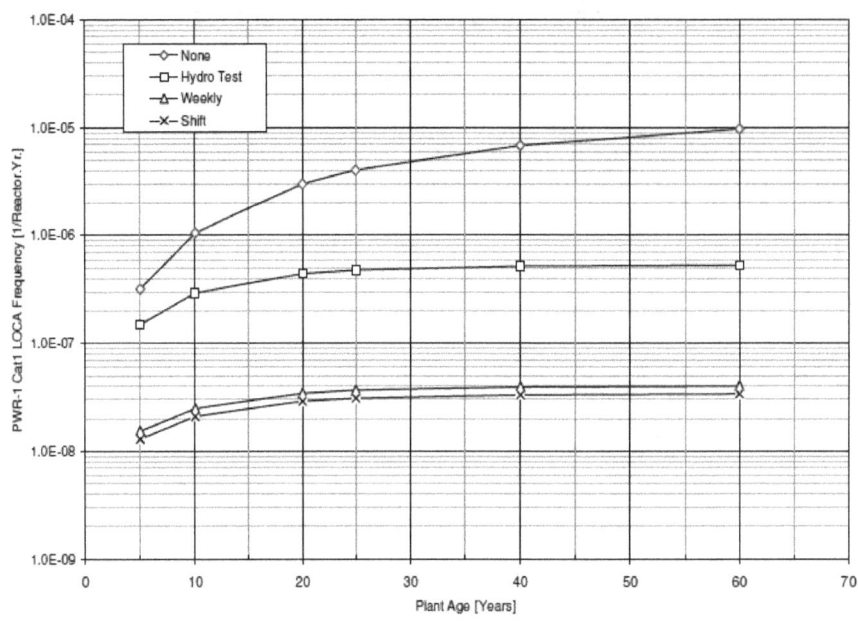

Figure D.34 Time-Dependent PWR-1 Cat 1 LOCA Frequency Given 'Comprehensive ISI'

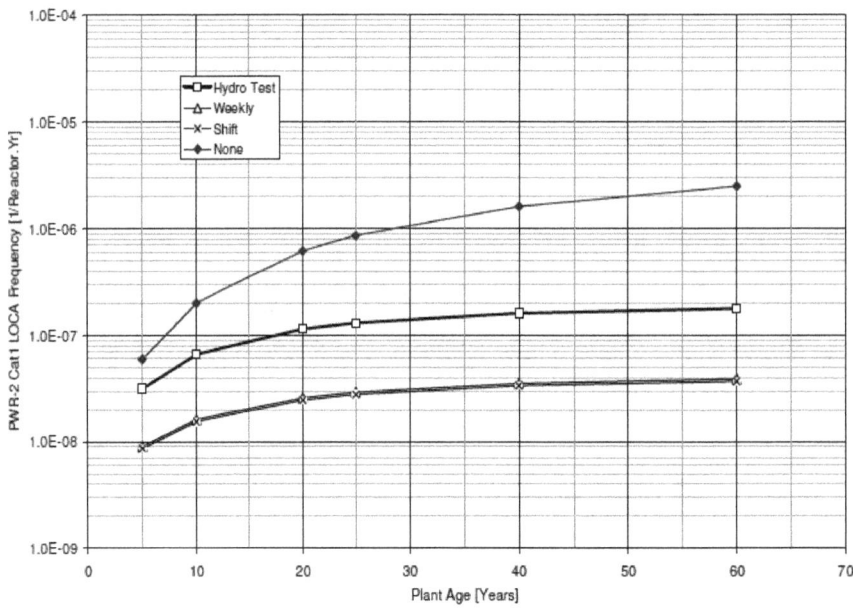

Figure D.35 Time-Dependent PWR-2 Cat 1 LOCA Frequency Given 'Cause-Based' ISI

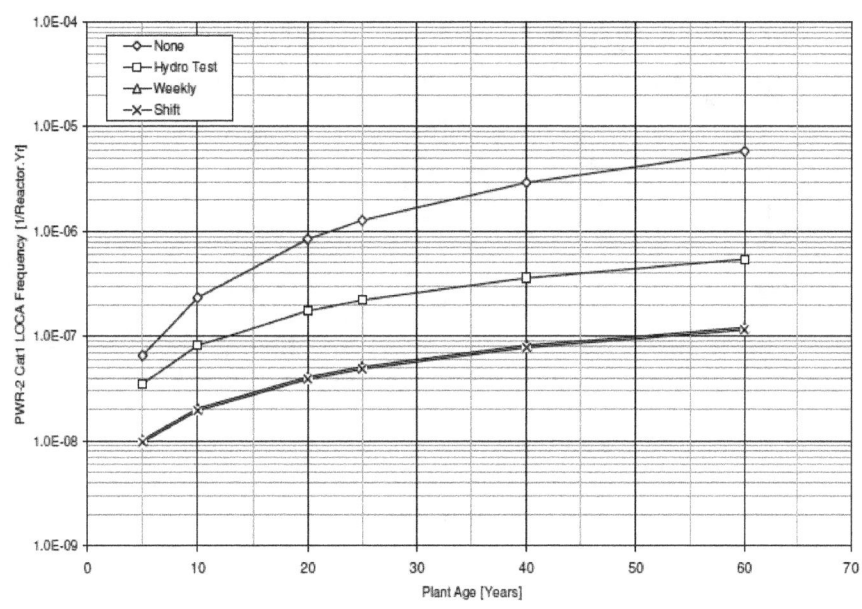

Figure D.36 Time-Dependent PWR-2 Cat 1 LOCA Frequency Assuming no ISI

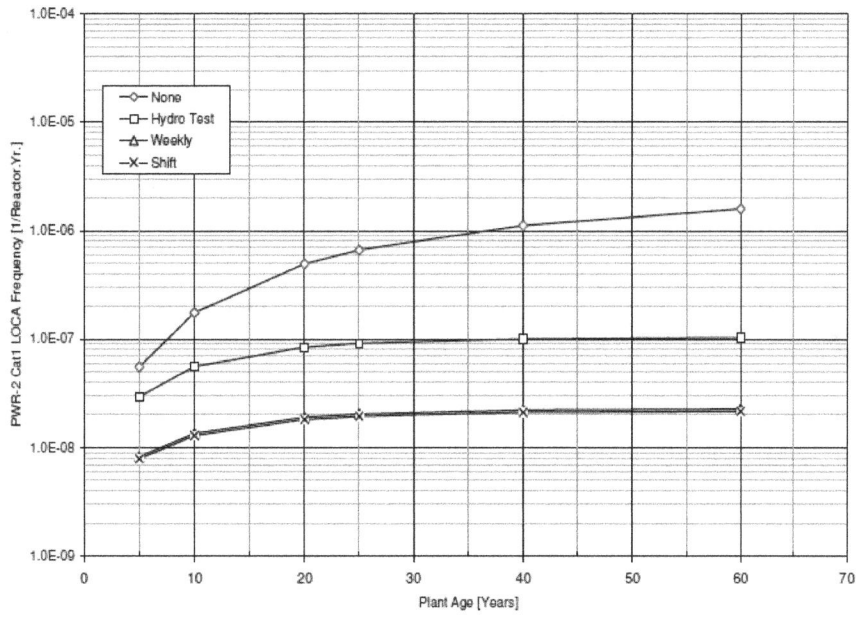

Figure D.37 Time-Dependent PWR-2 Cat 1 LOCA Frequency Given 'Comprehensive ISI'

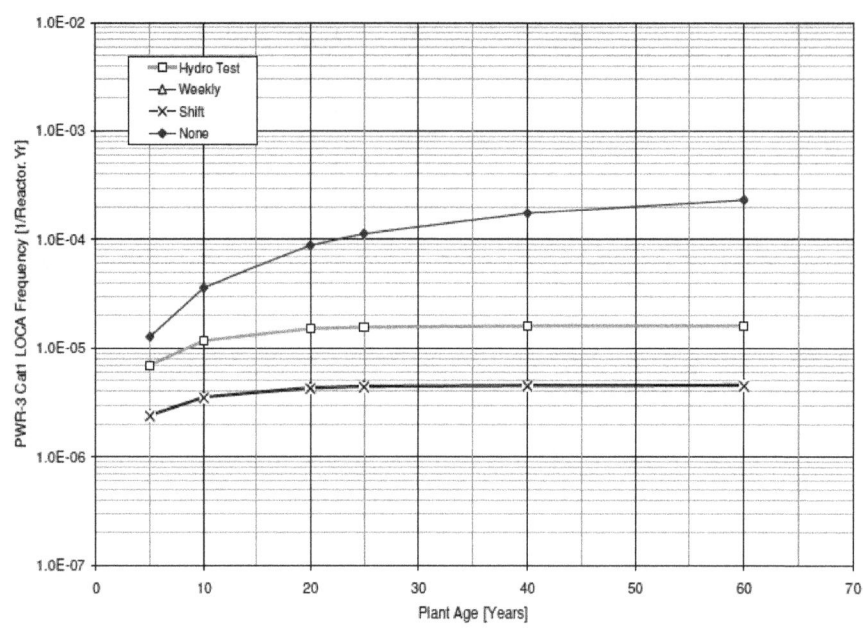

Figure D.38 Time-Dependent PWR-3 Cat 1 LOCA Frequency Given 'Cause-Based' ISI

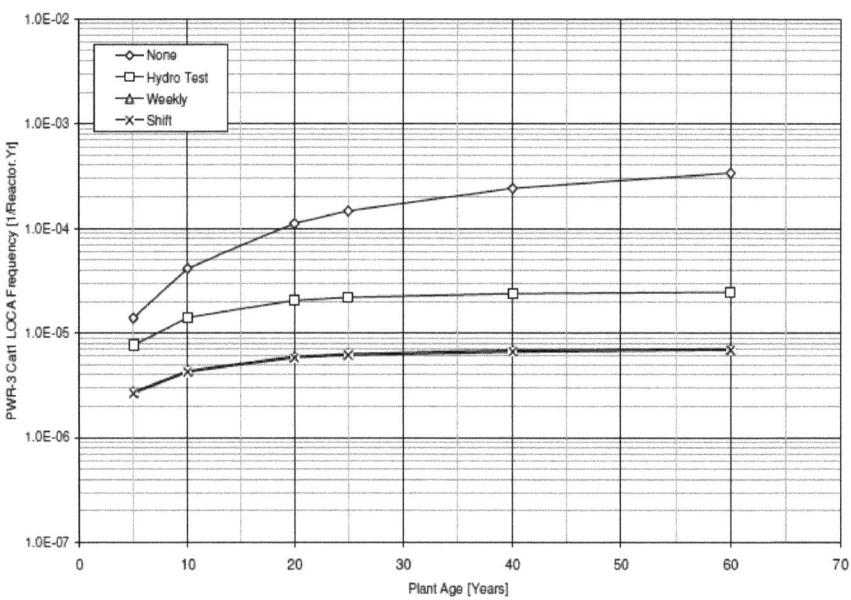

Figure D.39 Time-Dependent PWR-3 Cat 1 LOCA Frequency Assuming no ISI

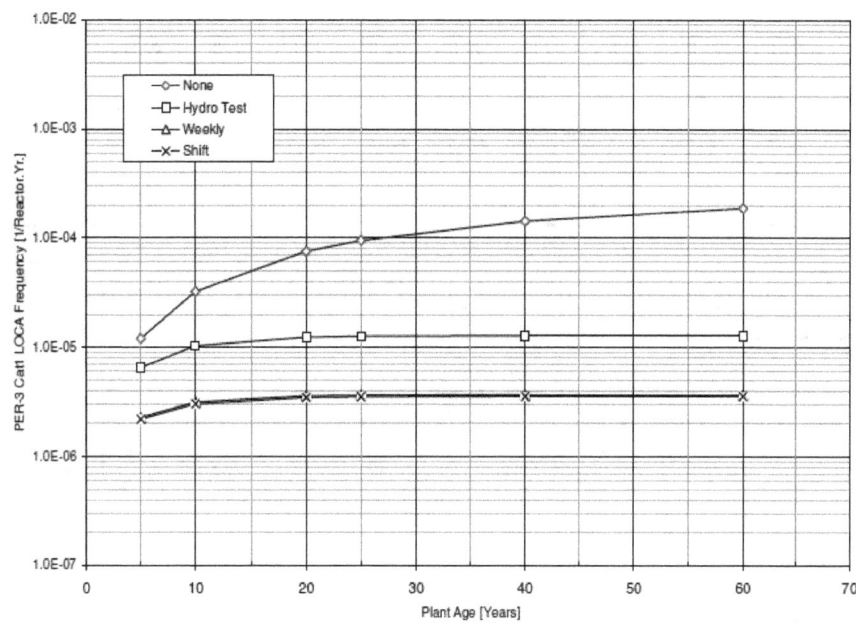

Figure D.40 Time-Dependent PWR-3 Cat 1 LOCA Frequency Given 'Comprehensive ISI'

D.6.4.2 Speculative LOCA Frequency at T = 40 & T = 60 - A retrospective evaluation is performed through a Bayesian update process whereby the exposure term in Equation 4.1 is modified to account for the longer exposure time. The analysis is performed by assuming that the service history at T = 40 and T = 60 years is known; zero (0) weld failures during the intervals $\Delta T = 15$ years (T40 to T25) and $\Delta T = 35$ years (T60 to T25). This is a purely speculative assumption implying that the ISI/NDE technologies and other piping reliability management programs remain at least as effective as at the present and that no unexpected material aging occurs. The extrapolated LOCA frequencies are summarized in Table D.20. Under the given assumptions the LOCA frequency would be expected to decrease with time.

Table D.20 Base Case LOCA Frequency Results (T = 25, 40 & 60 Years)

Base Case	LOCA Frequency – Statistical Mean [per Reactor-year]				
	Flow Rate Interval [gpm]				
	$100 < v \leq 1500$	$1500 < v \leq 5000$	$5000 < v \leq 25,000$	$25,000 < v \leq 100,000$	$100,000 < v \leq 500,000$
BWR-1, T = 25	8.24E-06	7.64E-07	3.07E-07	1.22E-07	3.05E-08
BWR-1; T = 40	2.67E-06	2.29E-07	9.14E-08	3.64E-08	1.45E-08
BWR-1; T = 60	2.44E-06	2.08E-07	8.38E-08	3.34E-08	1.34E-08
BWR-2, T = 25	2.21E-06	2.11E-07	8.40E-08	3.36E-08	7.33E-09
BWR-2, T = 40	2.07E-06	2.03E-07	8.05E-08	3.13E-08	6.61E-09
BWR-2, T = 60	1.87E-06	1.85E-07	7.35E-08	2.97E-08	6.09E-09
PWR-1, T = 25	6.65E-07	4.87E-08	1.83E-08	6.99E-09	2.55E-09
PWR-1, T = 40	2.14E-07	1.49E-08	6.10E-09	2.24E-09	8.14E-10
PWR-1, T = 60	1.19E-07	8.34E-09	3.38E-09	1.26E-09	4.62E-10
PWR-2, T = 25	1.14E-07	9.60E-09	3.84E-09	1.44E-09	5.31E-10
PWR-2, T = 40	1.07E-07	9.22E-09	3.68E-09	1.34E-09	4.79E-10
PWR-2, T = 60	9.67E-08	8.31E-09	3.36E-09	1.27E-09	4.41E-10
PWR-3, T = 25	1.60E-05	2.33E-06	9.22E-07	N/A	N/A
PWR-3, T = 40	1.08E-05	1.58E-06	6.31E-07	N/A	N/A
PWR-3, T = 60	8.23E-06	1.20E-06	4.81E-07	N/A	N/A

Note 1: PWR-1 in this table accounts for 3-of-3 hot legs.
Note 2: PWR-3 in this table accounts for 2-of-2 HPI/NMU lines.

D.6.5 Influence of Service Data on LOCA Frequency

The LOCA frequencies in this Base Case Report are derived from service data on Code Class 1 piping. In this section we investigate how the LOCA frequencies relate to two data issues: 1) completeness of the pipe failure data collection, and 2) data interpretations. The former remains an ever-present issue in probabilistic safety assessment. Completeness is addressed by having in place an active and rigorous data collection process (*c.f.* Appendix A). Two sensitivity cases (SC:s) are defined to demonstrate how changes in the input to the failure rate calculations affect the estimated LOCA frequency. The sensitivity cases are defined as:

1. <u>SC1</u>: A small leak (≤ T.S. limit for unidentified RCPB-leakage) is assumed to have occurred in a pipe-to-safe-end weld in a BWR NPS28 reactor recirculation pipe during the time period 1988 – 2002. This evidence is used to modify the posterior weld failure rates.

2. <u>SC2</u>: This sensitivity case is concerned with an assumed large leak (= Cat0 LOCA) in a NPS28 BWR reactor recirculation pipe. Again, the large leak is assumed to have occurred in the time period 1988 – 2002. This evidence is used to modify the posterior weld failure rates and the conditional failure probability.

The results of the sensitivity analysis are summarized in Table D.21. These sensitivity cases are hypothetical in that they do not account for effects on piping reliability by the anticipated industry and regulatory actions that invariably would arise in response to the results of root cause analysis to determine the reasons behind a significant RCPB degradation such as defined by SC1 or SC2.

Table D.21 BWR LOCA Frequency Sensitivity Analysis Results

Base Case	LOCA Frequency – Statistical Mean [per Reactor-year]				
	Flow Rate Interval [gpm]				
	Cat1 $100 < v \leq 1500$	Cat2 $1500 < v \leq 5000$	Cat3 $5000 < v \leq 25,000$	Cat4: $25,000 < v \leq 100,000$	Cat5: $100,000 < v \leq 500,000$
Base-1	8.24E-06	7.64E-07	3.07E-07	1.22E-07	3.05E-08
Base-1 – SC1	8.70E-06	8.07E-07	3.27E-07	1.29E-07	3.29E-08
Base-1 – SC2	1.30E-05	1.17E-06	4.77E-07	1.87E-07	5.63E-07

D.6.6 Service Data and Conditional Failure Probabilities

There is no service data associated with Cat0 LOCA events. Therefore, the estimation of conditional failure probabilities is based on zero-failure statistics. Since not all flaws propagate through-wall if left unattended, an alternative to the approach in Section D.5.2 (constrained noninformative prior) would be to use Jeffrey's noninformative prior and to assume all flaws (non-through wall and through wall) as pressure boundary integrity challenges. The result would be conditional failure probabilities that are closely approximated by the power law (Equation D.8), however. It is acknowledged that this is just one way of representing the current state-of-knowledge with respect to gross Code Class 1 pipe failure. It is not a physical model of flaw propagation given its interactions with certain loading conditions and pipe stresses.

D.7 Summary of Results

An application of a parametric attribute/influence method has yielded results as summarized in this section. Central to the method is the processing and interpretation of service data on Code Class 1 piping. A Markov model of piping reliability is used to develop time-dependent LOCA frequencies.

D.7.1 Discussion of Assumptions

A parametric attribute/influence method is applied to five base cases. Three types of assumptions are made in the analysis; global assumptions (applicable to all five base cases), BWR-specific assumptions and PWR-specific assumptions:

Global Assumptions

- Pipe failure results from observable degradation mechanisms and loading conditions. A statistical evaluation of service experience data therefore provides a sufficiently accurate basis for piping reliability analysis.

- The PIPExp database is of sufficient completeness and depth to support an application of the parametric attribute/influence methodology. This database addresses piping performance in response to both anticipated and unanticipated loading conditions.

- The effect on piping reliability from pressure, deadweight, weld residual stresses, thermal loading, and thermal stratification is implicitly accounted for in the PIPExp database. This database also accounts for the effects from inadvertent over-pressurization and relief valve actuation, water hammer and seismic[12] events.

---

[12] The database includes a single event involving the fracture of a small-diameter steam line due to seismic event (Fukushima-Daiichi Unit 6 on 07-21-2000).

- The BWR-specific LOCA frequencies are assumed to be representative of a plant with IGSCC Category D and E welds operating with normal water chemistry (NWC). The pipe failure database includes plants with hydrogen water chemistry (HWC) and NWC. This study did not differentiate between plants with weld overlays and HWC versus plants with weld overlays and NWC, however. This study shows improved water chemistry together with weld reinforcements to lower the weld failure rates by about a factor of ten (10).

- Because of service conditions and piping arrangements, flow accelerated corrosion (FAC) is not viewed as a significant degradation mechanism affecting Code Class 1 feedwater piping. Degradation involving wall thinning is therefore not viewed as having an effect on the time-dependent LOCA frequency.

PWR-Specific Assumptions

- The estimation of RC-HL weld failure rates is based on the assumption that the observed (in 4th quarter 2000) weld degradation at V.C. Summer is a circumferential flaw in the RPV nozzle-to-safe-end weld. This assumption is believed to result in an over-estimation of the actual weld failure rate.

- Relative to PWRs of Westinghouse design, the pipe failure database includes no records on through-wall flaws in large-diameter pressurizer surge line welds. The analysis assumes that the piping is susceptible to thermal fatigue of sufficient magnitude to potentially cause a flaw in the through-wall direction.

D.7.2 Summary of Input Data and Results

Tables D.22 and D.23 summarize the input data to the LOCA frequency calculation. Tables D.24 through D.26 give the results at T = 25, 40 and 60 years, respectively. Consistent with the LOCA frequency elicitation structure, Table D.27 is summary of mid values (MV, or 50th percentiles) at T = 25 years rather than mean values, however. Figure D.41 shows the time-dependent LOCA frequencies. Figure D.42 shows selected weld failure rates. Figures D.43 through D.46 show the contribution to LOCA frequency by respective Base Case. Note that Figure D.45 includes the contribution to LOCA frequency by PWR-1 (Reactor Coolant System Hot Legs; all 3 loops are accounted for in this figure) and PWR-2 (Pressurizer Surge Line). Note that the Base Case results used in Table 4.1 in the main body can be obtained from Tables D.16, D.17, and D.20 in this report.

Table D.22 Summary of Key BWR Base Case Input Data

Input Data	Base Case					
	BWR-1			BWR-2		
	NPS12	NPS22	NPS28	NPS12	NPS14	NPS20
Weld count	50	16	56	63	5	53
Weld failure rate Dominant [1/Reactor-yr.]	6.50E-05	1.54E-04	1.44E-04	2.20E-06	2.20E-06	1.58E-06
Weld failure rate Minimum [1/Reactor-yr.]	2.37E-05	3.32E-05	1.29E-05	1.77E-07	1.77E-07	1.73E-07

Table D.23 Summary of Key PWR Base Case Input Data

Input Data	Base Case		
	PWR-1	PWR-2	PWR-3
	NPS30	NPS14	NPS3-¾
Weld count	50	14	9
Weld failure rate Dominant [1/Reactor-yr.]	7.64E-05	1.56E-06	6.56E-04
Weld failure rate Minimum [1/Reactor-yr.]	1.05E-06	4.60E-08	1.58E-06

Table D.24 Calculated LOCA Frequencies (T = 25 Years)

Base Case	LOCA Frequency – Statistical Mean [per Reactor-year]					
	Flow Rate Threshold Value [gpm]					
	Cat1 $v > 100$	Cat2 $v > 1,500$	Cat3 $v > 5,000$	Cat4 $v > 25,000$	Cat5 $v > 100,000$	Cat6 $v > 500,000$
BWR-1[13]	9.46E-06	1.22E-06	4.60E-07	1.53E-07	3.05E-08	N/A[14]
BWR-2[15]	2.54E-06	3.36E-07	1.25E-07	4.09E-08	7.33E-09	N/A
PWR-1[16]	7.42E-07	7.62E-08	2.93E-08	1.09E-08	3.77E-09	1.26E-09
PWR-2	1.29E-07	1.50E-08	5.40E-09	1.56E-09	5.31E-10	N/A
PWR-3[17]	1.60E-05	2.32E-06	9.22E-07	N/A	N/A	N/A

Table D.25 Calculated LOCA Frequencies (T = 40 Years)

Base Case	LOCA Frequency – Statistical Mean [per Reactor-year]					
	Flow Rate Threshold Value [gpm]					
	Cat1 $v > 100$	Cat2 $v > 1,500$	Cat3 $v > 5,000$	Cat4 $v > 25,000$	Cat5 $v > 100,000$	Cat6 $v > 500,000$
BWR-1	1.14E-05	1.47E-06	5.54E-07	1.84E-07	3.78E-08	N/A
BWR-2	2.56E-06	3.39E-07	1.26E-07	4.13E-08	7.40E-09	N/A
PWR-1	8.96E-07	9.20E-08	3.54E-08	1.32E-08	4.55E-09	1.45E-09
PWR-2	1.60E-07	1.86E-08	6.70E-09	1.93E-09	6.59E-10	N/A
PWR-3	1.95E-05	3.30E-06	9.44E-07	N/A	N/A	N/A

Table D.26 Calculated LOCA Frequencies (T = 60 Years)

Base Case	LOCA Frequency – Statistical Mean [per Reactor-year]					
	Flow Rate Threshold Value [gpm]					
	Cat1 $v > 100$	Cat2 $v > 1,500$	Cat3 $v > 5,000$	Cat4 $v > 25,000$	Cat5 $v > 100,000$	Cat6 $v > 500,000$
BWR-1	1.88E-05	2.43E-06	9.16E-07	3.05E-07	6.07E-08	N/A
BWR-2	2.56E-06	3.39E-07	1.26E-07	4.13E-08	7.40E-09	N/A
PWR-1	9.74E-07	1.00E-07	3.85E-08	1.43E-08	4.95E-09	1.57E-09
PWR-2	1.77E-07	2.06E-08	7.41E-09	2.14E-09	7.29E-10	N/A
PWR-3	1.96E-05	3.32E-06	9.50E-07	N/A	N/A	N/A

[13] BWR-1 is the combination of RR Loop A and B.
[14] N/A = not applicable.
[15] The results are for FW Loop A and B.
[16] The results are for 3-of-3 RC hot legs.
[17] The results are for 2-of-2 HPI/NMU lines.

Table D.27 BWR and PWR LOCA Frequency Elicitation Anchor (MV) Values (T = 25 Years)

Base Case	Median (MV) LOCA Frequency [per Reactor-year]					
	Flow Rate Threshold Value [gpm]					
	Cat1 v > 100	Cat2 v > 1,500	Cat3 v > 5,000	Cat4 v > 25,000	Cat5 v > 100,000	Cat6 v > 500,000
BWR-1	8.23E-06	1.08E-06	4.03E-07	1.29E-07	2.19E-08	N/A
BWR-2	1.09E-06	1.35E-07	5.03E-08	1.65E-08	2.10E-09	N/A
PWR-1	1.54E-07	2.25E-08	8.33E-09	2.85E-09	8.53E-10	1.58E-10
PWR-2	1.37E-08	1.39E-09	5.15E-10	1.54E-10	5.46E-11	N/A
PWR-3	6.87E-06	1.15E-06	2.14E-07	N/A	N/A	N/A

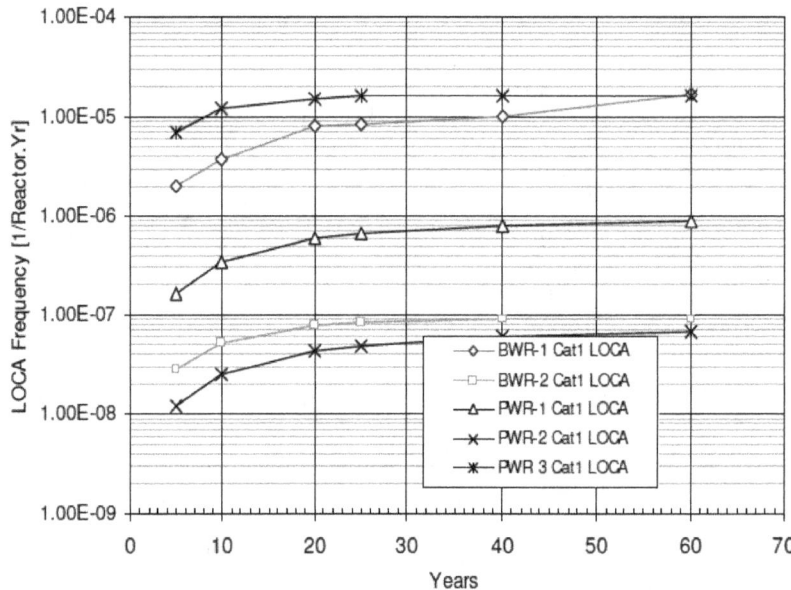

Figure D.41 Time-Dependent Cat 1 LOCA Frequencies

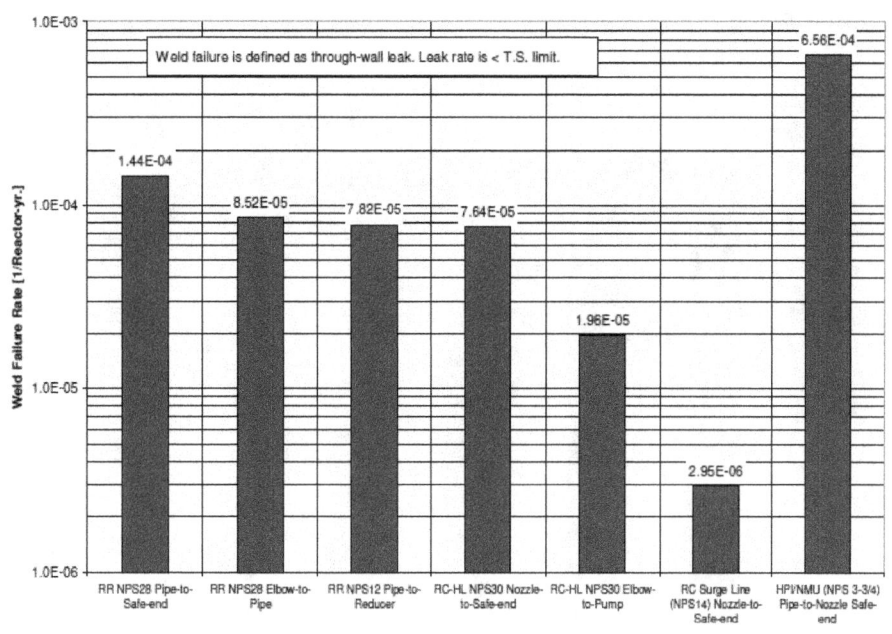

Figure D.42 Selected Base Case Weld Failure Rates

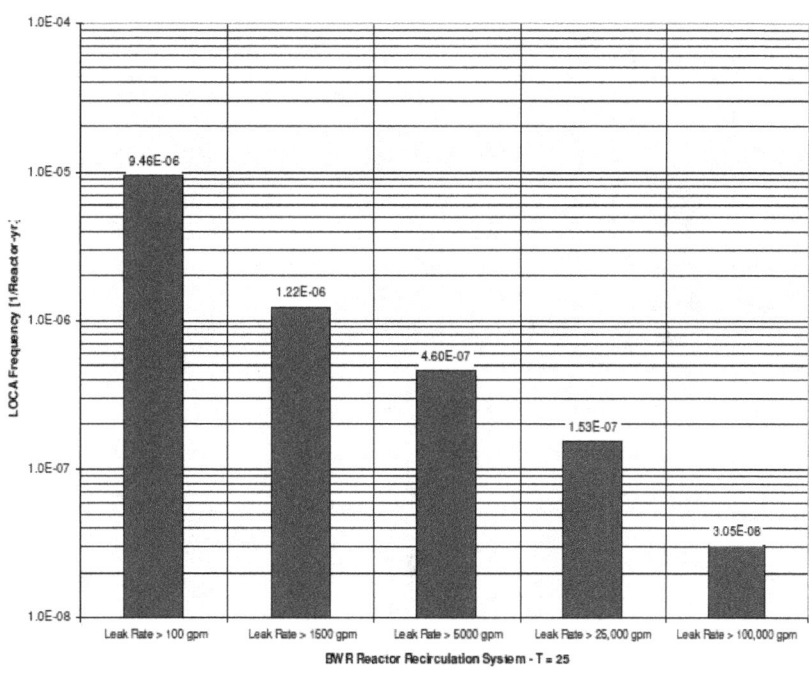

Figure D.43 BWR-1 LOCA Frequency

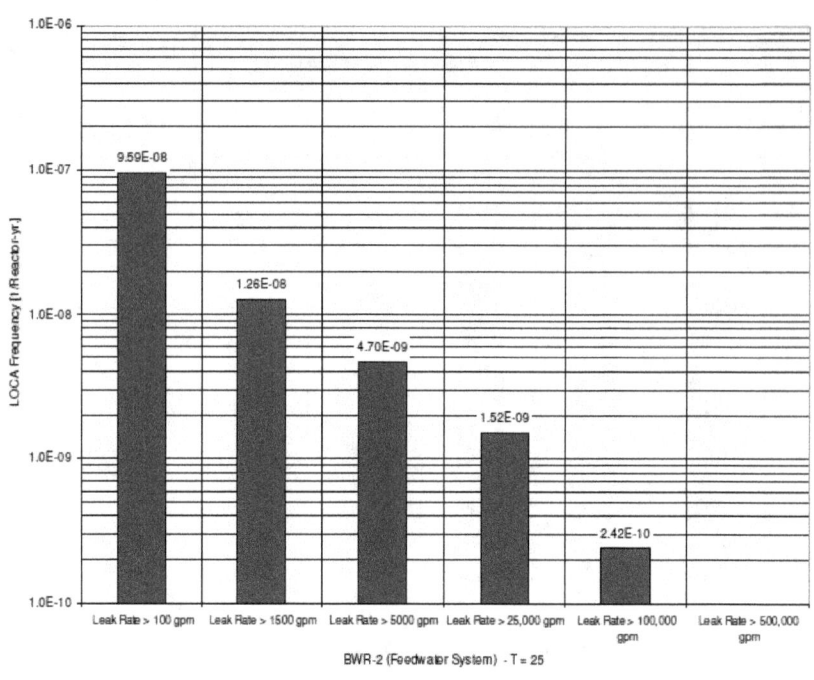

Figure D.44 BWR-2 LOCA Frequency (Feedwater System Loop A & B)

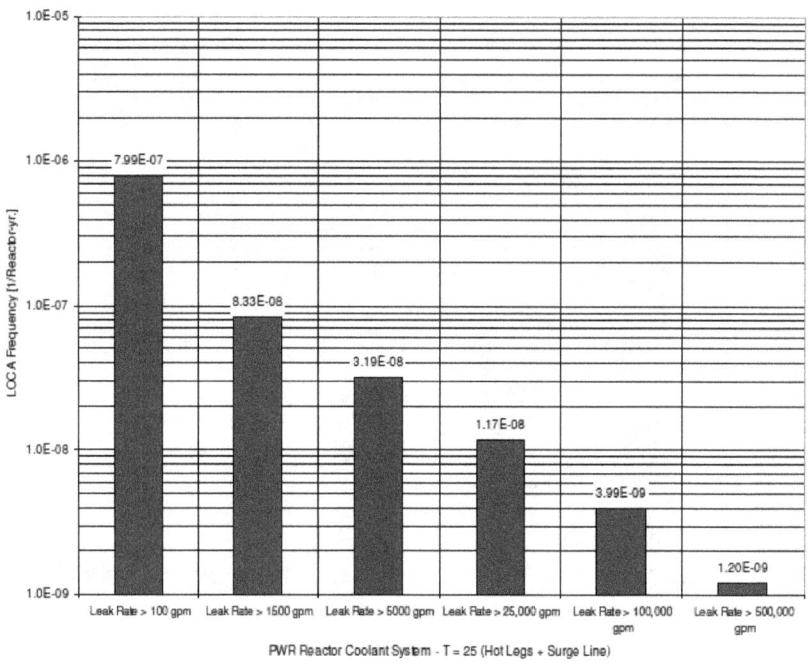

Figure D.45 PWR-1 and PWR-2 LOCA Frequency

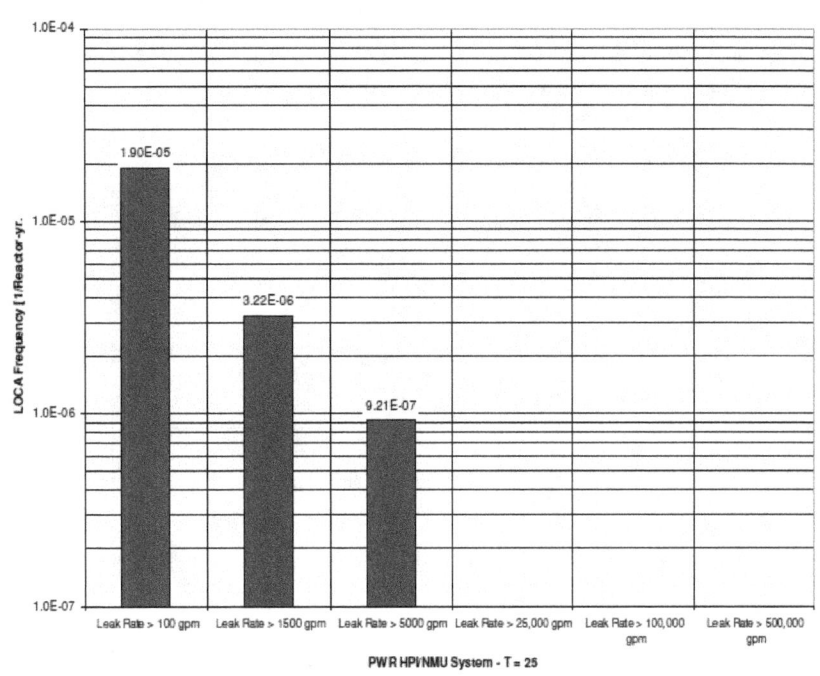

Figure D.46 PWR-3 LOCA Frequency (ASME Code Class 1 HPI/NMU System)

Figure D.47 displays the results of a sensitivity analysis associated with the BWR Base Cases. It is concerned with the influence by service data on the calculated LOCA frequency.

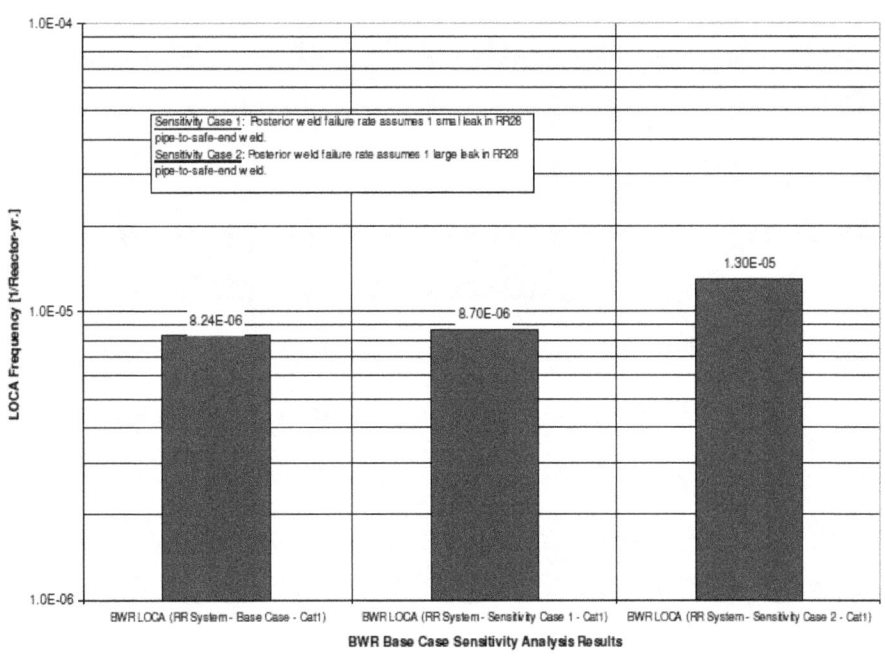

Figure D.47 Selected BWR Base Case Sensitivity Analysis Results – Cat 1 LOCA

Figures D.48 (BWR) and D.49 (PWR) show the influence of in-service inspection on the time-dependent LOCA frequency; no ISI and ISI with POD = 0.5 and 0.9, respectively.

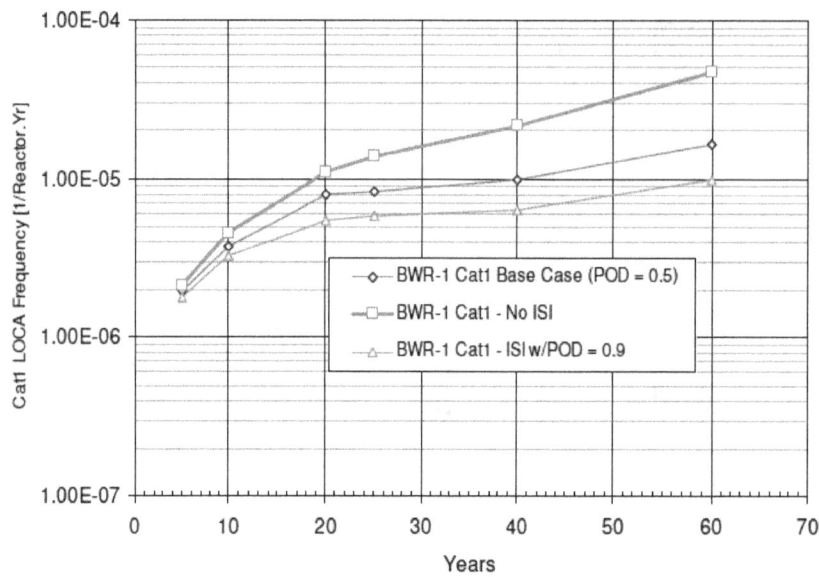

Figure D.48 Influence of ISI on Time-Dependent BWR-1 Cat 1 LOCA Frequency

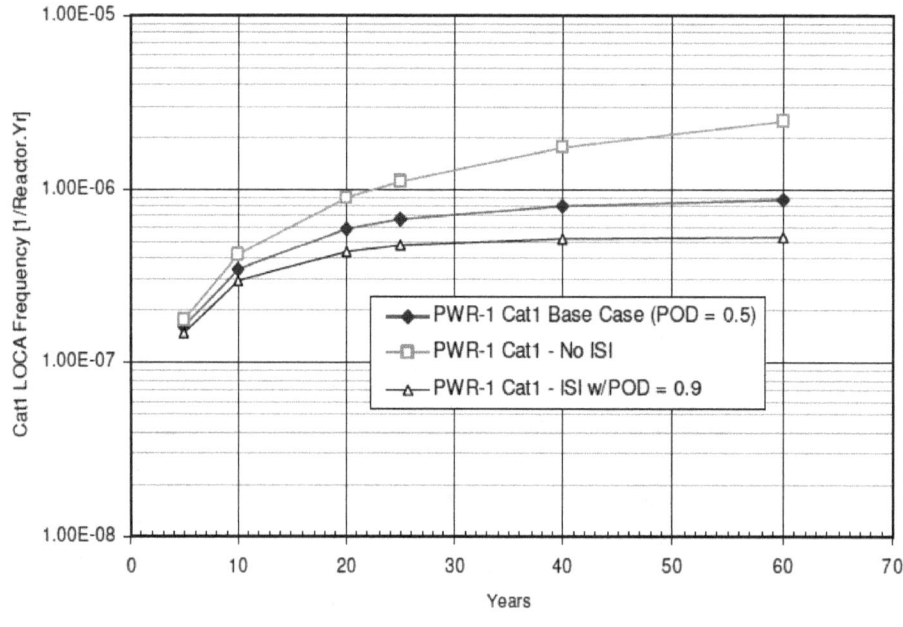

Figure D.49 Influence of ISI on Time-Dependent PWR-1 Cat 1 LOCA Frequency

D.7.3 Benchmarking

A limited scope benchmarking exercise was performed to compare predicted weld failure rates with the reported service experience. The benchmarking was limited to NPS12 BWR reactor recirculation welds susceptible to IGSCC. Probabilistic fracture mechanics (PFM) calculations using the WinPRAISE computer code generated predictions about the weld failure rate for different assumptions about the normal operating stresses (σ_{NO}).[18] Bayesian reliability analysis was used to derive weld failure rates from service experience data. Figure D.50 shows the results of the benchmarking exercise. Table D.28 includes a description of the different cases of the benchmarking exercise.

Table D.28 Benchmarking of WinPRAISE Versus Service Experience

Case	Definition
BOYL (PIPExp)	Table D.13 (this report). NPS12 Reactor Recirculation pipe-to-reducer weld with weld overlay. T = 25 years. This weld configuration has the highest predicted failure rate.
DOH-1 (D.O. Harris)	NPS12 reactor recirculation system weld with normal operating stress, σ_{NO} = 10 ksi[19]
DOH-2	NPS12 reactor recirculation system weld; σ_{NO} = 12 ksi
DOH-3	NPS12 reactor recirculation system weld; σ_{NO} = 15 ksi
DOH-4	NPS12 reactor recirculation system weld; σ_{NO} = 20 ksi

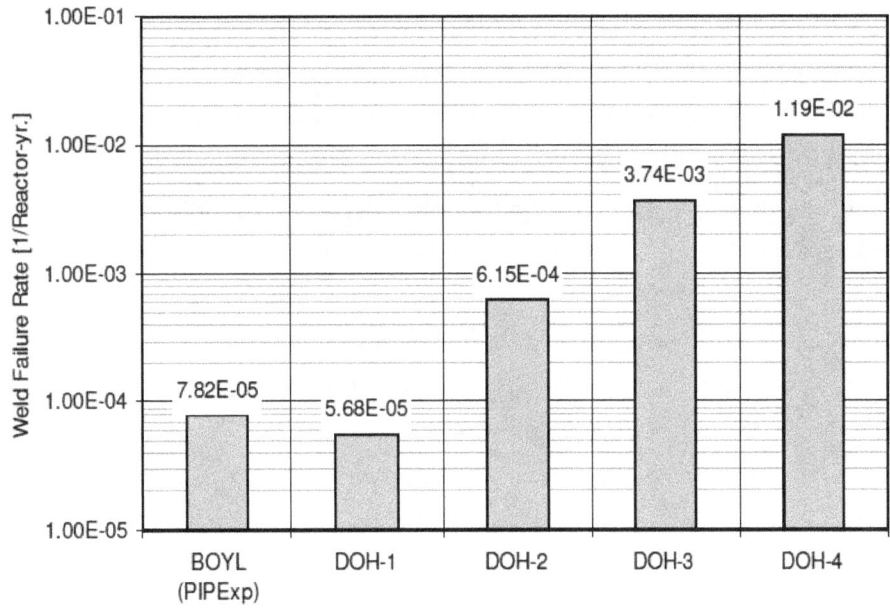

Figure D.50 Predicted (WinPRAISE) Versus Observed Weld Failure Rate (PIPExp)

---

[18] D.O. Harris, "Progress in Benchmarking SCC for 12 Inch Recirculation Line, July 1, 2003.

[19] 1 ksi = 6.9 MPa

D.7.4 Comparison to Historical LOCA Frequency Estimates

Figures D.51 and D.52 compare the Base Case results to historical LOCA frequency estimates. Direct (one-to-one) comparisons are not feasible due to different LOCA definitions and estimation approaches. Listed below are the selected BWR and PWR LOCA frequency references.

BWR Large (≥ Cat3) LOCA Frequency Estimates (Figure D. 51)

- SKI 98:30 (FW/RR); the displayed value range is taken from Reference [D.18]. It excludes contribution from thermal fatigue in Code Class 1 feedwater system piping. The feedwater system design is unique to the pilot plant in SKI Report 98:30 and it is therefore not applicable to BWR-2.

- NUREG/CR-5750 (Appendix J) provides recommended pipe LOCA frequencies. The given value range accounts for all Code Class 1 pipe failure contributions.

- GRS-98 is a probabilistic safety assessment of the German plant Gundremmingen; a BWR plant designed and built by Kraftwerk Union. This reactor design has no external recirculation loops; the given LOCA frequency value range accounts only for contributions from Code Class 1 feedwater pipe failure.

- BFN-1 (NUREG/CR-2802) is the 1982 probabilistic safety assessment of Browns Ferry Unit 1 performed as part of the NRC-sponsored Interim Reliability Evaluation Program. The given LOCA frequency value range accounts for Reactor Recirculation pump suction piping failure.

PWR Large (≥ Cat3) LOCA Frequency Estimates (Figure D. 52)

- NUREG/CR-5750 (Appendix J) provides recommended pipe LOCA frequencies. The given value range accounts for all Code Class 1 pipe failure contributions.

- Surry-1 (1990 Expert Elicitation). Surry-1 is a 3-loop Westinghouse reactor, similar to the PWR-1/PWR-2 reference design. The given LOCA frequency value range applies to RCS pipe failure and resulted from a NRC-sponsored expert elicitation.[20]

- EPRI TR-100380 (Piping Failures in U.S. Commercial Nuclear Power Plants, 1992) includes recommended BWR and PWR LOCA frequencies that are based on statistical analysis of service data. The given LOCA frequency value range accounts for all Code Class 1 pipe failure contributions.

---

[20] See for example T.V. Vo et al (1991). "Estimates of Rupture Probabilities for Nuclear Power Plant Components: Expert Judgment Elicitation," Nuclear Technology, **96**:259-270.

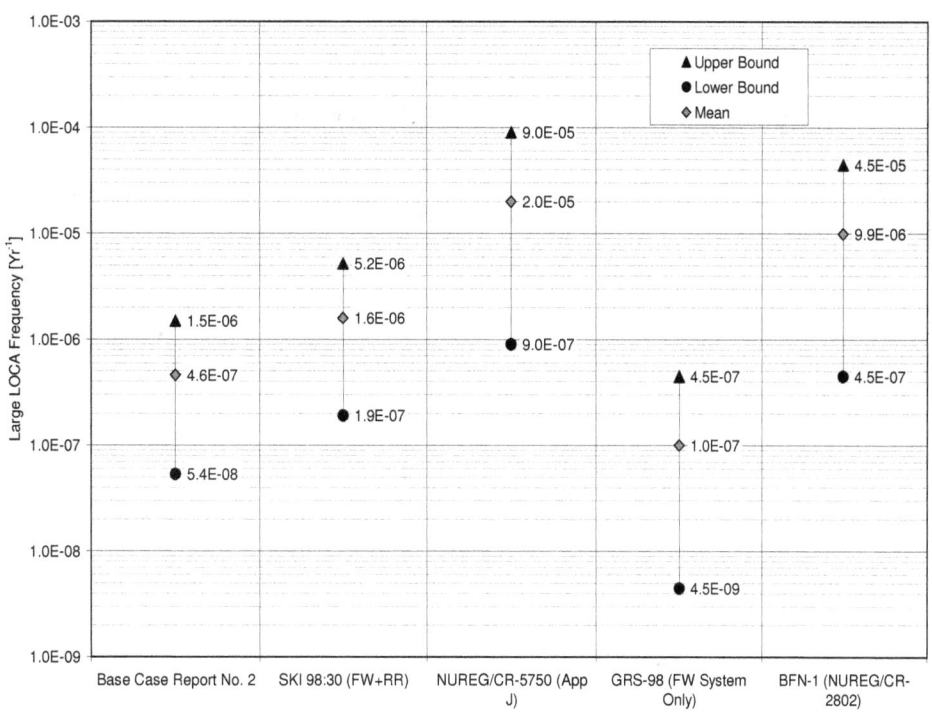

Figure D.51 Comparison of Selected BWR Large LOCA Frequency Estimates

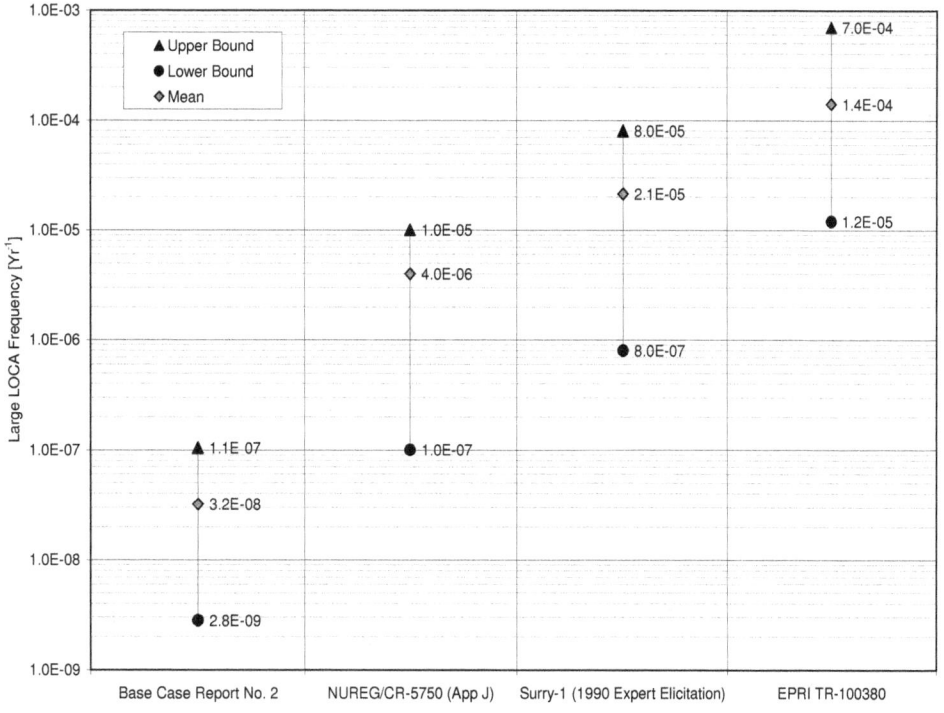

Figure D.52 Comparison of Selected PWR Large LOCA Frequency Estimates

D.8 References

D.1 Scott, P., 2003. Meeting Notes from U.S. NRC LOCA Elicitation Kick-Off Meeting, Rockville (MD), February 4-6, 2003.

D.2 Zigler, G. et al, 1995. Parametric Study of the Potential for BWR ECCS Strainer Blockage Due to LOCA Generated Debris, NUREG/CR-6224, U.S. Nuclear Regulatory Commission, Washington (DC).

D.3 U.S. Nuclear Regulatory Commission, 1988. NRC Position on IGSCC in BWR Austenitic Stainless Steel Piping, Generic Letter 88-01 (January 25, 1988), Washington (DC).

D.4 U.S. Nuclear Regulatory Commission, 1992. NRC Position on IGSCC in BWR Austenitic Stainless Steel Piping, Generic Letter 88-01, Supplement 1 (February 4, 1992), Washington (DC).

D.5 Structural Integrity Associates, Inc., 2001. Degradation Mechanisms Evaluation for Class 1 Piping at Plant B Nuclear Station, File No. EPRI-156-310.

D.6 Structural Integrity Associates, Inc., 2001. Degradation Mechanisms Evaluation for the Class 1 (Category B-J/B-F) Piping at Plant A Nuclear Station, File No. EPRI-156-310.

D.7 Shah, V.N. et al, 1998. Assessment of Pressurized Water Reactor Primary System Leaks, NUREG/CR-6582, U.S. Nuclear Regulatory Commission, Washington (DC).

D.8 Boman, B.L. et al, 2000. "Evaluation of Oconee-2 High Pressure Injection/Normal Makeup (HPI/NMU) Line Weld Failure," Assessment Methodologies for Preventing Failure: Service Experience and Environment Considerations, PVP-Vol. 410-2, American Society of Mechanical Engineers, New York (NY), ISBN: 0-7918-1891-8, pp 111-117.

D.9 Deardorff, A.F., 2001. Interim Thermal Fatigue Management Guideline (MRP-24), 1000701, Electric Power Research Institute, Palo Alto (CA).

D.10 Trolle, M., 1996. Status för gjutet rostfritt stål i äldre svenska kärnkraftverk, mars 1996 (Status of Cast Austenitic Stainless Steel in Older Swedish Nuclear Power Plants), SKI Report 96:26, Swedish Nuclear Power Inspectorate, Stockholm (Sweden).

D.11 Nyman, R. et al, 2003. Reliability of Piping System Components. Framework for Estimating Failure Parameters from Service Data, SKI Report 97:26 (2nd Edition), Swedish Nuclear Power Inspectorate, Stockholm (Sweden).

D.12 Fleming, K.N. and B.O.Y. Lydell, 2004. "Database Development and Uncertainty Treatment for Estimating Pipe Failure Rates and Rupture Frequencies," Reliability Engineering and System Safety, article in press.

D.13 Lydell, B.O.Y., E. Mathet and K. Gott, 2004. "Piping Service Life Experience in Commercial Nuclear Power Plants: Progress with the OECD Pipe Failure Data Exchange Project," Proc. ASME PVP-2004 Conf.: 2004 ASME Pressure Vessels and Piping Conference, American Society of Mechanical Engineers, New York (NY).

D.14 Fleming, K.N., 2004. "Markov Models for Evaluating Risk-informed In-service Inspection Strategies for Nuclear Power Plant Piping Systems," Reliability Engineering & System Safety, 83:27-45.

D.15 Lydell, B.O.Y., 1999. Failure Rates in Barsebäck-1 Reactor Coolant Pressure Boundary Piping. An Application of a Piping Failure Database, SKI Report 98:30, Swedish Nuclear Power Inspectorate, Stockholm (Sweden).

D.16 Lydell, B.O.Y., 1999. Failure Rates in Barsebäck-1 Reactor Coolant Pressure Boundary Piping. SKI 98:30 Appendix H: Barsebäck-1 Piping Reliability Database, RSA-R-99-01P (Proprietary), Prepared for Barsebäck Kraft AB (Sweden), RSA Technologies, Fallbrook (CA).

D.17 Lydell, B.O.Y., 1997. Strategies for Reactor Safety: Preventing Loss of Coolant Accidents, Nordic Nuclear Safety Research, Risø (Denmark), NKS/RAK-1(97)R10, ISBN: 87-7893-046-4.

D.18 Wachter, O. and G. Brümmer, 1997. "Experiences with Austenitic Steels in Boiling Water Reactors," Nuclear Engineering and Design, **168**:35-52.

D.19 U.S. Nuclear Regulatory Commission, 1992. Higher Than Predicted Erosion/Corrosion in Unisolable Reactor Coolant Pressure Boundary Piping Inside Containment at a Boiling Water Reactor, Information Notice 92-35 (May 6, 1992), Washington (DC).

D.20 International Incident Reporting System, 2000. Cracks in Weld Area of Safe-end to the Reactor Pressure Vessel, IRS No. 7397, International Atomic Energy Agency, Vienna (Austria).

D.21 Crowley, B. et al, 2001. Virgil C. Summer Nuclear Station – NRC Special Inspection Report No. 50-395/00-08, Exercise of Enforcement Discretion, U.S. Nuclear Regulatory Commission, Washington (DC).

D.22 U.S. Nuclear Regulatory Commission, 1988. Thermal Stresses in Piping Connected to Reactor Coolant Systems, Bulletin 88-08 (June 22, 1988), Washington (DC).

D.23 U.S. Nuclear Regulatory Commission, 1988. Pressurizer Surge Line Thermal Stratification, Bulletin 88-11 (December 20, 1988), Washington (DC).

D.24 OECD Nuclear Energy Agency, 1998. Experience with Thermal Fatigue in LWR Piping Caused by Mixing and Stratification, NEA/CSNI/R(98)8, Issy-les-Moulineaux (France).

D.25 Aaltonen, P., K. Saarinen and K. Simola, 1993. "The Correlation of IGSCC Propagation with the Power Plant Trainsient History," The International Journal of Pressure Vessels and Piping, **55**:149-162.

ATTACHMENT A TO APPENDIX D

SUMMARY OF PIPEXP DATABASE

The PIPExp database has evolved over a period of about ten years. A first database version was developed with financial support from the Swedish Nuclear Power Inspectorate (SKI). Since the conclusion of the initial R&D effort in 1998 an active maintenance program has supported the database.

D.A.1 Database Structure

Designed in Access, the database consists of searchable free-format text fields and a large number of data fields that are used as data filters in support of a range of data processing needs. Table D.A.1 is a summary of text and data fields.

D.A.2 Completeness and Quality Management

The completeness of the pipe failure data is addressed through a continuous database management program. Extracted from Monthly Summary Reports, Table D.A.2 provides snapshots of the database evolution from 1998 to the present. In PIPExp, each record is assigned a 'Quality Index' (Table D.A.1, item #4, and Table D.A.3) as one means of monitoring the completeness and technical accuracy of source information as well as the process of classifying and coding of the source information.

Table D.A.1 Description of Data Fields in PIPExp

Item No.	Field Name	Type	Description
1	UPDATE	Date	Date of the most recent update.
2	MER	Yes/No[21]	Multiple Events Report; some reports include information on more than one crack/leak in one system. Used to identify events where a discovery resulted in an investigation (e.g., augmented ISI) to identify further piping degradation due to a common cause. A new record is added if additional degradation is positively identified (by component socket).
3	DDA	Text	Data filter used to classify a record as either 'public' (= Licensee Event Report), 'restricted' or 'proprietary.'
4	QA-Index	Number	QA-Index of '1' signifies a data entry determined to be 'complete.' By contrast, a QA-Index of '6' signifies a database entry for which only a LER (or equivalent) abstract was available.
5	EVENT DATE	Date	Event date (MM/DD/YY); date of discovery (in case of ISI).
6	PLANT TYPE	Text	Plant type; e.g., BWR, PWR, used as data filter.
7	DESIGN	Text	NSSS design/design generation; keyword using generally accepted or standard nomenclature. This field is used as a data filter.
8	NSSS-VENDOR	Text	Reactor vendor; e.g., ABB-Atom, KWU/Siemens, Westinghouse; used as data filter.
9	PLANT NAME	Text	Plant name
10	COUNTRY	Text	Two-letter code based on the ISO 3166-1-alpha-2 code elements.
11	CONSTRUCTOR	Text	Name of company responsible for the original piping system design. The default name is the architect engineering firm. Used as data filter.
12	COD	Date	Date (MM/DD/YY) of commercial operation as default. If known, date of initial criticality. For U.S. data, based on NUREG-0020
13	PLANT	Text	Plant operational state (at the time of discovery); keyword using generally accepted or standard nomenclature. This field is used as a data

---

[21] A check box without check mark implies 'No' or 'Unknown/Pending.'

Table D.A.1 Description of Data Fields in PIPExp

Item No.	Field Name	Type	Description
	OPERATIONAL STATE - POS		filter. Pulldown menu with the following options: § CSD – Cold Shutdown § HSD – Hot Shutdown § HSB – Hot Standby § Refueling § Shutting Down § Starting Up § Power Operation
14	REFERENCE-1	Text	Primary reference
15	REFERENCE-2	Text	Secondary (or supplemental) reference
16	REFERENCE-3	Text	Tertiary (or supplemental) reference
17	LER-RO?	Yes/No	Check if the information source is a Licensee Event Report (or equivalent); i.e., from a regulatory reporting system.
18	EVENT TYPE	Text	Event type; 'Crack', 'Wall Thinning', 'P/H-leak' (P/H = pinhole), 'Leak', 'Severance', 'Rupture.' Used as data filter.
19	FAILURE-ON-DEMAND	Yes/No	Check if pipe failure occurred when a demand was placed on the affected system (e.g., standby system). Used as data filter.
20	SYNERGY	Yes/No	Check if the pipe failure was caused by multiple degradation mechanisms; e.g., crack initiation through IGSCC and crack propagation through thermal fatigue. Used as data filter.
21	DEGRADATION+ LOADING	Yes/No	Check if the pipe failure resulted from the combined effect of a degradation mechanism (e.g., flow-accelerated corrosion, FAC) and a severe (or unusual) loading condition. Used as data filter.
22	ECA	Text	Event Category. Used as data filter. This database field is used to characterize actual or potential impact on plant risk by a degradation or failure. The following options are available: § S-/M-/L-LOCA (implies that a pressure boundary failure resulted in ESF actuation); § S-/M-/L-LOCA Precursor (implies mitigation of a pressure boundary failure through prompt operator response; e.g., plant shutdown prior to reaching ESF actuation setpoint); § Internal Flooding (spill rate in excess of room/compartment floor drain capacity); § Internal Flooding Precursor (accumulation of large water volumes prevented through prompt operator response); § Common Cause Initiating (CCI) Event (pressure boundary failure results in spatial effects through spraying or steaming of safety equipment); § CCI Precursor (pressure boundary failure results in spraying or steaming but prompt operator action prevents safety equipment from being affected); § System Disabled (pressure boundary failure is large enough to incapacitate a system function); § System Degraded (default used for at-power events that result in an entry into a Technical Specification Action Statement).
23	CCC	Yes/No	Check if event is considered to be a 'common cause candidate' (CCC) event. Used as data filter.
24	CA	Text	Corrective Action. Used as data filter. The following types of corrective action are defined: § REPAIR (used in a generic sense); § REPLACEMENT § REPLACEMENT – IN-KIND § REPLACEMENT – NEW MATERIAL § TEMP. REPAIR (temporary repair to allow continued operation until next refueling outage or major maintenance outage at which time a Code-repair (which would require system isolation and

Table D.A.1 Description of Data Fields in PIPExp

Item No.	Field Name	Type	Description
			draining) or a replacement is performed. § WOR (= weld overlay repair); primarily applies to ASME Section XI Class 1 or 2 (or equivalent) piping.
25	ISS	Yes/No	Safety system actuation; check if pipe failure resulted automatic actuation of a make-up system or other safety system. Used as data filter.
26	IRT	Yes/No	Automatic reactor trip; check if pipe failure resulted in automatic reactor trip/turbine trip. Used as data filter.
27	IPO	Text	Impact of pipe failure on plant operation; e.g., power reduction, manual reactor trip. Used as data filter.
28	TTR	Number	Repair time in hours.
29	TTR-Class	Number	A data filter: 1: TTR ≤ 8 hours; 2: 8 < TTR ≤ 24 hours; 3: 24 < TTR ≤ 96 hours; 4: 96 < TTR ≤168 hours; 5: TTR > 168 hours.
30	NARRATIVE	Memo	Event narrative; includes details on plant condition prior to event and plant response during event, method of detection, corrective action plan. This field should include sufficient information to support independent verification of the event data classification.
31	LQT	Number	Quantity of process medium released [kg]
32	DOL	Text	Duration of release
33	LRT	Number	Leak rate [kg/s]
34	gpm	Number	Leak rate [U.S. gallons/minute]
35	LEAK CLASS	Number	A data filter: 1: Leak Rate (LR) ≤ 1 gpm; 2: 1 < LR ≤ 5 gpm; 3: 5 < LR ≤ 10 gpm; 4: 10 < LR ≤ 50 gpm; 5: LR > 50 gpm.
36	FLO	Text	Location of crack/leak/rupture; description of where in the piping system a degradation or failure occurred. Include sufficient detail to support the consequence evaluation/classification.
37	K1	Yes/No	Data filter; steamline break outside containment.
38	K2	Yes/No	Data filter; feedwater line break
39	K3	Yes/No	Data filter; steamline break inside containment.
40	IMPULSE-LINE	Yes/No	Check (= 'Yes') if affected line is a valve impulse line.
41	INSTR. LINE	Yes/No	Check (= 'Yes') if affected line is an instrument sensing line.
42	ISOMETRIC DRAWING #	Text	Isometric drawing number
43	P&ID #	Text	Piping and instrument drawing number.
44	MSA	Text	Name of the affected plant system
45	SHARED	Yes/No	Check (= 'Yes') if affected piping is shared by two reactor units. Mainly applies to support systems (e.g., Service Water, Instrument Air) where sections of a piping system may be shared by two reactor units; this is relatively common in the U.S.
46	OSA	Text	Name of other systems affected by the degradation or failure. Secondary effects of piping failure
47	S-TYPE	Text	Category of system affected by the degradation or failure. Used as data filter. The following types are used: • RCPB (Reactor Coolant Pressure Boundary); • SIR (Safety Injection & Recirculation); includes emergency core cooling systems & decay heat removal). • CS (Containment Spray) • AUX (Reactor Auxiliary Systems); includes component cooling water, chemical & volume control, reactor water cleanup, control rod drive, containment heat removal, standby liquid control, radwaste control, spent fuel pool cooling. • FWC (Feedwater & Condensate Systems) • STEAM (Main Steam System) • SUPPORT (Service Water & Instrument Air systems) • PCS (turbine generator) • FIRE (Fire Protection).

Table D.A.1 Description of Data Fields in PIPExp

Item No.	Field Name	Type	Description
48	ISO	Yes/No	Check if the affected pipe section can be isolated to prevent or mitigate direct/indirect impacts.
49	DET	Text	Method of detection; e.g., ISI, WT = walk-through inspection, leak detection system in combination with control room indication and/or alarm. Used as data filter. Pulldown menu with the following options: § Walk-through § UT-examination § Liquid penetrant testing § Hydrotesting § Leak detection § Containment/drywell inspection § Control Room Indication
50	DRYWELL ENTRY	Yes/No	BWR-specific data field. Checked for 'at-power', unidentified P/H-leak or leak requiring power reduction or reactor shutdown for containment drywell entry to determine leak source. Used as a data filter. Also, this data could be input to plant availability models.
51	CONTAINMENT ENTRY	Yes/No	Check if power reduction initiated to allow for containment entry to identify source of leakage. Used for other than BWR plants.
52	CRS	Text	Verbal description of crack morphology; orientation and size/ geometry of crack or fracture
53	CRACK-DEPTH	Number	Crack depth in percent of wall thickness (a/t-ratio)
54	AXIAL-LENGTH	Number	Axial crack length in [mm].
55	CRACK-LENGTH	Number	Circumferential crack length as percent of inside diameter
56	ASPECT-RATIO	Number	Ratio of crack depth (a) to flaw length (L)
57	WELD-CONFIG	Text	Configuration of the affected weld in a piping system; e.g., BP = bend-to-pipe weld, PP = pipe-to-pipe weld, etc.
58	INSIDE CONTAINMENT	Yes/No	Check if pipe failure located inside containment
59	AUXILIARY BUILDING	Yes/No	Check if pipe failure located in Auxiliary Building (PWR)
60	REACTOR BUILDING	Yes/No	Check if pipe failure located in Reactor Building (BWR)
61	TURBINE BUILDING	Yes/No	Check if pipe failure located in Turbine Building
62	Not used	N/A	N/A
63	Not used	N/A	N/A
64	CTA	Text	Component Type; pulldown menu with the following options: § Bend § Elbow § Elbow – 45-degree § Elbow – 90-degree § Elbow – LR (Long Radius) § Pipe § Reducer § Tee § Weld § Socket weld
65	ASME Class	Number	Differentiate between 1, 2, 3 and 4 (= non-Code Class)
66	BELOW-GRADE	Yes/No	Check if 'Yes'; Below Grade / Underground Piping. Used as data filter.
67	FIELD-WELD	Yes/No	Check if 'Yes'. Used as data filter.
68	SHOP-WELD	Yes/No	Check if 'Yes'. Used as data filter.
69	CONCRETE-LINED	Yes/No	Check if 'Yes'; could apply to essential or non-essential service water (or equivalent) system piping. Used as data filter.
70	REPLACEMENT	Yes/No	Check if piping replaced using new material.
71	REPL-DATE	Date/Time	Date of component (e.g., weld and spool piece) replacement. Used in hazard plotting.

D-77

Table D.A.1 Description of Data Fields in PIPExp

Item No.	Field Name	Type	Description
72	YOO	Number	Years of commercial operation when failure occurred. Used in aging analysis.
73	AGE	Number	Age of component socket [hours]. Used in hazard plotting. For additional information.
74	CLASS	Number	Based on diameter; events grouped in six diameter classes; $\underline{1}$ = ($\leq$ DN15), $\underline{2}$ = (15 < DN $\leq$ 25), $\underline{3}$ = (25 < DN $\leq$ 50), $\underline{4}$ = (50 < DN $\leq$ 100), $\underline{5}$ = (100 < DN $\leq$ 250), $\underline{6}$ = (> DN250). DN = nominal diameter in [mm]. This field is used as a data filter.
75	THOMAS	Number	Ratio of diameter and pipe wall thickness ([CSI/WTK]); for details, see the paper by H.M. Thomas (1981): "Pipe and Vessel Failure Probability," *Reliability Engineering*, **2**:83-124. This field is used as a data filter.
76	CSI	Number	Nominal diameter [DN] in [mm]. Used as data filter.
77	WTK	Number	Wall thickness [mm]
78	SCHEDULE	Number	Pipe schedule number
79	DIS-MET	Yes/No	Dissimilar metal weld; check if 'yes'. Used as data filter.
80	MTR	Text	Material; e.g., carbon steel, stainless steel, etc. Used as data filter.
81	MTR-DES	Text	Material designation according to national standard; e.g., AISI 304, SS2343, etc. Used as data filter.
82	PMD	Text	Process medium. Used as data filter.
83	RAW WATER	Text	Source of raw water (applies to Fire Protection and Service Water piping); differentiate between LAKE – RIVER – SEA-BRACKISH. Used as data filter.
84	STG	Yes/No	Normally stagnant process medium? Used as data filter.
85	HWC	Yes/No	For BWRs; hydrogen water chemistry; check if 'Yes'. Used as data filter (e.g., in factor-of-influence assessments).
86	HWC-START	Date/Time	Date when HWC was introduced
87	NMCA	Yes/No	Check if Noble Metal Chemical Addition. Used as data filter.
88	NMCA-Start	Date/Time	Date when NMCA started.
89	IHSI	Yes/No	Induction heat stress improvement; check if 'Yes'. Used as data filter.
90	IHSI-DATE	Date/Time	Date when IHSI was performed
91	MSIP	Yes/No	Check if Mechanical Stress Improvement Process applied to weld. Used as data filter.
92	MSIP-Date	Date/Time	Date of MSIP application
93	S-A	Number	Stress intensity allowance; ratio of the critical stress intensity factor to the assessed stress intensity factor given a flaw. This information is extracted from fracture mechanics evaluations.
94	OPA	Number	Operating temperature [°C]
95	DPA	Number	Design temperature [°C]
96	OPB	Number	Operating pressure [MPa]
97	DPB	Number	Design pressure [MPa]
98	OPC	Text	Process medium chemistry (for primary system); e.g., NWC = normal water chemistry, HWC = hydrogen water chemistry. Used as data filter.
99	MPR	Text	Method of fabrication; e.g., cold formed, hot formed. Used as data filter.
100	SYS	Yes/No	Systematic failure? Used as data filter to enable queries that address the effectiveness of remedial actions (e.g., preventing recurring failures).
101	RFL	Text	Description of the extent and nature of a systematic failure
102	REST	Yes/No	Failure due to deficient system restoration?; e.g., no venting prior to fill procedure, etc. Used as data filter.
103	CEA	Text	Apparent cause of failure; e.g., IGSCC, PWSCC, TGSCC, etc. Used as data filter. Pull down menu with the following options: • B/A SCC (B/A = boric acid) • Corrosion (general, pitting or crevice corrosion) • Corrosion-fatigue • Erosion

Table D.A.1 Description of Data Fields in PIPExp

Item No.	Field Name	Type	Description
			• Erosion-cavitation • Flow accelerated corrosion (FAC) • Fretting • HF: Construction/installation error • HF: Human error • HF: Repair/maintenance error • HF: Welding error • HPSCC (High Potential SCC) • IGSCC • MIC (Microbiologically induced corrosion) • Overpressurization • PWSCC • Severe overloading (other than water hammer) • SICC (Strain-rate induced SCC) • TGSCC • Thermal fatigue • Unknown • Unreported • Vibration • Vibration-fatigue • Water hammer
104	EPRI-CODE	Text	Failure code as used in the EPRI '97 / EPRI '98 databases (see for example EPRI TR-110161 (Piping System Reliability and Failure Rate Estimation Models for Use in Risk-Informed In-Service Inspection Applications, December 1998). Used as data filter. Pull down menu with the following options: § CF – Corrosion-fatigue § COR – General corrosion, microbiologically induced corrosion (MIC), pitting corrosion § COR-EXT – external corrosion § D&C – Design & Construction errors § E-C – Erosion-cavitation § E/C – Erosion-corrosion § FRET – Fretting § HE – Human Error § OVP – Overpressurization § SCC – Stress corrosion cracking § TF – Thermal Fatigue § UNK – Unknown § VF – Vibration-fatigue § WH – Water hammer
105	RC1	Text	Contributing factor number 1
106	RC2	Text	Contributing factor number 2
107	CEC	Memo	Description of events and causal factors. Include sufficient technical detail from the root cause analysis process so that recurrence may be prevented.
108	CMT	Memo	Any other information of relevance to understanding of underlying causal factors. Also, information on the type and extent of repair/replacement. The purpose of this free-format database field is to facilitate future applications, for example, by codifying the information on piping replacements.
109	ISI	Yes/No	Deficient ISI; e.g., ISI not performed, or ISI failed to detect a flaw. Used as data filter to identify events caused by ISI program deficiencies (e.g., affected component should have been included in program) or an inspection prior to failure missed a degradation that propagated in the

Table D.A.1 Description of Data Fields in PIPExp

Item No.	Field Name	Type	Description
			through-wall direction.
110	ISI-CMT	Memo	Comments on ISI history; e.g., date of last inspection, details on examination technique(s).
111	mCF	Yes/No	Check if there are multiple circumferential flaws in a weld.
112	Number of Flaws	Number	
113	D0-1	Number	Distance from 12 o'clock position to the first circumferential flaw; this field is repeated for up to nine flaws.
114	CF-1	Number	Length of the first circumferential flaw (counted from the 12 o'clock position

Table D.A.2 Summary of PIPExp Database Development

Database as of 12-31-1998 — No. Damage Reports by QA-Index[22]

Plant Type	Totals	1	2	3	4	5	6
BWR	673	210	66	3	74	7	277
PHWR	100	30	3	--	56	1	10
PWR	1376	386	123	6	152	84	746
RBMK	57	3	6	--	19	28	1
	2291	629	198	9	301	120	1034

Database as of 12-31-1999 — No. Damage Reports by QA-Index

Plant Type	Totals	1	2	3	4	5	6
BWR	1595	1000	168	2	146	53	226
PHWR	100	30	3	--	56	1	10
PWR	1656	645	176	5	211	208	411
RBMK	66	7	6	--	22	31	--
	3417	1682	253	7	435	243	647

Database as of 12-31-2000 — No. Damage Reports by QA-Index

Plant Type	Totals	1	2	3	4	5	6
BWR	1711	1111	164	2	175	58	201
PHWR	95	43	1	--	41	10	--
PWR	1748	696	181	5	260	209	397
RBMK	125	12	5	1	77	30	--
	3679	1862	351	8	553	307	598

Database as of 12-31-2001 — No. Damage Reports by QA-Index

Plant Type	Totals	1	2	3	4	5	6
BWR	1784	1172	166	2	197	63	184
PHWR	96	44	1	--	41	10	--
PWR	1952	811	194	5	329	221	392
RBMK	125	12	5	1	77	30	--
	3957	2039	366	8	644	324	576

Database as of 12-31-2002 — No. Damage Reports by QA-Index

Plant Type	Totals	1	2	3	4	5	6
BWR	1872	1216	174	12	219	75	176
GCR, HWLWR	--	--	--	--	--	--	--
PHWR	106	51	2	--	42	11	--
PWR	2077	1011	198	6	351	233	278
RBMK	160	48	--	--	18	81	--
	4215	2290	379	22	721	349	454

Database as of 06-01-2004 — No. Damage Reports by QA-Index

Plant Type	Totals	1	2	3	4	5	6
BWR	2033	1370	189	77	233	93	71
GCR, HWLWR	12	12	--	--	--	--	--
PHWR	101	47	1	--	41	12	--
PWR	2280	1213	218	103	344	270	132
RBMK	160	12	5	4	109	30	--
	4586	2654	413	184	727	405	203

[22] See Table A-3 for a definition of QA-Index

Table D.A.3 Definition of QA Index for Database Management

QA-Index	Definition
1	Validated – all source data has been accessed & reviewed – no further action required
2	Validated – source data may be missing some, non-critical information – no further action anticipated
3	Validated – incomplete source data – assumptions made about material grade and/or exact flaw location – no further action anticipated
4	Validation based on incomplete information – depending on application requirements, further action may be necessary
5	Validation based on available, incomplete information – further action expected (e.g., retrieval of additional source data)
6	Not validated – validation is pending, or record is subject to deletion from database

Table D.A.4 Summary of Through-Wall Cracks in BWR Reactor Recirculation Piping[23]

EID	Date of Detection	Plant	NPS	Comment
1978	06-14-1978	Duane Arnold	10	LER 50-331/78-030. Active, at power leak
1067	01-26-1983	Brunswick-1	12	Weld B32-RR-12-BR-H4
2419	02-12-1980	Santa Maria de Garona	10	Active, 0.8 gpm leak
1397	02-21-1985	Browns Ferry-2	12	Riser-to-manifold weld
404	11-02-1982	Monticello	12	LER 50-220/82-013; Riser-to-safe-end weld
7414	01-15-1986	Hatch-1	12	Weld 1B31-1RC-12BR-B-3
3231	11-22-1982	Monticello	12	Leak detected during hydrostatic testing
2681	04-01-1985	Quad Cities-2	12	Weld 02M-F7
1848	11-05-1986	Quad Cities-2	12	Weld 02K-S3
2085	10-20-1982	Monticello	22	Pipe-to-safe-end weld
443	10-31-1982	Monticello	22	Pipe-to-safe-end weld
2850	11-06-1982	Hatch-1	22	LER 50-321/82-089; End cap-to-manifold
3217	03-10-1985	Duane Arnold	22	Weld RHB-J1
7543	03-23-1982	Nine Mile Point-1	28	LER 50-220/82-009; Weld P32-FW-17-W
7542	03-23-1982	Nine Mile Point-1	28	Weld P32-FW-22-W
7538	04-15-1982	Nine Mile Point-1	28	Weld P32-FW-27-W
437	04-15-1985	Nine Mile Point-1	28	Weld P32-FW-42-W
3224	03-10-1985	Duane Arnold	28	Weld RCB-J36
2839	07-18-1985	Brunswick-1	28	Weld 1B32-RR-28-A-4
2838	07-18-1985	Brunswick-1	28	Weld 1B32-RR-28-A-15
2837	07-18-1985	Brunswick-1	28	Weld 1B32-RR-28-B-8
2836	07-18-1985	Brunswick-1	28	Weld 1B32-RR-28-B-4
1711	07-18-1985	Brunswick-1	28	Weld 1B32-RR-28-A-14
3183	01-09-1986	Brunswick-2	28	Weld 2B32-RR-28-B-3
3182	01-09-1986	Brunswick-2	28	Weld 2B32-RR-28-B-4
3181	01-09-1986	Brunswick-2	28	Weld 2B32-RR-28-B-5
3180	01-09-1986	Brunswick-2	28	Weld 2B32-RR-28-B-11
1723	01-09-1986	Brunswick-2	28	Weld 2B32-RR-28-A-4

[23] Basis for Table D.5 (Section D.4).

ATTACHMENT B TO APPENDIX D

BASIS FOR LOCA FREQUENCY MODELS

Attached as Tables D.B.1 (RR System Loop B) and D.B.2 (FW System Loop B) are the Excel spreadsheets on which the BWR LOCA frequency model is based. Attached as Tables D.B.3 (RC-HL), D.B.4 (RC Surge Line), and D.B.5 (HPI/NMU Line) are the spreadsheets on which the PWR LOCA frequency model is based.

The input to the calculation of a selection of posterior weld failure rates is summarized in Tables D.B.6 (BWR-1, NPS28), D.B.7 (BWR-2, NPS12), D.B.8 (PWR-1), D.B.9 (PWR-2), and D.B.10 (PWR-3).

Table D.B.1 BWR-1 – RR System Loop B Weld List

System ID	Loop	Exam Category	Category Item	Line Number	Weld Order	Component ID	NPS (in)	Wall Thk (in) or Schedule	Configuration	Description
B31	B	B-F	B5.10	5358-5(2)	5	101-304E	12	Schd. 80	Nozzle-safe-end	RRI Nozzle-to-Safe End Butt Weld (N2E)
B31	B	B-F	B5.10	5358-5(5)	5	2-303A	12	Schd. 80	Nozzle-safe-end	RRI Nozzle-to-Safe End Butt Weld (N2A)
B31	B	B-F	B5.10	5358-5(4)	5	2-303B	12	Schd. 80	Nozzle-safe-end	RRI Nozzle-to-Safe End Butt Weld (N2B)
B31	B	B-F	B5.10	5358-5(6)	4	2-303C	12	Schd. 80	Nozzle-safe-end	RRI Nozzle-to-Safe End Butt Weld (N2C)
B31	B	B-F	B5.10	5358-5(3)	5	2-303D	12	Schd. 80	Nozzle-safe-end	RRI Nozzle-to-Safe End Butt Weld (N2D)
B31	B	B-J	B9.11	5358-5(5)	1	FW-RD-2-B10	12	Schd. 80	Pipe-sweepolet	B31-Reactor Recirc - Loop B Circ Weld
B31	B	B-J	B9.11	5358-5(4)	1	FW-RD-2-B11	12	Schd. 80	Pipe-sweepolet	B31-Reactor Recirc - Loop B Circ Weld
B31	B	B-J	B9.11	5358-5(6)	0.5	FW-RD-2-B12	12	Schd. 80	Pipe-reducer	B31-Reactor Recirc - Loop B Circ Weld
B31	B	B-J	B9.11	5358-5(3)	1	FW-RD-2-B13	12	Schd. 80	Pipe-sweepolet	B31-Reactor Recirc - Loop B Circ Weld
B31	B	B-J	B9.11	5358-5(2)	1	FW-RD-2-B14	12	Schd. 80	Pipe-sweepolet	B31-Reactor Recirc - Loop B Circ Weld
B31	B	B-J	B9.11	5358-5(5)	4	FW-RD-2-B15	12	Schd. 80	Pipe-safe-end	B31-Reactor Recirc - Loop B Circ Weld
B31	B	B-J	B9.11	5358-5(4)	4	FW-RD-2-B16	12	Schd. 80	Pipe-safe-end	B31-Reactor Recirc - Loop B Circ Weld
B31	B	B-J	B9.11	5358-5(6)	3	FW-RD-2-B17	12	Schd. 80	Pipe-safe-end	
B31	B	B-J	B9.11	5358-5(3)	4	FW-RD-2-B18	12	Schd. 80	Pipe-safe-end	B31-Reactor Recirc - Loop B Circ Weld
B31	B	B-J	B9.11	5358-5(2)	4	FW-RD-2-B19	12	Schd. 80	Pipe-safe-end	B31-Reactor Recirc - Loop B Circ Weld
B31	B	B-J	B9.11	5358-5(5)	2	SW-RD-2-B4-W1	12	Schd. 80	Pipe-elbow	B31-Reactor Recirc - Loop B Circ Weld
B31	B	B-J	B9.11	5358-5(5)	3	SW-RD-2-B4-W2	12	Schd. 80	Elbow-pipe	B31-Reactor Recirc - Loop B Circ Weld
B31	B	B-J	B9.11	5358-5(4)	2	SW-RD-2-B5-W1	12	Schd. 80	Pipe-elbow	
B31	B	B-J	B9.11	5358-5(4)	3	SW-RD-2-B5-W2	12	Schd. 80	Elbow-pipe	B31-Reactor Recirc - Loop B Circ Weld
B31	B	B-J	B9.11	5358-5(3)	2	SW-RD-2-B6-W1	12	Schd. 80	Pipe-elbow	
B31	B	B-J	B9.11	5358-5(3)	3	SW-RD-2-B6-W2	12	Schd. 80	Elbow-pipe	
B31	B	B-J	B9.11	5358-5(2)	2	SW-RD-2-B7-W1	12	Schd. 80	Pipe-elbow	
B31	B	B-J	B9.11	5358-5(2)	3	SW-RD-2-B7-W2	12	Schd. 80	Elbow-pipe	
B31	B	B-J	B9.11	5358-5(6)	1	SW-RD-2-B8-W1	12	Schd. 80	Pipe-elbow	B31-Reactor Recirc - Loop B Circ Weld
B31	B	B-J	B9.11	5358-5(6)	2	SW-RD-2-B8-W2	12	Schd. 80	Elbow-safe-end	B31-Reactor Recirc - Loop B Circ Weld
B31	B	B-J	B9.31	5358-5(1)	2	SW-RD-2-B3-W1	22	Schd. 80	Pipe-sweepolet	
B31	B	B-J	B9.31	5358-5(1)	3	SW-RD-2-B3-W2	22	Schd. 80	Pipe-sweepolet	
B31	B	B-J	B9.31	5358-5(1)	6	SW-RD-2-B3-W3	22	Schd. 80	Pipe-sweepolet	
B31	B	B-J	B9.31	5358-5(1)	7	SW-RD-2-B3-W4	22	Schd. 80	Pipe-sweepolet	B31-Reactor Recirc - Loop B Circ Weld
B31	B	B-J	B9.11	5358-5(1)	4	SW-RD-2-B3-W6	22	Schd. 80	Pipe-cross	B31-Reactor Recirc - Loop B Circ Weld
B31	B	B-J	B9.11	5358-5(1)	5	SW-RD-2-B3-W7	22	Schd. 80	Pipe-cross	B31-Reactor Recirc - Loop B Circ Weld

Table D.B.1 BWR-1 – RR System Loop B Weld List

System ID	Loop	Exam Category	Category Item	Line Number	Weld Order	Component ID	NPS (in)	Wall Thk (in) or Schedule	Configuration	Description
B31	B	B-J	B9.11	5358-5(1)	1	SW-RD-2-B3-W8	22	Schd. 80	Pipe-end-cap	B31-Reactor Recirc - Loop B Circ Weld
B31	B	B-J	B9.11	5358-5(1)	8	SW-RD-2-B3-W9	22	Schd. 80	Pipe-end-cap	
B31	B	B-F	B5.10	5359-5(S)	1	4-303B	28	Schd. 80	Nozzle-safe-end	RRS Nozzle-to-Safe End Butt Welds (N1B)
B31	B	B-J	B9.11	5359-5(D)-5358-5(6)	3	FW-RD-2-B1-W1	28	Schd. 80	Pipe-pipe	B31-Reactor Recirc - Loop B Circ Weld
B31	B	B-J	B9.11	5359-5(D)-5358-5(6)	8	FW-RD-2-B2-W2	28	Schd. 80	Pipe-sweepolet	B31-Reactor Recirc - Loop B Circ Weld - 28 x 4 inch SWOL
B31	B	B-J	B9.11	5359-5(D)-5358-5(6)	1	FW-RD-2-B6	28	Schd. 80	Pipe-pump	RR Pump discharge
B31	B	B-J	B9.11	5359-5(D)-5358-5(6)	4	FW-RD-2-B7	28	Schd. 80	Pipe-valve	B31-Reactor Recirc - Loop B Circ Weld
B31	B	B-J	B9.11	5359-5(D)-5358-5(6)	5	FW-RD-2-B8	28	Schd. 80	Pipe-valve	
B31	B	B-J	B9.11	5359-5(D)-5358-5(6)	11	FW-RD-2-B9	28	Schd. 80	Cross-tee	
B31	B	B-J	B9.11	5359-5(S)	2	FW-RS-2-B1	28	Schd. 80	Pipe-safe-end	RPV Nozzle area
B31	B	B-J	B9.11	5359-5(S)	5	FW-RS-2-B2	28	Schd. 80	Pipe-pipe	B31-Reactor Recirc - Loop B Circ Weld
B31	B	B-J	B9.11	5359-5(S)	9	FW-RS-2-B3	28	Schd. 80	Pipe-valve	
B31	B	B-J	B9.11	5359-5(S)	10	FW-RS-2-B4	28	Schd. 80	Pipe-valve	B31-Reactor Recirc - Loop B Circ Weld
B31	B	B-J	B9.11	5359-5(S)	14	FW-RS-2-B5	28	Schd. 80	Pipe-pump	RR Pump suction
B31	B	B-J	B9.31	5359-5(D)-5358-5(6)	2	SW-RD-2-B1-W1	28	Schd. 80	Pipe-pipe	RPV Nozzle area
B31	B	B-J	B9.11	5359-5(D)-5358-5(6)	6	SW-RD-2-B2-W1	28	Schd. 80	Elbow-pipe	
B31	B	B-J	B9.11	5359-5(D)-5358-5(6)	10	SW-RD-2-B2-W2	28	Schd. 80	Pipe-tee	B31-Reactor Recirc - Loop B Circ Weld
B31	B	B-J	B9.11	5359-5(D)-5358-5(6)	9	SW-RD-2-B2-W2O	28	Schd. 80	Pipe-pipe	B31-Reactor Recirc - Loop B Circ Weld
B31	B	B-J	B9.31	5359-5(D)-5358-5(6)	7	SW-RD-2-B2-W3	28	Schd. 80	Pipe-pipe	
B31	B	B-J	B9.11	5359-5(D)-5358-5(6)	99	SW-RD-2-B3-W5	28	Schd. 80	Cross-reducer	B31-Reactor Recirc - Loop B Circ Weld
B31	B	B-J	B9.11	5359-5(S)	3	SW-RS-2-B1-W1	28	Schd. 80	Elbow-pipe	B31-Reactor Recirc - Loop B Circ Weld
B31	B	B-J	B9.11	5359-5(S)	4	SW-RS-2-B1-W2	28	Schd. 80	Elbow-pipe	B31-Reactor Recirc - Loop B Circ Weld
B31	B	B-J	B9.11	5359-5(S)	8	SW-RS-2-B2-W1	28	Schd. 80	Elbow-pipe	
B31	B	B-J	B9.11	5359-5(S)	7	SW-RS-2-B2-W10A	28	Schd. 80	Pipe-tee	B31-Reactor Recirc - Loop B Circ Weld
B31	B	B-J	B9.11	5359-5(S)	6	SW-RS-2-B2-W2	28	Schd. 80	Pipe-tee	

Table D.B.1 BWR-1 – RR System Loop B Weld List

System ID	Loop	Exam Category	Category Item	Line Number	Weld Order	Component ID	NPS (in)	Wall Thk (in) or Schedule	Configuration	Description
B31	B	B-J	B9.11	5359-5(S)	13	SW-RS-2-B3-W1	28	Schd. 80	Elbow-pipe	B31-Reactor Recirc - Loop B Circ Weld
B31	B	B-J	B9.31	5359-5(S)	12	SW-RS-2-B3-W2	28	Schd. 80	Pipe-pipe	B31-Reactor Recirc - Loop B Circ Weld
B31	B	B-J	B9.31	5359-5(S)	11	SW-RS-2-B3-W3	28	Schd. 80	Pipe-pipe	
B31	B	B-J	B9.11	5359-5(S)	11.1	SW-RS-2-B3-W4	28	Schd. 80	Pipe-pipe	
B31	B	B-J	B9.11	5359-5(S)	11.2	SW-RS-2-B3-W5	28	Schd. 80	Pipe-pipe	

Table D.B.2 BWR-2 – FW System Loop B Weld List

System ID	Loop	Exam Category	Category Item	Line Number	Weld Order	Component ID	NPS (in)	Wall Thk (in) or Schedule	Configuration	Description
N21	B	B-J	B9.11	3537-5(4)	2	SW-N21-2336-19WF	12	Schd. 100	Elbow-pipe	
N21	B	B-J	B9.11	3537-5(4)	3	SW-N21-2336-19WG	12	Schd. 100	Elbow-pipe	
N21	B	B-J	B9.11	3537-5(4)	4	FW-N21-2336-19W20	12	Schd. 100	Elbow-pipe	
N21	B	B-J	B9.11	3537-5(4)	7	FW-N21-2336-20WF4	12	Schd. 100	Elbow-pipe	
N21	B	B-J	B9.11	3537-5(4)	8	SW-N21-2336-20WM	12	Schd. 100	Elbow-pipe	
N21	B	B-J	B9.11	3537-5(5)	2	SW-N21-2336-17WB	12	Schd. 100	Elbow-pipe	
N21	B	B-J	B9.11	3537-5(5)	3	SW-N21-2336-17WD	12	Schd. 100	Elbow-pipe	
N21	B	B-J	B9.11	3537-5(5)	4	FW-N21-2336-17W18	12	Schd. 100	Elbow-pipe	
N21	B	B-J	B9.11	3537-5(5)	6	SW-N21-2336-18WP	12	Schd. 100	Elbow-pipe	
N21	B	B-J	B9.11	3537-5(5)	7	SW-N21-2336-18WQ	12	Schd. 100	Elbow-pipe	
N21	B	B-J	B9.11	3537-5(5)	8	FW-N21-2336-18W0	12	Schd. 100	Elbow-pipe	Transition piece
N21	B	B-J	B9.11	3537-5(6)	2	SW-N21-2336-14WB	12	Schd. 100	Elbow-pipe	
N21	B	B-J	B9.11	3537-5(6)	6	SW-N21-2336-15WP	12	Schd. 100	Elbow-pipe	N21-2336-Feedwater Loop A Circ Weld
N21	B	B-J	B9.11	3537-5(6)	7	FW-N21-2336-15WF2	12	Schd. 100	Elbow-pipe	
N21	B	B-J	B9.11	3537-5(5)	1	FW-N21-2336-16W17	12	Schd. 100	Elbow-reducing-tee	
N21	B	B-J	B9.11	3537-5(6)	1	FW-N21-2336-13W14	12	Schd. 100	Elbow-reducing-tee	N21-2336-Feedwater Loop A Circ Weld
N21	B	B-J	B9.11	3537-5(4)	11	3-316C	12	Schd. 100	Nozzle-safe-end	N21-2336-Feedwater Loop A Circ Weld
N21	B	B-J	B9.11	3537-5(5)	10	3-316B	12	Schd. 100	Nozzle-safe-end	N21-2336-Feedwater Loop A Circ Weld
N21	B	B-J	B9.11	3537-5(6)	10	3-316A	12	Schd. 100	Nozzle-safe-end	N21-2336-Feedwater Loop A Circ Weld
N21	B	B-J	B9.11	3537-5(4)	5	FW-N21-2336-20WF2	12	Schd. 100	Pipe-pipe	
N21	B	B-J	B9.11	3537-5(4)	6	FW-N21-2336-20WF3	12	Schd. 100	Pipe-pipe	
N21	B	B-J	B9.11	3537-5(5)	5	FW-N21-2336-18WF1	12	Schd. 100	Pipe-pipe	
N21	B	B-J	B9.11	3537-5(6)	3	FW-N21-2336-14WF1	12	Schd. 100	Pipe-pipe	N21-2336-Feedwater Loop A Circ Weld
N21	B	B-J	B9.11	3537-5(6)	4	FW-N21-2336-14W15	12	Schd. 100	Pipe-pipe	N21-2336-Feedwater Loop A Circ Weld
N21	B	B-J	B9.11	3537-5(6)	5	FW-N21-2336-15WF1	12	Schd. 100	Pipe-pipe	
N21	B	B-J	B9.11	3537-5(4)	1	FW-N21-2336-16W19	12	Schd. 100	Pipe-reducer	N21-2336-Feedwater Loop A Circ Weld
N21	B	B-J	B9.11	3537-5(4)	9	FW-N21-2336-20W0	12	Schd. 100	Pipe-safe-end	

Table D.B.2 BWR-2 – FW System Loop B Weld List

System ID	Loop	Exam Category	Category Item	Line Number	Weld Order	Component ID	NPS (in)	Wall Thk (in) or Schedule	Configuration	Description
N21	B	B-J	B9.11	3537-5(6)	8	FW-N21-2336-15W0	12	Schd. 100	Pipe-safe-end	N21-2336-Feedwater Loop A Circ Weld
N21	B	B-J	B9.11	3537-5(4)	10	N4-C	12	Schd. 100	Safe-end-Safe-end	
N21	B	B-J	B9.11	3537-5(5)	9	N4-B	12	Schd. 100	Safe-end-Safe-end	
N21	B	B-J	B9.11	3537-5(6)	9	N4A	12	Schd. 100	Safe-end-Safe-end	N21-2336-Feedwater Loop A Circ Weld
N21	B	B-J	B9.11	3537-5(2)	3	SW-N21-2336-11WD	20	Schd. 100	Elbow-pipe	
N21	B	B-J	B9.11	3537-5(3)	8	FW-N21-2336-2WB	20	Schd. 100	Elbow-pipe	
N21	B	B-J	B9.11	3537-5(2)	4	SW-N21-2336-11WE	20	Schd. 100	Elbow-pipe	
N21	B	B-J	B9.11	3537-5(3)	12	SW-N21-2336-13WC	20	Schd. 100	Elbow-reducing-tee	N21-2336-Feedwater Loop A Circ Weld
N21	B	B-J	B9.11	3537-5(3)	7	FW-N21-2336-0W02	20	Schd. 100	Elbow-valve	
N21	B	B-J	B9.11	3537-5(3)	11	SW-N21-2336-13WB	20	Schd. 100	Elbow-valve	
N21	B	B-J	B9.11	3537-5(3)	5	SW-X9A-W1	20	Schd. 100	Penetration	Longitudinal weld
N21	B	B-J	B9.11	3537-5(3)	4	FW-N21-2336-11W01	20	Schd. 100	Pipe-penetration	
N21	B	B-J	B9.11	3537-5(2)	2	FW-N21-2336-11WF1	20	Schd. 100	Pipe-pipe	
N21	B	B-J	B9.11	3537-5(2)	5	FW-N21-2336-11WF2	20	Schd. 100	Pipe-pipe	
N21	B	B-J	B9.11	3537-5(2)	6	FW-N21-2336-11WF3	20	Schd. 100	Pipe-pipe	
N21	B	B-J	B9.11	3537-5(3)	14	FW-N21-2336-3AW13	20	Schd. 100	Pipe-pipe	
N21	B	B-J	B9.11	3537-5(3)	15	FW-N21-2336-3W03A	20	Schd. 100	Pipe-pipe	
N21	B	B-J	B9.11	3537-5(3)	16	FW-N21-2336-3W16	20	Schd. 100	Pipe-pipe	
N21	B	B-J	B9.11	3537-5(2)	1	SW-N21-2336-11WB	20	Schd. 100	Pipe-reducer	
N21	B	B-J	B9.11	3537-5(3)	13	SW-N21-2336-13WE	20	Schd. 100	Pipe-reducing-tee	N21-2336-Feedwater Loop A Circ Weld
N21	B	B-J	B9.11	3537-5(3)	17	SW-N21-2336-16WC	20	Schd. 100	Pipe-reducing-tee	
N21	B	B-J	B9.11	3537-5(2)	7	SW-N21-2336-11WN	20	Schd. 100	Pipe-tee	
N21	B	B-J	B9.11	3537-5(3)	3	SW-N21-2336-11WH	20	Schd. 100	Pipe-tee	
N21	B	B-J	B9.11	3537-5(3)	1	FW-N21-2336-1VW11	20	Schd. 100	Pipe-valve	
N21	B	B-J	B9.11	3537-5(3)	6	FW-N21-2336-1VW12	20	Schd. 100	Pipe-valve	
N21	B	B-J	B9.11	3537-5(3)	9	FW-N21-2336-2W0	20	Schd. 100	Pipe-valve	N21-2336-Feedwater Loop A Circ Weld
N21	B	B-J	B9.11	3537-5(3)	10	FW-N21-2336-0W13	20	Schd. 100	Pipe-valve	N21-2336-Feedwater Loop A Circ Weld
N21	B	B-J	B9.11	3537-5(3)	18	SW-N21-2336-16WE	20	Schd. 100	Reducer-reducing-tee	
N21	B	B-J	B9.11	3537-5(3)	2	FW-N21-2336-0W11	20	Schd. 100	Tee-valve	

Table D.B.3 PWR-1 – Reactor Coolant System Hot Leg

Examination Category	Category Item	Component ID	Configuration	NPS (in)	Wall Thk (in)	Weld Material
B-J	B9.11	1-4100A- 8	ELBOW TO PIPE	31.00	2.600	304N/351CF
B-J	B9.11	1-4100A- 9	ELBOW TO PIPE	31.00	2.625	SS
B-J	B9.11	1-4100A- 10	ELBOW TO PIPE	31.00	2.625	SS
B-J	B9.11	1-4100A- 11	PIPE TO ELBOW	31.00	2.625	SS
B-J	B9.11	1-4100A- 12	ELBOW TO PUMP	31.00	2.625	SS
B-J	B9.11	1-4100A- 13	PIPE TO PUMP	27.50	2.375	304N
B-J	B9.11	1-4100A- 14	PIPE TO ELBOW	27.50	2.375	SS
B-J	B9.11	1-4100A- 15	BIMETAL (INCONEL) WELD.ELBOW TO SAFE END	27.50	2.375	SS/INCONEL
B-F	B5.10	1-4100A- 16(DM)	BIMETAL (INCONEL) WELD. R.V. LOOP A INLET NOZZLE TO SAFE END	27.50	2.375	CS/SS

Table D.B.4 PWR-2 – Pressurizer Surge Line

ASME XI Examination Category	Component ID	Description / Configuration	NPS [inch]	Wall Thk [inch]	Weld Material
B-J	1-4100A- 19BC	14" Branch (CGE-1-4500) Connection to 29-inch pipe	14.00	2.350	304N/SA182
B-F	1-4500A- 1(DM)	Bimetal (INCONEL) weld; Pressurizer Surge Line Nozzle to Safe End	14.00	1.406	SS
B-J	1-4500A- 2	Bimetal (INCONEL) Weld: Safe End to Pipe	14.00	1.406	SS
B-J	1-4500A- 3	Pipe-to-Elbow	14.00	1.406	SS
B-J	1-4500A- 4	Pipe-to-Elbow	14.00	1.406	SS
B-J	1-4500A- 5	Pipe-to-Elbow	14.00	1.406	SS
B-J	1-4500A- 6	Pipe-to-Elbow	14.00	1.406	SS
B-J	1-4500A- 7	Pipe-to-Elbow	14.00	1.406	SS
B-J	1-4500A- 8	Pipe-to-Elbow	14.00	1.406	SS
B-J	1-4500A- 9	Pipe-to-Elbow	14.00	1.406	SS
B-J	1-4500A- 10	Pipe-to-Elbow	14.00	1.406	SS
B-J	1-4500A- 11	Pipe-to-Elbow	14.00	1.406	SS
B-J	1-4500A- 12	Pipe-to-Elbow	14.00	1.406	SS
B-J	1-4500A- 13	Pipe-to-Branch Connection	14.00	1.406	SS

Table D.B.5 PWR-5 – HPI/NMU Line

Weld ID	Configuration	Inside Diameter
17-MU-23-73AR	Elbow-valve	2.50
17-MU-23-21-055	Elbow-pipe	2.50
17-MU-23-21-057	Elbow-pipe	2.50
17-MU-23-21-058	Elbow-pipe	2.50
17-MU-23-21-059	Elbow-valve	2.50
17-MU-23-21-059B	Elbow-valve	2.50
17-MU-23-21-061	Pipe-pipe	2.50
17-MU-23-21-062	Elbow-pipe	2.50
17-MU-23-21-063	Elbow-nozzle	2.50

Table D.B.6 BWR-1 Weld Failure Rate Calculation Sheet

Description	Mean (Prior)[24]	Range Factor	Failures (Evidence)	Exposure [Weld-Yr]	Weld Count[25]	Mean	5th	50th	95th	Range Factor	Calc Date
NPS28 E-P Low	7.19E-04	10	0	4,725	5	1.28E-04	1.41E-05	8.78E-05	3.70E-04	5.1	5/13/2003 15:16
NPS28 E-P Medium	7.19E-04	10	0	9,450	10	8.37E-05	1.10E-05	6.09E-05	2.29E-04	4.6	5/13/2003 15:16
NPS28 E-P High	7.19E-04	10	0	14,175	15	6.42E-05	9.28E-06	4.81E-05	1.70E-04	4.3	5/13/2003 15:16
NPS28 P-V Low	2.25E-04	10	0	3,780	4	8.20E-05	6.29E-06	4.84E-05	2.66E-04	6.5	5/13/2003 15:16
NPS28 P-V Medium	2.25E-04	10	0	7,560	8	5.86E-05	5.41E-06	3.75E-05	1.80E-04	5.8	5/13/2003 15:16
NPS28 P-V High	2.25E-04	10	0	11,340	12	4.71E-05	4.84E-06	3.15E-05	1.40E-04	5.4	5/13/2003 15:16
NPS28 N-Se Low	1.50E-04	10	0	1,890	2	8.34E-05	4.89E-06	4.27E-05	2.93E-04	7.7	5/13/2003 15:16
NPS28 N-Se Medium	1.50E-04	10	0	3,780	4	6.49E-05	4.49E-06	3.64E-05	2.18E-04	7.0	5/13/2003 15:16
NPS28 N-Se High	1.50E-04	10	0	5,670	6	5.46E-05	4.20E-06	3.23E-05	1.78E-04	6.5	5/13/2003 15:16
NPS28 P-Se Low	6.74E-04	10	0	1,890	2	2.03E-04	1.74E-05	1.26E-04	6.39E-04	6.1	5/13/2003 15:16
NPS28 P-Se Medium	6.74E-04	10	0	3,780	4	1.41E-04	1.45E-05	9.43E-05	4.19E-04	5.4	5/13/2003 15:16
NPS28 P-Se High	6.74E-04	10	0	5,670	6	1.12E-04	1.28E-05	7.76E-05	3.20E-04	5.0	5/13/2003 15:16
NPS28 P-P Low	4.99E-05	20	0	8,505	9	1.75E-05	3.99E-07	6.56E-06	7.03E-05	13.3	5/13/2003 15:16
NPS28 P-P Medium	4.99E-05	20	0	17,010	18	1.27E-05	3.64E-07	5.45E-06	4.91E-05	11.6	5/13/2003 15:16
NPS28 P-P High	4.99E-05	20	0	25,515	27	1.04E-05	3.39E-07	4.77E-06	3.88E-05	10.7	5/13/2003 15:16
NPS28 P-Pu Low	2.25E-04	10	0	1,890	2	1.09E-04	7.01E-06	5.88E-05	3.73E-04	7.3	5/13/2003 15:16
NPS28 P-Pu Medium	2.25E-04	10	0	3,780	4	8.20E-05	6.29E-06	4.84E-05	2.66E-04	6.5	5/13/2003 15:16
NPS28 P-Pu High	2.25E-04	10	0	5,670	6	6.78E-05	5.79E-06	4.20E-05	2.13E-04	6.1	5/13/2003 15:16
NPS28 P-T Low	1.25E-04	10	0	1,890	2	7.36E-05	4.15E-06	3.68E-05	2.61E-04	7.9	5/13/2003 15:16
NPS28 P-T Medium	1.25E-04	10	0	3,780	4	5.81E-05	3.84E-06	3.18E-05	1.98E-04	7.2	5/13/2003 15:16
NPS28 P-T High	1.25E-04	10	0	5,670	6	4.93E-05	3.61E-06	2.85E-05	1.63E-04	6.7	5/13/2003 15:16
NPS28 P-X Low	7.49E-05	20	0	945	1	4.63E-05	6.74E-07	1.27E-05	1.95E-04	17.0	5/13/2003 15:16
NPS28 P-X Medium	7.49E-05	20	0	1,890	2	3.79E-05	6.51E-07	1.17E-05	1.58E-04	15.6	5/13/2003 15:16
NPS28 P-X High	7.49E-05	20	0	2,835	3	3.29E-05	6.33E-07	1.10E-05	1.36E-04	14.6	5/13/2003 15:16
NPS28 R-X Low	7.49E-05	20	0	945	1	4.63E-05	6.74E-07	1.27E-05	1.95E-04	17.0	5/13/2003 15:16
NPS28 R-X Medium	7.49E-05	20	0	1,890	2	3.79E-05	6.51E-07	1.17E-05	1.58E-04	15.6	5/13/2003 15:16
NPS28 R-X High	7.49E-05	20	0	2,835	3	3.29E-05	6.33E-07	1.10E-05	1.36E-04	14.6	5/13/2003 15:16

---

[24] From Table 11 (RR NPS28)
[25] Weld count (medium value) is taken from PlantBWelds xls (see Section 1 3); Table B-1 is an excerpt from this Excel-file

Table D.B.7 BWR-2 Weld Failure Rate Calculation Sheet

Description	Mean (Prior)	Range Factor	Failures (Evidence)	Exposure [Weld-Yr]	Welds	Mean	5th	50th	95th	Range Factor	Calc Date
NPS12 N-Se Low	1.26E-04	10	0	2,835	3	6.47E-05	4.00E-06	3.42E-05	2.24E-04	7.5	5/13/2003 15:05
NPS12 N-Se Medium	1.26E-04	10	0	5,670	6	4.95E-05	3.63E-06	2.86E-05	1.63E-04	6.7	5/13/2003 15:05
NPS12 N-Se High	1.26E-04	10	0	8,505	9	4.12E-05	3.36E-06	2.50E-05	1.32E-04	6.3	5/13/2003 15:05
NPS12 E-P Low	6.73E-06	20	0	13,230	14	3.92E-06	6.00E-08	1.11E-06	1.64E-05	16.6	5/13/2003 15:05
NPS12 E-P Medium	6.73E-06	20	0	26,460	28	3.15E-06	5.77E-08	1.02E-06	1.31E-05	15.1	5/13/2003 15:05
NPS12 E-P High	6.73E-06	20	0	39,690	42	2.71E-06	5.58E-08	9.51E-07	1.11E-05	14.1	5/13/2003 15:05
NPS12 P-P Low	1.05E-05	15	0	5,670	6	7.48E-06	1.71E-07	2.45E-06	3.01E-05	13.3	5/13/2003 15:05
NPS12 P-P Medium	1.05E-05	15	0	11,340	12	6.33E-06	1.66E-07	2.29E-06	2.53E-05	12.3	5/13/2003 15:05
NPS12 P-P High	1.05E-05	15	0	17,010	18	5.61E-06	1.62E-07	2.17E-06	2.21E-05	11.7	5/13/2003 15:05
NPS12 P-R Low	3.14E-05	15	0	945	1	2.54E-05	5.23E-07	7.64E-06	1.02E-04	13.9	5/13/2003 15:05
NPS12 P-R Medium	3.14E-05	15	0	1,890	2	2.24E-05	5.14E-07	7.34E-06	9.04E-05	13.3	5/13/2003 15:05
NPS12 P-R High	3.14E-05	15	0	2,835	3	2.05E-05	5.06E-07	7.09E-06	8.22E-05	12.7	5/13/2003 15:05
NPS12 P-Se Low	8.80E-06	20	0	4,725	5	6.16E-06	8.05E-08	1.55E-06	2.57E-05	17.9	5/13/2003 15:05
NPS12 P-Se Medium	8.80E-06	20	0	9,450	10	5.22E-06	7.86E-08	1.47E-06	2.19E-05	16.7	5/13/2003 15:05
NPS12 P-Se High	8.80E-06	20	0	14,175	15	4.63E-06	7.71E-08	1.40E-06	1.94E-05	15.9	5/13/2003 15:05
NPS12 E-Rt Low	9.43E-06	20	0	1,890	2	7.63E-06	8.76E-08	1.72E-06	3.12E-05	18.9	5/13/2003 15:05
NPS12 E-Rt Medium	9.43E-06	20	0	3,780	4	6.80E-06	8.65E-08	1.67E-06	2.83E-05	18.1	5/13/2003 15:05
NPS12 E-Rt High	9.43E-06	20	0	5,670	6	6.24E-06	8.56E-08	1.63E-06	2.62E-05	17.5	5/13/2003 15:05

Table D.B.8 PWR-1 Weld Failure Rate Calculation Sheet

Description	Mean (Prior)[26]	Range Factor	Failures (Evidence)	Exposure [Weld-Yr]	Weld Count	Mean	5th	50th	95th	Range Factor	Calc Date
NPS30 E-P Low	3.65E-06	100	0	6,720	20	1.49E-06	7.05E-10	6.85E-08	5.61E-06	89.2	5/2/2003 9 25
NPS30 E-P Medium	3.65E-06	100	0	13,440	40	1.19E-06	6.96E-10	6.65E-08	4.88E-06	83.7	5/2/2003 9 25
NPS30 E-P High	3.65E-06	100	0	20,160	60	1.02E-06	6.89E-10	6.49E-08	4.40E-06	79.9	5/2/2003 9 25
NPS30 P-Pu Low	4.06E-05	100	0	504	2	1.75E-05	7.86E-09	7.67E-07	6.42E-05	90.4	5/2/2003 9 25
NPS30 P-Pu Medium	4.06E-05	100	0	1,008	3	1.40E-05	7.77E-09	7.47E-07	5.66E-05	85.3	5/2/2003 9 25
NPS30 P-Pu High	4.06E-05	100	0	1,512	5	1.22E-05	7.71E-09	7.30E-07	5.14E-05	81.7	5/2/2003 9 25
NPS30 N-Se Low	8.12E-04	100	0	504	2	1.06E-04	1.42E-07	1.19E-05	5.07E-04	59.8	5/2/2003 9 25
NPS30 N-Se Medium	8.12E-04	100	0	1,008	3	7.35E-05	1.33E-07	1.04E-05	3.53E-04	51.4	5/2/2003 9 25
NPS30 N-Se High	8.12E-04	100	0	1,512	5	5.85E-05	1.28E-07	9.41E-06	2.80E-04	46.8	5/2/2003 9 25
NPS30 E-Se Low	8.12E-05	100	0	504	2	2.81E-05	1.55E-08	1.49E-06	1.13E-04	85.3	5/2/2003 9 25
NPS30 E-Se Medium	8.12E-05	100	0	1,008	3	2.18E-05	1.53E-08	1.43E-06	9.49E-05	78.8	5/2/2003 9 25
NPS30 E-Se High	8.12E-05	100	0	1,512	5	1.85E-05	1.51E-08	1.39E-06	8.34E-05	74.3	5/2/2003 9 25
NPS30 E-Pu Low	5.07E-05	100	0	504	2	2.04E-05	9.78E-09	9.51E-07	7.74E-05	89.0	5/2/2003 9 25
NPS30 E-Pu Medium	5.07E-05	100	0	1,008	3	1.62E-05	9.66E-09	9.22E-07	6.72E-05	83.4	5/2/2003 9 25
NPS30 E-Pu High	5.07E-05	100	0	1,512	5	1.40E-05	9.57E-09	8.99E-07	6.04E-05	79.4	5/2/2003 9 25

[26] From Table 12 (RC Hot Leg)

Table D.B.9 PWR-2 Weld Failure Rate Calculation Sheet

Description	Mean (Prior)	Range Factor	Failures (Evidence)	Exposure	Welds Count	Mean	5th	50th	95th	Range Factor	Calc Date
NPS14 E-P Low	6.70E-07	100	0	18,105	5	3.33E-07	1.30E-10	1.28E-08	1.13E-06	93.3	5/2/2003 10 14
NPS14 E-P Medium	6.70E-07	100	0	36,210	10	2.74E-07	1.29E-10	1.26E-08	1.03E-06	89.3	5/2/2003 10 14
NPS14 E-P High	6.70E-07	100	0	54,315	15	2.41E-07	1.29E-10	1.24E-08	9.57E-07	86.3	5/2/2003 10 14
NPS14 N-Se Low	3.75E-05	100	0	1,811	1	1.08E-05	7.10E-09	6.70E-07	4.62E-05	80.7	5/2/2003 10 14
NPS14 N-Se Medium	3.75E-05	100	0	3,621	1	8.18E-06	6.94E-09	6.34E-07	3.72E-05	73.2	5/2/2003 10 14
NPS14 N-Se High	3.75E-05	100	0	5,432	2	6.84E-06	6.81E-09	6.06E-07	3.19E-05	68.4	5/2/2003 10 14
NPS14 P-Se Low	3.35E-06	100	0	1,811	1	1.96E-06	6.55E-10	6.48E-08	6.03E-06	96.0	5/2/2003 10 14
NPS14 P-Se Medium	3.35E-06	100	0	3,621	1	1.67E-06	6.52E-10	6.41E-08	5.67E-06	93.3	5/2/2003 10 14
NPS14 P-Se High	3.35E-06	100	0	5,432	2	1.49E-06	6.49E-10	6.35E-08	5.39E-06	91.1	5/2/2003 10 14
NPS14 B-HL Low	4.69E-05	100	0	1,811	1	1.24E-05	8.82E-09	8.25E-07	5.42E-05	78.4	5/2/2003 10 14
NPS14 B-HL Medium	4.69E-05	100	0	3,621	1	9.28E-06	8.60E-09	7.74E-07	4.28E-05	70.6	5/2/2003 10 14
NPS14 B-HL High	4.69E-05	100	0	5,432	2	7.71E-06	8.42E-09	7.37E-07	3.63E-05	65.7	5/2/2003 10 14
NPS14 B-P Low	3.35E-06	100	0	1,811	1	1.96E-06	6.55E-10	6.48E-08	6.03E-06	96.0	5/2/2003 10 14
NPS14 B-P Medium	3.35E-06	100	0	3,621	1	1.67E-06	6.52E-10	6.41E-08	5.67E-06	93.3	5/2/2003 10 14
NPS14 B-P High	3.35E-06	100	0	5,432	2	1.49E-06	6.49E-10	6.35E-08	5.39E-06	91.1	5/2/2003 10 14

Table D.B.10 PWR-3 Weld Failure Rate Calculation Sheet

Description	Mean (Prior)	Range Factor	Failures (Evidence)	Exposure	Weld Count	Mean	5th	50th	95th	Range Factor	Calc Date
NPS4 E-P Low	1.90E-06	97	0	140	4	1.69E-06	4.05E-10	3.93E-08	3.79E-06	96.7	5/2/2003 14 14
NPS4 E-P Medium	1.90E-06	97	0	280	8	1.61E-06	4.05E-10	3.92E-08	3.77E-06	96.5	5/2/2003 14 14
NPS4 E-P High	1.90E-06	97	0	420	12	1.54E-06	4.05E-10	3.92E-08	3.75E-06	96.3	5/2/2003 14 14
NPS4 E-V Low	1.31E-06	98	0	105	3	1.21E-06	2.69E-10	2.64E-08	2.59E-06	98.2	5/2/2003 14 14
NPS4 E-V Medium	1.31E-06	98	0	210	6	1.16E-06	2.69E-10	2.64E-08	2.59E-06	98.1	5/2/2003 14 14
NPS4 E-V High	1.31E-06	98	0	315	9	1.13E-06	2.69E-10	2.64E-08	2.58E-06	98.0	5/2/2003 14 14
NPS4 P-P Low	6.48E-06	100	0	35	1	5.81E-06	1.28E-09	1.27E-07	1.27E-05	99.6	5/2/2003 14 14
NPS4 P-P Medium	6.48E-06	100	0	70	2	5.52E-06	1.28E-09	1.27E-07	1.26E-05	99.4	5/2/2003 14 14
NPS4 P-P High	6.48E-06	100	0	105	3	5.32E-06	1.28E-09	1.27E-07	1.26E-05	99.2	5/2/2003 14 14
NPS4 E-N Low	9.86E-04	16	0	35	1	7.59E-04	1.37E-05	2.17E-04	3.09E-03	15.0	5/2/2003 14 15
NPS4 E-N Medium	9.86E-04	16	0	70	2	6.61E-04	1.34E-05	2.07E-04	2.70E-03	14.2	5/2/2003 14 15
NPS4 E-N High	9.86E-04	16	0	105	3	5.97E-04	1.32E-05	2.00E-04	2.43E-03	13.6	5/2/2003 14 15

ATTACHMENT C TO APPENDIX D

BASIS FOR LOCA FREQUENCY MODELS

Attached as Table D.C.1 is an Excel spreadsheet used to calculate time-dependent LOCA frequencies. Table D.C.1 includes the parameters input to the BWR-1 Cat1 LOCA frequency calculation.

Table D.C.1 Application of Markov Model to BWR-1 Cat 1 LOCA Frequency

Config. (Count)	CASE	ISI Inspection Coverage	Leak Inspection Interval	P_LI	TSUBLI	MU	PHI	RHO_F	RHO_L	Lambda	T_FI	TSUBR	P_FI	P_FD	OMEGA	Haz[T]
Cross-to-reducer (2)	1	None	None	0.00	1.50	0.0000	3.34E-03	5.96E-07	2.00E-02	4.40E-04	10.00	0.023	0.000	0.000	0.000	7.62E-06
	2	None	RF	0.90	1.50	0.5910	3.34E-03	5.96E-07	2.00E-02	4.40E-04	10.00	0.023	0.000	0.000	0.000	1.12E-06
	3	None	Wk	0.90	1.92E-02	21.3971	3.34E-03	5.96E-07	2.00E-02	4.40E-04	10.00	0.023	0.000	0.000	0.000	8.01E-08
	4	None	Shift	0.90	9.13E-04	37.9048	3.34E-03	5.96E-07	2.00E-02	4.40E-04	10.00	0.023	0.000	0.000	0.000	6.59E-08
	5	Secondary	None	0.00	1.50	0.0000	3.34E-03	5.96E-07	2.00E-02	4.40E-04	10.00	0.023	1.000	0.500	0.050	5.18E-06
	6	Secondary	RF	0.90	1.50	0.5910	3.34E-03	5.96E-07	2.00E-02	4.40E-04	10.00	0.023	1.000	0.500	0.050	6.66E-07
	7	Secondary	Wk	0.90	1.92E-02	21.3971	3.34E-03	5.96E-07	2.00E-02	4.40E-04	10.00	0.023	1.000	0.500	0.050	4.62E-08
	8	Secondary	Shift	0.90	9.13E-04	37.9048	3.34E-03	5.96E-07	2.00E-02	4.40E-04	10.00	0.023	1.000	0.500	0.050	3.80E-08
	9	Primary	None	0.00	1.50	0.0000	3.34E-03	5.96E-07	2.00E-02	4.40E-04	10.00	0.023	1.000	0.900	0.090	4.02E-06
	10	Primary	RF	0.90	1.50	0.5910	3.34E-03	5.96E-07	2.00E-02	4.40E-04	10.00	0.023	1.000	0.900	0.090	4.74E-07
	11	Primary	Wk	0.90	1.92E-02	21.3971	3.34E-03	5.96E-07	2.00E-02	4.40E-04	10.00	0.023	1.000	0.900	0.090	3.24E-08
	12	Primary	Shift	0.90	9.13E-04	37.9048	3.34E-03	5.96E-07	2.00E-02	4.40E-04	10.00	0.023	1.000	0.900	0.090	2.67E-08
Cross-to-tee (2)	1	None	None	0.00	1.50	0.0000	3.29E-03	5.88E-07	2.00E-02	4.34E-04	10.00	0.023	0.000	0.000	0.000	7.43E-06
	2	None	RF	0.90	1.50	0.5910	3.29E-03	5.88E-07	2.00E-02	4.34E-04	10.00	0.023	0.000	0.000	0.000	1.09E-06
	3	None	Wk	0.90	1.92E-02	21.3971	3.29E-03	5.88E-07	2.00E-02	4.34E-04	10.00	0.023	0.000	0.000	0.000	7.81E-08
	4	None	Shift	0.90	9.13E-04	37.9048	3.29E-03	5.88E-07	2.00E-02	4.34E-04	10.00	0.023	0.000	0.000	0.000	6.42E-08
	5	Secondary	None	0.00	1.50	0.0000	3.29E-03	5.88E-07	2.00E-02	4.34E-04	10.00	0.023	1.000	0.500	0.050	5.05E-06
	6	Secondary	RF	0.90	1.50	0.5910	3.29E-03	5.88E-07	2.00E-02	4.34E-04	10.00	0.023	1.000	0.500	0.050	6.49E-07
	7	Secondary	Wk	0.90	1.92E-02	21.3971	3.29E-03	5.88E-07	2.00E-02	4.34E-04	10.00	0.023	1.000	0.500	0.050	4.50E-08
	8	Secondary	Shift	0.90	9.13E-04	37.9048	3.29E-03	5.88E-07	2.00E-02	4.34E-04	10.00	0.023	1.000	0.500	0.050	3.71E-08
	9	Primary	None	0.00	1.50	0.0000	3.29E-03	5.88E-07	2.00E-02	4.34E-04	10.00	0.023	1.000	0.900	0.090	3.92E-06
	10	Primary	RF	0.90	1.50	0.5910	3.29E-03	5.88E-07	2.00E-02	4.34E-04	10.00	0.023	1.000	0.900	0.090	4.62E-07
	11	Primary	Wk	0.90	1.92E-02	21.3971	3.29E-03	5.88E-07	2.00E-02	4.34E-04	10.00	0.023	1.000	0.900	0.090	3.16E-08
	12	Primary	Shift	0.90	9.13E-04	37.9048	3.29E-03	5.88E-07	2.00E-02	4.34E-04	10.00	0.023	1.000	0.900	0.090	2.60E-08
Elbow-to-pipe (10)	1	None	None	0.00	1.50	0.0000	7.36E-03	1.32E-06	2.00E-02	9.71E-04	10.00	0.023	0.000	0.000	0.000	3.57E-05
	2	None	RF	0.90	1.50	0.5910	7.36E-03	1.32E-06	2.00E-02	9.71E-04	10.00	0.023	0.000	0.000	0.000	5.19E-06
	3	None	Wk	0.90	1.92E-02	21.3971	7.36E-03	1.32E-06	2.00E-02	9.71E-04	10.00	0.023	0.000	0.000	0.000	3.69E-07
	4	None	Shift	0.90	9.13E-04	37.9048	7.36E-03	1.32E-06	2.00E-02	9.71E-04	10.00	0.023	0.000	0.000	0.000	3.04E-07
	5	Secondary	None	0.00	1.50	0.0000	7.36E-03	1.32E-06	2.00E-02	9.71E-04	10.00	0.023	1.000	0.500	0.050	2.45E-05
	6	Secondary	RF	0.90	1.50	0.5910	7.36E-03	1.32E-06	2.00E-02	9.71E-04	10.00	0.023	1.000	0.500	0.050	3.11E-06
	7	Secondary	Wk	0.90	1.92E-02	21.3971	7.36E-03	1.32E-06	2.00E-02	9.71E-04	10.00	0.023	1.000	0.500	0.050	2.15E-07
	8	Secondary	Shift	0.90	9.13E-04	37.9048	7.36E-03	1.32E-06	2.00E-02	9.71E-04	10.00	0.023	1.000	0.500	0.050	1.77E-07
	9	Primary	None	0.00	1.50	0.0000	7.36E-03	1.32E-06	2.00E-02	9.71E-04	10.00	0.023	1.000	0.900	0.090	1.91E-05
	10	Primary	RF	0.90	1.50	0.5910	7.36E-03	1.32E-06	2.00E-02	9.71E-04	10.00	0.023	1.000	0.900	0.090	2.23E-06
	11	Primary	Wk	0.90	1.92E-02	21.3971	7.36E-03	1.32E-06	2.00E-02	9.71E-04	10.00	0.023	1.000	0.900	0.090	1.52E-07
	12	Primary	Shift	0.90	9.13E-04	37.9048	7.36E-03	1.32E-06	2.00E-02	9.71E-04	10.00	0.023	1.000	0.900	0.090	1.25E-07
Nozzle-to-	1	None	None	0.00	1.50	0.0000	5.66E-03	1.01E-06	2.00E-02	7.47E-04	10.00	0.023	0.000	0.000	0.000	2.15E-05

Table D.C.1 Application of Markov Model to BWR-1 Cat 1 LOCA Frequency

Config. (Count)	CASE	ISI Inspection Coverage	Leak Inspection Interval	P_{LI}	TSUBLI	MU	PHI	RHO_F	RHO_L	Lambda	T_{FI}	TSUBR	P_{FI}	P_{FD}	OMEGA	Haz(T)
safe-end (2)	2	None	RF	0.90	1.50	0.5910	5.66E-03	1.01E-06	2.00E-02	7.47E-04	10.00	0.023	0.000	0.000	0.000	3.13E-06
	3	None	Wk	0.90	1.92E-02	21.3971	5.66E-03	1.01E-06	2.00E-02	7.47E-04	10.00	0.023	0.000	0.000	0.000	2.23E-07
	4	None	Shift	0.90	9.13E-04	37.9048	5.66E-03	1.01E-06	2.00E-02	7.47E-04	10.00	0.023	0.000	0.000	0.000	1.84E-07
	5	Secondary	None	0.00	1.50	0.0000	5.66E-03	1.01E-06	2.00E-02	7.47E-04	10.00	0.023	1.000	0.500	0.050	1.47E-05
	6	Secondary	RF	0.90	1.50	0.5910	5.66E-03	1.01E-06	2.00E-02	7.47E-04	10.00	0.023	1.000	0.500	0.050	1.87E-06
	7	Secondary	Wk	0.90	1.92E-02	21.3971	5.66E-03	1.01E-06	2.00E-02	7.47E-04	10.00	0.023	1.000	0.500	0.050	1.30E-07
	8	Secondary	Shift	0.90	9.13E-04	37.9048	5.66E-03	1.01E-06	2.00E-02	7.47E-04	10.00	0.023	1.000	0.500	0.050	1.07E-07
	9	Primary	None	0.00	1.50	0.0000	5.66E-03	1.01E-06	2.00E-02	7.47E-04	10.00	0.023	1.000	0.900	0.090	1.14E-05
	10	Primary	RF	0.90	1.50	0.5910	5.66E-03	1.01E-06	2.00E-02	7.47E-04	10.00	0.023	1.000	0.900	0.090	1.34E-06
	11	Primary	Wk	0.90	1.92E-02	21.3971	5.66E-03	1.01E-06	2.00E-02	7.47E-04	10.00	0.023	1.000	0.900	0.090	9.15E-08
	12	Primary	Shift	0.90	9.13E-04	37.9048	5.66E-03	1.01E-06	2.00E-02	7.47E-04	10.00	0.023	1.000	0.900	0.090	7.52E-08
Pipe-to-pipe (18)	1	None	None	0.00	1.50	0.0000	1.11E-03	1.99E-07	2.00E-02	1.47E-04	10.00	0.023	0.000	0.000	0.000	8.70E-07
	2	None	RF	0.90	1.50	0.5910	1.11E-03	1.99E-07	2.00E-02	1.47E-04	10.00	0.023	0.000	0.000	0.000	1.29E-07
	3	None	Wk	0.90	1.92E-02	21.3971	1.11E-03	1.99E-07	2.00E-02	1.47E-04	10.00	0.023	0.000	0.000	0.000	9.23E-09
	4	None	Shift	0.90	9.13E-04	37.9048	1.11E-03	1.99E-07	2.00E-02	1.47E-04	10.00	0.023	0.000	0.000	0.000	7.59E-09
	5	Secondary	None	0.00	1.50	0.0000	1.11E-03	1.99E-07	2.00E-02	1.47E-04	10.00	0.023	1.000	0.500	0.050	5.89E-07
	6	Secondary	RF	0.90	1.50	0.5910	1.11E-03	1.99E-07	2.00E-02	1.47E-04	10.00	0.023	1.000	0.500	0.050	7.62E-08
	7	Secondary	Wk	0.90	1.92E-02	21.3971	1.11E-03	1.99E-07	2.00E-02	1.47E-04	10.00	0.023	1.000	0.500	0.050	5.29E-09
	8	Secondary	Shift	0.90	9.13E-04	37.9048	1.11E-03	1.99E-07	2.00E-02	1.47E-04	10.00	0.023	1.000	0.500	0.050	4.35E-09
	9	Primary	None	0.00	1.50	0.0000	1.11E-03	1.99E-07	2.00E-02	1.47E-04	10.00	0.023	1.000	0.900	0.090	4.55E-07
	10	Primary	RF	0.90	1.50	0.5910	1.11E-03	1.99E-07	2.00E-02	1.47E-04	10.00	0.023	1.000	0.900	0.090	5.40E-08
	11	Primary	Wk	0.90	1.92E-02	21.3971	1.11E-03	1.99E-07	2.00E-02	1.47E-04	10.00	0.023	1.000	0.900	0.090	3.70E-09
	12	Primary	Shift	0.90	9.13E-04	37.9048	1.11E-03	1.99E-07	2.00E-02	1.47E-04	10.00	0.023	1.000	0.900	0.090	3.04E-09
Pipe-to-pump (4)	1	None	None	0.00	1.50	0.0000	7.23E-03	1.29E-06	2.00E-02	9.53E-04	10.00	0.023	0.000	0.000	0.000	3.44E-05
	2	None	RF	0.90	1.50	0.5910	7.23E-03	1.29E-06	2.00E-02	9.53E-04	10.00	0.023	0.000	0.000	0.000	5.00E-06
	3	None	Wk	0.90	1.92E-02	21.3971	7.23E-03	1.29E-06	2.00E-02	9.53E-04	10.00	0.023	0.000	0.000	0.000	3.56E-07
	4	None	Shift	0.90	9.13E-04	37.9048	7.23E-03	1.29E-06	2.00E-02	9.53E-04	10.00	0.023	0.000	0.000	0.000	2.93E-07
	5	Secondary	None	0.00	1.50	0.0000	7.23E-03	1.29E-06	2.00E-02	9.53E-04	10.00	0.023	1.000	0.500	0.050	2.36E-05
	6	Secondary	RF	0.90	1.50	0.5910	7.23E-03	1.29E-06	2.00E-02	9.53E-04	10.00	0.023	1.000	0.500	0.050	3.00E-06
	7	Secondary	Wk	0.90	1.92E-02	21.3971	7.23E-03	1.29E-06	2.00E-02	9.53E-04	10.00	0.023	1.000	0.500	0.050	2.08E-07
	8	Secondary	Shift	0.90	9.13E-04	37.9048	7.23E-03	1.29E-06	2.00E-02	9.53E-04	10.00	0.023	1.000	0.500	0.050	1.71E-07
	9	Primary	None	0.00	1.50	0.0000	7.23E-03	1.29E-06	2.00E-02	9.53E-04	10.00	0.023	1.000	0.900	0.090	1.84E-05
	10	Primary	RF	0.90	1.50	0.5910	7.23E-03	1.29E-06	2.00E-02	9.53E-04	10.00	0.023	1.000	0.900	0.090	2.15E-06
	11	Primary	Wk	0.90	1.92E-02	21.3971	7.23E-03	1.29E-06	2.00E-02	9.53E-04	10.00	0.023	1.000	0.900	0.090	1.47E-07
	12	Primary	Shift	0.90	9.13E-04	37.9048	7.23E-03	1.29E-06	2.00E-02	9.53E-04	10.00	0.023	1.000	0.900	0.090	1.21E-07
Pipe-to-valve (8)	1	None	None	0.00	1.50	0.0000	5.15E-03	9.20E-07	2.00E-02	6.80E-04	10.00	0.023	0.000	0.000	0.000	1.78E-05
	2	None	RF	0.90	1.50	0.5910	5.15E-03	9.20E-07	2.00E-02	6.80E-04	10.00	0.023	0.000	0.000	0.000	2.61E-06
	3	None	Wk	0.90	1.92E-02	21.3971	5.15E-03	9.20E-07	2.00E-02	6.80E-04	10.00	0.023	0.000	0.000	0.000	1.86E-07

Table D.C.1 Application of Markov Model to BWR-1 Cat 1 LOCA Frequency

Config. (Count)	CASE	ISI Inspection Coverage	Leak Inspection Interval	P_LI	TSUBLI	MU	PHI	RHO_F	RHO_L	Lambda	T_FI	TSUBR	P_FI	P_FD	OMEGA	Haz(T)
	4	None	Shift	0.90	9.13E-04	37.9048	5.15E-03	9.20E-07	2.00E-02	6.80E-04	10.00	0.023	0.000	0.000	0.000	1.53E-07
	5	Secondary	None	0.00	1.50	0.0000	5.15E-03	9.20E-07	2.00E-02	6.80E-04	10.00	0.023	0.500	0.500	0.050	1.22E-05
	6	Secondary	RF	0.90	1.50	0.5910	5.15E-03	9.20E-07	2.00E-02	6.80E-04	10.00	0.023	1.000	0.500	0.050	1.56E-06
	7	Secondary	Wk	0.90	1.92E-02	21.3971	5.15E-03	9.20E-07	2.00E-02	6.80E-04	10.00	0.023	1.000	0.500	0.050	1.08E-07
	8	Secondary	Shift	0.90	9.13E-04	37.9048	5.15E-03	9.20E-07	2.00E-02	6.80E-04	10.00	0.023	1.000	0.500	0.050	8.88E-08
	9	Primary	None	0.00	1.50	0.0000	5.15E-03	9.20E-07	2.00E-02	6.80E-04	10.00	0.023	1.000	0.900	0.090	9.46E-06
	10	Primary	RF	0.90	1.50	0.5910	5.15E-03	9.20E-07	2.00E-02	6.80E-04	10.00	0.023	1.000	0.900	0.090	1.11E-06
	11	Primary	Wk	0.90	1.92E-02	21.3971	5.15E-03	9.20E-07	2.00E-02	6.80E-04	10.00	0.023	1.000	0.900	0.090	7.61E-08
	12	Primary	Shift	0.90	9.13E-04	37.9048	5.15E-03	9.20E-07	2.00E-02	6.80E-04	10.00	0.023	1.000	0.900	0.090	6.26E-08
Pipe-to-safe-end (2)	1	None	None	0.00	1.50	0.0000	1.24E-02	2.22E-06	2.00E-02	1.64E-03	10.00	0.023	0.000	0.000	0.000	9.72E-05
	2	None	RF	0.90	1.50	0.5910	1.24E-02	2.22E-06	2.00E-02	1.64E-03	10.00	0.023	0.000	0.000	0.000	1.39E-05
	3	None	Wk	0.90	1.92E-02	21.3971	1.24E-02	2.22E-06	2.00E-02	1.64E-03	10.00	0.023	0.000	0.000	0.000	9.84E-07
	4	None	Shift	0.90	9.13E-04	37.9048	1.24E-02	2.22E-06	2.00E-02	1.64E-03	10.00	0.023	0.000	0.000	0.000	8.10E-07
	5	Secondary	None	0.00	1.50	0.0000	1.24E-02	2.22E-06	2.00E-02	1.64E-03	10.00	0.023	1.000	0.500	0.050	6.72E-05
	6	Secondary	RF	0.90	1.50	0.5910	1.24E-02	2.22E-06	2.00E-02	1.64E-03	10.00	0.023	1.000	0.500	0.050	8.44E-06
	7	Secondary	Wk	0.90	1.92E-02	21.3971	1.24E-02	2.22E-06	2.00E-02	1.64E-03	10.00	0.023	1.000	0.500	0.050	5.83E-07
	8	Secondary	Shift	0.90	9.13E-04	37.9048	1.24E-02	2.22E-06	2.00E-02	1.64E-03	10.00	0.023	1.000	0.500	0.050	4.79E-07
	9	Primary	None	0.00	1.50	0.0000	1.24E-02	2.22E-06	2.00E-02	1.64E-03	10.00	0.023	1.000	0.900	0.090	5.27E-05
	10	Primary	RF	0.90	1.50	0.5910	1.24E-02	2.22E-06	2.00E-02	1.64E-03	10.00	0.023	1.000	0.900	0.090	6.11E-06
	11	Primary	Wk	0.90	1.92E-02	21.3971	1.24E-02	2.22E-06	2.00E-02	1.64E-03	10.00	0.023	1.000	0.900	0.090	4.16E-07
	12	Primary	Shift	0.90	9.13E-04	37.9048	1.24E-02	2.22E-06	2.00E-02	1.64E-03	10.00	0.023	1.000	0.900	0.090	3.42E-07
Pipe-to-tee (6)	1	None	None	0.00	1.50	0.0000	5.00E-03	8.92E-07	2.00E-02	6.59E-04	10.00	0.023	0.000	0.000	0.000	1.68E-05
	2	None	RF	0.90	1.50	0.5910	5.00E-03	8.92E-07	2.00E-02	6.59E-04	10.00	0.023	0.000	0.000	0.000	2.46E-06
	3	None	Wk	0.90	1.92E-02	21.3971	5.00E-03	8.92E-07	2.00E-02	6.59E-04	10.00	0.023	0.000	0.000	0.000	1.75E-07
	4	None	Shift	0.90	9.13E-04	37.9048	5.00E-03	8.92E-07	2.00E-02	6.59E-04	10.00	0.023	0.000	0.000	0.000	1.44E-07
	5	Secondary	None	0.00	1.50	0.0000	5.00E-03	8.92E-07	2.00E-02	6.59E-04	10.00	0.023	1.000	0.500	0.050	1.15E-05
	6	Secondary	RF	0.90	1.50	0.5910	5.00E-03	8.92E-07	2.00E-02	6.59E-04	10.00	0.023	1.000	0.500	0.050	1.47E-06
	7	Secondary	Wk	0.90	1.92E-02	21.3971	5.00E-03	8.92E-07	2.00E-02	6.59E-04	10.00	0.023	1.000	0.500	0.050	1.02E-07
	8	Secondary	Shift	0.90	9.13E-04	37.9048	5.00E-03	8.92E-07	2.00E-02	6.59E-04	10.00	0.023	1.000	0.500	0.050	8.37E-08
	9	Primary	None	0.00	1.50	0.0000	5.00E-03	8.92E-07	2.00E-02	6.59E-04	10.00	0.023	1.000	0.900	0.090	8.91E-06
	10	Primary	RF	0.90	1.50	0.5910	5.00E-03	8.92E-07	2.00E-02	6.59E-04	10.00	0.023	1.000	0.900	0.090	1.05E-06
	11	Primary	Wk	0.90	1.92E-02	21.3971	5.00E-03	8.92E-07	2.00E-02	6.59E-04	10.00	0.023	1.000	0.900	0.090	7.16E-08
	12	Primary	Shift	0.90	9.13E-04	37.9048	5.00E-03	8.92E-07	2.00E-02	6.59E-04	10.00	0.023	1.000	0.900	0.090	5.89E-08
Pipe-to-socket-weld (capped bypass) (2)	1	None	None	0.00	1.50	0.0000	1.11E-03	1.99E-07	2.00E-02	1.47E-04	10.00	0.023	0.000	0.000	0.000	8.70E-07
	2	None	RF	0.90	1.50	0.5910	1.11E-03	1.99E-07	2.00E-02	1.47E-04	10.00	0.023	0.000	0.000	0.000	1.29E-07
	3	None	Wk	0.90	1.92E-02	21.3971	1.11E-03	1.99E-07	2.00E-02	1.47E-04	10.00	0.023	0.000	0.000	0.000	9.23E-09
	4	None	Shift	0.90	9.13E-04	37.9048	1.11E-03	1.99E-07	2.00E-02	1.47E-04	10.00	0.023	0.000	0.000	0.000	7.59E-09
	5	Secondary	None	0.00	1.50	0.0000	1.11E-03	1.99E-07	2.00E-02	1.47E-04	10.00	0.023	1.000	0.500	0.050	5.89E-07

Table D.C.1 Application of Markov Model to BWR-1 Cat 1 LOCA Frequency

Config. (Count)	INSPECTION CASE			INSPECTION INDEPENDENT PARAMETERS									INSPECTION DEPENDENT PARAMETERS			
	CASE	ISI Inspection Coverage	Leak Inspection Interval	P_{LI}	TSUBLI	MU	PHI	RHO_F	RHO_L	Lambda	T_{FI}	TSUBR	P_{FI}	P_{FD}	OMEGA	Haz(T)
	6	Secondary	RF	0.90	1.50	0.5910	1.11E-03	1.99E-07	2.00E-02	1.47E-04	10.00	0.023	1.000	0.500	0.050	7.62E-08
	7	Secondary	Wk	0.90	1.92E-02	21.3971	1.11E-03	1.99E-07	2.00E-02	1.47E-04	10.00	0.023	1.000	0.500	0.050	5.29E-09
	8	Secondary	Shift	0.90	9.13E-04	37.9048	1.11E-03	1.99E-07	2.00E-02	1.47E-04	10.00	0.023	1.000	0.500	0.050	4.35E-09
	9	Primary	None	0.00	1.50	0.0000	1.11E-03	1.99E-07	2.00E-02	1.47E-04	10.00	0.023	1.000	0.900	0.090	4.55E-07
	10	Primary	RF	0.90	1.50	0.5910	1.11E-03	1.99E-07	2.00E-02	1.47E-04	10.00	0.023	1.000	0.900	0.090	5.40E-08
	11	Primary	Wk	0.90	1.92E-02	21.3971	1.11E-03	1.99E-07	2.00E-02	1.47E-04	10.00	0.023	1.000	0.900	0.090	3.70E-09
	12	Primary	Shift	0.90	9.13E-04	37.9048	1.11E-03	1.99E-07	2.00E-02	1.47E-04	10.00	0.023	1.000	0.900	0.090	3.04E-09

ATTACHMENT D TO APPENDIX D

SIGNIFICANT FAILURES OF SAFETY RELATED PIPING

Table D.D.1 is a list of selected significant pipe failures during the period 1970 – 2003. The list includes failures of Code Class 1 and 2 piping systems inside the containment/drywell and auxiliary/reactor building structures of commercial nuclear power plants. The technical information has been extracted from the OPDE database (Attachment A).

Table D.D.1 Selected Historical Pipe Failure Information

Event Date	Plant	Country	Estimated Peak Leak/Flow Rate [gpm]	Description
12/14/02	Brunsbüttel (BWR)	Germany	-- (see Description)	Rupture of reactor head cooling pipe (the rupture occurred in section of pipe that was separated from the RPV through an isolation valve – no RPV steam leakage observed
11/7/02	Hamaoka-1 (BWR)	Japan	>> 50	Rupture of pipe in High Pressure Coolant Injection system; the rupture occurred during a functional system test
7/12/99	Tsuruga-2 (PWR)	Japan	16	Thermal fatigue induced fracture of elbow connected to regenerative heat exchanger
5/12/98	Civaux-1 (PWR)	France	131	Thermal fatigue induced fracture of seam welded elbow in the Residual Heat Removal System
5/27/97	Calvert Cliffs-1 (PWR)	USA	8.0	Fractured pressurizer instrument sensing line; attributed to vibration fatigue
4/21/97	Oconee-2 (PWR)	USA	12.0	Thermal fatigue induced fracture of weld connecting HPI/NMU pipe to RCS (see Base Case PWR-3)
12/21/96	Dampierre-1 (PWR)	France	0.6	Thermal-fatigue induced weld crack in straight section of Safety Injection line to RCS hot leg.
3/8/95	Borssele (PWR)	The Netherlands	65.8	While in hot standby prior to startup a weld fractured on the High Head Safety Injection common discharge header; attributed to vibration-fatigue
2/23/95	Biblis-B (PWR)	Germany	15.8	Thermal fatigue induced fracture of base metal of pipe in Chemical and Volume Control system
3/3/94	Kola-2 (PWR)	Russia	S-LOCA	Soviet-designed PWR of type WWER-440/230; full circumferential fracture of NPS2 makeup pipe while shutting down for maintenance outage. Event resulted in High Pressure Safety Injection system actuation; a beyond-design basis accident.
9/20/92	Dampierre-2 (PWR)	France	3.2	Non-isolable, thermal fatigue induced weld fracture in Safety Injection System.
6/18/88	Tihange-1	Belgium	6.3	Thermal fatigue induced fracture of base

Table D.D.1 Selected Historical Pipe Failure Information

Event Date	Plant	Country	Estimated Peak Leak/Flow Rate [gpm]	Description
	(PWR)			metal of Safety Injection line to RCS hot leg
12/9/87	Farley-2 (PWR)	USA	0.7	Thermal fatigue induced weld fracture in Safety Injection line to RCS cold leg
8/16/87	McGuire-1 (PWR)	USA	39.5	Fracture (80% of circumference) of 1-inch socket weld in drain line off of letdown line inside containment. The weld fracture occurred during startup operations (8% reactor power)
5/31/86	Obrigheim (PWR)	Germany	0.32	Thermal fatigue induced weld fracture in makeup line to RCS.
7/29/85	Sequoyah-2 (PWR)	USA	60.0	Fractured sample line in Chemical and Volume Control system; attributed to vibration-fatigue
8/6/84	McGuire-2 (PWR)	USA	8.0	Water hammer induced fracture of socket weld in letdown line
1/25/83	Maine Yankee (PWR)	USA	100	Fractured main feedwater pipe adjacent to weld joining pipe and steam generator safe end; attributed to severe water hammer.
1/21/82	Crystal River-3 (PWR)	USA	1	140-degree circumferential crack in makeup line near valve-to-safe end weld; attributed to thermal fatigue
2/12/80	Santa Maria de Garona (BWR)	Spain	0.8	IGSCC induced through-wall flaw in Reactor Recirculation nozzle-to-safe end weld
8/29/80	TVO-1 (BWR)	Finland	315	Thermal fatigue induced fracture of tee in Reactor Water Cleanup system. The fracture occurred during the commissioning of this reactor unit.
6/14/78	Duane Arnold (BWR)	USA	3.0	IGSCC induced through-wall flaw in Reactor Recirculation nozzle-to-safe end weld
11/13/73	Indian Point-2 (PWR)	USA	15.8	180-degree circumferential crack of 18-inch feedwater line weld inside containment
4/28/70	H.B. Robinson-2 (PWR)	USA	>> 50[27]	360-degree break in 6-inch branch line between No. 3 steam generator main steam line and safety valve. The failure occurred during the final stages of hot functional testing

---

[27] At the time of the pipe break the primary system was at 278 C (533 F) and 15.3 MPa (2,225 psi) primary system pressure with a secondary system pressure of 6.2 MPa (900 psi)

APPENDIX E

PIPING BASE CASE RESULTS OF

WILLIAM GALYEAN

APPENDIX E

PIPING BASE CASE RESULTS OF WILLIAM GALYEAN

E.1 Summary

In this base case study, LOCA frequencies are calculated using a "top-down" approach. Specifically, a total LOCA frequency is calculated using U.S. commercial nuclear power plant (NPP) operating experience. This total frequency is then allocated to the LOCA size categories, RCS subsystems and components, and degradation mechanisms. This allocation is performed using data on primary system leaks and cracks from both U.S. and foreign PWR and BWR reactors.

E.2 Assumptions and Observations

As with all analyses, there are a number of implicit assumptions associated with this approach. First is that past performance is representative of future performance. The common scenario for the occurrence of a LOCA starts with postulating the existence of a flaw or defect in the primary reactor coolant boundary. This flaw is then subjected to a stress that results in the catastrophic failure of the primary pressure boundary, producing a LOCA. The U.S. LWR operating experience to date consists of approximately 100 reactors with an average age of about 23 years. During this time the RCS of these plants have experience numerous transients and loads, which have produce a wide range of stresses. Whether these plants operate for 40 years (or 60 years with license extensions) this available operating experience represents a significant portion of the average plants lifetime. It is therefore reasonable to assume that the stresses that have already occurred are representative of those that will occur in the future. Similarly, various degradation mechanisms have affected RCS pipe, welds and components. However, when these degradation mechanisms have been detected, mitigation programs have subsequently been implemented (e.g., IGSCC in BWRs). Therefore, the number of flaws and defects in the RCS is likely to be cyclic over time. As the degradation mechanism manifests itself, the number of defects grows, as the degradation mechanism is addressed and mitigated, the number of defects is reduced. Again, the assumption here is that current 23 years of operating (on average, per reactor) are representative of the remaining operating life.

Another observation is the occurrence of zero LOCAs for both PWRs and BWRs. Although this does not prove that the LOCA frequencies are the same for both designs, it likewise does not support different LOCA frequencies. Therefore, for this analysis, the operating experience data (i.e., zero failures) will be pooled to generate a single LOCA frequency.

Furthermore, this analysis, just as every LOCA frequency estimate performed to date, assumes that the frequency of a LOCA decreases as pipe size increases. This might be attributable to a couple of issues. First, for small diameter pipe, some failure mechanisms exist that don't apply to larger diameter pipe (e.g., compression fitting failures and socket welds). Second, the same flaw in both a small diameter pipe and a large diameter pipe represents a large percentage of the pipe diameter in the small diameter pipe. Third, inspection is probably more thorough in larger diameter pipe so that the chance of a defect going undetected is less in the larger diameter pipe. For all of these reasons (and probably others), the total LOCA frequency is reduced as LOCA size category increases. The scaling factor of ½ order of magnitude (assuming a lognormal probability distribution on LOCA frequency) appears to be reasonably consistent with historical LOCA frequency estimates.

This assumption of a half-order of magnitude (i.e., approximately a factor of 3) decrease in frequency for each increase in LOCA size is an assumption based on the general practice employed in estimating LOCA frequencies over the past 30 years starting with the Reactor Safety Study (Ref. E.1). This assumption is further supported by work done by Beliczey and Schulz (Ref. E.2). In this study, a combination of operating experience and fracture mechanics is used to demonstrate that the conditional probability of a rupture, given a leak, decreases as pipe diameter increases. This conclusion is reached because the size of detectable cracks and leaks remains relatively constant as a function of pipe size. Therefore, the relative crack or leak size as a function of the pipe circumference decreases, and the safety margin increases, as the pipe diameter increases.

Additionally, Beliczey and Schulz developed a quantitative conditional failure probability --- which decreases by approximately ½ order of magnitude for each successively larger LOCA size --- that was based on the propensity of through-wall fatigue flaws to lead to successively larger LOCA sizes. Although the quantitative conditional failure probability is not applicable to all failure mechanisms and systems, this simple relationship has been extensively employed. This assumption was also employed in these analyses.

The final premise of this base case analysis is that the relative frequency of precursor data (i.e., leaks and cracks) is an indicator of the relative frequency of LOCA events. In the calculations that follow, the total LOCA frequency is allocated to the different RCS subsystems and components, and the different degradation mechanisms according to the relative frequency of observed leaks and cracks attributable to these subsystems and mechanism. Note that in order to determine the relative frequencies, complete crack and leak data are not needed, only consistent data that has not been biased by the over reporting of one attribute relative to another. Completeness in the data is neither required nor important, only consistency.

E.3 Total LOCA Frequency Estimates

The total LOCA frequency is calculated using U.S. NPP experience of zero Category-1 LOCAs (i.e., greater than 380 lpm [100 gpm]) in 2,647 LWR-years of operation (as of 4/24/2003). A Bayesian update of a non-informative prior-distribution was performed to produce a total LOCA frequency of 1.9E-4 per LWR-year.

Table E.1 Total LOCA Frequency (per LWR-Year) Including Uncertainty, Using a Non-Informative Prior and U.S. LWR Operating Experience

5%	50%	mean	95%
7.4E-07	8.6E-05	1.9E-04	7.3E-04

E.4 LOCA Frequency Allocation by RCS Pipe and Non-Pipe

The total LOCA frequency calculated above is first allocated between pipe and non-pipe passive components using data on primary system leaks and cracks collected from licensee event reports (LERs). These data records were collected, reviewed and categorized specifically for this effort. Since these data will only be used to ascribe a relative frequency between pipe and passive non-pipe components, complete data are not necessary, only data that have been reported consistently. These data and the resultant allocation are summarized in the table below. Steam generator tube ruptures are being assessed separately, and are therefore removed from this allocation.

Table E.2 Allocation of LOCA Frequency Between Pipe and Passive Non-Pipe Components

Reactor Coolant Pressure Boundary
failure (cracks or leaks) events
1990-2002 LERs

	LWR	PWR	BWR
Total number of failure events	448	388	60
Number of SG tube failure events	112	112	0
Total minus SG events	336	276	60
Number of pipe failure events	54	24	30

Exclude SG tube events since these can be estimated directly

Therefore

Number of non-pipe failure events	282	252	30

fraction of LOCA frequency attributed to

	LWR	PWR	BWR
pipes	0.16	0.09	0.50
non-pipes	0.84	0.91	0.50

total LOCA frequency = 1.9E-04

LOCA frequency attributable to

	LWR	PWR	BWR
pipes	3.0E-05	1.6E-05	9.4E-05
non-pipes	1.6E-04	1.7E-04	9.4E-05

E.5 LOCA Frequency by Size Category

The total LOCA frequencies calculated above are for Category-1 LOCAs. The simple approach taken here is that the LOCA frequency is reduced by ½ order of magnitude (assuming a lognormal distribution), for each step up in size category. There are a number of reasons for this approach. Between the smallest pipe size categories (i.e., < 2 inches, and > 2 inches) there is a significant difference in the failure mechanisms. For the smallest pipes, the operating experience includes failures of compression fittings and socket welds, which are not used in larger size pipe. Also, a number of studies on crack and leak events indicate a decrease in these precursor frequencies, as pipe diameter increases. Lastly, virtually every estimate of LOCA frequencies ever made has resulted in a reduced frequency for the large LOCA sizes.

E.6 LOCA Frequency by Degradation Mechanism and Subsystem

In addition to the LER data used to allocate the LOCA frequency between pipe and passive non-pipe components, the LOCA frequency was further allocated among the different degradation mechanisms observed and among the different RCS subsystems and components defined for this project. These allocations were based on data collected from both U.S. reactor operating experience (primarily LERs), and from foreign LWR operating experience (SLAP database). One complication to this approach is the IGSCC-related experience in U.S. BWR plants. IGSCC was an issue for BWRs in the early 1980's. Many U.S. BWRs implemented IGSCC mitigation programs in the mid-1980's, which have greatly reduced the occurrence of IGSCC. To avoid unrealistically over weighting the IGSCC mechanism, the

BWR experience was segregated and only the post 1985 experience was used for allocating the relative contribution to LOCA frequency by degradation mechanism.

Lastly, although the guidance for calculating base-case frequencies for this project included estimates for 25 years, 40 years and 60 years, this particular base-case calculation assumed that the frequencies were generally independent of plant life. This is based on the IGSCC experience that demonstrated that although degradation mechanisms are at work that can result in an increase in the LOCA frequency over time, so to are mitigation programs and general performance improvement programs (e.g., more effective inspections), that can result in a decrease in the LOCA frequency. Therefore, overall these competing effects are assumed to cancel each other out for a net zero effect on LOCA frequency. That is, the current LOCA frequency (approximately 25 year life) is assumed to be application for 40 and 60 years as well.

E.7 LOCA Frequency Tables

The following tables display the detailed results of Base Case #1 calculations. The legend of degradation categories is shown in Table E.3. LOCA data are presented on the tables listed below:

PWR pipe	Table E.4
PWR passive non-pipe	Table E.5
BWR pipe	Table E.6
BWR passive non-pipe	Table E.7

Table E.3 Degradation Categories

Deg Mech	DM Description
MA	Material Aging
FDR	Fabrication Defect and Repair
SCC	Stress Corrosion Cracking
LC	Local Corrosion
MF	Mechanical Fatigue
TF	Thermal Fatigue
FS	Flow Sensitive (includes FAC and E/C)
UNK	Unknown

E.8 References

E.1 "Reactor Safety Study: An Assessment of Accident Risks in U.S. Commercial Nuclear Power Plants," WASH-1400, U.S. Nuclear Regulatory Commission, October 1975.

E.2 Beliczey, S., and Schulz, H., "Comments on Probabilities of Leaks and Breaks of Safety-Related Piping in PWR Plants," *International Journal of Pressure Vessel and Piping*, Vol. 43, pp. 219 – 227, 1990.

Table E.4 PWR LOCA Frequency (for Pipes) Allocated by System, Degradation Mechanism, and Size Category

SLAP event counts by system (for PWRs)
Only includes RCPB (S-type)

LOCA System	Event Counts	Deg. Mech. Fractional Contribution	Est. LOCA Fraction by System	Min (mm)(in)	Max (mm)(in)	# of Welds Dom. Total	greater than Total LOCA Freq 1.6E-05	Cat-1 100 / 0.4 / 1	Cat-2 1500 / 1.7 / 0.3	Cat-3 5000 / 3.0 / 0.1	Cat-4 25000 / 6.8 / 0.03	Cat-5 100000 / 14.0937 / 0.01	Cat-6 500000 / 31.5146 / 0.003 gpm / dia.(in.) / probabilty
CRDM	4.6		0.01	70 / 2.8	100 / 3.9	1	1.4E-07	1.4E-07	4.1E-08	1.4E-08	0.0	0.0	0.0
CRDM	*4*	*1*					*1.4E-07*	*1.4E-07*	*4.1E-08*	*1.4E-08*	--	--	--
UA		*0.00*					*0.0E+00*	*0.0E+00*	*0.0E+00*	*0.0E+00*	--	--	--
MA		*0.00*					*0.0E+00*	*0.0E+00*	*0.0E+00*	*0.0E+00*	--	--	--
LC		*0.00*					*0.0E+00*	*0.0E+00*	*0.0E+00*	*0.0E+00*	--	--	--
FDR		*0.00*					*0.0E+00*	*0.0E+00*	*0.0E+00*	*0.0E+00*	--	--	--
SCC	*4*	*1.00*					*1.4E-07*	*1.4E-07*	*4.1E-08*	*1.4E-08*	--	--	--
MF		*0.00*					*0.0E+00*	*0.0E+00*	*0.0E+00*	*0.0E+00*	--	--	--
TF		*0.00*					*0.0E+00*	*0.0E+00*	*0.0E+00*	*0.0E+00*	--	--	--
FS		*0.00*					*0.0E+00*	*0.0E+00*	*0.0E+00*	*0.0E+00*	--	--	--
UK		*0.00*					*0.0E+00*	*0.0E+00*	*0.0E+00*	*0.0E+00*	--	--	--
CRDM pipe	5.0		0.01	60 / 2.4	100 / 3.9	1	1.5E-07	1.5E-07	4.5E-08	1.5E-08	0.0	0.0	0.0
CRDM pipe	*5*	*1*					*1.5E-07*	*1.5E-07*	*4.5E-08*	*1.5E-08*	--	--	--
UA		*0.00*					*0.0E+00*	*0.0E+00*	*0.0E+00*	*0.0E+00*	--	--	--
MA		*0.00*					*0.0E+00*	*0.0E+00*	*0.0E+00*	*0.0E+00*	--	--	--
LC		*0.00*					*0.0E+00*	*0.0E+00*	*0.0E+00*	*0.0E+00*	--	--	--
FDR	*1*	*0.20*					*3.0E-08*	*3.0E-08*	*9.0E-09*	*3.0E-09*	--	--	--
SCC	*2*	*0.40*					*6.0E-08*	*6.0E-08*	*1.8E-08*	*6.0E-09*	--	--	--
MF		*0.00*					*0.0E+00*	*0.0E+00*	*0.0E+00*	*0.0E+00*	--	--	--
TF		*0.00*					*0.0E+00*	*0.0E+00*	*0.0E+00*	*0.0E+00*	--	--	--
FS		*0.00*					*0.0E+00*	*0.0E+00*	*0.0E+00*	*0.0E+00*	--	--	--
UK	*2*	*0.40*					*6.0E-08*	*6.0E-08*	*1.8E-08*	*6.0E-09*	--	--	--
CVCS	140.0		0.26	13 / 0.5	250 / 9.8	1	4.2E-06	4.2E-06	1.3E-06	4.2E-07	1.3E-07	0.0	0.0
CVCS	*140*	*1*					*4.2E-06*	*4.2E-06*	*1.3E-06*	*4.2E-07*	*1.3E-07*	--	--
UA		*0.00*					*0.0E+00*	*0.0E+00*	*0.0E+00*	*0.0E+00*	*0.0E+00*	--	--
MA		*0.00*					*0.0E+00*	*0.0E+00*	*0.0E+00*	*0.0E+00*	*0.0E+00*	--	--
LC		*0.00*					*0.0E+00*	*0.0E+00*	*0.0E+00*	*0.0E+00*	*0.0E+00*	--	--
FDR	*12*	*0.09*					*3.6E-07*	*3.6E-07*	*1.1E-07*	*3.6E-08*	*1.1E-08*	--	--
SCC	*19*	*0.14*					*5.7E-07*	*5.7E-07*	*1.7E-07*	*5.7E-08*	*1.7E-08*	--	--
MF	*94*	*0.67*					*2.8E-06*	*2.8E-06*	*8.5E-07*	*2.8E-07*	*8.5E-08*	--	--
TF	*7*	*0.05*					*2.1E-07*	*2.1E-07*	*6.3E-08*	*2.1E-08*	*6.3E-09*	--	--
FS	*6*	*0.04*					*1.8E-07*	*1.8E-07*	*5.4E-08*	*1.8E-08*	*5.4E-09*	--	--

SLAP event counts by system (for PWRs)
Only includes RCPB (S-type)

								Cat-1	Cat-2	Cat-3	Cat-4	Cat-5	Cat-6	gpm
								100	1500	5000	25000	100000	500000	
								0.4	1.7	3.0	6.8	14.0937	31.5146	dia. (in.)
		Deg. Mech.	Est.	Min	Max	# of	greater than	1	0.3	0.1	0.03	0.01	0.003	probabilty
	Event	Fractional	LOCA Fraction	(mm) (in)	(mm) (in)	Welds Dom.	Total LOCA Freq							
LOCA System	Counts	Contribution	by System			Total	1.6E-05							
UK	2	0.01					6.0E-08	6.0E-08	1.8E-08	6.0E-09	1.8E-09	--	--
Drain Lines	58.5		0.11	10 / 0.4	80 / 3.1								
Drain Lines	46	1				1	1.8E-06	1.8E-06	5.3E-07	1.8E-07	0.0	0.0	0.0
							1.8E-06	1.8E-06	5.3E-07	1.8E-07	--	--	--
UA		0.00					0.0E+00	0.0E+00	0.0E+00	0.0E+00	--	--	--
MA		0.00					0.0E+00	0.0E+00	0.0E+00	0.0E+00	--	--	--
LC		0.00					0.0E+00	0.0E+00	0.0E+00	0.0E+00	--	--	--
FDR	9	0.20					3.4E-07	3.4E-07	1.0E-07	3.4E-08	--	--	--
SCC	5	0.11					1.9E-07	1.9E-07	5.7E-08	1.9E-08	--	--	--
MF	27	0.59					1.0E-06	1.0E-06	3.1E-07	1.0E-07	--	--	--
TF	2	0.04					7.7E-08	7.7E-08	2.3E-08	7.7E-09	--	--	--
FS	2	0.04					7.7E-08	7.7E-08	2.3E-08	7.7E-09	--	--	--
UK	1	0.02					3.8E-08	3.8E-08	1.1E-08	3.8E-09	--	--	--
In-Core Instr.	16.6		0.03	10 / 0.4	25 / 1.0								
In-Core Instr.	13	1				1	5.0E-07	5.0E-07	0.0	0.0	0.0	0.0	0.0
							5.0E-07	5.0E-07	--	--	--	--	--
UA		0.00					0.0E+00	0.0E+00	--	--	--	--	--
MA		0.00					0.0E+00	0.0E+00	--	--	--	--	--
LC	4	0.31					1.5E-07	1.5E-07	--	--	--	--	--
FDR	2	0.15					7.7E-08	7.7E-08	--	--	--	--	--
SCC	2	0.15					7.7E-08	7.7E-08	--	--	--	--	--
MF	2	0.15					7.7E-08	7.7E-08	--	--	--	--	--
TF		0.00					0.0E+00	0.0E+00	--	--	--	--	--
FS		0.00					0.0E+00	0.0E+00	--	--	--	--	--
UK	3	0.23					1.2E-07	1.2E-07	--	--	--	--	--
Instr. lines	151.8		0.28	9 / 0.4	200 / 7.9								
Instr. Lines	119	1				1	4.6E-06	4.6E-06	1.4E-06	4.6E-07	1.4E-07	0.0	0.0
							4.6E-06	4.6E-06	1.4E-06	4.6E-07	1.4E-07	--	--
UA		0.00					0.0E+00	0.0E+00	0.0E+00	0.0E+00	0.0E+00	--	--
MA	1	0.01					3.8E-08	3.8E-08	1.2E-08	3.8E-09	1.2E-09	--	--
LC	1	0.01					3.8E-08	3.8E-08	1.2E-08	3.8E-09	1.2E-09	--	--
FDR	11	0.09					4.2E-07	4.2E-07	1.3E-07	4.2E-08	1.3E-08	--	--
SCC	19	0.16					7.3E-07	7.3E-07	2.2E-07	7.3E-08	2.2E-08	--	--
MF	74	0.62					2.8E-06	2.8E-06	8.5E-07	2.8E-07	8.5E-08	--	--
TF	2	0.02					7.7E-08	7.7E-08	2.3E-08	7.7E-09	2.3E-09	--	--
FS	1	0.01					3.8E-08	3.8E-08	1.2E-08	3.8E-09	1.2E-09	--	--
UK	10	0.08					3.8E-07	3.8E-07	1.2E-07	3.8E-08	1.2E-08	--	--

SLAP event counts by system (for PWRs)
Only includes RCPB (S-type)

			Est.			# of	greater than	Cat-1	Cat-2	Cat-3	Cat-4	Cat-5	Cat-6	gpm
		Deg. Mech.	LOCA	Min	Max	Welds	Total	100	1500	5000	25000	100000	500000	dia. (in.)
LOCA System	Event Counts	Fractional Contribution by System	Fraction by System	(mm)(in)	(mm)(in)	Dom. Total	LOCA Freq 1.6E-05	0.4 / 1	1.7 / 0.3	3.0 / 0.1	6.8 / 0.03	14.0937 / 0.01	31.5146 / 0.003	probabilty
Pressurizer	6.4													
Pressurizer	*5*	*1*	0.01	25 / 1.0	25 / 1.0	*1*	1.9E-07	1.9E-07	5.8E-08	0.0	0.0	0.0	0.0	
UA		*0.00*					0.0E+00	0.0E+00	0.0E+00	--	--	--	--	
MA		*0.00*					0.0E+00	0.0E+00	0.0E+00	--	--	--	--	
LC		*0.00*					0.0E+00	0.0E+00	0.0E+00	--	--	--	--	
FDR	*2*	*0.40*					7.7E-08	7.7E-08	2.3E-08	--	--	--	--	
SCC	*2*	*0.40*					7.7E-08	7.7E-08	2.3E-08	--	--	--	--	
MF		*0.00*					0.0E+00	0.0E+00	0.0E+00	--	--	--	--	
TF	*1*	*0.20*					3.8E-08	3.8E-08	1.2E-08	--	--	--	--	
FS		*0.00*					0.0E+00	0.0E+00	0.0E+00	--	--	--	--	
UK		*0.00*					0.0E+00	0.0E+00	0.0E+00	--	--	--	--	
Pzr Spray Lines	6.4													
Pzr Spray Lines	*5*	*1*	0.01	20 / 0.8	75 / 3.0	*1*	1.9E-07	1.9E-07	5.8E-08	1.9E-08	0.0	0.0	0.0	
UA		*0.00*					0.0E+00	0.0E+00	0.0E+00	0.0E+00	--	--	--	
MA		*0.00*					0.0E+00	0.0E+00	0.0E+00	0.0E+00	--	--	--	
LC		*0.00*					0.0E+00	0.0E+00	0.0E+00	0.0E+00	--	--	--	
FDR		*0.00*					0.0E+00	0.0E+00	0.0E+00	0.0E+00	--	--	--	
SCC	*1*	*0.20*					3.8E-08	3.8E-08	1.2E-08	3.8E-09	--	--	--	
MF	*3*	*0.60*					1.2E-07	1.2E-07	3.5E-08	1.2E-08	--	--	--	
TF		*0.00*					0.0E+00	0.0E+00	0.0E+00	0.0E+00	--	--	--	
FS		*0.00*					0.0E+00	0.0E+00	0.0E+00	0.0E+00	--	--	--	
UK	*1*	*0.20*					3.8E-08	3.8E-08	1.2E-08	3.8E-09	--	--	--	
RCP cold-leg	1.3													
RCP cold-leg	*1*	*1*	0.002			*1*	3.8E-08	3.8E-08	1.2E-08	3.8E-09	1.2E-09	3.8E-10	1.2E-10	
UA		*0.00*					0.0E+00	0.0E+00	0.0E+00	0.0E+00	0.0E+00	0.0E+00	0.0E+00	
MA		*0.00*					0.0E+00	0.0E+00	0.0E+00	0.0E+00	0.0E+00	0.0E+00	0.0E+00	
LC		*0.00*					0.0E+00	0.0E+00	0.0E+00	0.0E+00	0.0E+00	0.0E+00	0.0E+00	
FDR		*0.00*					0.0E+00	0.0E+00	0.0E+00	0.0E+00	0.0E+00	0.0E+00	0.0E+00	
SCC	*1*	*1.00*					3.8E-08	3.8E-08	1.2E-08	3.8E-09	1.2E-09	3.8E-10	1.2E-10	
MF		*0.00*					0.0E+00	0.0E+00	0.0E+00	0.0E+00	0.0E+00	0.0E+00	0.0E+00	
TF		*0.00*					0.0E+00	0.0E+00	0.0E+00	0.0E+00	0.0E+00	0.0E+00	0.0E+00	
FS		*0.00*					0.0E+00	0.0E+00	0.0E+00	0.0E+00	0.0E+00	0.0E+00	0.0E+00	
UK		*0.00*					0.0E+00	0.0E+00	0.0E+00	0.0E+00	0.0E+00	0.0E+00	0.0E+00	
RCS hot-leg	5.1													
RCP hot-leg -	*4.5*	*1*	0.009	25 / 1.0	650 / 25.6	*1*	1.5E-07	1.5E-07	4.6E-08	1.5E-08	4.6E-09	1.5E-09	0.0E+00	

SLAP event counts by system (for PWRs)
Only includes RCPB (S-type)

LOCA System	Event Counts	Deg. Mech. Fractional Contribution by System	Est. LOCA Fraction by System	Min (mm)(in)	Max (mm)(in)	# of Welds Dom. Total	greater than Total LOCA Freq 1.6E-05	Cat-1 100	Cat-2 1500	Cat-3 5000	Cat-4 25000	Cat-5 100000	Cat-6 500000
								0.4	1.7	3.0	6.8	14.0937	31.5146	gpm dia. (in.)
								1	0.3	0.1	0.03	0.01	0.003	probabilty
BC						40.0							
UA		0.00					0.0E+00	0.0E+00	0.0E+00	0.0E+00	0.0E+00	0.0E+00	--
MA		0.00					0.0E+00	0.0E+00	0.0E+00	0.0E+00	0.0E+00	0.0E+00	--
LC		0.00					0.0E+00	0.0E+00	0.0E+00	0.0E+00	0.0E+00	0.0E+00	--
FDR	2	0.44					6.8E-08	6.8E-08	2.1E-08	6.8E-09	2.1E-09	6.8E-10	--
SCC	0.5	0.11					1.7E-08	1.7E-08	5.1E-09	1.7E-09	5.1E-10	1.7E-10	--
MF	1	0.22					3.4E-08	3.4E-08	1.0E-08	3.4E-09	1.0E-09	3.4E-10	--
TF	1	0.22					3.4E-08	3.4E-08	1.0E-08	3.4E-09	1.0E-09	3.4E-10	--
FS		0.00					0.0E+00	0.0E+00	0.0E+00	0.0E+00	0.0E+00	0.0E+00	--
UK		0.00					0.0E+00	0.0E+00	0.0E+00	0.0E+00	0.0E+00	0.0E+00	--
Rx-Head	1.3		0.002				3.8E-08	3.8E-08	1.2E-08	3.8E-09	1.2E-09	3.8E-10	1.2E-10
Rx-Head	*1*	*1*				*1*	3.8E-08	3.8E-08	1.2E-08	3.8E-09	1.2E-09	3.8E-10	1.2E-10
UA		0.00					0.0E+00	0.0E+00	0.0E+00	0.0E+00	0.0E+00	0.0E+00	0.0E+00
MA		0.00					0.0E+00	0.0E+00	0.0E+00	0.0E+00	0.0E+00	0.0E+00	0.0E+00
LC		0.00					0.0E+00	0.0E+00	0.0E+00	0.0E+00	0.0E+00	0.0E+00	0.0E+00
FDR		0.00					0.0E+00	0.0E+00	0.0E+00	0.0E+00	0.0E+00	0.0E+00	0.0E+00
SCC		0.00					0.0E+00	0.0E+00	0.0E+00	0.0E+00	0.0E+00	0.0E+00	0.0E+00
MF		0.00					0.0E+00	0.0E+00	0.0E+00	0.0E+00	0.0E+00	0.0E+00	0.0E+00
TF		0.00					0.0E+00	0.0E+00	0.0E+00	0.0E+00	0.0E+00	0.0E+00	0.0E+00
FS		0.00					0.0E+00	0.0E+00	0.0E+00	0.0E+00	0.0E+00	0.0E+00	0.0E+00
UK	*1*	1.00					3.8E-08	3.8E-08	1.2E-08	3.8E-09	1.2E-09	3.8E-10	1.2E-10
RHR	49.0		0.09	13 0.5	500 19.7		1.5E-06	1.5E-06	4.4E-07	1.5E-07	4.4E-08	1.5E-08	0.0
RHR	*49*	*1*				*1*	1.5E-06	1.5E-06	4.4E-07	1.5E-07	4.4E-08	1.5E-08	--
UA		0.00					0.0E+00	0.0E+00	0.0E+00	0.0E+00	0.0E+00	0.0E+00	--
MA		0.00					0.0E+00	0.0E+00	0.0E+00	0.0E+00	0.0E+00	0.0E+00	--
LC		0.00					0.0E+00	0.0E+00	0.0E+00	0.0E+00	0.0E+00	0.0E+00	--
FDR	9	0.18					2.7E-07	2.7E-07	8.1E-08	2.7E-08	8.1E-09	2.7E-09	--
SCC	9	0.18					2.7E-07	2.7E-07	8.1E-08	2.7E-08	8.1E-09	2.7E-09	--
MF	23	0.47					6.9E-07	6.9E-07	2.1E-07	6.9E-08	2.1E-08	6.9E-09	--
TF	1	0.02					3.0E-08	3.0E-08	9.0E-09	3.0E-09	9.0E-10	3.0E-10	--
FS	4	0.08					1.2E-07	1.2E-07	3.6E-08	1.2E-08	3.6E-09	1.2E-09	--
UK	3	0.06					9.0E-08	9.0E-08	2.7E-08	9.0E-09	2.7E-09	9.0E-10	--
SIS Accum	14.3		0.03	15 0.6	100 3.9		4.3E-07	4.3E-07	1.3E-07	4.3E-08	0.0	0.0	0.0
SIS Accum	*14*	*1*				*1*	4.3E-07	4.3E-07	1.3E-07	4.3E-08	--	--	--
UA		0.00					0.0E+00	0.0E+00	0.0E+00	0.0E+00	0.0E+00		

SLAP event counts by system (for PWRs)
Only includes RCPB (S-type)

									Cat-1	Cat-2	Cat-3	Cat-4	Cat-5	Cat-6	gpm
			Est.			# of			100	1500	5000	25000	100000	500000	dia. (in.)
		Deg. Mech.	LOCA	Min	Max	Welds		greater than	0.4	1.7	3.0	6.8	14.0937	31.5146	probabilty
	Event	Fractional	Fraction	(mm)	(mm)	Dom.		Total	1	0.3	0.1	0.03	0.01	0.003	
LOCA System	Counts	Contribution by System	by System	(in)	(in)	Total		LOCA Freq 1.6E-05							
MA		0.00						0.0E+00	0.0E+00	0.0E+00	0.0E+00	--	--	--	
LC		0.00						0.0E+00	0.0E+00	0.0E+00	0.0E+00	--	--	--	
FDR	2	0.14						6.1E-08	6.1E-08	1.8E-08	6.1E-09	--	--	--	
SCC	5	0.36						1.5E-07	1.5E-07	4.6E-08	1.5E-08	--	--	--	
MF	6	0.43						1.8E-07	1.8E-07	5.5E-08	1.8E-08	--	--	--	
TF	1	0.07						3.1E-08	3.1E-08	9.2E-09	3.1E-09	--	--	--	
FS		0.00						0.0E+00	0.0E+00	0.0E+00	0.0E+00	--	--	--	
UK		0.00						0.0E+00	0.0E+00	0.0E+00	0.0E+00	--	--	--	
SIS inj	78.0		0.14	9 / 0.4	600 / 23.6	1		2.3E-06	2.3E-06	7.0E-07	2.3E-07	7.0E-08	2.3E-08	0.0	
SIS Inj	*78*	*1*				*114*		*2.3E-06*	*2.3E-06*	*7.0E-07*	*2.3E-07*	*7.0E-08*	*2.3E-08*		
UA		*0.00*						*0.0E+00*	*0.0E+00*	*0.0E+00*	*0.0E+00*	*0.0E+00*	*0.0E+00*		
MA		*0.00*						*0.0E+00*	*0.0E+00*	*0.0E+00*	*0.0E+00*	*0.0E+00*	*0.0E+00*		
LC		*0.00*						*0.0E+00*	*0.0E+00*	*0.0E+00*	*0.0E+00*	*0.0E+00*	*0.0E+00*		
FDR	*5*	*0.06*						*1.5E-07*	*1.5E-07*	*4.5E-08*	*1.5E-08*	*4.5E-09*	*1.5E-09*		
SCC	*28*	*0.36*						*8.4E-07*	*8.4E-07*	*2.5E-07*	*8.4E-08*	*2.5E-08*	*8.4E-09*		
MF	*26*	*0.33*						*7.8E-07*	*7.8E-07*	*2.3E-07*	*7.8E-08*	*2.3E-08*	*7.8E-09*		
TF	*9*	*0.12*						*2.7E-07*	*2.7E-07*	*8.1E-08*	*2.7E-08*	*8.1E-09*	*2.7E-09*		
FS	*4*	*0.05*						*1.2E-07*	*1.2E-07*	*3.6E-08*	*1.2E-08*	*3.6E-09*	*1.2E-09*		
UK	*6*	*0.08*						*1.8E-07*	*1.8E-07*	*5.4E-08*	*1.8E-08*	*5.4E-09*	*1.8E-09*		
SRV lines	6.4		0.01	50 / 2.0	75 / 3.0	1		1.9E-07	1.9E-07	5.8E-08	1.9E-08	0.0	0.0	0.0	
SRV lines	*6*	*1*						*1.9E-07*	*1.9E-07*	*5.8E-08*	*1.9E-08*				
UA		*0.00*						*0.0E+00*	*0.0E+00*	*0.0E+00*	*0.0E+00*				
MA		*0.00*						*0.0E+00*	*0.0E+00*	*0.0E+00*	*0.0E+00*				
LC		*0.00*						*0.0E+00*	*0.0E+00*	*0.0E+00*	*0.0E+00*				
FDR	*1*	*0.17*						*3.2E-08*	*3.2E-08*	*9.6E-09*	*3.2E-09*				
SCC	*3*	*0.50*						*9.6E-08*	*9.6E-08*	*2.9E-08*	*9.6E-09*				
MF	*1*	*0.17*						*3.2E-08*	*3.2E-08*	*9.6E-09*	*3.2E-09*				
TF		*0.00*						*0.0E+00*	*0.0E+00*	*0.0E+00*	*0.0E+00*				
FS		*0.00*						*0.0E+00*	*0.0E+00*	*0.0E+00*	*0.0E+00*				
UK	*1*	*0.17*						*3.2E-08*	*3.2E-08*	*9.6E-09*	*3.2E-09*				
Surge Line	0.5		0.001		10			1.5E-08	1.5E-08	4.5E-09	1.5E-09	4.5E-10	0.0E+00	0.0E+00	
Surge Line - BC	*0.5*	*1*				*13*		*1.5E-08*	*1.5E-08*	*4.5E-09*	*1.5E-09*	*4.5E-10*			
UA	*0.0*	*0.00*						*0.0E+00*	*0.0E+00*	*0.0E+00*	*0.0E+00*	*0.0E+00*			
MA	*0.0*	*0.00*						*0.0E+00*	*0.0E+00*	*0.0E+00*	*0.0E+00*	*0.0E+00*			
LC		*0.00*						*0.0E+00*	*0.0E+00*	*0.0E+00*	*0.0E+00*	*0.0E+00*			

SLAP event counts by system (for PWRs)
Only includes RCPB (S-type)

LOCA System	Event Counts	Deg. Mech. Fractional Contribution by System	Est. LOCA Fraction by System	Min (mm) (in)	Max (mm) (in)	# of Welds	Dom. Total	Total LOCA Freq greater than 1.6E-05	Cat-1 100 0.4 1	Cat-2 1500 1.7 0.3	Cat-3 5000 3.0 0.1	Cat-4 25000 6.8 0.03	Cat-5 100000 14.0937 0.01	Cat-6 500000 31.5146 0.003 gpm dia. (in.) probability
FDR		0.00						0.0E+00	0.0E+00	0.0E+00	0.0E+00	0.0E+00	--	--
SCC	0.3	0.50						7.5E-09	7.5E-09	2.3E-09	7.5E-10	2.3E-10	--	--
MF		0.00						0.0E+00	0.0E+00	0.0E+00	0.0E+00	0.0E+00	--	--
TF	0.3	0.50						7.5E-09	7.5E-09	2.3E-09	7.5E-10	2.3E-10	--	--
FS		0.00						0.0E+00	0.0E+00	0.0E+00	0.0E+00	0.0E+00	--	--
UK		0.00						0.0E+00	0.0E+00	0.0E+00	0.0E+00	0.0E+00	--	--
total	545		1					1.64E-05	1.64E-05	4.78E-06	1.57E-06	3.86E-07	4.06E-08	2.31E-10

Table E.5 PWR Passive, Non-Pipe Component LOCA Frequency

PWR Non-Pipe LOCA Contributors
Total non-pipe event count = 252

						gpm	100	1500	5000	25000	100000	500000
						dia. (in.)	0.5	1.8	3.3	7.3	18.4	41.2
						probability	1	0.3	0.1	0.03	0.01	0.003
Non-Pipe LOCA System	Degradation Mechanism	Counts	Non-Pipe LOCA Fract (252)	Fract Contr	PWR Non-Pipe LOCA Freq 1.7E-04	Cat-1	Cat-2	Cat-3	Cat-4	Cat-5	Cat-6
LIV		3	0.01	**1.0**	**2.05E-06**	**2.05E-06**	**6.16E-07**	**2.05E-07**	**6.16E-08**	**2.05E-08**	**6.16E-09**
	FDR			0.00	0.00E+00	0.00E+00	0.00E+00	0.00E+00	0.00E+00	0.00E+00	0.00E+00
	FS			0.00	0.00E+00	0.00E+00	0.00E+00	0.00E+00	0.00E+00	0.00E+00	0.00E+00
	LC			0.00	0.00E+00	0.00E+00	0.00E+00	0.00E+00	0.00E+00	0.00E+00	0.00E+00
	MA	2		0.67	1.37E-06	1.37E-06	4.10E-07	1.37E-07	4.10E-08	1.37E-08	4.10E-09
	MF			0.00	0.00E+00	0.00E+00	0.00E+00	0.00E+00	0.00E+00	0.00E+00	0.00E+00
	SCC	1		0.33	6.84E-07	6.84E-07	2.05E-07	6.84E-08	2.05E-08	6.84E-09	2.05E-09
	TF			0.00	0.00E+00	0.00E+00	0.00E+00	0.00E+00	0.00E+00	0.00E+00	0.00E+00
	UNK			0.00	0.00E+00	0.00E+00	0.00E+00	0.00E+00	0.00E+00	0.00E+00	0.00E+00
PIV		1	0.00	**1.0**	**6.84E-07**	**6.84E-07**	**2.05E-07**	**6.84E-08**	**2.05E-08**	**6.84E-09**	**2.05E-09**
	FDR			0.00	0.00E+00	0.00E+00	0.00E+00	0.00E+00	0.00E+00	0.00E+00	0.00E+00
	FS			0.00	0.00E+00	0.00E+00	0.00E+00	0.00E+00	0.00E+00	0.00E+00	0.00E+00
	LC			0.00	0.00E+00	0.00E+00	0.00E+00	0.00E+00	0.00E+00	0.00E+00	0.00E+00
	MA	1		1.00	2.05E-06	2.05E-06	6.16E-07	2.05E-07	6.16E-08	2.05E-08	6.16E-09
	MF			0.00	0.00E+00	0.00E+00	0.00E+00	0.00E+00	0.00E+00	0.00E+00	0.00E+00
	SCC			0.00	0.00E+00	0.00E+00	0.00E+00	0.00E+00	0.00E+00	0.00E+00	0.00E+00
	TF			0.00	0.00E+00	0.00E+00	0.00E+00	0.00E+00	0.00E+00	0.00E+00	0.00E+00
	UNK			0.00	0.00E+00	0.00E+00	0.00E+00	0.00E+00	0.00E+00	0.00E+00	0.00E+00
Pzr		65	0.26	**1.0**	**4.45E-05**	**4.45E-05**	**1.33E-05**	**4.45E-06**	**1.33E-06**	**4.45E-07**	**1.33E-07**
	FDR	4		0.06	1.26E-07	1.26E-07	3.79E-08	1.26E-08	3.79E-09	1.26E-09	3.79E-10
	FS			0.00	0.00E+00	0.00E+00	0.00E+00	0.00E+00	0.00E+00	0.00E+00	0.00E+00
	LC	1		0.02	3.16E-08	3.16E-08	9.47E-09	3.16E-09	9.47E-10	3.16E-10	9.47E-11
	MA	1		0.02	3.16E-08	3.16E-08	9.47E-09	3.16E-09	9.47E-10	3.16E-10	9.47E-11
	MF			0.00	0.00E+00	0.00E+00	0.00E+00	0.00E+00	0.00E+00	0.00E+00	0.00E+00
	SCC	59		0.91	1.86E-06	1.86E-06	5.59E-07	1.86E-07	5.59E-08	1.86E-08	5.59E-09
	TF			0.00	0.00E+00	0.00E+00	0.00E+00	0.00E+00	0.00E+00	0.00E+00	0.00E+00

PWR Non-Pipe LOCA Contributors
Total non-pipe event count = 252

252

Non-Pipe LOCA System	Degradation Mechanism Counts	Fract Contr	Non-Pipe LOCA Fract	PWR Non-Pipe LOCA Freq	Cat-1	Cat-2	Cat-3	Cat-4	Cat-5	Cat-6	gpm
					100	1500	5000	25000	100000	500000	dia. (in.)
					0.5	1.8	3.3	7.3	18.4	41.2	
				1.7E-04	1	0.3	0.1	0.03	0.01	0.003	probability
UNK			0.00	0.00E+00	0.00E+00	0.00E+00	0.00E+00	0.00E+00	0.00E+00	0.00E+00	
RCP	**4**	**1.0**	**0.02**	**2.74E-06**	**2.74E-06**	**8.21E-07**	**2.74E-07**	**8.21E-08**	**2.74E-08**	**8.21E-09**	
FDR	3	0.75		1.54E-06	1.54E-06	4.62E-07	1.54E-07	4.62E-08	1.54E-08	4.62E-09	
FS		0.00		0.00E+00	0.00E+00	0.00E+00	0.00E+00	0.00E+00	0.00E+00	0.00E+00	
LC		0.00		0.00E+00	0.00E+00	0.00E+00	0.00E+00	0.00E+00	0.00E+00	0.00E+00	
MA		0.00		0.00E+00	0.00E+00	0.00E+00	0.00E+00	0.00E+00	0.00E+00	0.00E+00	
MF		0.00		0.00E+00	0.00E+00	0.00E+00	0.00E+00	0.00E+00	0.00E+00	0.00E+00	
SCC		0.00		0.00E+00	0.00E+00	0.00E+00	0.00E+00	0.00E+00	0.00E+00	0.00E+00	
TF		0.00		0.00E+00	0.00E+00	0.00E+00	0.00E+00	0.00E+00	0.00E+00	0.00E+00	
UNK	1	0.25		5.13E-07	5.13E-07	1.54E-07	5.13E-08	1.54E-08	5.13E-09	1.54E-09	
RPV	**173**	**1.00**	**0.69**	**1.18E-04**	**1.18E-04**	**3.55E-05**	**1.18E-05**	**3.55E-06**	**1.18E-06**	**3.55E-07**	
FDR	6	0.03		7.12E-08	7.12E-08	2.14E-08	7.12E-09	2.14E-09	7.12E-10	2.14E-10	
FS		0.00		0.00E+00	0.00E+00	0.00E+00	0.00E+00	0.00E+00	0.00E+00	0.00E+00	
LC	7	0.04		8.30E-08	8.30E-08	2.49E-08	8.30E-09	2.49E-09	8.30E-10	2.49E-10	
MA		0.00		0.00E+00	0.00E+00	0.00E+00	0.00E+00	0.00E+00	0.00E+00	0.00E+00	
MF		0.00		0.00E+00	0.00E+00	0.00E+00	0.00E+00	0.00E+00	0.00E+00	0.00E+00	
SCC	153	0.88		1.82E-06	1.82E-06	5.45E-07	1.82E-07	5.45E-08	1.82E-08	5.45E-09	
TF		0.00		0.00E+00	0.00E+00	0.00E+00	0.00E+00	0.00E+00	0.00E+00	0.00E+00	
UNK	7	0.04		8.30E-08	8.30E-08	2.49E-08	8.30E-09	2.49E-09	8.30E-10	2.49E-10	
SG	**6**	**1.00**	**0.02**	**4.10E-06**	**4.10E-06**	**1.23E-06**	**4.10E-07**	**1.23E-07**	**4.10E-08**	**1.23E-08**	
FDR		0.00		0.00E+00	0.00E+00	0.00E+00	0.00E+00	0.00E+00	0.00E+00	0.00E+00	
FS	1	0.17		3.42E-07	3.42E-07	1.03E-07	3.42E-08	1.03E-08	3.42E-09	1.03E-09	
LC		0.00		0.00E+00	0.00E+00	0.00E+00	0.00E+00	0.00E+00	0.00E+00	0.00E+00	
MA		0.00		0.00E+00	0.00E+00	0.00E+00	0.00E+00	0.00E+00	0.00E+00	0.00E+00	
MF	1	0.17		3.42E-07	3.42E-07	1.03E-07	3.42E-08	1.03E-08	3.42E-09	1.03E-09	
SCC	2	0.33		6.84E-07	6.84E-07	2.05E-07	6.84E-08	2.05E-08	6.84E-09	2.05E-09	
TF		0.00		0.00E+00	0.00E+00	0.00E+00	0.00E+00	0.00E+00	0.00E+00	0.00E+00	
UNK	2	0.33		6.84E-07	6.84E-07	2.05E-07	6.84E-08	2.05E-08	6.84E-09	2.05E-09	

Table E.6 BWR LOCA Frequency (for Pipes) Allocated by System, Degradation Mechanism, and Size Category

SLAP event counts by system (for BWRs) Only includes RCPB (S-type)

LOCA System	Counts (Event Counts)	Contribution	Deg. Mech. Fraction	Est. LOCA Fraction by System	Min (mm) (in)	Max (mm) (in)	# of Welds Dom. Total	greater than Total LOCA Freq 9.4E-05	Cat-1 100 / 0.5 / 1	Cat-2 1500 / 1.8 / 0.3	Cat-3 5000 / 3.3 / 0.1	Cat-4 25000 / 7.3 / 0.03	Cat-5 100000 / 18.4 / 0.01	Cat-6 500000 / 41.2 / 0.003
					gpm / dia.(in.) / probability →									
CRD Piping	16			0.09	10 / 0.4	550 / 21.7	1	8.0E-06	8.0E-06	2.4E-06	8.0E-07	2.4E-07	8.0E-08	0.0
CRD Piping	*16*	*1*						*8.0E-06*	*8.0E-06*	*2.4E-06*	*8.0E-07*	*2.4E-07*	*8.0E-08*	*--*
UA			*0.00*					*0.0E+00*	*0.0E+00*	*0.0E+00*	*0.0E+00*	*0.0E+00*	*0.0E+00*	*--*
MA			*0.00*					*0.0E+00*	*0.0E+00*	*0.0E+00*	*0.0E+00*	*0.0E+00*	*0.0E+00*	*--*
LC	*0*		*0.00*					*0.0E+00*	*0.0E+00*	*0.0E+00*	*0.0E+00*	*0.0E+00*	*0.0E+00*	*--*
FDR	*3*		*0.19*					*1.5E-06*	*1.5E-06*	*4.5E-07*	*1.5E-07*	*4.5E-08*	*1.5E-08*	*--*
SCC	*12*		*0.75*					*6.0E-06*	*6.0E-06*	*1.8E-06*	*6.0E-07*	*1.8E-07*	*6.0E-08*	*--*
MF	*0*		*0.00*					*0.0E+00*	*0.0E+00*	*0.0E+00*	*0.0E+00*	*0.0E+00*	*0.0E+00*	*--*
TF	*0*		*0.00*					*0.0E+00*	*0.0E+00*	*0.0E+00*	*0.0E+00*	*0.0E+00*	*0.0E+00*	*--*
FS			*0.00*					*0.0E+00*	*0.0E+00*	*0.0E+00*	*0.0E+00*	*0.0E+00*	*0.0E+00*	*--*
UK	*1*		*0.06*					*5.0E-07*	*5.0E-07*	*1.5E-07*	*5.0E-08*	*1.5E-08*	*5.0E-09*	*--*
Drain Lines	11			0.06	19 / 0.7	250 / 9.8	1	5.5E-06	5.5E-06	1.7E-06	5.5E-07	1.7E-07	0.0	0.0
Drain Lines	*11*	*1*						*5.5E-06*	*5.5E-06*	*1.7E-06*	*5.5E-07*	*1.7E-07*	*--*	*--*
UA			*0.00*					*0.0E+00*	*0.0E+00*	*0.0E+00*	*0.0E+00*	*0.0E+00*	*--*	*--*
MA			*0.00*					*0.0E+00*	*0.0E+00*	*0.0E+00*	*0.0E+00*	*0.0E+00*	*--*	*--*
LC			*0.00*					*0.0E+00*	*0.0E+00*	*0.0E+00*	*0.0E+00*	*0.0E+00*	*--*	*--*
FDR	*0*		*0.00*					*0.0E+00*	*0.0E+00*	*0.0E+00*	*0.0E+00*	*0.0E+00*	*--*	*--*
SCC	*2*		*0.18*					*1.0E-06*	*1.0E-06*	*3.0E-07*	*1.0E-07*	*3.0E-08*	*--*	*--*
MF	*8*		*0.73*					*4.0E-06*	*4.0E-06*	*1.2E-06*	*4.0E-07*	*1.2E-07*	*--*	*--*
TF	*0*		*0.00*					*0.0E+00*	*0.0E+00*	*0.0E+00*	*0.0E+00*	*0.0E+00*	*--*	*--*
FS	*1*		*0.09*					*5.0E-07*	*5.0E-07*	*1.5E-07*	*5.0E-08*	*1.5E-08*	*--*	*--*
UK			*0.00*					*0.0E+00*	*0.0E+00*	*0.0E+00*	*0.0E+00*	*0.0E+00*	*--*	*--*
Feedwater	8			0.04	8 / 0.3	300 / 11.8	1	4.0E-06	4.0E-06	1.2E-06	4.0E-07	1.2E-07	0.0	0.0
Feedwater - BC	*8*	*1*						*4.0E-06*	*4.0E-06*	*1.2E-06*	*4.0E-07*	*1.2E-07*	*--*	*--*

SLAP event counts by system (for BWRs)
Only includes RCPB (S-type)

LOCA System	Event Counts	Deg. Mech. Fractional Contribution by System	Est. LOCA Fraction	Pipe Sizes Min (mm)/(in)	Pipe Sizes Max (mm)/(in)	# of Welds Dom.	# of Welds Total	greater than Total LOCA Freq 9.4E-05	Cat-1 100 / 0.5 / 1	Cat-2 1500 / 1.8 / 0.3	Cat-3 5000 / 3.3 / 0.1	Cat-4 25000 / 7.3 / 0.03	Cat-5 100000 / 18.4 / 0.01	Cat-6 500000 / 41.2 / 0.003 (gpm / dia.(in) / probabilty)
UA		0.00					123	0.0E+00	0.0E+00	0.0E+00	0.0E+00	0.0E+00	--	--
MA		0.00						0.0E+00	0.0E+00	0.0E+00	0.0E+00	0.0E+00	--	--
LC		0.00						0.0E+00	0.0E+00	0.0E+00	0.0E+00	0.0E+00	--	--
FDR	0	0.00						0.0E+00	0.0E+00	0.0E+00	0.0E+00	0.0E+00	--	--
SCC	0	0.00						0.0E+00	0.0E+00	0.0E+00	0.0E+00	0.0E+00	--	--
MF	0	0.00						0.0E+00	0.0E+00	0.0E+00	0.0E+00	0.0E+00	--	--
TF	7	0.88						3.5E-06	3.5E-06	1.1E-06	3.5E-07	1.1E-07	--	--
FS	1	0.13						5.0E-07	5.0E-07	1.5E-07	5.0E-08	1.5E-08	--	--
UK		0.00						0.0E+00	0.0E+00	0.0E+00	0.0E+00	0.0E+00	--	--
ECCS	20		0.11	10 / 0.4	500 / 19.7	1	60	1.0E-05	1.0E-05	3.0E-06	1.0E-06	3.0E-07	1.0E-07	0.0
ECCS	20	1						1.0E-05	1.0E-05	3.0E-06	1.0E-06	3.0E-07	1.0E-07	--
UA		0.00						0.0E+00	0.0E+00	0.0E+00	0.0E+00	0.0E+00	0.0E+00	--
MA		0.00						0.0E+00	0.0E+00	0.0E+00	0.0E+00	0.0E+00	0.0E+00	--
LC	1	0.05						5.0E-07	5.0E-07	1.5E-07	5.0E-08	1.5E-08	5.0E-09	--
FDR	0	0.00						0.0E+00	0.0E+00	0.0E+00	0.0E+00	0.0E+00	0.0E+00	--
SCC	9	0.45						4.5E-06	4.5E-06	1.4E-06	4.5E-07	1.4E-07	4.5E-08	--
MF	5	0.25						2.5E-06	2.5E-06	7.5E-07	2.5E-07	7.5E-08	2.5E-08	--
TF	2	0.10						1.0E-06	1.0E-06	3.0E-07	1.0E-07	3.0E-08	1.0E-08	--
FS	3	0.15						1.5E-06	1.5E-06	4.5E-07	1.5E-07	4.5E-08	1.5E-08	--
UK	0	0.00						0.0E+00	0.0E+00	0.0E+00	0.0E+00	0.0E+00	0.0E+00	--
Inst	20		0.11	10 / 0.4	150 / 5.9	1		1.0E-05	1.0E-05	3.0E-06	1.0E-06	0.0	0.0	0.0
Inst	20	1						1.0E-05	1.0E-05	3.0E-06	1.0E-06	--	--	--
UA		0.00						0.0E+00	0.0E+00	0.0E+00	0.0E+00	0.0E+00	0.0E+00	--
MA		0.00						0.0E+00	0.0E+00	0.0E+00	0.0E+00	0.0E+00	0.0E+00	--
LC		0.00						0.0E+00	0.0E+00	0.0E+00	0.0E+00	0.0E+00	0.0E+00	--
FDR	5	0.25						2.5E-06	2.5E-06	7.5E-07	2.5E-07	0.0E+00	0.0E+00	--
SCC	4	0.20						2.0E-06	2.0E-06	6.0E-07	2.0E-07	0.0E+00	0.0E+00	--
MF	11	0.55						5.5E-06	5.5E-06	1.7E-06	5.5E-07	0.0E+00	0.0E+00	--
TF	0	0.00						0.0E+00	0.0E+00	0.0E+00	0.0E+00	0.0E+00	0.0E+00	--
FS		0.00						0.0E+00	0.0E+00	0.0E+00	0.0E+00	0.0E+00	0.0E+00	--
UK	0	0.00						0.0E+00	0.0E+00	0.0E+00	0.0E+00	0.0E+00	0.0E+00	--
RCIC	2		0.01	20 / 0.8	150 / 5.9	1		1.0E-06	1.0E-06	3.0E-07	1.0E-07	0.0	0.0	0.0
RCIC	2	1						1.0E-06	1.0E-06	3.0E-07	1.0E-07	--	--	--

SLAP event counts by system (for BWRs)
Only includes RCPB (S-type)

LOCA System	Event Counts	Deg. Mech. Fractional Contribution	Est. LOCA Fraction	Fraction by System	Pipe Sizes Min (mm / in)	Pipe Sizes Max (mm / in)	# of Welds Dom. Total	greater than Total (LOCA Freq 9.4E-05)	Cat-1 (100 gpm / 0.5 in / 1)	Cat-2 (1500 gpm / 1.8 in / 0.3)	Cat-3 (5000 gpm / 3.3 in / 0.1)	Cat-4 (25000 gpm / 7.3 in / 0.03)	Cat-5 (100000 gpm / 18.4 in / 0.01)	Cat-6 (500000 gpm / 41.2 in / 0.003 probability)
UA		0.00					16	0.0E+00	0.0E+00	0.0E+00	0.0E+00	--	--	--
MA		0.00						0.0E+00	0.0E+00	0.0E+00	0.0E+00	--	--	--
LC		0.00						0.0E+00	0.0E+00	0.0E+00	0.0E+00	--	--	--
FDR		0.00						0.0E+00	0.0E+00	0.0E+00	0.0E+00	--	--	--
SCC		0.00						0.0E+00	0.0E+00	0.0E+00	0.0E+00	--	--	--
MF	2	1.00						1.0E-06	1.0E-06	3.0E-07	1.0E-07	--	--	--
TF	0	0.00						0.0E+00	0.0E+00	0.0E+00	0.0E+00	--	--	--
FS	0	0.00						0.0E+00	0.0E+00	0.0E+00	0.0E+00	--	--	--
UK		0.00						0.0E+00	0.0E+00	0.0E+00	0.0E+00	--	--	--
Recirc	48							2.4E-05	2.4E-05	7.2E-06	2.4E-06	7.2E-07	2.4E-07	0.0E+00
Recirc - ave	48		0.26		20 / 0.8	700 / 27.6	1	2.4E-05	2.4E-05	7.2E-06	2.4E-06	7.2E-07	2.4E-07	0.0
Recirc - old (BC)	45	1					121	5.3E-05	5.3E-05	1.6E-05	5.3E-06	1.6E-06	5.3E-07	--
UA		0.00						0.0E+00	0.0E+00	0.0E+00	0.0E+00	0.0E+00	0.0E+00	--
MA		0.00						0.0E+00	0.0E+00	0.0E+00	0.0E+00	0.0E+00	0.0E+00	--
LC		0.00						0.0E+00	0.0E+00	0.0E+00	0.0E+00	0.0E+00	0.0E+00	--
FDR	1	0.022						1.2E-06	1.2E-06	3.5E-07	1.2E-07	3.5E-08	1.2E-08	--
SCC	43	0.96						5.0E-05	5.0E-05	1.5E-05	5.0E-06	1.5E-06	5.0E-07	--
MF	1	0.02						1.2E-06	1.2E-06	3.5E-07	1.2E-07	3.5E-08	1.2E-08	--
TF		0.00						0.0E+00	0.0E+00	0.0E+00	0.0E+00	0.0E+00	0.0E+00	--
FS		0.00						0.0E+00	0.0E+00	0.0E+00	0.0E+00	0.0E+00	0.0E+00	--
UK		0.00						0.0E+00	0.0E+00	0.0E+00	0.0E+00	0.0E+00	0.0E+00	--
Recirc - new	3	1						2.4E-06	2.4E-06	7.2E-07	2.4E-07	7.2E-08	2.4E-08	0.0
UA		0.00						0.0E+00	0.0E+00	0.0E+00	0.0E+00	0.0E+00	0.0E+00	--
MA		0.00						0.0E+00	0.0E+00	0.0E+00	0.0E+00	0.0E+00	0.0E+00	--
LC		0.00						0.0E+00	0.0E+00	0.0E+00	0.0E+00	0.0E+00	0.0E+00	--
FDR	2	0.67						1.6E-06	1.6E-06	4.8E-07	1.6E-07	4.8E-08	1.6E-08	--
SCC	1	0.33						8.0E-07	8.0E-07	2.4E-07	8.0E-08	2.4E-08	8.0E-09	--
MF	0	0.00						0.0E+00	0.0E+00	0.0E+00	0.0E+00	0.0E+00	0.0E+00	--
TF		0.00						0.0E+00	0.0E+00	0.0E+00	0.0E+00	0.0E+00	0.0E+00	--
FS		0.00						0.0E+00	0.0E+00	0.0E+00	0.0E+00	0.0E+00	0.0E+00	--
UK		0.00						0.0E+00	0.0E+00	0.0E+00	0.0E+00	0.0E+00	0.0E+00	--

SLAP event counts by system (for BWRs)
Only includes RCPB (S-type)

LOCA System	Deg. Mech. Event	Fraction Counts	Fractional Contribution	Est. LOCA Fraction by System	Pipe Sizes Min (mm)(in)	Max (mm)(in)	# of Welds Dom.	Total	greater than Total LOCA Freq 9.4E-05	Cat-1 100 0.5 1	Cat-2 1500 1.8 0.3	Cat-3 5000 3.3 0.1	Cat-4 25000 7.3 0.03	Cat-5 100000 18.4 0.01	Cat-6 500000 41.2 0.003 gpm dia.(in) probabilty
RHR	39			0.21	19 / 0.7	600 / 23.6	1	74	2.0E-05	2.0E-05	5.9E-06	2.0E-06	5.9E-07	2.0E-07	0.0
RHR	*39*	*1*							*2.0E-05*	*2.0E-05*	*5.9E-06*	*2.0E-06*	*5.9E-07*	*2.0E-07*	*---*
UA			*0.00*						*0.0E+00*	*0.0E+00*	*0.0E+00*	*0.0E+00*	*0.0E+00*	*0.0E+00*	*---*
MA			*0.00*						*0.0E+00*	*0.0E+00*	*0.0E+00*	*0.0E+00*	*0.0E+00*	*0.0E+00*	*---*
LC			*0.00*						*0.0E+00*	*0.0E+00*	*0.0E+00*	*0.0E+00*	*0.0E+00*	*0.0E+00*	*---*
FDR		*2*	*0.05*						*1.0E-06*	*1.0E-06*	*3.0E-07*	*1.0E-07*	*3.0E-08*	*1.0E-08*	*---*
SCC		*33*	*0.85*						*1.7E-05*	*1.7E-05*	*5.0E-06*	*1.7E-06*	*5.0E-07*	*1.7E-08*	*---*
MF		*2*	*0.05*						*1.0E-06*	*1.0E-06*	*3.0E-07*	*1.0E-07*	*3.0E-08*	*1.0E-08*	*---*
TF		*2*	*0.05*						*1.0E-06*	*1.0E-06*	*3.0E-07*	*1.0E-07*	*3.0E-08*	*1.0E-08*	*---*
FS			*0.00*						*0.0E+00*	*0.0E+00*	*0.0E+00*	*0.0E+00*	*0.0E+00*	*0.0E+00*	*---*
UK			*0.00*						*0.0E+00*	*0.0E+00*	*0.0E+00*	*0.0E+00*	*0.0E+00*	*0.0E+00*	*---*
RWCU	17			0.09	10 / 0.4	400 / 15.7	1	72	8.5E-06	8.5E-06	2.6E-06	8.5E-07	2.6E-07	0.0	0.0
RWCU	*17*	*1*							*8.5E-06*	*8.5E-06*	*2.6E-06*	*8.5E-07*	*2.6E-07*	*---*	*---*
UA			*0.00*						*0.0E+00*	*0.0E+00*	*0.0E+00*	*0.0E+00*	*0.0E+00*	*---*	*---*
MA			*0.00*						*0.0E+00*	*0.0E+00*	*0.0E+00*	*0.0E+00*	*0.0E+00*	*---*	*---*
LC		*0*	*0.00*						*0.0E+00*	*0.0E+00*	*0.0E+00*	*0.0E+00*	*0.0E+00*	*---*	*---*
FDR		*1*	*0.06*						*5.0E-07*	*5.0E-07*	*1.5E-07*	*5.0E-08*	*1.5E-08*	*---*	*---*
SCC		*6*	*0.35*						*3.0E-06*	*3.0E-06*	*9.0E-07*	*3.0E-07*	*9.0E-08*	*---*	*---*
MF		*3*	*0.18*						*1.5E-06*	*1.5E-06*	*4.5E-07*	*1.5E-07*	*4.5E-08*	*---*	*---*
TF		*6*	*0.35*						*3.0E-06*	*3.0E-06*	*9.0E-07*	*3.0E-07*	*9.0E-08*	*---*	*---*
FS		*0*	*0.00*						*0.0E+00*	*0.0E+00*	*0.0E+00*	*0.0E+00*	*0.0E+00*	*---*	*---*
UK		*1*	*0.06*						*5.0E-07*	*5.0E-07*	*1.5E-07*	*5.0E-08*	*1.5E-08*	*---*	*---*
SLC	3			0.02	20 / 0.8	75 / 3.0	1		1.5E-06	1.5E-06	4.5E-07	0.0	0.0	0.0	0.0
SLC	*3*	*1*							*1.5E-06*	*1.5E-06*	*4.5E-07*	*---*	*---*	*---*	*---*
UA			*0.00*						*0.0E+00*	*0.0E+00*	*0.0E+00*	*0.0E+00*	*0.0E+00*	*---*	*---*
MA			*0.00*						*0.0E+00*	*0.0E+00*	*0.0E+00*	*0.0E+00*	*0.0E+00*	*---*	*---*
LC			*0.00*						*0.0E+00*	*0.0E+00*	*0.0E+00*	*0.0E+00*	*0.0E+00*	*---*	*---*
FDR			*0.00*						*0.0E+00*	*0.0E+00*	*0.0E+00*	*0.0E+00*	*0.0E+00*	*---*	*---*
SCC		*3*	*1.00*						*1.5E-06*	*1.5E-06*	*4.5E-07*	*---*	*---*	*---*	*---*
MF			*0.00*						*0.0E+00*	*0.0E+00*	*0.0E+00*	*0.0E+00*	*0.0E+00*	*---*	*---*
TF			*0.00*						*0.0E+00*	*0.0E+00*	*0.0E+00*	*0.0E+00*	*0.0E+00*	*---*	*---*
FS			*0.00*						*0.0E+00*	*0.0E+00*	*0.0E+00*	*0.0E+00*	*0.0E+00*	*---*	*---*
UK			*0.00*						*0.0E+00*	*0.0E+00*	*0.0E+00*	*0.0E+00*	*0.0E+00*	*---*	*---*

E-16

SLAP event counts by system (for BWRs)
Only includes RCPB (S-type)

										Cat-1	Cat-2	Cat-3	Cat-4	Cat-5	Cat-6	
										100	1500	5000	25000	100000	500000	gpm
		Deg. Mech.	Est.		Pipe Sizes			# of Welds	greater than	0.5	1.8	3.3	7.3	18.4	41.2	dia. (in.)
	Event Counts	Fractional Contribution	LOCA Fraction by System	Min (mm) (in)	Max (mm) (in)	Dom.	Total		Total	1	0.3	0.1	0.03	0.01	0.003	probabilty
LOCA System									LOCA Freq 9.4E-05							
SRV Lines	3		0.02	15	25				1.5E-06	1.5E-06	0.0	0.0	0.0	0.0	0.0	
SRV Lines	3	1		0.6	1.0		1		1.5E-06	1.5E-06	--	--	--	--	--	
UA		0.00							0.0E+00	0.0E+00	--	--	--	--	--	
MA		0.00							0.0E+00	0.0E+00	--	--	--	--	--	
LC		0.00							0.0E+00	0.0E+00	--	--	--	--	--	
FDR		0.00							0.0E+00	0.0E+00	--	--	--	--	--	
SCC	2	0.67							1.0E-06	1.0E-06	--	--	--	--	--	
MF	1	0.33							5.0E-07	5.0E-07	--	--	--	--	--	
TF	0	0.00							0.0E+00	0.0E+00	--	--	--	--	--	
FS		0.00							0.0E+00	0.0E+00	--	--	--	--	--	
UK		0.00							0.0E+00	0.0E+00	--	--	--	--	--	
Steam Lines	1		0.01	20	100				5.0E-07	5.0E-07	1.5E-07	5.0E-08	0.0	0.0	0.0	
Steam Lines	1	1		0.8	28.0	113.0	1		5.0E-07	5.0E-07	1.5E-07	5.0E-08	1.5E-08	5.0E-09	--	
UA		0.00							0.0E+00	0.0E+00	0.0E+00	0.0E+00	0.0E+00	0.0E+00	--	
MA		0.00							0.0E+00	0.0E+00	0.0E+00	0.0E+00	0.0E+00	0.0E+00	--	
LC		0.00							0.0E+00	0.0E+00	0.0E+00	0.0E+00	0.0E+00	0.0E+00	--	
FDR		0.00							0.0E+00	0.0E+00	0.0E+00	0.0E+00	0.0E+00	0.0E+00	--	
SCC	0	0.00							0.0E+00	0.0E+00	0.0E+00	0.0E+00	0.0E+00	0.0E+00	--	
MF	1	1.00							5.0E-07	5.0E-07	1.5E-07	5.0E-08	1.5E-08	5.0E-09	--	
TF		0.00							0.0E+00	0.0E+00	0.0E+00	0.0E+00	0.0E+00	0.0E+00	--	
FS		0.00							0.0E+00	0.0E+00	0.0E+00	0.0E+00	0.0E+00	0.0E+00	--	
UK		0.00							0.0E+00	0.0E+00	0.0E+00	0.0E+00	0.0E+00	0.0E+00	--	
	188		1.00						9.4E-05	9.4E-05	2.8E-05	9.1E-06	2.4E-06	6.2E-07	0.0E+00	

Table E.7 BWR Passive Non-Pipe LOCA Frequency

BWR Non-Pipe LOCA Contributors
Total non-pipe event count = 30

Non-Pipe LOCA System	Degradation Mechanism	Counts	Fract Contr	Non-Pipe LOCA Fract	Non-Pipe LOCA Freq 9.4E-05	Cat-1 100 0.5 1	Cat-2 1500 1.8 0.3	Cat-3 5000 3.3 0.1	Cat-4 25000 7.3 0.03	Cat-5 100000 18.4 0.01	Cat-6 500000 41.2 0.003 gpm dia. (in.) probability
LIV		1	1	0.03	3.15E-06	3.15E-06	9.44E-07	3.15E-07	9.44E-08	3.15E-08	0.00E+00
	FDR		0.00		0.00E+00	0.00E+00	0.00E+00	0.00E+00	0.00E+00	0.00E+00	0.00E+00
	FS		0.00		0.00E+00	0.00E+00	0.00E+00	0.00E+00	0.00E+00	0.00E+00	0.00E+00
	LC		0.00		0.00E+00	0.00E+00	0.00E+00	0.00E+00	0.00E+00	0.00E+00	0.00E+00
	MA	1	1.00		3.15E-06	3.15E-06	9.44E-07	3.15E-07	9.44E-08	3.15E-08	0.00E+00
	MF		0.00		0.00E+00	0.00E+00	0.00E+00	0.00E+00	0.00E+00	0.00E+00	0.00E+00
	SCC		0.00		0.00E+00	0.00E+00	0.00E+00	0.00E+00	0.00E+00	0.00E+00	0.00E+00
	TF		0.00		0.00E+00	0.00E+00	0.00E+00	0.00E+00	0.00E+00	0.00E+00	0.00E+00
	UNK		0.00		0.00E+00	0.00E+00	0.00E+00	0.00E+00	0.00E+00	0.00E+00	0.00E+00
RecP		5	1	0.17	1.57E-05	1.57E-05	4.72E-06	1.57E-06	4.72E-07	1.57E-07	0.00E+00
	FDR	3	0.60		1.89E-06	1.89E-06	5.66E-07	1.89E-07	5.66E-08	1.89E-08	0.00E+00
	FS		0.00		0.00E+00	0.00E+00	0.00E+00	0.00E+00	0.00E+00	0.00E+00	0.00E+00
	LC		0.00		0.00E+00	0.00E+00	0.00E+00	0.00E+00	0.00E+00	0.00E+00	0.00E+00
	MA	2	0.40		1.26E-06	1.26E-06	3.78E-07	1.26E-07	3.78E-08	1.26E-08	0.00E+00
	MF		0.00		0.00E+00	0.00E+00	0.00E+00	0.00E+00	0.00E+00	0.00E+00	0.00E+00
	SCC		0.00		0.00E+00	0.00E+00	0.00E+00	0.00E+00	0.00E+00	0.00E+00	0.00E+00
	TF		0.00		0.00E+00	0.00E+00	0.00E+00	0.00E+00	0.00E+00	0.00E+00	0.00E+00
	UNK		0.00		0.00E+00	0.00E+00	0.00E+00	0.00E+00	0.00E+00	0.00E+00	0.00E+00
RPV		24	1	0.80	7.55E-05	7.55E-05	2.27E-05	7.55E-06	2.27E-06	7.55E-07	2.27E-07
	FDR	1	0.04		1.31E-07	1.31E-07	3.93E-08	1.31E-08	3.93E-09	1.31E-09	3.93E-10
	FS		0.00		0.00E+00	0.00E+00	0.00E+00	0.00E+00	0.00E+00	0.00E+00	0.00E+00
	LC		0.00		0.00E+00	0.00E+00	0.00E+00	0.00E+00	0.00E+00	0.00E+00	0.00E+00
	MA		0.00		0.00E+00	0.00E+00	0.00E+00	0.00E+00	0.00E+00	0.00E+00	0.00E+00
	MF		0.00		0.00E+00	0.00E+00	0.00E+00	0.00E+00	0.00E+00	0.00E+00	0.00E+00
	SCC	23	0.96		3.02E-06	3.02E-06	9.05E-07	3.02E-07	9.05E-08	3.02E-08	9.05E-09

BWR Non-Pipe LOCA Contributors
Total non-pipe event count = 30

Non-Pipe LOCA System	Degradation Mechanism Counts	Fract Contr	Non-Pipe LOCA Fract	Non-Pipe LOCA Freq 9.4E-05	Cat-1 100 0.5 1	Cat-2 1500 1.8 0.3	Cat-3 5000 3.3 0.1	Cat-4 25000 7.3 0.03	Cat-5 100000 18.4 0.01	Cat-6 500000 41.2 0.003	gpm dia. (in.) probabilty
TF		0.00			0.00E+00	0.00E+00	0.00E+00	0.00E+00	0.00E+00	0.00E+00	
UNK		0.00			0.00E+00	0.00E+00	0.00E+00	0.00E+00	0.00E+00	0.00E+00	

APPENDIX F

PIPING BASE CASE RESULTS OF

DAVID HARRIS

APPENDIX F

PIPING BASE CASE RESULTS OF DAVID HARRIS

**Probabilistic Fracture Mechanics Analyses
Performed in Support of LOCA Frequency
Re-evaluation Effort**

**D.O. Harris
Engineering Mechanics Technology, Inc.
San Jose, California**

F.1 Introduction

The purpose of this document is to report the procedures used and the results obtained in probabilistic fracture mechanics analyses of the base case systems considered in the LOCA Re-evaluation effort performed by use of expert elicitation by the Nuclear Regulatory Commission in the period February 2003 – March 2004. The base case systems, which were defined in the kick-off meeting of the expert panel that was held in Rockville, Maryland in February 2003, consisted of the following:

Pressurized Water Reactor
-     hot leg   (cast austenitic stainless steel)
-     surge line   (austenitic stainless steel)
-     HPI makeup nozzle   (austenitic stainless steel)

Boiling Water Reactor
-     recirculation line   (austenitic stainless steel)
-     feedwater   (carbon steel)

These were identified as key systems that could serve as benchmarks for use by members of the expert panel in their estimation of LOCA frequencies.

Piping isometrics of the base case systems and other systems identified in the kick-off meeting as important to estimations of flow rate probabilities were included in the FTP site that was set up for the use of panel members. Times in this appendix are in reactor-years (1 calendar year ~ 0.8 reactor years).

F.2 Software

The following discussion provides only a brief review of the PRAISE software. The references cited give the details. The results reported here were generated by use of the PRAISE software, which was developed with NRC support over a period of some 20 years. PRAISE is based on deterministic fracture mechanics, with some of the inputs considered as random variables. This allows the statistical distribution of lifetime to be computed, rather than a single deterministic failure time. The probability of failure (leaks of various sizes) is obtained from the computed lifetime distribution.

Several versions of PRAISE were employed, depending on the nature of the problem. The original version of PRAISE [F.1] considers fatigue crack growth from crack-like weld defects introduced during

fabrication. Semi-elliptical interior surface cracks are considered, usually circumferentially oriented.
The initial crack size and fatigue crack growth properties are the major random variables, and Monte
Carlo simulation is used to generate numerical results. Stratified sampling of crack depth and aspect ratio
is employed to allow very small probabilities to be obtained without excessive computer time. Figure F.1
is a schematic representation of the probabilistic fracture mechanics procedures used in the original
version of PRAISE. The cumulative probability of a flow (leak) rate exceeding a specified size is
generated by PRAISE as a function of time. If the stress history is specified in reactor-years, then the
PRAISE results are also in reactor years.

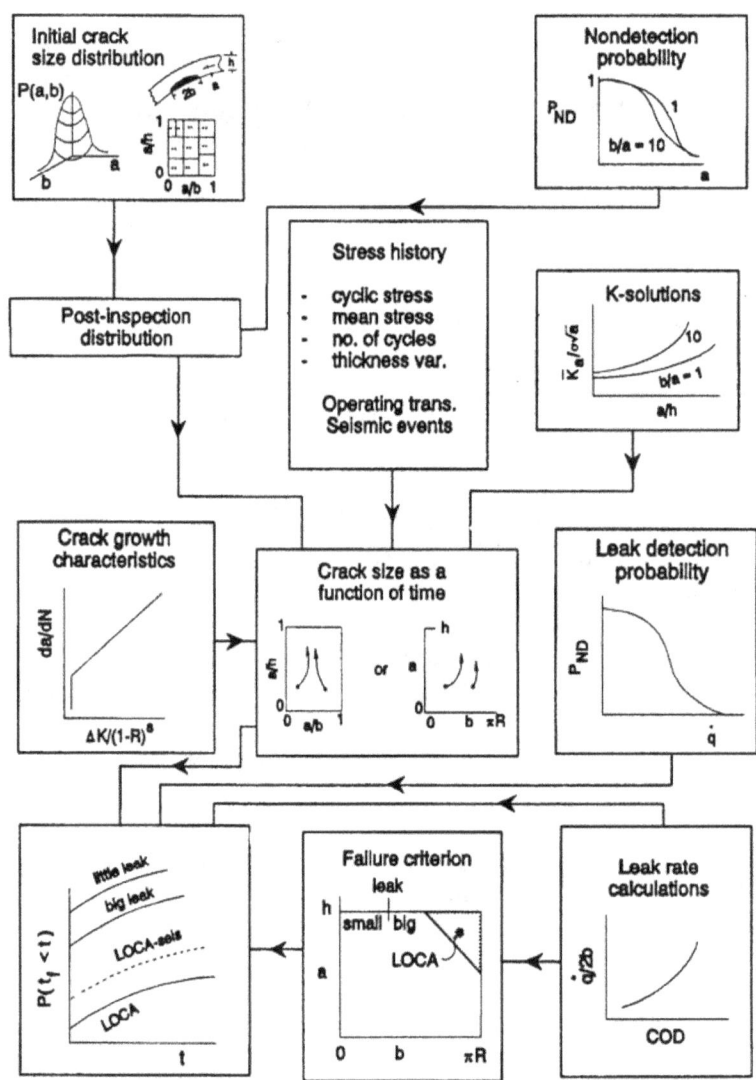

Figure F.1 Overview of PRAISE Methodology for Probabilistic Analysis of Fatigue Crack Growth

A later version of PRAISE was used the analyze initiation and growth of stress corrosion cracks [F.2].
Both the fatigue crack growth and stress corrosion initiation and growth capabilities are included in
pcPRAISE [F.3] and also in WinPRAISE [F.4], which is a Windows version that is much easier to use
than the earlier DOS versions. WinPRAISE gives the same results as pcPRAISE for the same problem
with the same inputs.

In some cases, the cyclic stresses were such that fatigue crack initiation is the expected dominant degradation mechanism, and it was necessary to use a later version of PRAISE. This version was developed and used in Reference F.5, with additional capabilities described in Reference F.6 (such as the ability to consider detailed circumferential variations of the stresses). The fatigue crack initiation analyses employed probabilistic strain-life relations developed by Argonne National Laboratory [F.7, F.8]. Once a fatigue crack initiates, the original growth analysis capabilities in PRAISE are used for the crack growth portion of the lifetime. The depth of an initiated fatigue crack is taken to be 3.0 mm (0.12 inches), in accordance with the ANL correlations, with a random surface length. In some instances, modifications to the source code of PRAISE were made to provide results of particular use for the problem at hand. These instances are discussed in the specific base case problem where they were employed.

In cases where inspection was considered, the nondetection probability was represented by the expression

$$P_{ND}(a) = \varepsilon + \frac{1}{2}(1+\varepsilon)\,\mathrm{erfc}\left[\upsilon \ln \frac{a}{a^*}\right] \qquad [F.1]$$

In this expression, ε is the probability of not detecting a crack no matter how deep it is, a^* is the crack depth having about a 50% chance of being detected, and υ controls the slope of the P_{ND} curve. "Good" and "outstanding" detection capabilities were considered [F.9], with the parameters given in Table F.1.

Table F.1 Parameters in Non-Detection Relation

		Ferritic	Austenitic	
		Fatigue Cracks	Fatigue Cracks	SCC Cracks
outstanding				
	a^*	0.05h	0.05h	0.05h
	υ	1.6	1.6	1.6
	ε	0.005	0.005	0.005
good				
	a^*	0.15h	0.15h	0.4h
	υ	1.6	1.6	1.6
	ε	0.02	0.02	0.10

The operating and hydro pressures of Table F.2 were considered.

Table F.2 Operating and Hydro Pressures, psi

	Operating Pressure	Hydro Pressure
PWR	2250	3125
BWR	1250	1560

A leak detection capability of 19 lpm (5 gpm) was considered, which means that any through-wall crack with a leak rate greater than 19 lpm (5 gpm) was immediately detected and removed from service.

F.3 Post-Processing of PRAISE Results

PRAISE analyses are performed on a piping location-by-location basis to provide the cumulative failure probability as a function of time, whereas the desired end result is the system failure frequency within various time frames. The time increments of interest are now (0-25 years), the near future (25-40 years) and the more distant future (40-60 years). If the reactor transients analyzed are per reactor-year (rather than calendar year), then the times considered are also per reactor-year. As an average, one calendar year corresponds to 0.8 reactor years, but no adjustments for this are made in this appendix. The system failure frequency is obtained from the failure frequency for individual locations by analyzing the most highly stressed location and multiplying by the number of such highly stressed locations in the system. The failure frequency for a flow rate exceeding $\dot{q}$ for a given time increment from t_1 to t_2 is obtained from the following relation

$$p_{\dot{q}}(t_2) = \frac{P_{\dot{q}}(t_2) - P_{\dot{q}}(t_1)}{t_2 - t_1} \qquad \text{[F.2]}$$

The values of $P_{\dot{q}}(t)$ are the output from PRAISE for the dominant location(s) in the system. The failure frequency for the system is then obtained by multiplying by the number of locations in the system that have the high stresses of the dominant location.

F.3.1 Material Properties
Numerous material properties enter into a PRAISE analysis, many of which are described in the references cited above. A Ramberg-Osgood stress-strain curve is used in the computation of the applied value of the J-integral, as represented by the following relation

$$\varepsilon = \frac{\sigma}{E} + \left(\frac{\sigma}{D}\right)^n \qquad \text{[F.3]}$$

Crack instability is governed by exceedance of a critical net section stress and/or a J-integral based tearing instability using a bilinear tearing resistance curve.

A set of default material tensile and fracture properties are provided in WinPRAISE [F.4], which are summarized in Table F.3. Unless otherwise stated, these properties are used in the base case analyses.

Table F.3 Summary of Default Material Properties

Material	Low Alloy	Carbon	Type 304	Type 316
Flow Stress (normal)				
Mean, ksi	50.	52.	44.9	44.9
Standard Deviation	2.1	2.2	1.9	1.9
Tearing Instability Data				
D, ksi	146	154	106	106
n	5	4	5	5
J_{Ic} kips/in	0.6	1.5	5	5
T_{mat}	60	45	300	300
E, ksi	27300	27300	25800	25800
ν	0.3	0.3	0.3	0.3
Yield Strength, ksi	30.8	28.3	19.4	19.4
Tensile Strength ksi	70.0	60.0	59.3	61.6
Fatigue Crack Growth Properties (random)	based on ASME Code		see Reference F.1 or F.4	

The properties of Table F.3 are generally somewhat conservative and are representative of undegraded materials. In some instances, degraded material properties are considered, as discussed at the particular component involved.

In the case of fatigue of initial cracks, the distributions of the initial crack depth and aspect ratio are probably the most important random variables. Unless otherwise stated, the depth distribution is taken from Reference F.10, which is the default in WinPRAISE and is also included in Reference F.9.

F.3.2 Hot Leg

The main coolant piping system is one of the base case systems. The failure probability of the large piping of this system is dominated by the hot leg to pressure vessel weld, because this location is at the highest temperature and sees the highest stress.

F.3.2.1 Dimensions and Welds - From the piping isometrics made available to the panel members, the hot leg has a 29 inch inner diameter, and a thickness of 63.5 mm (2.5 inch) (OD=34 inches), fabricated from SA-376 (which is an austenitic stainless steel). The example plant has two coolant loops. There are several welds in the hot leg, including shop and field welds and safe ends. There is a safe end and field weld at the pressure vessel.

F.3.2.2 Stresses and Cycles - Table 1-2, page 9 of Reference F.1 summarizes the deadweight, pressure and restraint of thermal expansion stresses for the 14 field welds in one loop of the large primary piping in the plant considered in that report. Joint 1 is the hot leg to pressure vessel joint, and it has the highest stresses. The seismic stresses are included in Table 1-3, page 10 [F.1]. They are generally quite low. The postulated list of transients is provided in Table 4-1, page 152 [F.1], and is the transients occurring over the 40 year design life, which corresponds to reactor years. The list contains 11 types of transients. There is sufficient information in Reference F.1 to consider all of these transients, but only the heat-up cool-down transient will be considered in this case, because it is the dominant transient contributing to fatigue crack growth [F.1]. The heat-up cool-down transient was postulated to occur 200 times in 40 years (5/year). This is excessive, and 3/year is used herein.

The seismic stresses are given in terms of the maximum load controlled stress (deadweight + pressure + max seismic), and the summary of the stress history needed for fatigue crack growth analysis is also provided. This summary, denoted as S, is the sum of the cyclic stresses as follows

$$S^4 = \sum_{seismic\,stress\,history} \sigma_{max}^2 \Delta\sigma^2 \qquad [F.4]$$

This is what controls the amount of fatigue crack growth during a seismic event for a fourth power crack growth law that includes R-ratio effects (See Reference F.1). Table F.4 summarizes the suggested stress history for the hot leg to pressure vessel joint.

Table F.4 Summary of Stress History for Hot Leg to Pressure Vessel Joint

deadweight stress = 2.08 ksi
pressure stress = 6.49 ksi (axial)
restraint of thermal expansion stress = 6.50 ksi
3 times per year

	max σ_{LC} ksi	S^4 ksi^4	$\Delta\sigma$, ksi
OBE	8.76	521.6	1.27
SSE	9.06	2958.3	1.96
3SSE	10.26	63430	4.22
5SSE	10.62	162000	5.33

The right-hand column in the above table is the cyclic stress if the seismic event contains 200 stress cycles all of the same amplitude, with a low minimum load. This column is derived from the value of S^4 and Equation F.4, with $\sigma_{max}=\Delta\sigma$, and is included just to provide an idea of the size of the seismic stresses. They are not large.

Residual stresses, when considered, are taken to be the default values for large lines, as reported in Reference F.2.

F.3.2.3 Results - WinPRAISE runs were made for the hot leg to pressure vessel weld using the above stresses and default material properties. Table F.5 summarizes the results. "Good" inspections at 0, 20 and 40 years were considered. These results are the cumulative leak probability. The left hand column gives the leak rate in gallons per minute. The next column in gives the time (25, 40 and 60 years), and the probabilities are directly from the PRAISE output for these times. For the Monte Carlo simulation, the crack size plane (*a/h – a/b*) was divided into 20 by 20 strata, with a maximum of 2000 trials drawn from each stratum. Sampling from a given stratum was stopped when 20 failures occurred in that stratum. Sampling began at the corner of the *a/h-a/b* plane corresponding to long deep cracks (1,0), and continued to shorter, then shallower cracks until no failures occurred in a stratum within 2000 trials. Sampling was then stopped. This procedure is referred to as automated stratification, and is a feature unique to WinPRAISE [F.4]. Earlier versions of the PRAISE software require the user to define each stratum and the sampling from each.

Table F.5 Cumulative Probability PRAISE Results for Hot Leg-Pressure Vessel Weld for Fatigue Crack Growth from Pre-Existing Defects

OD=34.0 inches, h=2.50 inches, cast austenitic, no σ_{DL}, times in reactor years

		Base			No hydro			Aging		
Hydro		yes			no			no		
Insp		good			good			good		
t_{insp}		0,20,40			0,20,40			0,20,40		
Aging		no			no			yes		
J_{IC}		5			5			1.5		
dJ/da		23.44			23.44			15		
		no EQ	SSE	5SSE	no EQ	SSE	5SSE	no EQ	SSE	5SSE
>0	25	1.20×10^{-18}	2.34×10^{-16}	2.45×10^{-16}	6.61×10^{-15}	7.04×10^{-15}	7.08×10^{-15}	1.43×10^{-14}	1.47×10^{-14}	1.47×10^{-14}
	40	1.29×10^{-18}	2.35×10^{-16}	2.45×10^{-16}	6.61×10^{-15}	7.04×10^{-15}	7.08×10^{-15}	1.43×10^{-14}	1.47×10^{-14}	1.47×10^{-14}
	60	1.29×10^{-18}	6.01×10^{-18}	6.19×10^{-18}	6.61×10^{-15}	6.62×10^{-15}	6.62×10^{-15}	1.43×10^{-14}	1.44×10^{-14}	1.44×10^{-14}
		HLA0			HLB0			HLC0		
>100	25	2.44×10^{-19}	2.53×10^{-18}	3.89×10^{-18}	2.93×10^{-17}	3.18×10^{-17}	3.22×10^{-17}	5.11×10^{-17}	5.52×10^{-17}	5.63×10^{-17}
	40	2.55×10^{-19}	2.60×10^{-18}	3.97×10^{-18}	2.94×10^{-17}	3.19×10^{-17}	3.22×10^{-17}	5.11×10^{-17}	5.53×10^{-17}	5.63×10^{-17}
	60	2.56×10^{-19}	3.04×10^{-19}	3.33×10^{-18}	2.94×10^{-17}	2.94×10^{-17}	2.9471^{-17}	2.12×10^{-17}	5.12×10^{-17}	5.13×10^{-17}
		HLA1			HLB1			HLC1		
>1500	25	1.20×10^{-20}	6.48×10^{-19}	1.31×10^{-18}	1.31×10^{-18}	1.99×10^{-18}	2.64×10^{-18}	2.72×10^{-18}	4.33×10^{-18}	5.99×10^{-18}
	40	1.26×10^{-20}	6.62×10^{-19}	1.32×10^{-18}	1.31×10^{-18}	2.00×10^{-18}	2.65×10^{-18}	2.72×10^{-18}	4.36×10^{-18}	5.97×10^{-18}
	60	1.27×10^{-20}	2.61×10^{-20}	3.93×10^{-20}	1.31×10^{-18}	1.33×10^{-18}	1.34×10^{-18}	2.72×10^{-18}	2.75×10^{-18}	2.79×10^{-18}
		HLA2			HLB2			HLC2		
>5000	25	1.19×10^{-20}	6.48×10^{-19}	1.31×10^{-18}	1.31×10^{-18}	1.99×10^{-18}	2.64×10^{-18}	2.72×10^{-18}	4.33×10^{-18}	5.99×10^{-18}
	40	1.26×10^{-20}	6.62×10^{-19}	1.32×10^{-18}	1.31×10^{-18}	2.00×10^{-18}	2.65×10^{-18}	2.72×10^{-18}	4.36×10^{-18}	5.97×10^{-18}
	60	1.27×10^{-20}	2.61×10^{-20}	3.93×10^{-20}	1.31×10^{-18}	1.33×10^{-18}	1.34×10^{-18}	2.72×10^{-18}	2.75×10^{-18}	2.79×10^{-18}
		HLA3			HLB3			HLC3		
>500000	25	1.20×10^{-20}	6.48×10^{-19}	1.31×10^{-18}	1.31×10^{-18}	1.99×10^{-18}	2.64×10^{-18}	2.72×10^{-18}	4.33×10^{-18}	5.99×10^{-18}
	40	1.26×10^{-20}	6.62×10^{-19}	1.32×10^{-18}	1.31×10^{-18}	2.00×10^{-18}	2.65×10^{-18}	2.72×10^{-18}	4.36×10^{-18}	5.97×10^{-18}
	60	1.27×10^{-20}	2.61×10^{-20}	3.93×10^{-20}	1.31×10^{-18}	1.33×10^{-18}	1.34×10^{-18}	2.72×10^{-18}	2.75×10^{-18}	2.79×10^{-18}
		HLA4			HLB4			HLC4		

noticeable effect of hydro
noticeable effect of seismic when hydro test is performed, less effect when no hydro
aging has about x2 effect
>1500 gpm same as DEPB

Runs were made with and without a hydro test, and it is seen that hydro testing has a noticeable effect. Moderate material degradation is considered, with the values of J_{Ic} and $(dJ/da)_{matl}$ identified in the table. The failure probabilities are all very small, even the leak probabilities. The influence or seismic events is seen to be quite small.

Table F.5 provides the base case results for the hot leg. Additional runs were made to study the following variables:

•   The effects of applying a load-controlled overload stress at a specified time were studied. This is called a design-limiting stress, and represents an overload event, such as water hammer or a seismic event even larger than the 5 SSE already considered for this component.

• The effects of the fatigue crack growth relation employed were studied. The fatigue crack growth relation in PRAISE for austenitic stainless steel is based on information available during the original software development. More recent crack growth relations have been suggested [F.11]. For the simple stress history in this case, it is possible to run PRAISE with a crack growth relation that is equivalent to the more recent relation.

• PWSCC crack initiation and growth has been identified in the control drive mechanisms (CRDM) in PWRs. This occurs in the Alloy 600 weldment. This alloy is also used in the safe end of the pressure vessel to main coolant piping welds, so is present in the hot leg to pressure vessel weld under consideration. In order to model the initiation and growth of PWSCC cracks, the initiation kinetics were assumed to be the same as for Type 316NG stainless steel as currently in PRAISE [F.2], but the crack growth kinetics were changed to be representative of Alloy 600. Based on information in Reference F.12, the crack growth kinetics is represented by the relation

$$\frac{da}{dt} = CK^m \qquad\qquad [F.5]$$

where m equals 1.16, and C is lognormally distributed with a median value that depends on the temperature and material (weld, base metal, etc.). The median value of C for a weld at 315 C (600°F) is 7.86×10^{-7}, when crack growth rates are in inches/hour and K is in ksi-in$^{1/2}$. Combining the within-heat and heat-to-heat variation in C, the second parameter of the lognormal distribution is 1.193 (standard deviation of $\ln C = 1.193$).

• The effects of more severe material degradation were studied, with the values of the degraded toughness given along with the results. Since PRAISE can not consider time-dependent material properties, the degraded material properties are present even in new pipe. The values of the degraded properties are from Reference F.13.

The results of these additional runs are summarized in Tables F.6 and F.7.

Table F.6 Cumulative PRAISE Results Additional Runs for Hot Leg Pressure Vessel Weld

		From Table F.5	Ref F.11 da/dN	σ_{DL} @ t-1	PWSCC Growth no σ_{res}	PWSCC Growth σ_{res}	PWSCC Initiation σ_{res}
Hydro		yes	yes	yes	yes	yes	--
Insp		good	good	good	good	good	good
t_{insp}		0, 20,40	0,20,40	0,20,40	0,20,40	0,20,40	20,40
Aging		no	no	no	no	no	no
J_{IC}		5	5	5	5	5	5
dJ/da		23.44	23.44	23.44	23.44	23.44	23.44
		no EQ	no EQ	no EQ	no EQ	no EQ	no EQ
>0	25	1.20×10^{-18}	2.20×10^{-18}	2.38×10^{-16}	0.923	0.916	0.001
	40	1.29×10^{-18}	2.42×10^{-18}	--	0.926	0.918	0.020
	60	1.29×10^{-18}	2.43×10^{-18}	--	0.926	0.919	0.068
			HLD0	HLE0			
>100	25	2.44×10^{-19}	2.61×10^{-19}	2.38×10^{-16}	7.97×10^{-7}	2.16×10^{-7}	1.0×10^{-5}
	40	2.55×10^{-19}	2.71×10^{-19}	--	7.97×10^{-7}	2.16×10^{-7}	2.69×10^{-4}
	60	2.56×10^{-19}	2.72×10^{-19}	--	7.97×10^{-7}	2.16×10^{-7}	1.78×10^{-3}
			HLD1	HLE1			
>1500	25	1.20×10^{-20}	1.63×10^{-20}	6.41×10^{-20}	9.68×10^{-10}	2.78×10^{-11}	$<10^{-4}$
	40	1.26×10^{-20}	1.65×10^{-20}	--	9.68×10^{-10}	2.78×10^{-11}	1.0×10^{-4}
	60	1.27×10^{-20}	1.66×10^{-20}	--	9.68×10^{-10}	2.78×10^{-11}	4.85×10^{-4}
			HLD2	HLE2			
>5000	25	1.20×10^{-20}	1.63×10^{-20}	6.41×10^{-20}	2.78×10^{-11}	4.66×10^{-11}	$<10^{-5}$
	40	1.26×10^{-20}	1.65×10^{-20}	--	2.78×10^{-11}	4.66×10^{-11}	9.0×10^{-5}
	60	1.27×10^{-20}	1.66×10^{-20}	--	2.78×10^{-11}	4.66×10^{-11}	3.77×10^{-4}
			HLD3	HLE3			
break	25	1.20×10^{-20}	1.63×10^{-20}	6.41×10^{-20}	2.19×10^{-14}	2.59×10^{-13}	$<10^{-5}$
	40	1.26×10^{-20}	1.65×10^{-20}	--	2.19×10^{-14}	2.59×10^{-13}	9.0×10^{-5}
	60	1.27×10^{-20}	1.66×10^{-20}	--	2.19×10^{-14}	2.59×10^{-13}	3.77×10^{-4}
			HLD4	HLE4	DEPB	DEPB	

The design limiting stress was 40.0 MPa (4.49 ksi). In the case of PWSCC growth, initial fabrication defects were considered with the default depth distribution discussed above. Both initiation and growth were considered for the column identified as PWSCC initiation. The higher large leak rates for the initiation relative to the PWSCC growth are due to the possibility of multiple initiation sites, whereas the growth considers only one initial crack.

Table F.7 Additional Hot Leg Pressure Vessel Runs Considering Material Aging

OD=34 inches
t=2.50 inches
σ_{dw}=2.08 ksi
σ_{te}=6.50 ksi

Good Inspection at 0, 20 ,40
3 HU-CD per year
Degraded Properties Used for All Times

Updated da/dN
No Hydro Unless Specified
Type 304 Stainless

		Base	no Hydro	A	B	C	D	E
J_{Ic} kips/in		5		1.11	0.67	1.72	0.75	0.20
dJ/da ksi			23.44	13.4	8.0	22.6	6.5	0.05
σ_{ys} ksi		19.4				29.2		
σ_{ult} ksi		--				76.7		
σ_{flo} ksi		44.9				53.0		
D ksi		106				104.5		
N		5				4.84		
>0	25	1.20×10^{-18}	6.61×10^{-15}	1.34×10^{-14}	1.96×10^{-14}	9.73×10^{-15}	2.07×10^{-14}	4.02×10^{-13}
	40	1.29×10^{-18}	6.61×10^{-15}	1.34×10^{-18}	1.96×10^{-14}	9.73×10^{-15}	2.07×10^{-14}	4.02×10^{-13}
	60	1.29×10^{-18}	6.61×10^{-15}	1.34×10^{-18}	1.96×10^{-14}	9.73×10^{-15}	2.07×10^{-14}	4.02×10^{-13}
>100	25	2.44×10^{-19}	2.93×10^{-17}	5.26×10^{-18}	6.76×10^{-17}	--	--	2.81×10^{-14}
	40	2.55×10^{-19}	2.94×10^{-17}	5.27×10^{-18}	6.77×10^{-17}	--	--	2.81×10^{-14}
	60	2.56×10^{-19}	2.94×10^{-17}	5.27×10^{-18}	6.77×10^{-17}	--	--	2.81×10^{-14}
break	25	1.20×10^{-20}	1.31×10^{-18}	2.67×10^{-18}	4.30×10^{-18}	1.48×10^{-18}	5.31×10^{-18}	2.81×10^{-14}
	40	1.26×10^{-20}	1.31×10^{-18}	2.67×10^{-18}	4.31×10^{-18}	1.48×10^{-18}	5.31×10^{-18}	2.81×10^{-14}
	60	1.27×10^{-20}	1.31×10^{-18}	2.67×10^{-18}	4.31×10^{-18}	1.48×10^{-18}	5.31×10^{-18}	2.81×10^{-14}
		earlier base case, default WinPRAISE properties, with hydro	no hydro	unaged weld metal J-T CF8M tensile	mult J-T by 0.6	all CF8M	more sensitive aged	extremely sensitive aged

The biggest effect in Table F.7 is not having a hydro test. This assumption is necessary, because when degraded material properties are used, everything that fails does so during the hydro test.

"Extremely sensitive aged" material properties are needed before degradation has a large effect.

F.3.3 Surge Line
The surge line is one of the base case systems.

F.3.3.1 Dimensions and Welds - From the piping isometric available to the panel members, the surge line is a 14 inch line (14 inch outer diameter) with a thickness of 35.7 mm (1.406 inches). The material is SA376 Type 304, which is an austenitic stainless steel. There are some 13 welds in the line.

F.3.3.2 Stresses and Cycles - The stresses at the surge line elbow are provided in Reference F.5, which is evidently the highest stressed location in the line. These stresses include seismic events and are given in Table F.8. The stress amplitude is contained in this table, which is one-half the stress range (peak-to-peak value).

Table F.8 Summary of Stress Cycles for Surge Line Elbow
(Stress Amplitudes with Seismic Stresses)

Load Pair	Amplitude (ksi)	Number/ 40 yr	Load Pair	Amplitude (ksi)	Number/ 40 yr
HYDRO-EXTREME	190.17	6	9D-LEAK TEST	52.20	50
8A-OBE	163.18	14	8G-LEAK TEST	52.20	65
9B-OBE	162.06	14	8G-UPSET3	51.00	30
8B-HYDRO	138.05	4	8G-12	50.96	90
8B-OBE	127.94	10	8G-16	50.93	90
9A-OBE	127.04	14	8E-8G	50.92	13
8C-OBE	64.76	68	8E-OBE	43.38	77
9F-OBE	64.17	68	9H-OBE	42.79	500
8F-18	63.40	68	8H-13	39.82	90
9C-11	63.38	68	8H-OBE	37.43	203
8D-OBE	54.02	72	8H-UPSET4	35.42	40
9G-OBE	53.42	400	8H-9E	33.94	90
8G-18	52.38	22	2A-8H	33.94	77
9D-11	52.35	22	3A-10A	33.10	4120
8G-17	52.35	90	6-10A	33.10	200
9D-LEAK TEST	52.20	50	3B-10A	33.10	4120
8G-LEAK TEST	52.20	65	7-10A	33.10	4580
8G-UPSET3	51.00	30	2B-SLUG1	32.87	100
8G-12	50.96	90	2B-SLUG2	32.87	500
			4B-10A	29.90	17040

To estimate the influence of seismic events, it is necessary to also have the stress history without such events. It is not possible to remove seismic events knowing only the information in the above table. This information was provided in Reference F.14 and is summarized in Table F.9.

Table F.9 Summary of Stress Cycles for Surge Line Elbow
(Stress Amplitudes without Seismic Stresses)

Load Pair	Amplitude (ksi)	Number/ 40 yr	Load Pair	Amplitude (ksi)	Number/ 40 yr
HYDRO-EXTREME	190.17	6	8G-16	50.93	90
9B-HYDRO	149.86	4	8G-9H	50.92	128
8A-UPSET 4	140.42	14	2A-8E	40.10	90
9B-UPSET4	139.43	10	8H-9H	40.09	100
8B-UPSET4	105.89	14	9H-10A	40.09	272
9A-UPSET4	105.13	2	9E-13	39.82	90
9A-LEAK	103.86	12	3A-10A	33.10	4120
8F-18	63.40	68	6-10A	33.10	200
9C-11	63.38	68	3B-10A	33.10	4120
9F-LEAK	63.37	68	7-10A	33.10	4580
8C-LEAK	63.37	35	2B-SLUG1	32.87	100
2A-8C	62.30	33	2B-SLUG2	32.87	500
8G-18	52.38	22	5-10A	29.90	9400
8G-17	52.35	90	4A-10A	29.90	17040
9D-11	52.35	22	4B-10A	29.90	17040
2A-8D	51.20	72	2B-10A	20.60	14400
8H-9G	51.18	400	2A-10A	20.60	14805
8G-UPSET3	51.00	30	10A-UPSET1	20.59	70
9D-12	50.96	50	10A-UPSET5	20.59	30
8G-12	50.96	40	10A-UPSET6	20.59	5
			10A-UPSET2	20.59	95
			1B-10A	20.59	1533
			1B-10B	20.00	87710

The cyclic stress amplitudes of Tables F.8 and F.9 provide the information for the initiation analysis, but additional information is required for the growth portion of the analysis. The spatial gradient (primarily radial) is required. Also, when analyzing the stability of a through-wall crack, the steady normal operating stress is needed. This stress is considered to be the sum of the pressure, deadweight and restraint of thermal expansion stresses. The values of these latter two are given in Reference F.5 as

$$\sigma_{dw} = 0 \qquad \sigma_{te} = 102.6 \text{ MP (14.88 ksi)}.$$

Many of the high stress contributors in Tables F.8 and F.9 are from rapid excursions of the coolant temperature. The largest stress amplitude (half the peak-to-peak) is 1,310 MPa (190 ksi), so the stresses are large (but localized). These are the stresses at the peak stress location, which is not at weld. The spatial stress gradients (both along the surface and into the pipe wall) are required for a thorough analysis. The radial gradient (into the pipe wall) can be estimated by the procedure given in Section 5.3 of Reference F.5, i.e.,

The following specific rules were applied to assign stress to the uniform and gradient categories:
- Cyclic stresses associated with seismic loads were treated as 100 percent uniform stress.
- Cyclic stresses greater than 310 MPa (45 ksi) were treated as having a uniform component of 310 MPa (45 ksi), and the remainder were assigned to the gradient category.
- For those transients with more than 1000 cycles over a 40 year life, it was assumed that 50% of the stress was uniform stress and 50% a through-wall gradient stress. In addition, for these transients, the uniform stress component was not permitted to exceed 69 MPa (10 ksi).

The gradient stress mentioned above is assumed to vary through the thickness as

$$\sigma(\xi) = \sigma_o \left(1 - 3\xi + \frac{3}{2}\xi^2 \right)$$ [F.6]

In this equation, σ_o is the stress at the inner wall of the pipe, $\xi = x/h$, x is the distance into the pipe wall from the inner surface, and h is the wall thickness. The stresses and cycles are high enough that fatigue crack initiation is important, which has been considered in Reference F.5, which shows a probability of 0.981 of a leak in 40 years for this component. The LOCA probabilities will be less. The use of the gradient along the surface will reduce this.

A refined stress analysis was available as part of the efforts reported in Reference F.6. These stresses included details of the variation of the stress in the circumferential direction, and are referred to as the "refined stresses". The stresses used in the surge line evaluation were based on the actual stress analysis for a CE-designed plant in response to NRC Bulletin 88-11 dealing with surge line stratification. The loadings were based on the methods approved by the NRC staff in the CE Owner Group Report CEN 387-NP, "Pressurizer Surge Line Flow Stratification Evaluation," Rev. 1-NP, December 1991. Additional evaluations of the local stress distributions in the elbow were conducted to get the detailed stress distribution around the circumference of the elbow. The critically stressed location that produced the highest probability of cracking was the circumferential stresses in the side of the elbow due to stratification bending. Detailed stresses are not provided, because they belong to the plant that allowed us to use them.

F.3.3.3 Results - PRAISE runs were made using the versions that can treat fatigue crack initiation. No inspections were considered. Since crack initiation is considered, there will be no effect of a pre-service inspection. The results are summarized in Table F.10.

Table F.10 Cumulative PRAISE Results for the Surge Line Elbow

Condition		Ref. F.5	Table F.8 Stresses	Table F.9 Stresses	Refined Stresses
Seismic		yes	yes	no	yes
σ_{DL}		no	no	no	no
>0	25			0.372	0.233	
	40	0.982	0.772	0.587	8×10^{-7}
	60	0.998	0.968	0.882	3.3×10^{-5}
					CENC4H1
>100	25	--	1.6×10^{-5}	7.5×10^{-6}	$<10^{-7}$
	40	--	3.11×10^{-4}	7.1×10^{-5}	$<10^{-7}$
	60		1.33×10^{-3}	2.51×10^{-4}	$<10^{-7}$
		--	CENC4D01	CENC4A3	20 hrs
>1500	25	--	$<10^{-7}$	$<10^{-7}$	
	40	--	$<10^{-7}$	$<10^{-7}$	
	60	--	2.0×10^{-7}	1.0×10^{-7}	
			CENC4D15	CENC4A4	
			axisymmetric seismic	axisymmetric nonseismic	strain rates and bivariate stresses

It is seen that the seismic stresses do not have a large effect, roughly a factor of 3. The use of the refined stresses greatly reduces the calculated failure probabilities. The computer run for 380 lpm (100 gpm) took about 20 hours and resulted in no failures in 10^7 trials. The runs for > 5,700 lpm (1,500 gpm) with the stresses from Tables F.8 and F.9 had 2 and 1 failures in 10^7 trials, respectively, and these runs each took many hours. Hence, it is evident that the Monte Carlo simulation with multiple fatigue crack initiation sites does not allow definition of the small probabilities of large leaks in the surge line elbow, and an alternate procedure was developed. Stratified sampling is not used for fatigue crack initiation.

F.3.3.4 Alternate Procedure - In cases where the dominant degradation mechanism is fatigue crack initiation with subsequent growth, PRAISE currently has no way of generating low probability results other than conventional Monte Carlo simulation. This is the dominant mechanism for three of the base line components; the surge line elbow, the HPI make up nozzle and the BWR feedwater line elbow. Excessive computer time is needed to generate probabilities of various size leaks for these components, with some runs taking 4 days on a 3 GHz pc, with no leaks of even 380 lpm (100 gpm). An alternate procedure is needed to estimate leak probabilities for the large leaks of interest, and such a procedure is described below.

As part of a standard analysis, the PRAISE software computes the crack opening area and leak rate as functions of the length of through-wall cracks. Hence, this information is readily available, and can be used to determine the length of a through-wall crack needed to produce a given leak rate, such as 380 lpm (100 gpm), 5,700 lpm (1,500 gpm), etc. The probability of having a leak of a given magnitude is then the probability of having a through-wall crack exceeding that length. The half-crack length, b, is considered, which is a function of the desired leak rate, $\dot{q}$. Hence, $b(\dot{q})$ can be considered as known.

The probability of a double-ended-pipe-break (DEPB) is also of interest. In the cases of interest here, the critical net section stress failure criterion is used. For a through-wall crack, the value of b for a DEPB is given by the expression

$$\frac{b_{DEPB}}{\pi R_I} = 1 - \frac{\sigma_{LC}}{\sigma_{flo}} \qquad\qquad [F.7]$$

where R_I is the inside radius, σ_{flo} is the flow stress (average of yield and ultimate) and σ_{LC} is the load controlled stress, which is equal to the pressure plus deadweight stress.

The version of PRAISE that performs Monte Carlo simulation of fatigue crack initiation and growth commonly provides information on the probability of having any leak and a leak exceeding a given magnitude. In order to have a nonzero number for the latter, a leak exceeding that magnitude must occur during the simulation. The problem is that this often does not occur within a number of trials that can be reasonably performed. In order to overcome this, PRAISE was modified to print out the length of any crack resulting in a leak and the time at which it first became through-wall. This was then used to estimate the size distribution of through-wall cracks as a function of time. The complementary cumulative distribution, denoted as $P_b(>b)$, is concentrated upon. Then the probability of a leak greater than $\dot{q}$ is given by

$$P_{LK}(>\dot{q}) = P_b[>b(\dot{q})] \qquad\qquad [F.8]$$

Table F.11 summarizes the information from a PRAISE run using the stresses from Table F.9 for the crack opening area (A) and leak rate ($\dot{q}$) for a given half-length of a through-wall crack (b).

Table F.11 Half Crack Lengths and Areas for a Given Leak Rate
(Surge Line Elbow, Table F.9 Stresses)

Q, gpm	b, inches	$\dfrac{b}{\pi R_I}$	A, in^2	$\dfrac{A}{A_{pipe}}$
100	5.981	0.445	0.936	0.010
1500	10.379	0.591	10.028	0.143
5000	11.791	0.671	46.762	0.476
DEPB	15.95	0.907	--	--

A table of lengths of through-wall cracks was generated from the modified version of PRAISE using 10^4 trials using the stresses of Table F.9 (no seismic). In this run, there were 2,162 leaks within 25 years, 5,932 within 40 years and 8,890 within 60 years. Dividing these numbers by 10^4 provides leak probabilities that are nearly the same as obtained from the Monte Carlo simulation with 10^7 trials. Figure F.2 is the complementary cumulative distribution of leaking crack sizes for the three times of interest. The upper curve is for 60 years, because there is a higher probability of encountering a longer crack at this longer time. The lines in this figure are least squares curve fits, which are discussed later.

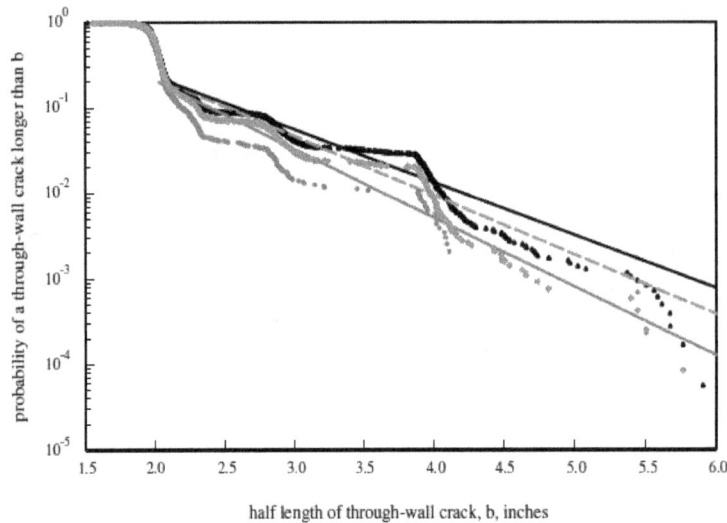

Figure F.2 Complementary Cumulative Distribution of Half-Crack Length of Through-Wall Cracks in Surge Line Elbow at 25, 40 and 60 Years

Figure F.2 shows changes in slope, and the "lumpiness" of the distribution is readily apparent. This "lumpiness" is representative of a multi-modal probability density function of crack length, which is most likely due to the fatigue crack initiation sites being taken as 50 mm (2 inches) in length. That is, each two-inch segment around the circumference is taken as an independent initiation site. The surface length of an initiated crack is also a random variable. Once a crack initiates, it grows, and can link with neighboring cracks. This growth and linking can lead to sudden increases in crack length (by linking) and evidently is responsible for the multi-modal nature of the probability density function of crack length. The multi-modal nature of the probability density function is clearly shown in Figure F.3.

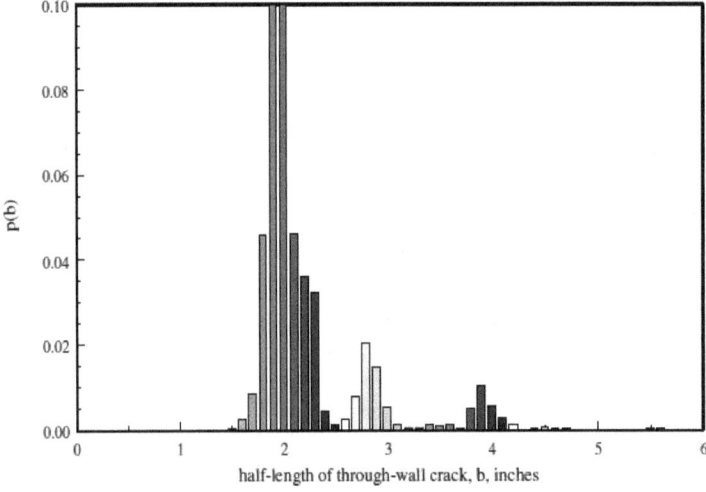

Figure F.3 Histogram of the Half-Length of Through-Wall Cracks at 60 Years for the Surge Line Elbow

Pleasing curve fits to the lines in Figure F.2 are not possible. A good fit could perhaps be obtained by assuming that the histogram of Figure F.3 consists of a sum of lognormals with medians to match the location of the modes and relative weights adjusted to match the relative heights of their modes. This is believed to be unwarranted, since the multi-modal nature of the density function is an artifact of the modeling assumption of a 50 mm (2 inch) long initiation site. It is better to just smooth out the cumulative distribution, and this was accomplished by a linear least squares curve fit to the cumulatives on log-linear scales. Since it is desired to represent the curve at long cracks, the fit was performed only for cracks corresponding to a probability less than 0.2. This eliminates the numerous cracks at higher probabilities that would skew the curve fit if included in the least squares calculations. The lines in Figure F.2 are the curve fits obtained in this manner.

The assumed functional form was

$$P(>b) = 0.2e^{-C(b-b_{0.2})}$$ [F.9]

The values of C and $b_{0.2}$ depend on the time. Once they are evaluated, the leak probabilities of a given size are obtained using the crack sizes in Table F.11. Table F.12 summarizes the results.

Table F.12 Summary of Results for Surge Line Elbow (Table F.9 Stresses)

Time, years	25	40	60
Number of cracks	2162	5932	8890
Cracks above 0.2	1730	4744	7109
$b_{0.2}$	2.072	2.089	2.108
C	1.876	1.597	1.425
P(>5.981)	1.31×10^{-4}	4.00×10^{-4}	8.02×10^{-4}
P(>10.379)	3.41×10^{-8}	3.56×10^{-7}	1.52×10^{-6}
P(>11.791)	2.42×10^{-9}	3.73×10^{-8}	2.04×10^{-7}
P(>15.95)	9.86×10^{-13}	4.86×10^{-11}	5.43×10^{-10}

Table F.13 summarizes the results along with corresponding ones obtained directly from the Monte Carlo simulation. The conventional Monte Carlo simulation used 10^6 trials for 380 lpm (100 gpm) and 10^7 trials for 5,700 lpm (1,500 gpm).

Table F.13 Cumulative PRAISE Results for the Surge Line Elbow as Obtained from the Alternate Procedure and Directly from Monte Carlo Simulation (Table F.9 Stresses)

		Direct Monte Carlo	Alternate Procedure
>0	25	0.233	0.216
	40	0.587	0.593
	60	0.882	0.889
>100	25	7.5×10^{-6}	1.31×10^{-4}
	40	7.1×10^{-5}	4.00×10^{-4}
	60	2.51×10^{-4}	8.02×10^{-4}
>1500	25	$<10^{-7}$	3.41×10^{-8}
	40	$<10^{-7}$	3.56×10^{-7}
	60	1.0×10^{-7}	1.52×10^{-6}
>5000	25	--	2.42×10^{-9}
	40	--	3.73×10^{-8}
	60	--	2.04×10^{-7}
DEPB	25	--	9.86×10^{-13}
	40	--	4.86×10^{-11}
	60	--	5.43×10^{-10}

Table F.13 shows that the alternate procedure is able to greatly extend the leak rates whose probabilities can be estimated. In cases where direct comparisons are possible, the alternate procedure gives higher leak probabilities. The direct Monte Carlo for 5,700 lpm (1,500 gpm) employed 10^7 trials and took 36 hours of computer time. The alternate procedure used 10^4 trials, so took about 2 minutes. Even in this era of fast cheap computer time, it would still be prohibitive to use direct Monte Carlo to generate the results obtained by the alternate procedure. It would take 10^{10} trials to produce the DEPB results in the above table. This translates to 36,000 hours of computer time, or about 4 years.

F.3.4 HPI Makeup Nozzle

An HPI/makeup nozzle safe end from a B&W plant type was selected as one of the base case systems.

F.3.4.1 Dimensions and Welds - This type of component was considered in Reference F. 5, which identifies the component as 2 ½ inch schedule 160 pipe fabricated from Type 304 austenitic stainless steel. The location considered in Reference F.5 is in the safe end at the nozzle, which has a thickness of 11.1 mm (0.4375 inches) and a mean radius of 32.5 mm (1.28 inches) at the location of high stresses.

F.3.4.2 Stresses and Cycles - As shown in Reference F.5, the cyclic stress history is dominated by two types of transients, with the amplitudes and frequencies shown in Table F.14.

Table F.14 Stress History for HPI/Make Up Nozzle from NUREG/CR-6674 [F.5]

Name	Stress Amplitude ksi	Number in 40 years
HPI actuation A/B	221.24	33
Test Null	169.31	7

The deadweight and restraint of thermal expansion stresses for this location under normal operation that were used in Reference F.5 are

$$\sigma_{dw}=0$$
$$\sigma_{te}= 63.1 \text{ MPa (9.16 ksi)}$$

As discussed above, these stresses were composed of 310 MPa (45 ksi) uniform and the remainder the generic gradient of Equation F.6. These stresses are believed to be very conservative and are for the thermal sleeve being intact.

F.3.4.3 Results - The version of PRAISE that considers fatigue crack initiation was run for the HPI/make up nozzle. The stresses of Table F.14 were taken to be axisymmetric. Due to the small line size, only 4 initiation sites around the circumference were considered. Table F.15 summarizes the results.

Table F.15 Cumulative Probability PRAISE Results for HPI/Make Up Nozzle
(Intact Thermal Sleeve)

Condition		From Reference F.5	Here
>0	25		1.004×10^{-5}
	40	0.00210	6.08×10^{-4}
	60	0.0309	1.04×10^{-2}
			Inel4a2
>100	25	--	4.5×10^{-8}
	40	--	4.9×10^{-7}
	60	--	1.79×10^{-5}
			Inel4a1
>1500	25	--	2.0×10^{-8}
	40	--	2.10×10^{-7}
	60	--	4.56×10^{-6}
			Inel4a2

Table F.15 shows a cumulative leak probability of 10^{-5} in 25 years, which is quite low. However, leaks in this component have been observed in service, in which case the thermal sleeve in the component was failed. The results of Table F.15 use the stresses for an intact sleeve, and the stresses will be altered if the sleeve fails. A failed thermal sleeve is now considered.

F.3.4.4 Failed Thermal Sleeve - There is a thermal sleeve at the HPI nozzle, and the results in Table F.15 are for the case of the thermal sleeve not failing. The thermal sleeve has been observed to fail in service, which changes the stresses in the component.

In order to model the failure of the thermal sleeve, the following steps were taken:

1. Once the thermal sleeve fails, assume that a crack of the "initiation size" immediately appears. This size is a depth of 3.0 mm (0.12 inches). The WinPRAISE default distribution of the aspect ratio is used, as in other components.

2. A WinPRAISE run with this initial crack is performed, with the stresses that were present before the crack initiated (Table F.14), plus a uniform cyclic stress cycling each hour of sufficient amplitude to result in a high leak probability at not long times. This defines the uniform stress.

3. Use WinPRAISE to compute the leak frequencies for larger leak rates.

This procedure provides the results shown in Table F.16.

**Table F.16 Cumulative PRAISE Results for HPI/Make Up Nozzle with
Failed Thermal Sleeve and Additional Uniform Cyclic Stress, σ_u**

		Intact Sleeve	With Initial Crack and Original Stresses, σ_u=0	With Initial Crack and σ_u = 8 ksi	With Initial Crack and σ_u = 12 ksi	With Initial Crack and σ_u = 25 ksi
>0	5	--	5.67×10^{-5}	$<10^{-2}$	0.047	0.18
	25	1.004×10^{-5}	3.69×10^{-3}	0.032	0.14	0.727
	40	6.08×10^{-4}	1.26×10^{-2}	0.129	0.33	0.909
	60	1.04×10^{-2}	2.98×10^{-2}	0.161	0.47	0.909
>100	25	4.5×10^{-8}	6.49×10^{-4}			$<10^{-5}$
	40	4.9×10^{-7}	2.68×10^{-3}			$<10^{-5}$
	60	1.79×10^{-5}	5.31×10^{-3}			$<10^{-5}$
>1500	25	2.0×10^{-8}	--			
	40	2.10×10^{-7}	--			
	60	4.56×10^{-6}	--			
break	25	--	6.49×10^{-4}			
	40	--	2.68×10^{-3}			
	60	--	5.31×10^{-3}			

Table F.16 shows that a uniform stress of some 170 MPa (25 ksi) is needed to result in an appreciable leak probability within 25 years. However, the frequency of larger leak rates is actually reduced by imposing the uniform stress that is necessary to produce the high leak probabilities seen in service. This uniform stress grows cracks to leaks, so that the larger leak rate frequencies are reduced. The least favorable condition for larger leaks is a failed thermal sleeve with the original stresses ($\sigma_u = 0$).

F.3.5 Recirculation Line – 12 inch
The recirculation line is one of the base case systems for a BWR. This system has developed leaks in the past due to intergranular stress corrosion cracking (IGSCC). The 12 inch line has some of the highest stresses, so is considered here. The recirculation system also has 28 inch lines, which can contribute to larger flow rate failures than possible from a 12 inch line. Hence, the 28 inch line is also considered in subsequent sections.

F.3.5.1 Dimensions and Welds - The layout of the recirculation system is given in isometrics made available to panel members. There are two recirculation loops, which are very similar to one another. There are 121 welds in this system, including field welds, shop welds and safe ends. The piping is fabricated from A-358 Class 1 Type 304, and the piping is of diameters 12, 22 and 28 inches – all schedule 80.

F.3.5.2 Stresses and Cycles - IGSCC will be the dominant degradation mechanism. Hence, time at stress is of major concern, and the number of stress cycles is of secondary importance. Estimated stresses at the highest stressed locations for the two pipe sizes of interest are given in Table F.17.

Table F.17 Stress Information for Two Recirculation Joints

OD, inch	Thickness, inch	σ_{NO}, ksi	Seismic σ, ksi
12.75	0.687	20.41	20.41
28	1.201	9.48	10.60

The normal operating stress (σ_{NO}) is the sum of the pressure stress, deadweight stress and restraint of thermal expansion stress. A value of 14 MPa (2 ksi) for the deadweight stress is assumed. The seismic stress is the normal operating stress plus the seismic-induced stress. Note that the seismic stresses are small in this case. The magnitude of the seismic event is unknown.

The time at stress is important for this case, with the cycling frequency being of less importance. Consistent with what is used for the PWR, the cycling is considered to be composed of heat up and cool down at 3 per year. The parameters related to stress corrosion cracking are summarized in Table F.18.

Table F.18 Stress Corrosion Cracking Parameters

Oxygen at startup (PPM) = 8.0
Oxygen at steady state (PPM) = 0.20
Heat up (100-550F) time (hrs) = 5.00
Coolant conductivity (μs/cm) = 0.20
Degree of sensitization (C/cm^2) = 7.04

Residual stresses will be important, and the default residual stress distributions in pcPRAISE, which are documented in Reference F.2, are used when no remedial treatments are performed. In order to include remedial treatments that have been performed in service, a weld overlay at 20 years will be considered. This alters the thickness, crack growth kinetics (post-treatment analyses use Type 316NG crack growth defaults in PRAISE) and residual stresses. The axisymmetric through-wall residual stress distribution of Figure F.4 is employed. This figure is from Reference F.15. PRAISE can not treat the actual gradient, so the linear approximation in this figure is used. The linear gradient employed underestimates the beneficial effect of the weld overlay.

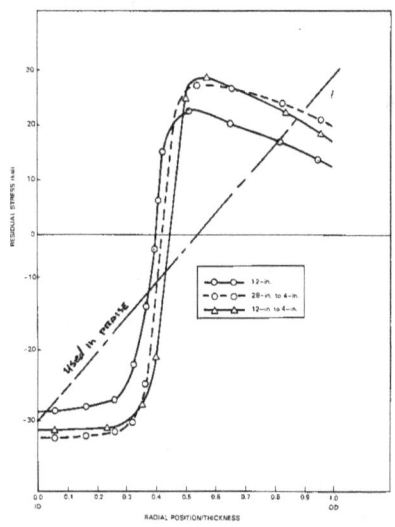

Figure F.4 Through-Wall Residual Axial Stress Distribution from Weld Overlay [F.15]

F.3.5.3 Results - Table F.19 summarizes the results obtained for the 12 inch weld in the recirculation system.

Table F.19 Cumulative Probability PRAISE Results for the 12 inch Recirculation Line Weld, with and without Weld Overlay at 20 Years (σ_{no} = 141 MPa [20.41 ksi])

OD=12.75 inches, h=0.687 inches, wrought austenitic,
stress corrosion crack initiation and growth

		Base	Overlay at 20 years	Overlay & σ_{DL} @ 39 years
>0	25	0.3674	0.2967	0.2968
	40	0.5986	0.3803	0.3872
	60	0.7435	0.4241	0.4253
>100	25	0.1682	0.1427	0.1429
	40	0.2452	0.1622	0.1632
	60	0.2872	0.1693	0.1708
>1500	25	0.1529	0.1066	0.1078
	40	0.2193	0.1250	0.1276
	60	0.2534	0.1312	0.1343
break	25	0.1529	0.0490	0.0502
	40	0.2193	0.0674	0.0700
	60	0.2535	0.0736	0.0767

5000 trials   304 full residual stress
σ_{dw}=2.0 ksi σ_{te}=13.32 ksi σ_{DL}=11.67ksi 3 HU-CD/yr p=1125 psi

The beneficial effect of the weld overlay at 20 years is not readily apparent from the results in Table F.19; such benefits are shown more clearly in Figure F.5, which provides a plot of the cumulative probability of a leak exceeding 380 lpm (100 gpm) as a function of time.

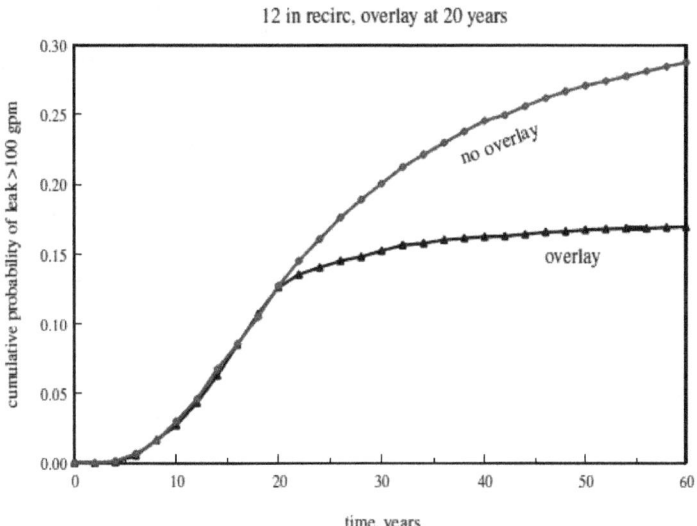

12 in recirc, overlay at 20 years

Figure F.5 Cumulative Probability of a Leak Exceeding 100 gpm as Functions of Time for the 12 inch Recirculation Line Weld with and without Weld Overlay at 20 Years

The slopes of the lines in Figure F.5 are the leak frequencies, and the slope at 40 years with no overlay is about 7 times that with overlay.

F.3.5.4 Summary of Observations from Service - Leak frequencies due to IGSCC in recirculation lines were estimated from service experience and reported in Reference F.16. Figure F.6 is Figure F.12 from that reference. With some exceptions, the results in Figure F.6 are between 10^{-4} and 10^{-3} per weld-year. The results are for times up to 15 years and do not include remedial actions. No strong dependencies on time or line size are apparent, but the smaller diameter lines appear to have a somewhat higher failure frequency.

Table F.20, which is from Charts 2 and 3 of Reference F.17, summarizes the depth distribution of observed cracks per weld-year for various pipe sizes in recirculation lines in BWRs. The remedial action of Reference F.17 is considered to consist of a weld overlay at 20 years. Observed crack sizes without remedial action, as reported in Reference F.16, are shown in Figure F.7.

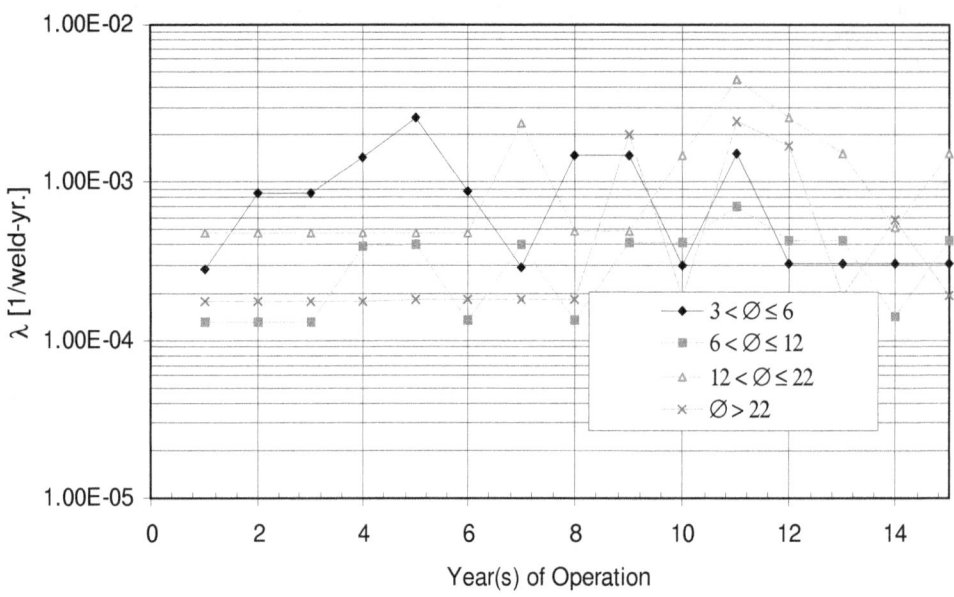

Figure F.6 Leak Frequencies as a Function of Time and Pipe Size (from Reference F.16)

Table F.20 Observed Crack Depth Frequencies in Various Line Sizes in Recirculation Lines as Percentages of the Wall Thickness (from Reference F.17)

No Remedial Action

Size	> 10%	> 20%	> 30%	> 40%	> 50%	> 60%	> 70%	> 80%	> 90%
NPS12	2.06E-03	1.62E-03	7.28E-04	3.64E-04	2.00E-04	1.46E-04	1.09E-04	7.28E-05	3.64E-05
NPS22	1.63E-03	1.11E-03	6.48E-04	3.21E-04	1.90E-04	1.24E-04	9.81E-05	6.54E-05	3.27E-05
NPS28	2.12E-03	1.50E-03	1.04E-03	5.99E-04	2.57E-04	1.84E-04	6.12E-05	3.67E-05	1.22E-05

With Remedial Action

Size	> 10%	> 20%	> 30%	> 40%	> 50%	> 60%	> 70%	> 80%	> 90%
NPS12	1.95E-04	1.60E-04	1.04E-04	8.31E-05	6.73E-05	4.61E-05	2.78E-05	1.90E-05	1.03E-05
NPS22	3.29E-04	2.74E-04	1.70E-04	1.32E-04	1.01E-04	8.62E-05	4.43E-05	2.95E-05	1.48E-05
NPS28	3.95E-04	2.84E-04	1.77E-04	9.15E-05	6.66E-05	3.82E-05	2.08E-05	1.24E-05	3.95E-06

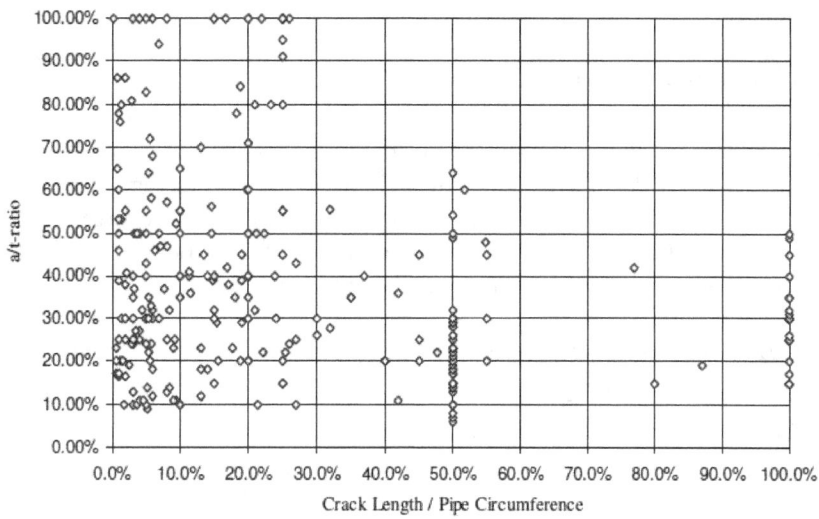

Figure F.7 Observed Crack Sizes as Reported in Reference F.16

F.3.5.5 Comparisons with PRAISE - The normal operating stress in Table F.17 of 20.41 ksi is for the highest stressed joint in the 12 inch recirculation line, whereas the observations are for all joints, including lower stressed locations. In order to generate PRAISE results that would be more representative of the population, runs were made with various stresses. Table F.21 summarizes the results.

**Table F.21 Cumulative PRAISE Results for a 12 inch Recirculation Line Weld
for Various Normal Operating Stresses (Remedial Action at 20 Years)**

		Cumulative			
	Mean σ_{no}	10	12	15	20
	COV	0.0	0.3	0.0	0.0
	Mean σ_{te}	3.32	5.32	8.32	13.32
>0	25	1.42×10^{-3}	1.54×10^{-2}	9.36×10^{-2}	0.2967
	40	1.46×10^{-3}	1.89×10^{-2}	0.1473	0.3803
	60	1.46×10^{-3}	2.08×10^{-2}	0.1781	0.4241
>100	25	4.90×10^{-4}	7.59×10^{-3}	3.90×10^{-2}	0.1427
	40	4.90×10^{-4}	8.86×10^{-3}	5.35×10^{-2}	0.1622
	60	4.90×10^{-4}	9.48×10^{-3}	6.11×10^{-2}	0.1693
>5000	25	3.50×10^{-4}	5.80×10^{-3}	3.19×10^{-2}	0.1066
	40	3.50×10^{-4}	7.06×10^{-3}	4.53×10^{-2}	0.1250
	60	3.50×10^{-4}	7.684×10^{-3}	5.27×10^{-2}	0.1312
DEPB	25	1.00×10^{-4}	2.70×10^{-3}	2.12×10^{-2}	0.0490
	40	1.00×10^{-4}	2.96×10^{-3}	3.46×10^{-2}	0.0674
	60	1.00×10^{-4}	4.58×10^{-3}	4.20×10^{-2}	0.0736

The results for 138 MPa (20 ksi) correspond to those in Table F.19. The normal operating stress was taken to be deterministic, except for the case of 83 MPa (12 ksi), in which case the normal operating stress is normally distributed with a mean of 83 MPa (12 ksi) and a standard deviation of 0.3x(36.7+13.8) MPa (0.3x(5.32+2.00) ksi) = 15.2 MPa (2.20 ksi).

The cumulative results from Table F.21 can be compared with the observed frequencies in Table F.20 by converting the cumulative results to a frequency by dividing by the time increment involved. In the current case, the increase in the cumulative following the remedial action is relatively small, as seen from Figure F.5. Hence, the cumulative results at 25 years from Table F.21 should be divided by 20 to provide frequencies for comparison purposes. This provides the results in Table F.22.

Table F.22 Estimated Leak Frequencies Prior to Remedial Action, from Table F.21

Mean σ_{no}	10	12	15	20
COV σ_{no}	0.0	0.3	0.0	0.0
Mean σ_{te}	3.32	5.32	8.32	13.32
Frequency	7.11×10^{-5}	7.69×10^{-4}	4.68×10^{-3}	1.48×10^{-2}

This table shows that the mean normal operating stress of 83 MPa (12 ksi) with some variance provides the best agreement with the results of Figure F.6. This is the case that will be used for benchmarking against observed cracks.

The following steps were followed in order to provide PRAISE results for comparison with observations of part-through cracks:

1. The WinPRAISE software was modified to print out the sizes of cracks present at each time step in the analysis. The depth and length of the deepest crack and the longest crack at that time step are printed into a file, along with the number of cracks present at that time. This file contains at most a number of lines equal to the number of Monte Carlo trials times the number of time steps (which can be a lot of lines).

2. The WinPRAISE file from step 1 is then processed to provide another file that includes only the sizes of part-through cracks present at the time of interest (25 years in this case). (Cracks of zero depth, leaks and other times are eliminated.)

3. The crack size file from step 2 is then loaded into a histogram, which provides the number of cracks present at 25 years that fall within a certain depth range.

4. Since Reference F.17 reports detected cracks, the detection probability (Equation F.1) must be accounted for. This is accomplished by multiplying the number of cracks in each bin by the detection probability for a crack of depth equal to the midpoint of the bin. This provides the number of detected cracks in this bin. The contents of each bin are then divided by the number of trials times the time (25 years) to provide the crack sizes per weld year.

5. The histogram is then converted to a complementary cumulative form, which is then directly comparable to results from Reference F.17.

Figure F.8 presents the crack size results for the benchmark case. Once again, not many deep cracks are observed. A pattern is observed in Figure F.8 which shows a preponderance of cracks below about 2.5 mm (0.1 inches). This pattern is due to cracks growing to a depth of 2.5 mm (0.1 inch) and then slowing down or arresting, which is most likely due to the transitioning from growth of "initiating cracks" to "fracture mechanics cracks" that occurs in the PRAISE modeling of initiation and growth. The transitioning criteria are discussed on page 42 of Reference F.2, and one of the criteria is "If the depth of the crack is greater than 2.5 mm (0.1 inch), its growth will always be by fracture mechanics velocity".

Figure F.9 is a plot of the predicted complementary cumulative number of observed cracks for the benchmark case, along with a comparison with reported observations. The outstanding inspection parameters of Table F.1 were employed. In this figure, Reference F.17 results from Chart 1 (prior) and Chart 2 (posterior) are both shown, since the analysis mixed with and without remedial action (weld overlay at 20 years). The analysis results fall midway between the two results, except for shallow cracks.

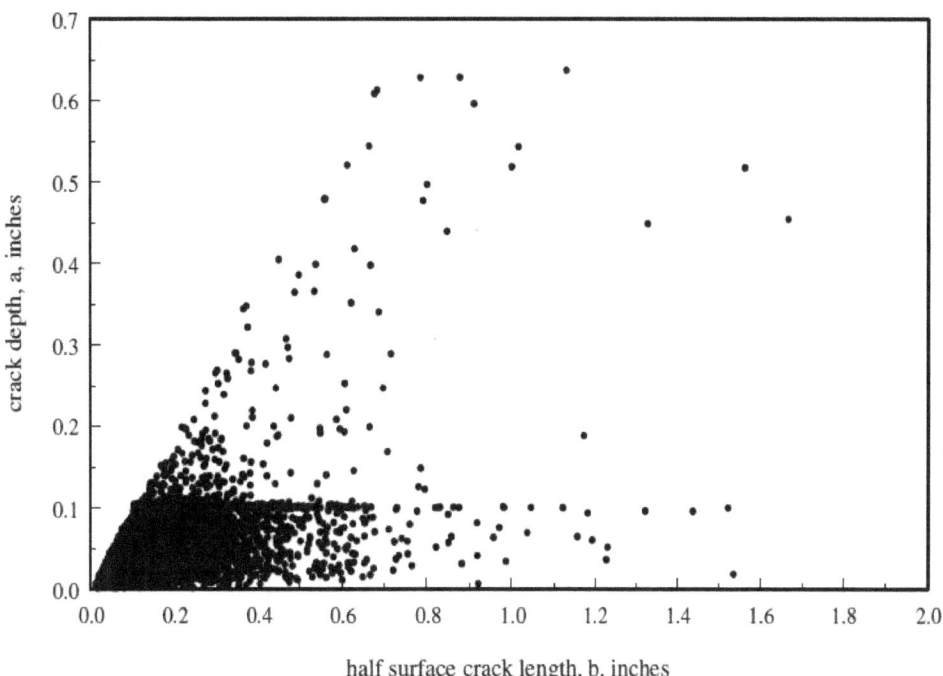

Figure F.8 Crack Sizes After 25 Years for the Benchmark Case (Mean σ_{NO} = 12 ksi) with Weld Overlay at 20 Years

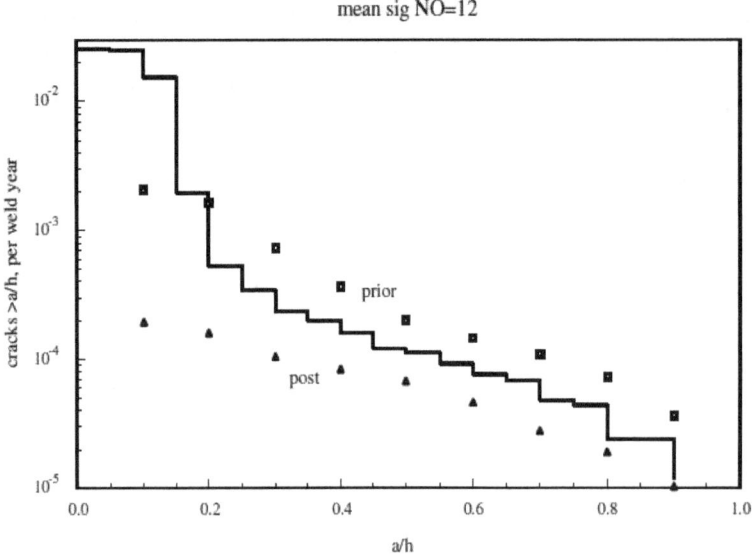

Figure F.9 Comparison of Results for the Benchmark Case Predicted for Outstanding Inspection Quality with Reported Prior and Post Observations [F.17]

The agreement shown in Figure F.9 is felt to be quite good, and indicates that the PRAISE model best fits the observed crack depths when the mean stress of 83 MPa (12 ksi) is used. Figure F.10 shows that the stress has an important effect, because the agreement is not so good when a stress of 103 MPa (15 ksi) is employed.

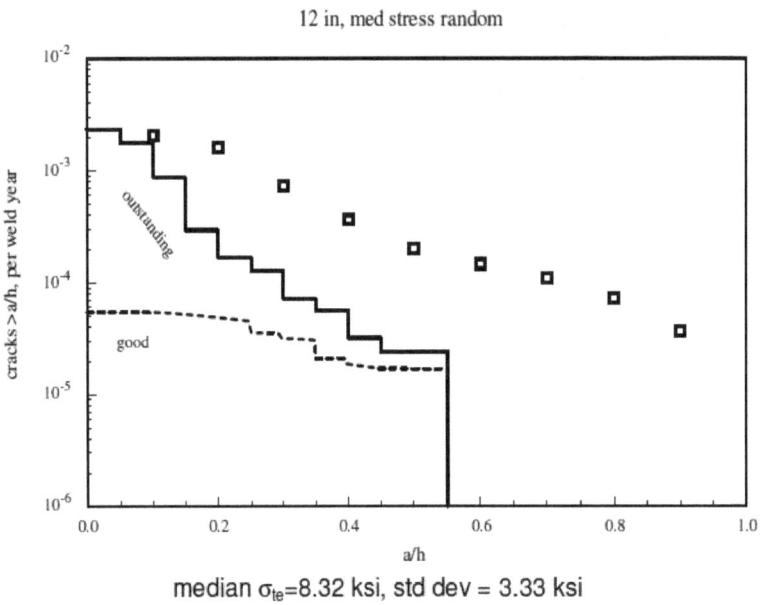

Figure F.10 Comparison of Results for the High-Stress Case Predicted for Two Inspection Qualities with Reported Observations (the Square Data Points are the Observations [F.17], no Weld Overlay)

The results of Figure F.7 are for observed (detected) cracks, whereas Figure F.8 has not had the nondetection probability applied. A direct comparison is therefore not possible, and the two figures are plotted on much different scales. However, the PRAISE predictions contain a much greater proportion of short and shallow cracks than the observations. This could be somewhat affected by the nondetection probabilities, but the differences would not be removed by applying the nondetection probabilities to the cracks predicted by PRAISE.

The question immediately arises regarding the results to be used in the estimation of the recirculation system reliability; the results of Table F.20 for the highly stressed joint or the results of Table F.21 for the benchmarked stress of 83 MPa (12 ksi). Interestingly, the system and average weld leak frequencies are nearly the same whether the 83 MPa (12 ksi) or 140 MPa (20 ksi) weld is used, because the number of joints involved also depends on the stress. Table F.23 summarizes this comparison. This table is in terms of the leak frequency per year, which is obtained from the cumulative results given above by use of Equation F.2.

The system frequency and average frequency per weld are nearly the same for both cases.

**Table F.23 System and Average Weld Leak Frequencies for Two Cases
of the 12 inch Recirculation Line**

Mean σ_{NO}, ksi	12	20
Mean σ_{te}, ksi	5.32	13.32
COV	0.3	0
Std Dev of σ_{te}, ksi	1.6	0
	per weld joint	
0 - 25 years	6.15×10^{-4}	1.19×10^{-2}
25 – 40 years	2.36×10^{-4}	5.57×10^{-3}
40 – 60 years	9.25×10^{-4}	2.19×10^{-3}
Number dominant. joints	49	2
Number in system	49	49
	System (times number dom. joints)	
0-25	3.01×10^{-2}	2.38×10^{-2}
25-40	1.16×10^{-2}	1.11×10^{-2}
40-60	4.53×10^{-3}	4.38×10^{-3}
	Average per joint ($\div 49$)	
0-25	6.15×10^{-4}	4.86×10^{-4}
25-40	2.36×10^{-4}	2.24×10^{-4}
40-60	9.25×10^{-4}	8.94×10^{-5}

Of the two cases in Table F.23, the case of a mean stress of 83 MPa (12 ksi) and coefficient of variation of 0.3 (on $\sigma_{te} + \sigma_{dw}$) is more representative of the population of joints as a whole, so is preferred for comparisons with observations of part-through cracks.

F.3.6 Recirculation Line– 28 inch
Stresses and dimensions are given in the corresponding sections for the 12 inch line. IGSCC crack initiation and growth are the dominant degradation mechanisms. Table F.24 summarizes the results for this weld.

**Table F.24 Cumulative PRAISE Results for the Weld
in the 28 inch Recirculation Line**

OD=28 inches   t = 1.201 inches

	Time	Probability
>0	25	6.23×10^{-3}
	40	1.02×10^{-2}
	60-	1.46×10^{-2}
>100	25	6×10^{-4}
	40	8×10^{-4}
	60-	8×10^{-4}
>1500	25	6.66×10^{-5}
	40	6.66×10^{-5}
	60-	1.25×10^{-4}
>5000	25	6.00×10^{-5}
	40	6.87×10^{-5}
	60-	9.79×10^{-5}
break	25	3.3×10^{-5}
	40	3.3×10^{-5}
	60-	6.7×10^{-5}

σ_{dw}=2.0 ksi
σ_{te}=1.75 ksi
P = 1,125 psi
Type 304 full
residual stress
3 HU-CD/yr

F.3.7 Feedwater Elbow

The feedwater elbow is one of the base case systems. This system is subject to flow accelerated corrosion (FAC), which can be a serious degradation mechanism if left unchecked. PRAISE can not model FAC, but some analyses are provided for fatigue crack initiation and growth.

F.3.7.1 Dimensions and Welds - The layout of the feedwater system is given in the piping isometrics made available to the panel members. There are some 123 welds in the two loops of the feedwater systems, all but 6 of them in 12 and 20 inch piping. The 12 inch lines are schedule 100 (17.4 mm [0.687 inches] thick) and the 20 inch lines are schedule 80 (32.5 mm [1.281 inches] thick). The material is A-333 Grade 6 (which is a carbon steel).

F.3.7.2 Stresses and Cycles - The feedwater line elbow is considered in Reference F.5, so this is evidently the high stress point in the system. Note that there are at least 6 such elbows in a feedwater system. (There are many more elbows, but they are likely to not be so highly stressed). The degradation mechanism is fatigue and flow accelerated corrosion (FAC). Stresses do not contribute to FAC, so are not needed for this mechanism. For fatigue, there are a considerable number of cycles of high stress amplitude. They are available from Reference F.5. Table F.25, which (except for the column of temperatures) is page A.25 of Reference F.5, summarizes the stresses. These stresses are "decomposed" according to the procedure discussed above for the surge line.   The analysis reported in Reference F.5

used a temperature of 590°F (310°C), as indicated in the text at the top of Table F.25. However, Table 5-123 of Reference F.18 provides the temperatures for these transients, and it is suggested that these temperatures be used, because their use is more realistic and less conservative. They are included as the right-hand column of Table F.25. The temperature influences the strain-life curve, and has a noticeable effect on the computed failure probabilities because of its influence on the initiation probabilities.

The values of the deadweight and restraint of thermal expansion under normal operation that Reference F.5 uses for this location are

$$\sigma_{dw} = 0$$

$$\sigma_{te} = 115 \text{ MPa } (16.68 \text{ ksi}).$$

The stress history in Table F.25 most likely contains seismic events. It is not possible to eliminate them from the list using information currently available, but their influence on the calculated failure probabilities is expected to be minimal.

**Table F.25 Summary of Stress Cycles for Feedwater Line Elbow
(from Page A.25 of NUREG/CR-6674 [F.5])**

```
NAME OF PLANT           =    GE-NEW
NAME OF COMPONENT       =    FEEDWATER LINE ELBOW
NUM OF LOAD PAIRS       =    28
MATERIAL                =    LAS
WALL THICK (INCH)       =    1.000
INNER DIAMETER          =    12.000
AIR/WATER               =    WATER
TEMPERATURE(F)          =    590.000
SULFUR(WHT%)            =    .015
DISOL O2 (PPM)          =    .100
STR RATE (%/SEC)        =    0.00100
USEAGE(DETERM.)         =    3.68800
P-INITIATION@40         =    1.59E-01
P-INITIATION@60         =    3.65E-01
      P-TWC @40         =    1.01E-03
      P-TWC @60         =    1.46E-02
```

LOAD PAIR	AMP(KSI)	NUM/40 YR	EDOT(%/S)	USEAGE	TEMP, °C
HIGH 18/LOW 21	106.040	5.0	.117000	.025000	200
HIGH 18/LOW 21	103.960	5.0	.114000	.024000	200
HIGH 18/LOW 21	102.610	5.0	.113000	.024000	200
HIGH 14/LOW 17	91.590	8.0	.001000	.123000	200
HIGH 8/LOW 17	89.400	10.0	.095000	.037000	200
HIGH 3/LOW 16	88.270	5.0	.094000	.018000	200
HIGH 8/HIGH 7	83.760	126.0	.041000	.519000	200
HIGH 7/HIGH 7	81.430	10.0	.086000	.033000	215
HIGH 7/LOW 13	67.930	97.0	.001000	.740000	200
HIGH 7/LOW 13	66.710	14.0	.001000	.101000	200
HIGH 7/LOW 15	61.290	6.0	.001000	.035000	200
HIGH 7/LOW 15	61.160	64.0	.001000	.451000	212
HIGH 8/LOW 12	55.500	92.0	.001000	.391000	200
HIGH 3/LOW 12	46.630	88.0	.001000	.254000	215
HIGH 7/LOW 22	42.880	15.0	.001000	.029000	212
HIGH 3/HIGH 7	39.440	212.0	.001000	.315000	215
HIGH 3/HIGH 7	38.130	69.0	.001000	.104000	224
HIGH 3/LOW 20	36.800	11.0	.001000	.014000	224
HIGH 4/LOW 20	34.320	60.0	.001000	.053000	215
LOW 11/LOW 20	32.950	203.0	.001000	.122000	200
HIGH 7/LOW 11	32.530	360.0	.001000	.203000	200
HIGH 6/LOW 11	29.770	222.0	.025000	.035000	200
HIGH 2/HIGH 19	26.090	30.0	.028000	.003000	212
HIGH 5/HIGH 19	26.040	81.0	.028000	.007000	200
HIGH 5/HIGH 9	21.640	96.0	.001000	.012000	212
HIGH 1/HIGH 11	20.560	40.0	.001000	.003000	200
LOW 10/LOW 11	14.180	30.0	.001000	.001000	200
HIGH 5/LOW 11	11.220	11515.0	.001000	.008000	200

F.3.7.3 Results - PRAISE runs for this component were made using the version that can treat fatigue crack initiation with details of the circumferential variation of the stresses. The feedwater system is relatively more likely to experience water hammer, so the influence of an overload event with a stress of $0.42\sigma_{flo} = 128$ MPa (18.5 ksi) above that normally present was considered. This stress is denoted as σ_{DL}, and results were generated for one cycle of this stress at 24, 39, or 59 years. The results are summarized in Table F.26, which includes the effects of σ_{DL} (columns D & F).

Table F.26 Cumulative PRAISE Results for Feedwater Line Elbow

		A	B	C	D	E	F	G
Stresses		Ref. F.5	Table F.25	Table F.25	Table F.25	Table F.25	Table F.25	80% of Table F.25
Failure Criterion		σ_{flow}	σ_{flow}	σ_{flow}	σ_{flow}	σ_{flow} & J-T	σ_{flow} & J-T	σ_{flow}
σ_{DL}		no	no	no	$\sigma_{DL}@(t-1)$	no	$\sigma_{DL}@(t-1)$	no
>0	25	--	--	$<10^{-8}$	2.5×10^{-8}	1.0×10^{-7}	3.10×10^{-6}	$<10^{-7}$
	40	0.001	2×10^{-6}	5.69×10^{-6}	7.19×10^{-6}	1.54×10^{-5}	1.43×10^{-4}	$<10^{-7}$
	60	0.0146	1.8×10^{-4}	2.57×10^{-4}	2.59×10^{-4}	$\sim5\times10^{-4}$	2.9×10^{-3}	4.6×10^{-7}
			Ref F.6 Table 4-8	10^8 trials				GEN6TWA4
>100	25	--		$<10^{-8}$	1.5×10^{-6}*	$<10^{-7}$	1.70×10^{-6}*	
	40	--		$<10^{-8}$	1.5×10^{-6}*	$<10^{-7}$	1.70×10^{-6}*	
	60	--		$<10^{-8}$	1.50×10^{-6}*	--	2.1×10^{-6}*	
				GENC6TW4				
>1500	25	--	--					$<10^{-7}$
	40	--	--					$<10^{-7}$
	60	--	--					$<10^{-7}$
			axi-symmetric actual T					reduced stresses

* also a break

Case A is directly from Reference F.5, and Case B is directly from Table 4-8 of Reference F.6. Case C is Case B rerun with 10^8 trials. Cases D-G are variations of C with different failure criteria, overloads and reduction of stresses. The results for various failure criteria (critical net section stress only or critical net section stress and tearing instability) show that consideration of tearing instability noticeably increases the computed failure probability (compare, for instance, cases C&E). Consideration of an overload event also has a noticeable effect (E&F). The use of lower stresses markedly reduces the computed failure probabilities (G & C). In the case of an overload event, the probability of a 100 gpm failure is the same as a complete pipe break.

F.3.7.4 Alternate Procedure - The results of Table F.26 show that the probability of a large leak was obtainable from the Monte Carlo procedure only when a large overload occurred. When this did not occur, there were no leaks of even 380 lpm (100 gpm) in 10^7 or 10^8 trials. In order to obtain estimates for the larger leak probabilities, the alternate procedure discussed for the surge line was also applied to Case C of Table F.26 for the feedwater elbow.

As before, the crack length for a given leak rate, $b(\dot{q})$, was obtained from a pcPRAISE run, along with the half-crack length of any cracks that become through-wall. Figure F.11 provides a plot of the leak rate as a function of b for the feedwater elbow.

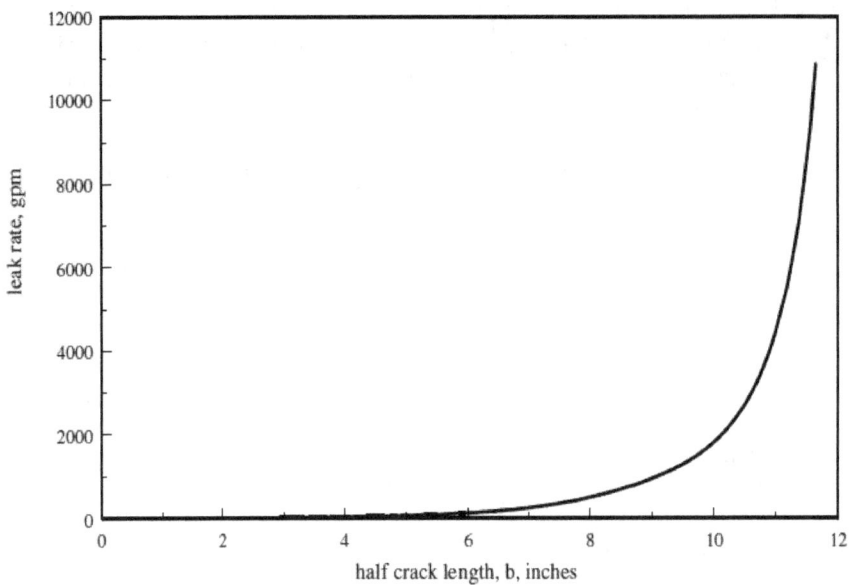

Figure F.11 Leak Rate as a Function of Half Crack Length for Feedwater Elbow Base Case C

The results in Table F.27 are obtained from this figure and the corresponding pcPRAISE results. This table also includes the portion of the circumference that is cracked and the proportion of the crack opening area to the flow area of the pipe. It is seen that the opening area of the crack is nearly equal to the flow area of the pipe when the leak rate is 19,000 lpm (5,000 gpm). The value of b for a complete pipe break, as obtained from Equation E.7 is also included. Table F.29 defines $b(\dot{q})$.

Table F.27 Half Crack Lengths and Areas for a Given Leak Rate
(Feedwater Elbow Base Case C)

$\dot{q}$, gpm	b, inches	$\dfrac{b}{\pi R_I}$	A, in^2	$\dfrac{A}{A_{pipe}}$
100	5.737	0.32	1.837	0.02
1500	9.743	0.55	27.554	0.27
5000	11.095	0.62	90.877	0.91
DEGB	15.925	0.89	--	--

As before, the next step is to estimate the probability of having a through-wall crack exceeding a given length as a function of time. The modified version of pcPRAISE was used to generate a table of values of b and the time at which the leak first occurred. A run was made with 10^7 trials, with 2,607 cracks becoming through-wall within 60 years. This corresponds to a leak probability of 2.607×10^{-4} at 60 years, which agrees closely with the leak probability obtained earlier. Of these 2,607 cracks, none appeared before 25 years, and 64 occurred between 25 and 40 years. The statistical distribution of these 64 cracks at 40 years provides the probability of having a through-wall crack greater than a given length within 40 years. Extrapolation is required to obtain results for the crack lengths included in Table F.27. Figure F.12 shows the complementary cumulative distribution of b at 40 years, along with the curve fit of Equation F.10.

$$P(>b) = e^{-5.34(b-1)} \qquad (40 \text{ years}) \qquad [F.10]$$

Note that the plot starts at a half-crack length of 25 mm (1 inch), and that the data are closely approximated by a straight line on log-linear scales.

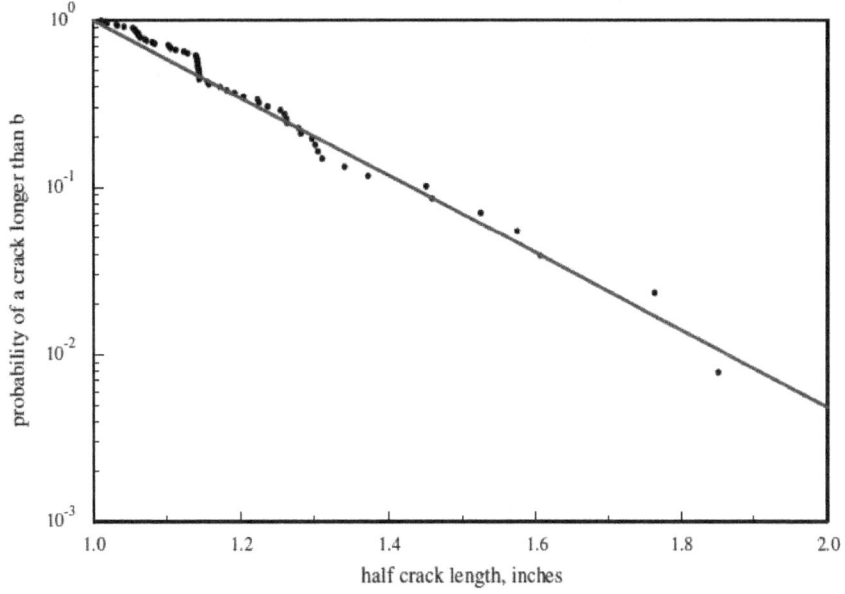

Figure F.12 Complementary Cumulative Distribution of Half-Crack Length of Through-Wall Cracks in Feedwater Elbow within 40 Years, Along with Fit

Figure F.13 provides a similar plot for the 2,607 through-wall cracks that occurred within 60 years. Equation F.11 is the fit of the distribution at 60 years within the range of interest.

$$P(>b) = 0.0274e^{-2.25(b-1)} \qquad \text{(60 years)} \qquad\qquad [F.11]$$

Note that in this case the data appear bilinear and are not well approximated by a straight line on log-linear scales. To represent the data at the longer crack lengths of interest, a straight line was assumed beyond a crack length of 50 mm (2 inches). This corresponds to a probability below about 0.003.

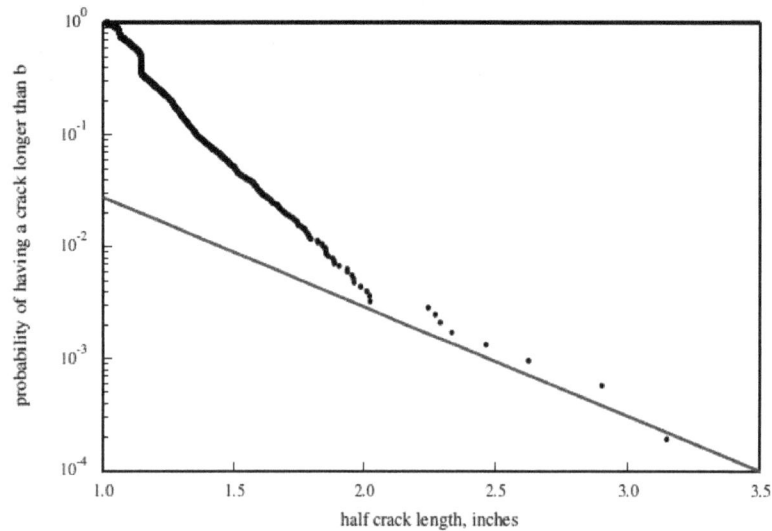

Figure F.13 Complementary Cumulative Distribution of Half-Crack Length of Through-Wall Cracks in Feedwater Elbow within 60 Years, Along with Fit

The probability of a leak exceeding a given size within 40 and 60 years is then obtained by taking using the value of *b* for a given leak rate from Table F.27 in conjunction with Equations F.10 and F.11, respectively. Table F.28 summarizes the results.

Table F.28 Cumulative Results for Feedwater Elbow Case C

	time years	$P(>\dot{q})$
>0	25	$<10^{-8}$
	40	5.69×10^{-6}
	60	2.57×10^{-4}
>100	25	--
	40	1.03×10^{-11}
	60	6.44×10^{-7} *
>1500	25	--
	40	5.29×10^{-21}
	60	7.84×10^{-11}
>5k	25	--
	40	3.88×10^{-24}
	60	3.74×10^{-12}
DEPB	25	--
	40	2.44×10^{-35}
	60	7.14×10^{-17}

* direct Monte Carlo gave $<10^{-8}$

The leak (>0) results in Table F.28 came directly from the Monte Carlo simulation. With 10^8 trials, no leaks exceeding 380 lpm (100 gpm) were obtained. Hence, the Monte Carlo simulation predicts $<10^{-8}$ probability of a leak exceeding 380 lpm (100 gpm) within 60 years. The alternative procedure gave a corresponding value of 6.44×10^{-7}. This suggests that the alternative procedure overestimates the probability of a given leak, as was also the case for the surge line elbow.

F.4 Selection of Reference Cases and Extension to System Frequencies

The earlier sections of this document contain many sets of results for each base case component. The multiple cases were generated primarily as a series of sensitivity studies. For these results are to be useful in the LOCA elicitation, a reference case must be selected for each component as being representative for that component. This section briefly discusses which case for each component is suggested as the reference case, and system leak frequencies are presented for each reference case.

The joint frequency is calculated from the cumulative results reported above by use of Equation F.2. The system frequencies are then obtained by multiplying by the number of highly stressed joints in the system (this approximation works because the failure probabilities are generally small).

Each component is discussed, with a summary table provided after all components are discussed.

F.4.1 Hot Leg Pressure Vessel
As shown in Tables F.5 – F.7, the large leak (> 380 lpm [100 gpm] and larger) probabilities for this component varied considerably, depending on the crack growth mechanism (cycle-dependent fatigue or time-dependent stress corrosion cracking [PWSCC]), and whether crack initiation or growth from pre-existing defects was considered. The fatigue crack growth results (Table F.5) were very low ($\sim 10^{-18}$), and the PWSCC crack initiation results (Table F.6) were quite large ($\sim 10^{-5}$). Since it is expected that this component will totally dominate the very large (> 380,000 lpm [100,000 gpm]) leak category, the selection of the reference case is critical for very large leak estimates. The PWSCC with fabrication defects has intermediate failure probability results ($\sim 10^{-10}$), and is recommended as the reference case. The case without residual stresses is selected. Table F.6 shows that residual stresses do not have a large influence. The time dependency of the large leak cumulative probability is very small, which suggests that the leak frequency is very small. For estimation purposes, the leak frequencies are estimated by taking the value of the cumulative at 60 years, dividing it by 60, and assuming the value to be applicable independently of time. This will overestimate the leak frequency at long time and underestimate it at short time.

For extension to system failure frequency, it is assumed that there are three comparably stressed joints in the large main coolant piping.

F.4.2 Surge Line Elbow
The surge line elbow result identified as "axisymmetric nonseismic" in Table F.12 is suggested as the reference case. Table F.13 summarizes the cumulative results for the larger flow rates, which were obtained by the alternative procedure.

Two of these highly stresses elbows are considered to be present in the surge line system

F.4.3 HPI Makeup Nozzle
Probability analyses were performed with and without failure of the thermal sleeve, which has been observed to fail in service. The least favorable large leak probabilities were for a failed thermal sleeve,

which immediately resulting in fatigue crack initiation, but with the same stresses as before. This is suggested as the reference case, with the column labeled $\sigma_u = 0$ in Table F.16 being the results of interest.

Three such locations are considered to be present in the system.

F.4.4 Recirculation Line – 12 inch

Analyses were performed for this component for a range of applied stresses, with predictions compared to field experience of leaks and observed part-through cracks. Analyses were performed for no remedial action, and for a weld overlay at 20 years. The weld overlay at 20 years is considered to be the most realistic. Comparisons with experience led to an estimate of stresses that were considerably below the peak value used in the original analysis. However, when compensated for the number of weld joints involved, the system leak frequencies were nearly the same whether 49 joints with a random stress (mean $\sigma_{NO} = 83$ MPa [12 ksi]) or 2 joints with a high stress ($\sigma_{NO} = 140$ MPa [20 ksi]) were considered (see Table F.25). The case of weld overlay at 20 years with the high stress is recommended as the reference case. Table F.19 contains the cumulative results.

Two of the highly stresses joints are considered to be present in the recirculation system.

F.4.5 Recirculation Line – 28 inch

The recirculation line with no remedial action and a high stress representing the dominant joints was the only case considered, and is summarized in Table F.24.

Two such joints are considered to be present in the system.

F.4.6 Feedwater Elbow

Case C in Table F.26 is suggested as the reference case. Results for > 380 lpm (100gpm) and larger were generated by the alternative procedure, and are summarized in Table F.28.

Four such locations were considered to be present in the system.

F.4.7 Summary Table

Table F.29 provides an overall summary of the leak flow rate frequencies for the reference cases of the base case systems.

Table F.29 Summary of Results for Reference Systems

			Hot Leg	Surge Line	HPI	Recirculation 12	Recirculation 28	Feedwater
		OD, in	34	14	3.44	12.75	28	12.75
		t, in	2.5	1.406	0.4375	0.687	1.201	0.687
		A, in^2	661	98.3	5.167	102	515	102
		Q_{max}	423	63	3.6	38	193	38
		matl	cast SS	SS	SS	SS	SS	CS
		Degr Mech	PWSCC growth	fatigue init&gro	fatigue	SCC init&gro	SCC init&gro	fatigue init&gro
		Table	F.6	F.12	F.16	F.21	F.26	F.26, F.28
		Case	PWSCC no σ_{res}	Table F.9 stresses	failed slv $\sigma_u=0$	overlay @ 20 yrs		C
		Insp	0,20,40	none	none	0,20,40	0,20,40	none
dominant joint freq	>0	0-25	--	9.3×10^{-3}	1.48×10^{-4}	1.19×10^{-2}	2.5×10^{-4}	$<4 \times 10^{-10}$
		25-40	--	0.024	5.94×10^{-4}	5.57×10^{-3}	2.6×10^{-4}	3.8×10^{-7}
		40-60	--	0.015	8.60×10^{-4}	2.19×10^{-3}	2.2×10^{-4}	1.3×10^{-5}
	>0.1	0-25	1.33×10^{-8}	3.0×10^{-7}	2.60×10^{-5}	5.71×10^{-3}	2.4×10^{-5}	--
		25-40	1.33×10^{-8}	4.2×10^{-6}	1.35×10^{-4}	1.30×10^{-3}	1.3×10^{-5}	6.9×10^{-13}
		40-60	1.33×10^{-8}	9.0×10^{-6}	1.32×10^{-4}	3.55×10^{-4}	$<5 \times 10^{-6}$	3.2×10^{-8}
	>1.5	0-25	1.6×10^{-11}	1.4×10^{-9}	2.60×10^{-5}	4.26×10^{-3}	2.7×10^{-6}	--
		25-40	1.6×10^{-11}	2.2×10^{-8}	1.35×10^{-4}	1.23×10^{-3}	--	3.5×10^{-22}
		40-60	1.6×10^{-11}	5.8×10^{-8}	1.32×10^{-4}	3.10×10^{-4}	3.0×10^{-6}	3.9×10^{-12}
	>5	0-25	4.6×10^{-13}	9.7×10^{-11}		3×10^{-3}	2.4×10^{-6}	--
		25-40	4.6×10^{-13}	2.3×10^{-9}		1.23×10^{-3}	5.8×10^{-7}	2.6×10^{-25}
		40-60	4.6×10^{-13}	8.3×10^{-9}		3.10×10^{-4}	1.5×10^{-6}	1.9×10^{-13}
	>25	0-25	4.6×10^{-13}	3.9×10^{-14}		1.96×10^{-3}	1.3×10^{-6}	--
		25-40	4.6×10^{-13}	3.2×10^{-12}		1.23×10^{-3}	$\sim 2 \times 10^{-6}$	1.6×10^{-36}
		40-60	4.6×10^{-13}	2.5×10^{-11}		3.10×10^{-4}	1.7×10^{-6}	3.6×10^{-18}
	>100**	0-25	3.6×10^{-16}				1.6×10^{-6}	
		25-40	3.6×10^{-16}				$\sim 2 \times 10^{-6}$	
		40-60	3.6×10^{-16}				1.7×10^{-6}	
		field	22	3		20	22	29
		shop	12	9		20	30	22
		safe end	16	1		9	3	12
		dominant	**3**	**2**	**3**	**2**	**2**	**4**
system frequencies	>0	0-25	--	0.019	4.44×10^{-4}	2.43×10^{-2}		$<1.6 \times 10^{-9}$
		25-40	--	0.048	1.78×10^{-3}	1.17×10^{-2}		1.5×10^{-6}
		40-60	--	0.030	2.58×10^{-3}	4.82×10^{-3}		5.2×10^{-5}
	>0.1	0-25	4.0×10^{-8}	6.0×10^{-7}	7.80×10^{-5}	1.15×10^{-2}		--
		25-40	4.0×10^{-8}	8.5×10^{-6}	4.05×10^{-4}	2.62×10^{-3}		2.8×10^{-12}
		40-60	4.0×10^{-8}	1.8×10^{-5}	3.96×10^{-4}	7.10×10^{-4}		1.3×10^{-7}
	>1.5	0-25	4.8×10^{-11}	2.8×10^{-9}	7.80×10^{-5}	8.52×10^{-3}		--
		25-40	4.8×10^{-11}	4.4×10^{-8}	4.05×10^{-4}	2.46×10^{-3}		1.4×10^{-21}
		40-60	4.8×10^{-11}	1.2×10^{-7}	3.96×10^{-4}	6.20×10^{-4}		1.6×10^{-11}
	>5	0-25	1.4×10^{-12}	1.9×10^{-10}		6×10^{-3}		--
		25-40	1.4×10^{-12}	4.6×10^{-9}		2.46×10^{-3}		1.0×10^{-24}
		40-60	1.4×10^{-12}	1.7×10^{-8}		6.20×10^{-4}		7.6×10^{-13}
	>25	0-25	1.4×10^{-12}	7.9×10^{-14}		3.92×10^{-3}		--
		25-40	1.4×10^{-12}	6.4×10^{-12}		2.46×10^{-3}		6.5×10^{-36}
		40-60	1.4×10^{-12}	5.0×10^{-11}		6.20×10^{-4}		1.4×10^{-17}
	>100**	0-25	1.1×10^{-15}			2.6×10^{-6}		
		25-40	1.1×10^{-15}			4×10^{-6}		
		40-60	1.1×10^{-15}			3.7×10^{-6}		

times in reactor years, 1 calendar year ~ 0.8 reactor years
shaded areas are estimates based on alternative procedure
leak rates in thousands of gallons per minute
cross-hatched cells are beyond maximum leak capability for that pipe size
** also applicable to > 1,900,000 lpm (500 kgpm) for hot leg if sufficient diameter

F.5 References

F.1. D. O. Harris, E. Y. Lim and D. Dedhia, *Probability of Pipe Fracture in the Primary Coolant Loop of a PWR Plant, Vol. 5: Probabilistic Fracture Mechanics Analysis*, U.S. Nuclear Regulatory Commission Report NUREG/CR-2189, Vol. 5, Washington, D.C., August 1981

F.2. D. O. Harris, D. Dedhia, E.D. Eason and S.D. Patterson, *Probability of Failure in BWR Reactor Coolant Piping: Probabilistic Treatment of Stress Corrosion Cracking in 304 and 316NG BWR Piping Weldments*, U.S. Nuclear Regulatory Commission Report NUREG/CR-4792, Vol. 3, Washington, D.C., December 1986

F.3. D. O. Harris, D. Dedhia and S. C. Lu, *Theoretical and User's Manual for pc-PRAISE, A Probabilistic Fracture Mechanics Code for Piping Reliability Analysis*, U.S. Nuclear Regulatory Commission Report NUREG/CR-5864, Washington, D.C., July 1992

F.4. D.O. Harris and D. Dedhia, *WinPRAISE: PRAISE Code in Windows*, Engineering Mechanics Technology, Inc. San Jose, California, Technical Report TR-98-4-1, 1998

F.5. M.A. Khaleel, F.A. Simonen, H.K. Phan, D.O. Harris and D. Dedhia, *Fatigue Analysis of Components for 60-Year Plant Life*, U.S. Nuclear Regulatory Commission Report NUREG/CR-6674, Washington, D.C., June 2000

F.6. A. Deardorff, D. Harris and D. Dedhia, *Materials Reliability Program: Re-Evaluation of Results in NUREG/CR-6774 for Carbon and Low-Alloy Steel Components"*, Electric Power Research Institute Report 1003667, Palo Alto, California, 2002

F.7. J. Keisler, O.K. Chopra and W.J. Shack, *Fatigue Strain-Life behavior of Carbon, Low-Alloy Steels, Austenitic Stainless Steels, and Alloy 600 in LWR Environments*, U.S. Nuclear Regulatory Commission Report NUREG/CR-6335, Washington, D.C., 1995

F.8. O.K. Chopra and W.J. Shack, *Effects of LWR Coolant Environments on Fatigue Design Curves of Carbon and Low-Alloy Steels*, U.S. Nuclear Regulatory Commission Report NUREG/CR-6583, Washington, D.C., March 1998

F.9. *Technical Elements of Risk-Informed Inservice Inspection Programs for Piping*, U.S. Nuclear Regulatory Commission Draft Report NUREG-1661, Washington, D.C., January 1999

F.10. M.A. Khaleel, O.J.V. Chapman, D.O. Harris and F.A. Simonen, "Flaw Size Distribution and Flaw Existence Frequencies in Nuclear Piping", *Probabilistic and Environmental Aspects of Fracture and Fatigue*, ASME PVP-Vol. 386, 1999, pp. 127-144

F.11. ASME Boiler and Pressure Vessel Code, Section XI, Appendix C, 1992

F.12. P. Ricardella, "Probabilistic Fracture Mechanics Analysis of CRDM Nozzles", presented at ACRS Meeting, Rockville, Maryland, June 5, 2002

F.13. e-mail from Gery Wilkowski to David Harris, "Material Property Inputs for Base Cases", June 10, 2003

F.14. Personal communication, Art Deardorff, Structural Integrity Associates, San Jose, California, to David Harris, Engineering Mechanics Technology, Inc., San Jose, California

F.15. T.C. Chapman, et al., *Assessment of Remedies for Degraded Piping*, Electric Power Research Institute Report NP-5881-LD, Palo Alto, California, 1988

F.16. B.O.Y. Lydell, *An Application of the Parametric Attribute/Influence Methodology to Determine Loss of Coolant Accident (LOCA) Frequency Distributions*, Document No. R2003-02, May 2003, provided to members of the NRC LOCA Frequency Expert Elicitation Panel.

F.17. Attachment (Action Item 45R1.xls) to e-mail from Bengt Lydell to base case panel members, June 20, 2003

F.18. A.G. Ware, D.K. Morton and M.E. Nitzel, *Application of NUREG/CR-5999 Interim Fatigue Curves to Selected Nuclear Power Plant Components,* U.S. Nuclear Regulatory Commission Report NUREG/CR-6260, Washington, D.C., 1995

APPENDIX G

PIPING BASE CASE RESULTS OF

VIC CHAPMAN

APPENDIX G

PIPING BASE CASE RESULTS OF VIC CHAPMAN

Summary of Benchmarking Analysis
Carried out Using 'RR-PRODIGAL'

G.1 General Background to RR-PRODIGAL

RR PRODIGAL is a basic fatigue failure probability model developed by Rolls Royce for the Naval Nuclear program. When analysing a weld, it first simulates the weld construction in order to determine a start of life defect distribution and density for both buried and surface breaking defects. A failure probability using standard linear elastic fracture mechanics methods is then evaluated for both the buried and surface breaking defects (assumptions about break through of buried defects to surface defects are based on the ASME criteria). Failure is achieved when the defect either exceeds the R6 failure criteria or simply grows through to the full thickness. The failure probability for all initial defects is then combined to form the total failure probability.

For non-weld areas, a probabilistic crack initiation analysis is carried out with a correlated crack growth analysis to failure. This correlation means that short times to crack initiation imply that a fast crack growth follows this initiation. There is no positive data to confirm or deny this proposition. It was chosen simply because it is pessimistic.

The modelling contains a routine to assess the growth of the defect around a welded pipe at the same time as the defect grows through the weld thickness, however, this part was not used in this assessment. RR-PRODIGAL does not, at present, contain a routine to evaluate the crack growth of a through wall defect around the outer surface of a pipe weld.

At failure, the model evaluates a critical through wall defect size based again on the R6 criteria.

At present RR-PRODIGAL does not contain a verified and validated assessment of the PWSCC degradation mechanism.

There are several publications that describe RR-PRODIGAL, which include a recent benchmarking exercise as part of a European initiative. References G.1 and G.2 should provide sufficient information for any readers wishing to obtain further information on this code.

G.2 Leak Rate Evaluation

When estimating RR-PRODIGAL leak rates through the final through wall defect in a pipe weld, evaluations were made using an elastic crack opening displacement (COD) analysis. However, it was felt that the uncertainties associated with assessing both the defect length around the pipe circumference as well as the COD needed for estimating the flow rate through the crack, were too great and too subject to ongoing development, to allow a suitable analysis of the leak rate. Thus, RR-PRODIGAL does not contain, within itself, a routine for evaluating the flow rate from the final defect size.

Instead, it was concluded that the leak rate from a through wall defect could be considered independently of the probability of the breach, i.e. the leak rate from the defect is not dependent on the probability of the defect cracking through the pipe wall. Note, however, that the COD, crack length, and hence leak rate is not independent of the mechanism that led to the failure, only the probability of the failure itself.

Within the Naval Nuclear program, computer programs have been developed to assess the leak rate from different defects based primarily on the 'SQUIRT' model. However, for consistency within this program, the data on leak rate against defect area provided by the USNRC were used, as shown in Figure G.1.

G.3 Procedure

The procedures used to develop the base case numbers are as follows:

1   Evaluate the basic fatigue failure probability using RR-PRODIGAL code using the transient data supplied[1].
2   Evaluate an elastic COD as a function of defect size.
3   Use expert judgement to extend this COD beyond the elastic limit.
4   Evaluate a mean defect cross-sectional area for a given defect size using its associated COD.
5   Evaluate the mean leak rate from a given defect size using the data supplied by the USNRC, see Figure G.1.

[Note for Steps 2, 3, 4 and 5 above a defect length is given. Thus, Steps 2, 3, 4 and 5 provide a mapping from a given defect size at failure to the mean leak rate in gpm, given this defect exists.]

6   Use expert judgement to assess the distribution of the defect length at failure.
7   Combine Steps 5 and 6 to obtain the conditional probability of a leak rate greater than the given leak rates for Categories 1 through 6. These categories being as follows;

Table G.1 Leak Category Leak Rates

	Leak Rate Greater than (gpm)	Log Leak Rate
Leak Category 1	100	2
Leak Category 2	1,500	3.2
Leak Category 3	5,000	3.7
Leak Category 4	25,000	4.4
Leak Category 5	100,000	5.0
Leak Category 6	500,000	5.7

8   Combining the conditional probability of Step 7 with the basic fatigue failure probability in Step 1 gives the required final probability of a leak greater than each of the categories.

G.4 Example Base Case Analysis

As a way of demonstrating the procedure given above, the results for the 14-inch Surge Line elbow are reproduced in this section. Two situations are considered, the elbow and the adjacent weld. The transients were based on data supplied and are reproduced in Attachment G.1

G.4.1 Probability of Failure Surge Line Elbow – Base Case
This is a failure from base material and so the analysis assumed a fatigue based crack initiation followed by crack growth to failure. As stated earlier, the crack initiation and crack growth are assumed to be positively correlated. This assumption assumes that if the properties of the base material are such as to lead to an early crack initiation, it is very possible that these same properties

---

[1] This information needed to be supplied because the transient experience for the Naval Nuclear program is a) confidential and b) not applicable to commercial plants.

could result in a subsequently fast crack growth rate. The results of this analysis are shown in the following table:

Table G.2 Results for PWR Surge Line Elbow Base Case Analysis

Time (years)	Cumulative Probability of Failure
25	6.1×10^{-6}
40	7.8×10^{-6}
60	9.4×10^{-6}

RR-PRODIGAL gave the critical through wall defect length, based on the R6 criterion, as 14 inches.

G.4.2 Probability of Failure Surge Line Weld

The surge line elbow weld was analysed at a 60-year life assuming the same cyclic conditions as for the elbow itself, but with the stresses factored down by 20 percent as suggested at the Elicitation Base Case Review Meeting on June 4 and 5, 2003 in Bethesda, Maryland. The two hydro cases were, however, maintained at their original values.

In this analysis, RR-PRODIGAL first simulates the weld construction, including any build inspections, to establish the start of life defect density and distribution for both buried and surface breaking defects. As stated earlier, conditional failure probabilities are assessed for both situations and combined to give the final failure probability.

The failure probability evaluated for this case was:

Table G.3 Results for PWR Surge Line Weld Analysis

Cumulative Probability of Failure at 60 years	1.3×10^{-4}

It can be seen that this failure probability is over an order of magnitude higher than the base case. This is due to the difference between having to initiate a defect and then grow this defect to failure, and having the probability of pre-existing defects in the weld. The base case values from the base material failure as reflected in Table G.2, i.e. crack initiation leading to failure, have been used in Table D.1 in the main body of this report. Note, however, that the values reported in Table G.2 are cumulative probabilities of failure in 25, 40, and 60 years whereas the values reported in Table D.1 of the main body are frequencies. Consequently, the Table G.2 values need to be divided by 25, 40, and 60 years, respectively to facilitate any comparisons. Furthermore, the values in Table D.1 are for leak rates greater than the threshold leak rates, i.e., 380 lpm (100 gpm) while the values in Table G.2 reflect the totals.

G.4.3 COD and Leak Rate for a Given Defect Size

Having established a basic failure probability, the COD can be evaluated independently of this probability. Once this is established, the leakage area of the defect follows, and given this leak area, the flow rate can be evaluated using the information from Figure G.1. A mean power law was then used to calculate the mean flow rate given a leakage area. The table below gives the elastic COD values evaluated for this case and the resultant flow rate.

Table G.4 Elastic COD and Resultant Leak Rates for a Given Defect Length

Defect Length (inches)	Elastic COD (inches)	Flow rate (gpm)
1.98	0.0025	17
3.96	0.0049	48
5.94	0.0074	92
7.41	0.01	145
9.89	0.012	200
11.87	0.015 (Invalid Result)	270

Interpolating between the results in Table G.4, it can be seen that a defect approximately 160 mm (6.2 inches) long, which is approximately 15 percent of the pipe circumference, results in the first leakage category of 380 lpm (100 gpm).

Clearly it is the behaviour of the defect beyond the elastic range that is of interest for the larger leak categories. If it were to be assumed that at the critical defect size the pipe would simply tear, in an unstable manner, to result in a Double Ended Guillotine Break (DEGB) failure, then the leak rate would simply jump from a Category 1 failure to the gpm associated with the DEGB. In this case that would be 250,000 lpm (65,000 gpm) or a Category 4 leak. The probability of a Category 2 leak rate would then be the same as a Category 3, which would be the same as the Category 4!

Such an assumption could be considered valid. However, in this work, it was assumed that the defect would continue opening in a stable, but plastic manner. Whilst models do exist to evaluate the plastic deformation of defective pipes, no such model was used in this analysis. Instead expert judgement was used to assess how the COD would develop beyond this elastic point, and at what defect size the pipe would finally tear into a DEGB failure. The results of this judgement are shown in Figure G.2. The area of leakage can then be calculated, and the leak rate, given a defect length also follows. The resulting gallon per minute flow rate, for this example, is shown in Figure G.3.

The failure probability gives the basic probability of a breach of the pressure boundary. Figure G.3 shows the leak rate in gallons per minute, given a defect of a given length. In order to obtain the probability of a leak rate greater than 'X' gallons per minute, it only remains to provide a distribution of the defect size at the moment of failure.

G.4.4 Defect Distribution and Leak Rate at Failure – No Leak Detection

First consider the case with no leak detection. For this case the instantaneous size of the defect, and its associated COD, at the moment of snap through to a breach of containment is required. As an example, if the aspect ratio were of the order of 8/1 at snap through, then given a pipe wall thickness of about 36 mm (1.4 inches), the defect length would be approximately ten or eleven inches long. If it were then pessimistically assumed that this was the full through wall defect length, then the instantaneous leak rate would be just above (actually about twice) our 'Category 1' failure criteria of 380 lpm (100 gpm). Thus, the probability of a leak rate greater than Category 1 becomes the basic probability of failure times the probability that the defect at snap through was greater than 250 mm (10 inches), i.e., the defect had an aspect ratio at snap through of about 8/1 or greater. It then follows, from Figure G.3, that in order to exceed the Category 2 leak rate, the instantaneous defect size at snap through would have to be greater than 380 or 405 mm (15 or 16 inches), i.e., the defect had an aspect ration of about 11/1. Furthermore, the defect snapped straight open to the fully plastic COD.

As stated earlier, RR-PRODIGAL has the capability of simulating the crack growth both around and through the pipe wall. However, this is not generally used as the solutions require a detailed knowledge of the stress distribution around the pipe, including any weld residual stress, and generally such knowledge is not well enough defined. Thus, expert judgement was again used. The expert judgement required is to generate a defect distribution at the moment the defect snaps to the COD of Figure G.3, assuming no leak detection.

This base case is for the surge line elbow and it has been assumed that most of the deformation and high stress will result from large bending moments at the elbow. It was felt that this would initiate a defect preferential on the hogging side of the elbow, and promote a crack to grow through the wall thickness on this side of the elbow. This would then imply that the crack growth around the pipe diameter would be restricted. Figure G.4 represents the distribution decided upon for this analysis. This distribution shows the most likely defect length to be up to about 250 mm (10 inches), which is about a quarter of the way around the pipe circumference. The probability of the defect being over halfway around the pipe is seen as a rare event, being about 0.025 or a 1 in 40 chance. If the loading were not dominated by bending, then this distribution would probably be judged to be flatter, with perhaps a 1 in 10 chance of being greater than halfway round the pipe circumference.

Combining Figures G.3 and G.4 gives the conditional probability of a leak greater than a given leak rate. This final plot is given in Figure G.5 and is combined with the basic failure probability to derive the values given in Table D.1 in Section D of the main body of this report.

G.4.5 Defect Distribution and Leak Rate at Failure – With Leak Detection

In the previous section it was assumed that the defect would instantaneously snap open to the full COD associated with its length at the moment the pressure boundary was breached. In reality this will probably not happen. Instead, the very large defects, which are those of interest, will probably grow to different through wall depths at different points around the length. Thus, much smaller surface defects would begin to breach the boundary at different points around the defect. The COD of these small defects would then remain elastic until the whole defect progressed to the surface. In this scenario, the leak from the defect would start very small and grow, slowly at first and then probably very quickly before snapping open to the fully plastic COD.

During this time of surface crack combination, the leak rate may exceed the value at which the operators shut the reactor down to a safe state in order to investigate the leak. Provided this occurs before the crack reaches a critical size, i.e., before the leak rate moves very quickly to the final leak state. Whilst the high leak rate may still occur, the plant would be in a safe condition. This can be seen as leak detection.

This probability of leak detection is almost certainly associated with the length of the defect that is itself related to the rate of leakage in the previous section. Thus, expert judgement was again used to introduce a factor, based on the leak rate, which would represent this probability of leak detection. Figure G.6 shows this plot as a function of leak rate.

From this plot it can be seen that the reduction factor for Category 1 (380 lpm [100 gpm]) is about five, rising to a factor of about fifty at Category 6 (1,900,000 lpm [500,000 gpm]).

G.5   Effect of In Service Inspection

An assessment of the effect of ISI was carried out for the surge line elbow weld, the defect distribution and density being those generated by RR-PRODIGAL, see section G.4.2. A Probability of Detection (POD) curve was defined by the following equation:

$$f_{POD} = \Phi\left(c_1 + c_2 \ln\left(\frac{a}{t}\right)\right) \qquad \text{where } c_1 = 1.526 \text{ and } c_2 = 0.533 \qquad (G.1)$$

This POD is shown in Figure G.7, and it can be seen that this sets the probability of detection at about 90 percent for defects 70 percent of the way through the wall thickness. This was felt to be representative of inspections carried out to date, but for future inspections that conform to modern standards, this POD could be much better.

The results are shown in the table below and in Figure G.8 for various ISI intervals.

Table G.5 Reduction Factors Due to ISI

ISI case	Cumulative Probability of Failure at 60 years	Factor for General Use
No ISI	1.3×10^{-4}	1
0 years (PSI)	4.2×10^{-5}	3
10 years	3.8×10^{-5}	3.4
10, 20 years	1.3×10^{-5}	10
10, 20, 30 years	6.5×10^{-6}	20
10, 20, 30, 40 years	4.8×10^{-6}	27
10, 20, 30, 40, 50 years	4.7×10^{-6}	28

These results suggest that even with this quite low inspection capability, and for a weld with a high failure probability, reductions of a decade can be achieved with two or three inspections during the life of the plant. It also indicates that going beyond three inspections gives little extra return.

An interesting conclusion from this figure would be that if a fourth inspection is carried out at the end of a forty year period, then, provided this inspection was clear, there would be little gain from an inspection at fifty years for a total life of sixty years! However, at this stage such a conclusion can only be taken as tentative and would require more investigation.

G.6 References

G.1 NUREG/CR-5505 PNNL-11898 'RR-PRODIGAL – A Model for Estimating the Probability of Defects in Reactor Pressure Vessel Welds.

G.2 NURBIM (Nuclear Risk-Based Inspection Methodology) WP4. Published by the European Commission under the EURATOM programme.

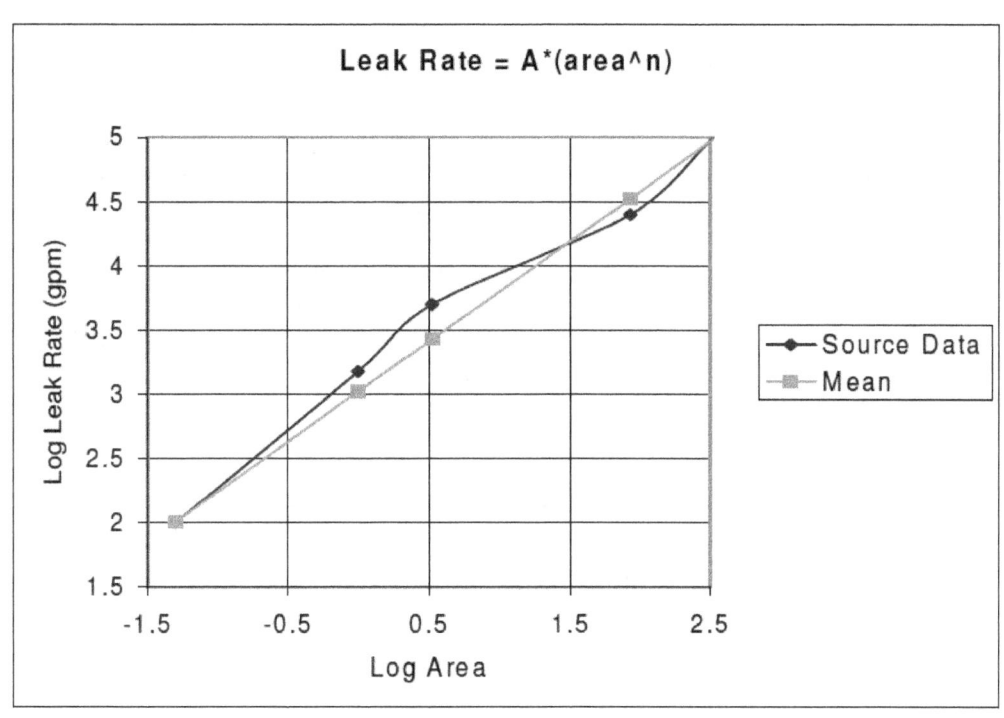

Figure G.1 Leak Rate as a Function of Leakage Area (Data Supplied by USNRC)

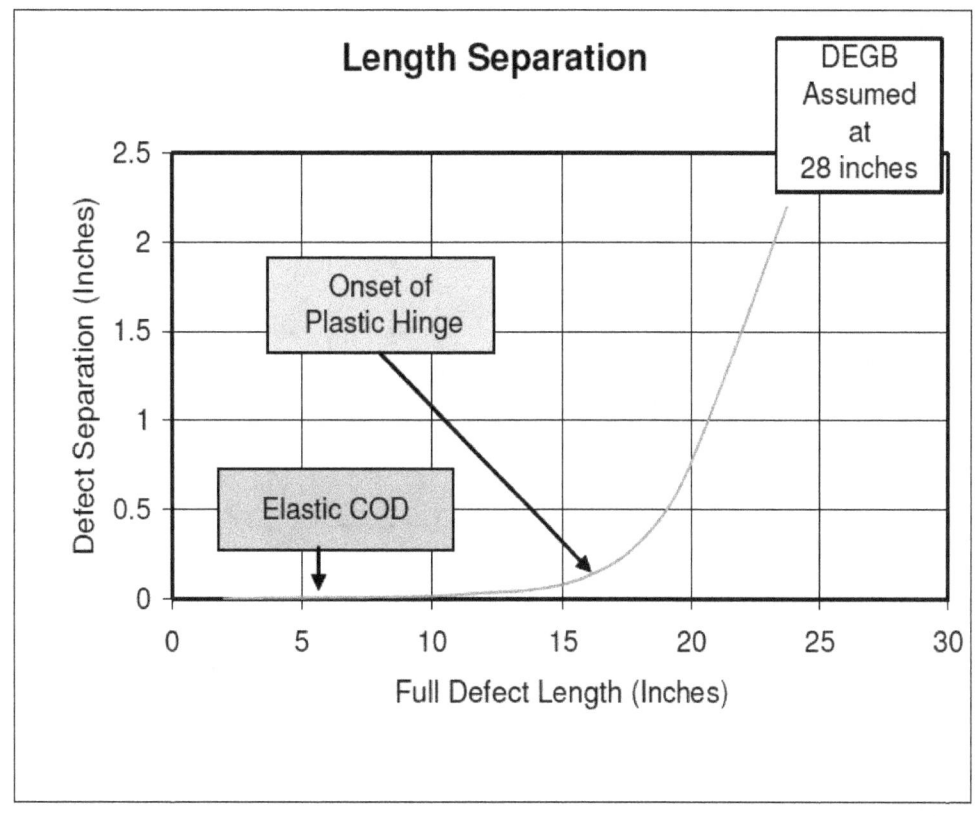

Figure G.2 Estimated Defect Separation (COD) Based on Expert Judgment
as a Function of Defect Length Assuming Plastic Deformation

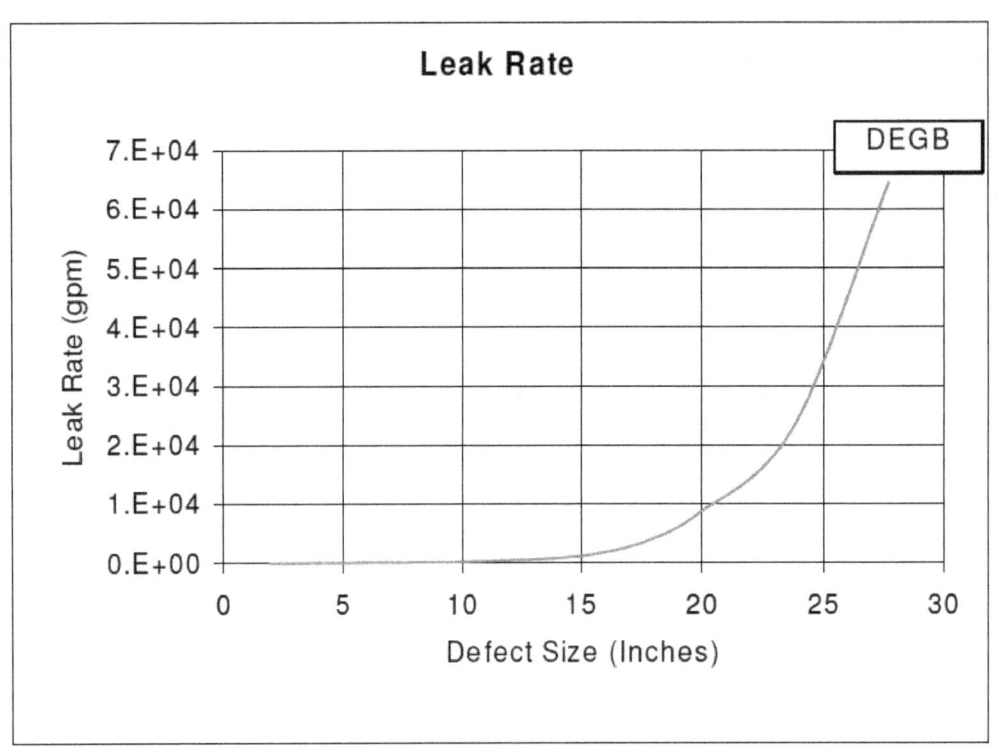

Figure G.3 Estimated Leak Rate Versus Defect Length Based on Expert Judgment

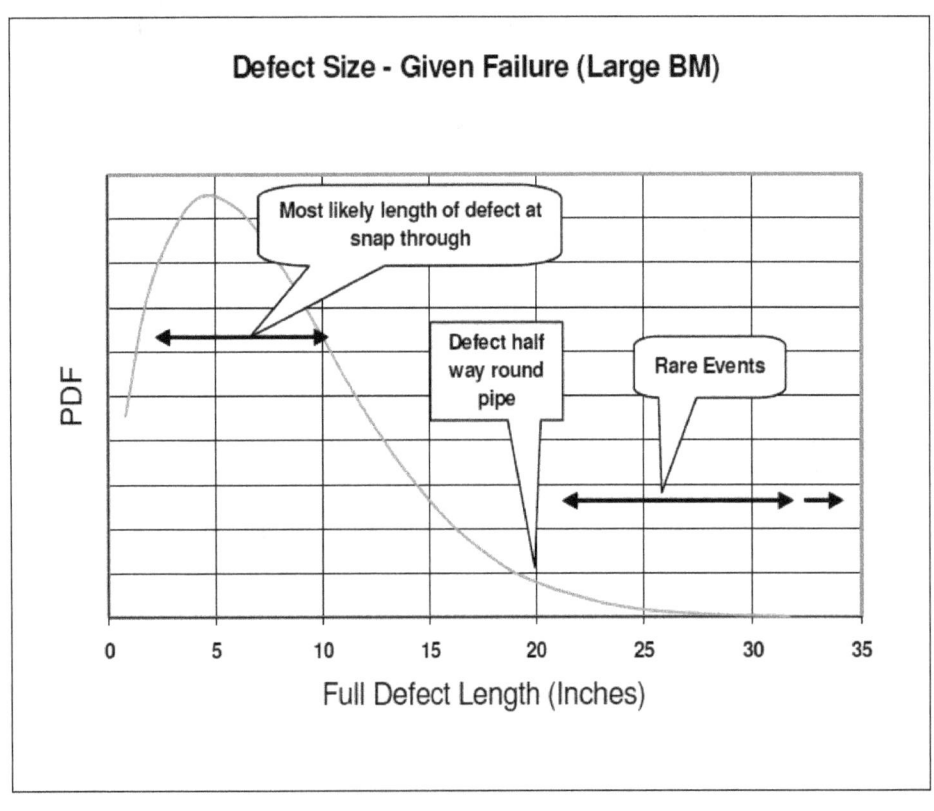

Figure G.4 Probability of the Existence of a Defect of a Certain Length for Surge Line Base Case

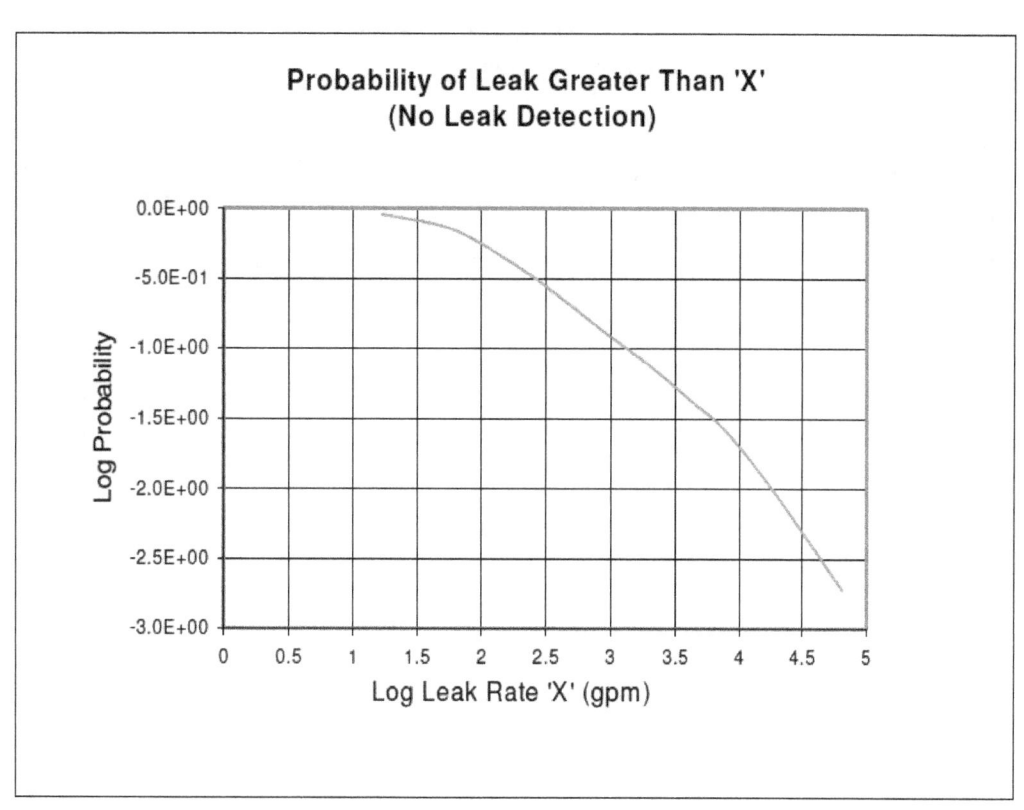

Figure G.5 Conditional Probability of a Leak of a Given Size

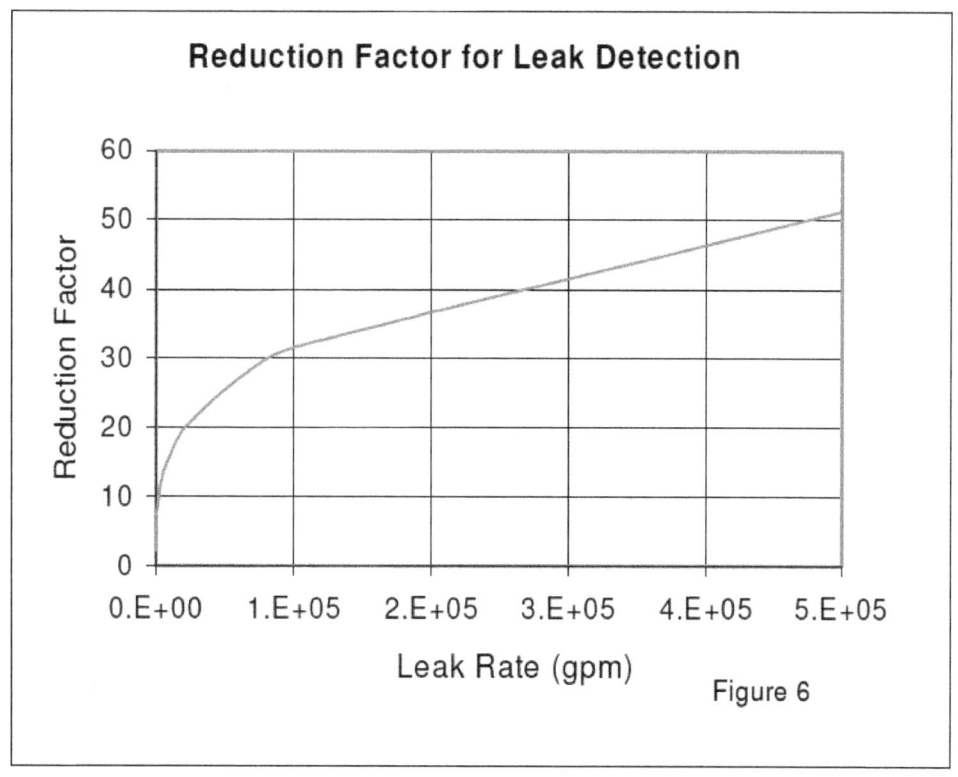

Figure G.6 Reduction Factors for Leak Detection Based on Expert Judgment

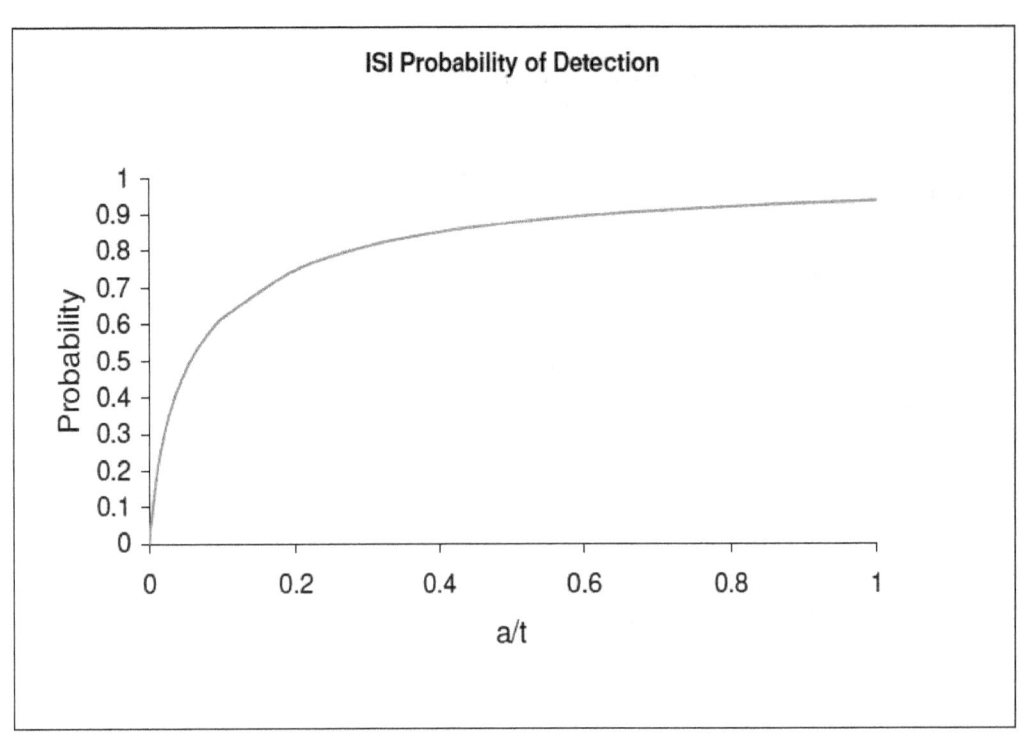

Figure G.7 Probability of Detection Curve

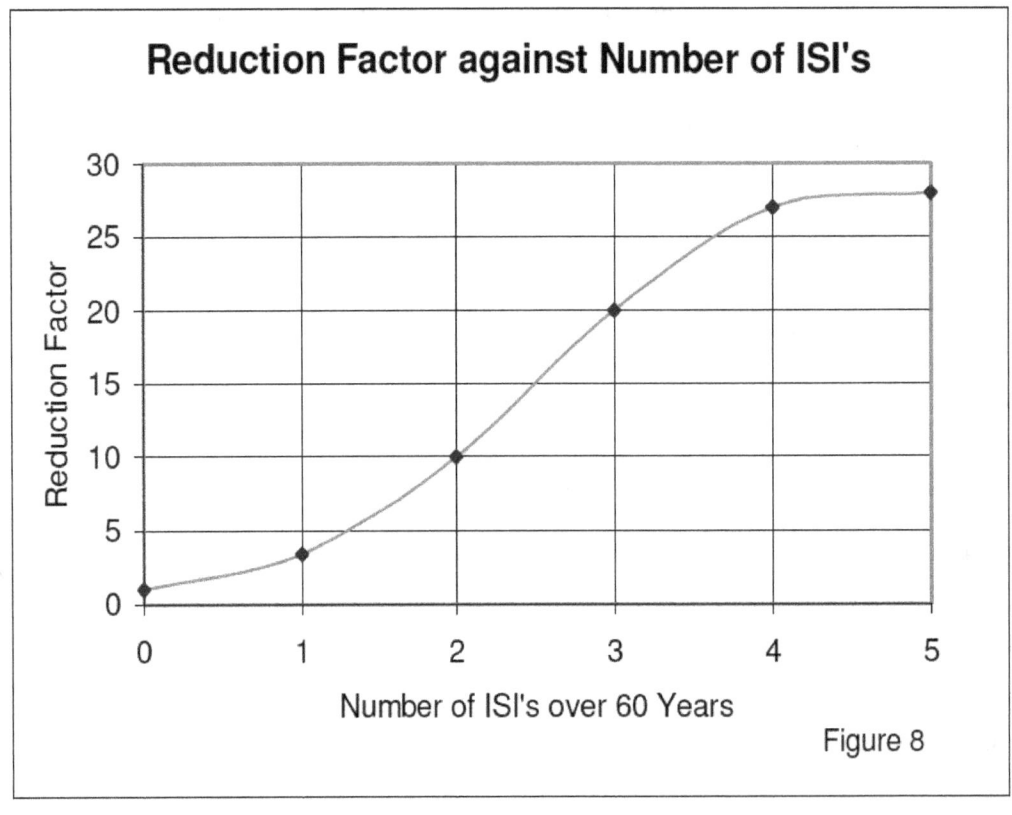

Figure G.8 Reduction Factors Due to ISI as a Function of the Number of Inspections

ATTACHMENT 1 TO APPENDIX G (ATTACHEMENT G.1) FROM NOTE 'SUMMARY OF BASE CASES STRESSES' APRIL 2003

Summary of Stress Cycles for Surge Line Elbow
(No seismic stresses)

Load Pair	Amplitude (ksi)	Number/40 years
HYDRO-EXTREME	190.17	6
9B-HYDRO	149.86	4
8A-UPSET 4	140.42	14
9B-UPSET4	139.43	10
8B-UPSET4	105.89	14
9A-UPSET4	105.13	2
9A-LEAK	103.86	12
8F-18	63.40	68
9C-11	63.38	68
9F-LEAK	63.37	68
8C-LEAK	63.37	35
2A-8C	62.30	33
8G-18	52.38	22
8G-17	52.35	90
9D-11	52.35	22
2A-8D	51.20	72
8H-9G	51.18	400
8G-UPSET3	51.00	30
9D-12	50.96	50
8G-12	50.96	40
8G-16	50.93	90
8G-9H	50.92	128
2A-8E	40.10	90
8H-9H	40.09	100
9H-10A	40.09	272
9E-13	39.82	90
3A-10A	33.10	4120
6-10A	33.10	200
3B-10A	33.10	4120
7-10A	33.10	4580
2B-SLUG1	32.87	100
2B-SLUG2	32.87	500
5-10A	29.90	9400
4A-10A	29.90	17040
4B-10A	29.90	17040
2B-10A	20.60	14400
2A-10A	20.60	14805
10A-UPSET1	20.59	70
10A-UPSET5	20.59	30
10A-UPSET6	20.59	5
10A-UPSET2	20.59	95
1B-10A	20.59	1533
1B-10B	20.00	87710

Many of the high stress contributors in Tables 2 and 3 are from rapid excursions of the coolant temperature. The largest stress amplitude (half the peak-to-peak) is 1,310 MPa (190 ksi), so the stresses are large (but localised). These are the stresses at the peak stress location, which is not at weld. The spatial stress gradients (both along the surface and into the pipe wall) are required for a thorough analysis. The radial gradient (into the pipe wall) can be estimated by the following procedure:

1 Cyclic stresses associated with seismic loads were treated as 100 percent uniform stress.

1 Cyclic stresses greater than 310 MPa (45 ksi) were treated as having a uniform component of 310 MPa (45 ksi), and the remainder were assigned to the gradient category.

1 For those transients with more than 1,000 cycles over a 40-year life, it was assumed that 50% of the stress was uniform stress and 50% a through-wall gradient stress. In addition, for these transients, the uniform stress component was not permitted to exceed 70 MPa (10 ksi).

The gradient stress mentioned above is assumed to vary through the thickness as

$$\sigma(\xi) = \sigma_o \left(1 - 3\xi + \frac{3}{2}\xi^2 \right) \qquad (G.2)$$

In this equation, σ_o is the stress at the inner wall of the pipe, $\xi = x/h$, x is the distance into the pipe wall from the inner surface, and h is the wall thickness. The stresses and cycles are high enough that fatigue crack initiation is important, which has been considered in Reference 6. Reference 6 shows a probability of 0.981 of a leak in 40 years for this component. The LOCA probabilities will be less. The use of the gradient along the surface will reduce this.

As mentioned above, the more thorough results that include a better estimate of the radial gradient and also consider the spatial gradient along the surface are available, and could be used for the base case calculations.

APPENDIX H

DESCRIPTION OF NON-PIPING DATABASE

APPENDIX H

DESCRIPTION OF NON-PIPING DATABASE

H.1 Background

The non-piping database has been compiled with the intention that it will serve as one source of information supporting the development of estimates of LOCA frequencies attributable to non-pipe components. The data has been obtained from two primary sources. First, a search of LERs was made to identify those instances where failures[1] of non-pipe components of the primary reactor coolant pressure boundary were reported to the NRC. The second source of information is data that has been incidentally collected on non-pipe components during the development and maintenance of the pipe-based OECD and SLAP databases. LER events compose the majority of the events in the database (see Attachment A of this appendix for a description of the LER reporting requirements).

The database is accessible in two formats, *Table* and *Forms*. The *Table* named "Failure Data" lists the data in a spreadsheet type of format where each line of the table contains one data record and each column contains the various fields that make-up the records. In the *Forms* format, only one record of the "Failure Data" is displayed at a time, but in a manner that allows all of the fields to be view at the same time. Both formats display the same data, only the presentation is different. Also, sorting and filtering of the data can be done in both views.

H.2 Approach

A search of LERs was performed (see Attachment B for the specific search criteria) using the SCSS. This search returned 1,036 LERs. Each LER was reviewed and coded in the Non-Pipe database. The database structure is based on information generated during the elicitation meetings. In particular, the component, piece part, and degradation mechanism are all identified using the tables documented in the elicitation meeting notes. Other fields of the database were developed and defined as judged appropriate.

The initial screening of the 1,036 LERs to remove those that were judged to not be applicable reduced the total number to 213. As discussed in Attachment B, the data search simply looked for leak and crack events associated with primary coolant systems. This conservative search included LERs that identified *potential* and *possible* leak and crack events (e.g., a problem with ECCS such that the plant would not respond as designed to a loss of coolant accident. Screening out the potential failures resulted in a reduction to 213 records. A further 34 records were removed since they identified problems with pipes or seals. Then 37 records were added that had been collected previously during the development of the OECD and SLAP databases. This results in a current total of 216 records.

H.3 Description of Data

This process results in 216 data records that document crack (both partial and full) and leak events associated with non-pipe primary coolant system components. This dataset can be considered complete for U.S. NPP operation from 1990 through 2002, inclusive, in as much as the LER reporting requirements (Attachment A) can be relied upon to generate complete reporting. Additionally, the dataset does include a limited amount of data from outside this time frame and from non-U.S. plants. Nevertheless, the dataset can be considered to be internally consistent, that is, the various components, failures and degradation

---

[1] Failures are classified using four categories: partial through-wall cracks, through-wall cracks without a significant leak rate (typically indicated by a boric acid deposit), leaks, and joint failures (i.e., non-welded connection).

mechanisms are believed to be represented equally such that relative ratios (if not the absolute frequencies) can be assumed to be reasonably accurate. Several of the database records represent multiple cases of degradation or failure. Attachment C includes a sample of multiple event records, including a discussion on how to estimate flaw frequencies from the observed events as recorded in the database.

The figure below (*Forms* view of the database) identifies the various fields maintained by the database. For additional detail on those records based on LERs, the LER hyperlink can be clicked to retrieve the full LER (internet access is require for this).

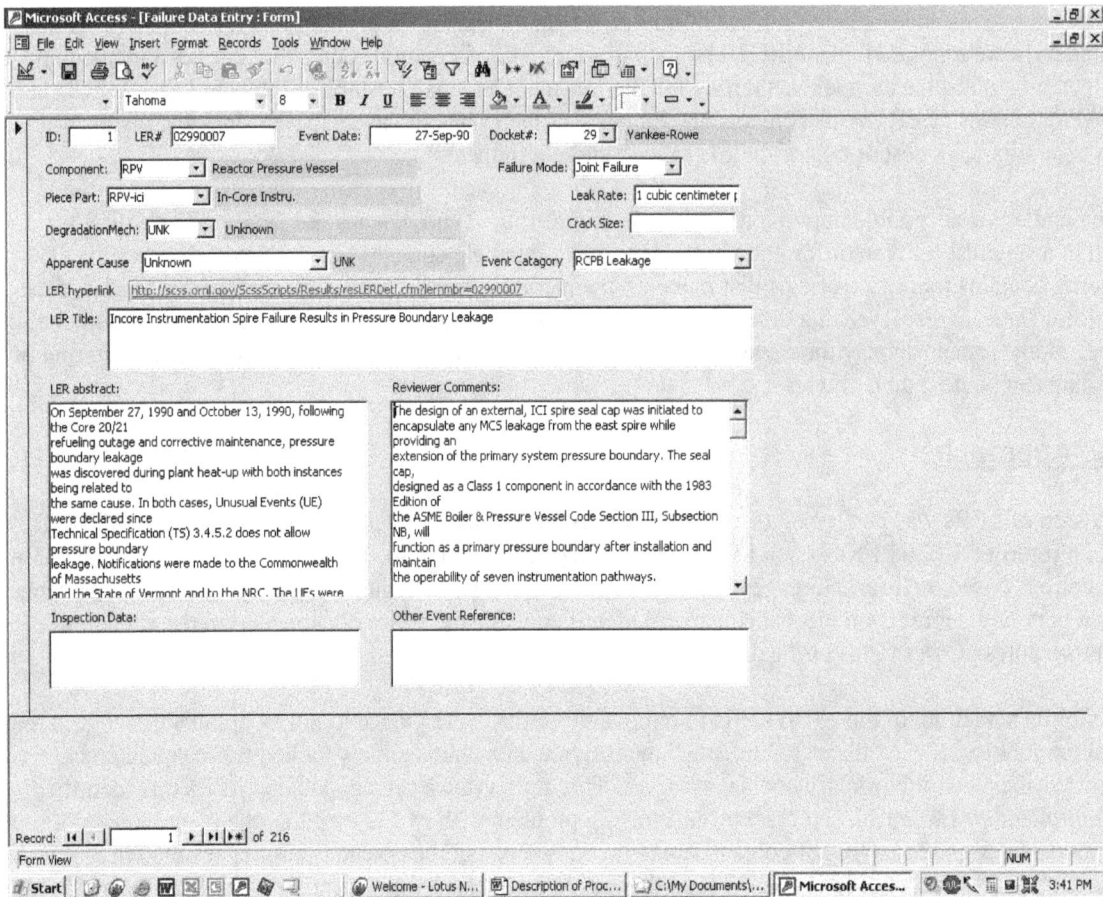

H.4 Limitations

As with many reliability databases, the completeness of the data is always an issue. While relative frequencies (e.g., percentage distribution of events by component or degradation mechanism) might be reasonably accurate, the accuracy of any absolute frequency (e.g., events per year) calculations will depend directly on how complete the data are. That is, have all events that have occurred been included in the database? In the current situation this question has two parts. First, have all relevant event been discovered? Second, of the discovered events, have they been reported (via LER)?

The completeness issue is probably more of an issue for the partial through-wall cracks than it is for the more severe failures. There are two causes for this concern. One is ambiguity in the interpretation of the LER reporting requirement (Attachment A), and the second and probably primary cause is simple lack of detection. While effort is made to make the LER reporting requirements as clear as possible, the "seriously degraded" aspect of 50.73(a)(2)(ii)(A) is difficult to quantify. How far does a crack have to extend to seriously degrade the primary pressure boundary? It is possible that some cracks are being detected and repaired (which might be considered normal plant maintenance rather than corrective action), without being reported as a LER. These events have not been captured in this search. However, detection likelihood is probably a bigger reason for coverage deficiencies of part-through wall flaws. A leak (or a non-leaking through-wall crack) is simply more likely to be detected. This issue is clearly illustrated by events in the data in which a detected leak prompted the plant to do a thorough inspection that found partial through-wall cracks. If the leak had not occurred and motivated the inspection, the partial through-wall cracks would not have been found.

H.5 Selected Non-Pipe Database Summary Tables

The following tables list some summary data from the non-pipe database.

Table H.1 Non-Pipe Event Counts by Component and Degradation Mechanism (for Each Plant Type)

Plant Type	Component		MA	FDR	SCC	LC	MF	TF	FS	UNK
					Degradation Mechanism (see legend)					
BWR	RPV	10		1	9					
		100%	0%	10%	90%	0%	0%	0%	0%	0%
	Valve	1					1			
		100%	0%	0%	0%	0%	100%	0%	0%	0%
	Pump	2		2						
		100%	0%	100%	0%	0%	0%	0%	0%	0%
	Totals	13	0	3	9	0	1	0	0	0
	Adjusted*	17	0.5	3.5	9.5	0.5	1.5	0.5	0.5	0.5
		100%	3%	21%	56%	3%	9%	3%	3%	3%
			MA	FDR	SCC	LC	MF	TF	FS	UNK
PWR	Pzr	28	1	3	23	1				
		100%	4%	11%	82%	4%	0%	0%	0%	0%
	RPV	42		5	27	5				5
		100%	0%	12%	64%	12%	0%	0%	0%	12%
	Valve	3	1		1	1				
		100%	33%	0%	33%	33%	0%	0%	0%	0%
	SG	124	2	29	85				3	5
		100%	2%	23%	69%	0%	0%	0%	2%	4%
	Pump	2		2						
		100%	0%	100%	0%	0%	0%	0%	0%	0%
	Instr nozzles	4			4					
		100%	0%	0%	100%	0%	0%	0%	0%	0%
	Totals	203	4	39	140	7	0	0	3	10
	Adjusted*	207	4.5	39.5	140.5	7.5	0.5	0.5	3.5	10.5
		100%	2%	19%	68%	4%	0%	0%	2%	5%

DM	DM Description
MA	Material Aging
FDR	Fabrication Defect and Repair
SCC	Stress Corrosion Cracking
LC	Local Corrosion
MF	Mechanical Fatigue
TF	Thermal Fatigue
FS	Flow Sensitive (includes FAC and E/C)
UNK	Unknown

* A half failure (0.5) was added to all degradation mechanism (DM) totals to force the representation of all DMs.

Table H.2 Summary of Non-Pipe Database by Plant Type and Piece Part (see Table H.3 for Legend of Piece Part Acronyms)

Plant Type	Component	Piece Part	No. Records	78	84	86	87	88	89	90	91	92	93	94	95	96	97	98	99	00	01	02	03
BWR	Pipe	Pipe-w	1								1												
BWR	RecP	RCP-bdy	1												1								
BWR	RecP	RecP-hx	1									1											
BWR	RPV	RPV-crc	4								1			1							1	1	
BWR	RPV	RPV-crd	4											1				2			1		
BWR	RPV	RPV-hbt	1								1												
BWR	RPV	RPV-noz	1																	1			
PWR	LIV	FLG-fbs	1						1														
PWR	LIV	LIV-bon	2																	1	1		
PWR	Pipe	Pipe-w	4												1				1	1	1		
PWR	Pzr	Pzr-bbs	4							1	1			1				1					
PWR	Pzr	Pzr-hsl	8				2		1	1				1				1		2		1	1
PWR	Pzr	Pzr-noz	16			1			1	1		2	3	1	1		1	2	1	1		1	
PWR	RCP	RCP-noz	2															2					
PWR	RPV	RPV-crc	13							1			1					2			7	2	
PWR	RPV	RPV-crd	13		1	1		1		1				1		1		1	1		3	1	1
PWR	RPV	RPV-hdt	3				1						1					1					
PWR	RPV	RPV-ici	2							1				1									
PWR	RPV	RPV-noz	2																				2
PWR	RPV	RPV-pen	9				1				1	1						1	1	1	1	1	1
PWR	SG	Pipe-w	1																1				
PWR	SG	SG-mwb	1					1															
PWR	SG	SG-noz	1																		1		
PWR	SG	SG-tub	3	1								1		1									
		Totals:	216	1	2	2	4	3	4	15	18	14	15	18	16	12	12	21	11	12	20	11	5

Table H.3 Piece Part Legend

PP-ID	Piece Part
FLG-fbs	Flange Bolts
LIV-bbs	Bonnet Bolts
LIV-bdy	Valve Body
LIV-bon	Bonnet
MSIV-bbs	Bonnet Bolts
MSIV-bdy	Valve Body
MSIV-bon	Bonnet
Pipe-w	Weld
PIV-bbs	Bonnet Bolts
PIV-bdy	Valve Body
PIV-bon	Bonnet
Pzr-bbs	Pzr valve bonnet bolts
Pzr-brv	Bolted Relief Valves
Pzr-hsl	Heater Sleeves
Pzr-mwb	Manway Bolts
Pzr-mwy	Manway
Pzr-noz	Pzr Nozzles
Pzr-rvb	Relief Valve Bolts
Pzr-shl	Shell
RCP-bdy	Pump Body
RCP-fwh	Flywheel
RCP-noz	Pump Nozzles
RCP-sel	Pump Seals
RecP-bbs	Bonnet Bolts
RecP-bdy	Pump Body
RecP-hx	Pump cooler
RecP-noz	Pump Nozzles
RecP-sel	Pump Seals
RPV-crc	CRDM connections
RPV-crd	CRDM
RPV-hbt	Head Bolts
RPV-hdb	Head (bottom)
RPV-hdt	Head (top)
RPV-ici	In-Core Instru.
RPV-noz	RPV Nozzles (incl. Instr.)
RPV-pen	Penetrations
SG-mwb	Manway Bolts
SG-mwy	Manway
SG-noz	SG Nozzles
SG-shl	Shell
SG-tbs	Tube Sheet
SG-tub	Tube

Table H.4 Summary of Non-Pipe Database by Plant Type, Piece Part and Failure Mode

Plant Type	Piece Part	Failure Mode[a]	No. Records	Calendar Year																			
				78	84	86	87	88	89	90	91	92	93	94	95	96	97	98	99	00	01	02	03
BWR	RCP-bdy	Crack-Part	1												1								
BWR	RPV-crc	Crack-Part	1								1												
BWR	RPV-crd	Crack-Part	1															1					
BWR	RPV-noz	Crack-Part	1																	1			
BWR		Subtotal:	4								1				1			1		1			
BWR	RPV-crc	Joint Failure	1											1									
BWR	RPV-hbt	Joint Failure	1								1												
BWR		Subtotal:	2								1			1									
BWR	Pipe-w	Leak	1								1												
BWR	RecP-hx	Leak	1									1											
BWR	RPV-crc	Leak	2																		1	1	
BWR	RPV-crd	Leak	3											1				1			1		
BWR		Subtotal:	7								1	1		1				1			2	1	
BWR		BWR Totals:	13								3	1		2	1			2		1	2	1	
PWR	RPV-crc	Crack-Part	1																		1		
PWR	RPV-hdt	Crack-Part	3				1						1									1	
PWR	RPV-pen	Crack-Part	2									1											1
PWR	SG-noz	Crack-Part	1									1											
PWR	SG-tub	Crack-Part	99							7	13	5	9	9	12	9	10	8	6	4	4	3	
PWR		Subtotal:	106				1			7	13	7	10	9	12	9	10	8	6	4	5	4	1
PWR	LIV-bon	Joint Failure	2											1						1			
PWR	Pzr-bbs	Joint Failure	2													1		1					
PWR	RPV-ici	Joint Failure	2							1				1									
PWR		Subtotal:	6							1				2		1		1		1			
PWR	Pipe-w	Crack-Full	4																2	1	1		
PWR	Pzr-hsl	Crack-Full	4											1	1					2			
PWR	Pzr-noz	Crack-Full	8									2		1	1			1	1		1	1	
PWR	RCP-noz	Crack-Full	1															1					
PWR	RPV-crc	Crack-Full	4																		4		
PWR	RPV-crd	Crack-Full	2											1							1		
PWR	RPV-pen	Crack-Full	3															1			1	1	

Plant Type	Piece Part	Failure Mode[a]	No. Records	Calendar Year																			
				78	84	86	87	88	89	90	91	92	93	94	95	96	97	98	99	00	01	02	03
PWR	SG-noz	Crack-Full	1																		1		
PWR	SG-tub	Crack-Full	28									3		3	3			3	3	3	7	3	
		Subtotal:	29									3		3	3			3	3	3	8	3	
PWR	FLG-fbs	Leak	1						1														
PWR	Pipe-w	Leak	1																	1			
PWR	Pzr-bbs	Leak	2							1	1												
PWR	Pzr-hsl	Leak	4				2		1														1
PWR	Pzr-noz	Leak	8			1			1	1			3				1	1					
PWR	RCP-noz	Leak	1															1					
PWR	RPV-crc	Leak	8							1			1					2			2	2	
PWR	RPV-crd	Leak	11		1	1	1	1		1						1		1	1	1	1		1
PWR	RPV-noz	Leak	2																				2
PWR	RPV-pen	Leak	4								1					1		1		1			
PWR	SG-tub	Leak	18	1	1			1	1	3		3	1	1		1	1	1	1		1	1	
PWR	SG-noz	Leak	1											1									
		Subtotal:	61	1	2	2	3	2	4	7	2	3	5	2		3	2	7	2	3	4	3	4
PWR	SG-mwb	NA	1	1																			
		PWR Totals:	203																				

a. Crack-Part – Partial through-wall crack
Crack-Full – Complete through-wall crack, but no active leak, typically indicated by boric acid deposit
Leak – Measurable leak
Joint Failure – Failure of a bolted connection

ATTACHMENT A TO APPENDIX H – LER REPORTING REQUIREMENTS

The database relies upon LERs submitted by plants under the requirement of 10 CFR 50.73. Of the LERs reviewed for this effort, the two most commonly cited reporting requirements (each LER must reference the requirement that necessitates the LER.) are 50.73(a)(2)(i)(B) and 50.73(a)(2)(ii)(A). These are described below.

50.73(a)(2)(i)(B) – Any operation or condition which was prohibited by the plant's Technical Specifications. Westinghouse Standard Tech Specs (NUREG-1431, Vol. 1, Rev. 2, June 2001, Section 3.4.13) related to RCS leakage are as follows.
RCS operational leakage shall be limited to:
   a. No pressure boundary leakage
   b. 1 gpm (3.8 lpm) unidentified leakage
   c. 10 gpm (38 lpm) identified leakage
   d. 1 gpm (3.8 lpm) total primary to secondary leakage through all steam generators (SGs), and
   e. 500 gallons (1,900 liters) per day primary to secondary leakage through any one SG
Pressure Boundary Leakage is defined as leakage through a non-isolable fault in an RCS component body, pipe wall, or vessel wall (except SG leakage). Leakage past seals and gaskets is not considered pressure boundary leakage.

50.73(a)(2)(ii) – Any event or condition that resulted in: **(A)** The condition of the nuclear power plant, including its principal safety barriers, being seriously degraded; or **(B)** The nuclear power plant being in an unanalyzed condition that significantly degraded plant safety.

NUREG-1022, Rev. 2 clarifies statement (A) as:
This criterion applies to material (e.g., metallurgical or chemical) problems that cause abnormal degradation of or stress upon the principal safety barriers (i.e., the fuel cladding, RCS pressure boundary, or the containment). Abnormal degradation of a barrier may be indicated by the necessity of taking corrective action to restore the barrier's capability . . .

PWR tech specs also contain reporting guidance (via LERs) associated with the plants SG tube surveillance program. Typically, this reporting requirement is triggered when an inspection reveals that greater than 1% of the tubes in a SG are found to be defective (i.e., greater than 40% thru-wall crack).

ATTACHMENT B TO APPENDIX H – SEQUENCE CODING AND SEARCH SYSTEM

The SCSS is an NRC-sponsored database maintained by Oak Ridge National Laboratory (ORNL). It is a web-accessible database of LERs that can execute searches using a variety of criteria. It can be accessed at:

http://scss.ornl.gov/

The following search criteria were used to generate the LER portions of the non-pipe database.

LER SYSTEM EVENT SEARCH CRITERIA

Primary System(s) =SAB, SAF, SAE, SAA, SAD, SAI, SAH
Interfacing System(s) =Any
Include Trains/Channels =Yes
Include Components =Yes
Happening(s) =Any
Event Cause(s) =Any
Event Effect(s) =BH, BF, BE, BI, DE, BN, BL, BK, BP, BX, BC, BB, BA, BD
Event Timing(s) =Any
Detection Methods(s) =Any
Nuclear Plant =Any
Beginning Event Date =01/01/1990
Ending Event Date =1/1/2003
Maximum LERs =2000

This search returned 1,036 LERs. Basically, this search criteria looks for any leaking or cracking event associated with any primary coolant related system. The above search criteria rely upon the coding effort performed by the staff at ORNL as part of the SCSS program. In that effort, each LER is reviewed and characterized for possible relevance to each related system. This characterization includes both actual and possible system failures. Therefore, these search criteria returned both pipe and non-pipe failures, as well as many "non-failure" events. Each of the returned 1,036 LERs was reviewed and approximately 80% (823 LERs) were judged to be non-failures and coded as not-applicable (NA). Most of these NA events were of the type where an engineering review or some other analysis was performed by the plant, and it was found that a pipe was inadequately (compared to the design requirements) constrained such that if an earthquake were to occur, there was an increased chance that the pipe might fail. Another common "non-failure" example is of a problem unrelated to the integrity of the primary coolant system, which would have adversely affected the ability of the plant to respond to a loss of coolant accident (i.e., a failure of the primary coolant system). These potential or possible issues were judged to not be actual failures and hence were deleted from the list. A further 34 LERs were removed from the set of LERs when they were found to document problems associated with pipe defects (or pipe-weld defects).

ATTACHMENT C TO APPENDIX H - ESTIMATION OF FLAW FREQUENCY

In general, a point estimate of the frequency of pipe failure (where 'failure' includes both small and large leaks and through-wall cracks, but excludes partial-through wall cracks), λ, is given by the following expression:

$$\lambda = \frac{n_F}{NT} \tag{H.C.1}$$

Where:

n_F = the number of failure events including both small and large leaks in the operational experience data;

T = the total time over which failure events were collected;

N = the number of components that provided the observed pipe failures.

A point estimate of the total frequency of flaws (cracks and leaks), ϕ, is given by the following expression:

$$\phi = \frac{n_C}{N \cdot T \cdot f \cdot P_{FD}} + \frac{n_F}{NT} = \frac{n_C}{N \cdot T \cdot f \cdot P_{FD}} + \lambda \tag{H.C.2}$$

Where:

n_C = the number of crack or flaw events

f = the fraction of welds inspected for cracks or flaws

P_{FD} = the probability that an inspected weld will find an existing flaw

Nearly all through-wall leaks are found from independent observations such as routine leak inspections and not from NDE inspections. However, part-through cracks are only typically found by NDE and thus the number is a function of the number of inspection locations. In Equation H.C.2 we account for the observed cracks in the data base and the fact that only a fraction (f) of the welds in the database are inspected according to ISI programs looking for cracks. The number of flaws actually discovered in ISI is subject to a finite NDE reliability, which is characterized by the factor P_{FD}.

If we now take the ratio of ϕ to λ, we get an expression for the factor by which to multiply the pipe failure rate to obtain the flaw (non-through wall crack) rate:

$$R_{C/F} = \frac{\phi}{\lambda} = \frac{n_C}{n_F \cdot f \cdot P_{FD}} + 1 \tag{H.C.3}$$

Where:

$R_{C/F}$ = Number of non-through wall cracks per leak event:

One approach to assess the $R_{C/F}$ ratio is to evaluate those records where both cracks and leaks were found during a single inspection of a component of interest. Ideally, the best data would be found in those instances where the component was 100% inspected. Without complete inspection, some assumptions about the inspection coverage, f, are required to assess this ratio. An example of this approach and the effect of the inspection coverage and POD is provided by analyzing the database for CRDM nozzle failures. For the component type 'CRDM Nozzles' in B&W PWR plants the database includes 6 LERs (= 6 database records) as identified in Table H.C.1. A detailed review of each of these LERs revealed multiple failures and degradation. Equation (C.3) together with an assumption about the inspection scope (f) makes it possible to estimate $R_{C/F}$.

Table H.C.1 B&W CRDM Nozzle Failures in 'Non-Pipe' Database

Plant	Date	LER Number	No. Components Leaking	No. Components Cracked	Population	Comment
Oconee-3	2/18/2001	2001-001	9	N/A[2]	69	Expanded inspection of an additional 9 nozzles. No recordable flaws.
Oconee-3	5/2/2003	2003-001	2	N/A	69	RVH replaced
Crystal River-3	10/1/2001	2001-004	1	N/A	69	5 flaws found in CRDM Nozzle #32. Expanded inspection of 8 nozzles found no flaws.
Three Mile Island-1	10/12/2001	2001-002	5	7	69	Inspection scope included 12 nozzles
ANO-1	3/24/2001	2001-002	1	N/A	69	Visual inspection only of remaining nozzles.
ANO-1	10/7/2002	2002-003	1	6	69	NDE of all nozzles
		Totals:	19	13		

Estimates of $R_{C/F}$ for the data set in Table H.C.1 are presented in Table H.C.2 for different assumptions about the POD and fraction of welds inspected. This analysis also assumes that the inspection criteria and the cracking characteristics of the events listed in Table H.C.1 are representative of the entire population. The fraction of welds inspected is a function of the ISI program requirements. In Table H.C.2, a LB for f is calculated using insights from piping reliability studies. This low f estimate results in the high $R_{C/F}$ estimate presented in the table.

The current ASME Section XI requirements are to inspect 25% of the Class 1 pipe welds and 7.5% of the Class 2 pipe welds. The current inspection practice for most if not all plants calls for the same welds to be inspected each inspection interval as opposed to randomly selecting a different set of welds for each interval. When cracks or significant flaws are found, the ASME code requires that an expanded search be made; however, the frequency of flaws and failures is so rare that this requirement adds very few additional inspections to the total population of inspected welds. Using data from an operating 4-loop Westinghouse PWR unit on the number of Class 1 and Class 2 welds of 1,605 and 1,800, the following estimate of the parameter f is obtained for Westinghouse PWR plants:

---

[2] N/A in Table H.C.1 means that a full-scope NDE was not pursued.

$$f = \frac{1,605(0.25) + 1,800(0.075)}{(1,605 + 1,800)} = 0.157 \qquad \text{(H.C.4)}$$

We assume this estimate of f to be representative of the non-pipe components. With additional assumptions about the reliability of the NDE we get the results as indicated in Table H.C.2.

Table H.C.2 Estimates of $R_{C/F}$ for B&W CRDM Nozzles

Parameter	Data Source	PWSCC		
		High Est.	*Median Est.*	*Low Est.*
Number of cracks	Table H.C.1	NA	13	NA
Number of leaks	Table H.C.1	NA	19	NA
Fraction of components inspected, f	Equation (H.C.4) – LB	1.0	0.5	0.157
$R_{C/F}$ with $P_{FD} = 0.5$	--	9.72	3.74	2.37
$R_{C/F}$ with $P_{FD} = 0.75$	--	6.81	2.82	1.91
$R_{C/F}$ with $P_{FD} = 0.9$	--	5.84	2.52	1.76

Hence, the relative number of flaws and leaks observed does not predict the relative frequency of flaws and leaks at a given weld. The estimate for the ratio of cracks to leaks obtained in Table H.C.2 reflects the degree to which components are exposed to PWSCC and inspection coverage.

APPENDIX I

REACTOR VESSEL LOCA PROBABILITY
BASE CASE ANALYSES
(BWR VESSELS AND PWR TOP HEAD NOZZLES)

APPENDIX I

REACTOR VESSEL LOCA PROBABILITY BASE CASE ANALYSES (BWR VESSELS AND PWR TOP HEAD NOZZLES)

I.1 Introduction

The LOCA expert panel elicitation team charter includes estimating the contribution to LOCA frequency from reactor vessels and other non-piping components. Extensive analyses were performed by members of the elicitation panel to develop LOCA frequencies for five piping "base cases" that were formulated by the panel in early meetings (documented as Appendices D, E, F and G to this NUREG). The piping base cases include failures on the piping side of vessel nozzles, including safe-ends. However, they do not include small diameter, partial penetration welded nozzles such as CRDM penetrations and other small nozzles, such as instrument nozzles, that aren't connected to piping systems. In addition, the piping base cases do not include consideration of a leak from or rupture of other regions of the reactor vessel, such as the irradiated reactor vessel beltline or the low alloy steel portions of large vessel nozzles. LOCA frequency estimates for these cases are presented in this appendix, based on prior PFM analyses performed for PWR top head nozzles [I.1, I.2], the BWR Reactor Vessel Beltline Region [I.3, I.4], and BWR reactor vessel feedwater nozzles [I.5]. These estimates are used to construct a complete set of LOCA frequency tables for BWR and PWR reactor vessels, for all LOCA categories defined in the elicitation, a comparison of them to the aforementioned piping base cases is also presented.

I.2 PWR Reactor Vessel Top Head Nozzles

Extensive PFM analyses have been conducted over the past several years to estimate the probability of leakage and rupture associated with the PWR CRDM penetration PWSCC problem [I.1, I.2]. The analysis model incorporates the following major elements:
- computation of applied stress intensity factors for circumferential cracks in various nozzle geometries as a function of crack length,
- determination of critical circumferential flaw sizes for nozzle failure,
- an empirical (Weibull) analysis of the probability of nozzle cracking or leakage as a function of operating time and temperature of the RPV head,
- statistical analysis of PWSCC crack growth rates in the PWR primary water environment as a function of applied stress intensity factor and service temperature, and
- modeling of the effects of inspections, including inspection type, frequency and effectiveness.

The model has been benchmarked with respect to field experience, considering the occurrence of cracking and leakage and of circumferential cracks of various sizes. Figures I.1 and I.2 illustrate the benchmarking. Figure I.1 presents a Weibull analysis of inspection results at thirty plants, of which 14 detected leakage or cracking (data points in the figure). The remaining plants that were inspected and found clean were treated as "suspended tests" according to standard Weibull analysis theory [I.2]. The data are plotted in terms of effective degradation years (EDYs) which are equivalent operating years at 600°F (315°C), using an activation energy (Arrhenius) model [I.1] to adjust for different head operating temperatures. For plants in which multiple cracked nozzles were detected in the inspections, the data were extrapolated back to the expected time of first cracking or leakage, using an assumed Weibull slope of 3. The straight line through the data represents a medium rank Weibull regression (also with a slope of 3) upon which the probability of leakage predictions in the model are based. Figure I.2 illustrates the benchmarking process used for the crack growth analysis algorithm in the model with respect to CRDM nozzles that exhibited circumferential cracks of various sizes. (Eleven (11) nozzles out of a total of 881 inspected nondestructively through the spring of 2003 exhibited circumferential cracking. No additional circumferential cracking has been detected in more recent inspections.) The figure shows that, when using original analysis parameters, the crack growth model under-predicted the probability of circumferential cracking somewhat, but after adjustment of selected analytical parameters, the PFM model was "benchmarked" so as to very accurately predict the field results, especially for the most important, larger crack sizes.

The benchmarked model was then used to evaluate the probabilities of nozzle failure and leakage in actual plants. A sample of the results is presented in Figures I.3 and I.4. Figure I.3 illustrates the probability of nozzle failure (ejection of a nozzle) for a head operating temperature of 580°F (304°C), the approximate average of U.S. PWRs. No inspections were assumed to be performed during the first 25 years of plant operation, resulting in the probability of nozzle failure constantly increasing with time during that period. The analysis then assumed that inspections begin after 25 years, at intervals and detection levels representative of current requirements [I.6]. It is seen from the figure that the current inspection regimen reduces the nozzle failure probability significantly.

Ejection of a 4 inch CRDM nozzle [2.75 inch (~70 mm) ID] due to a circumferential crack would yield a one-sided LOCA corresponding approximately to Category 2 LOCA [>1,500 gpm (5,700 lpm) but < 5,000 gpm (19,000 lpm)]. If periodic inspections are continued, with any nozzles in which leakage or cracking are detected repaired or the heads replaced (as is common practice), the nozzle ejection probability will be even lower in the future. Table I.1 below provides a summary of the average failure probabilities from Figure I.3, between 0 and 25 years, and from 25 to 40 years. The probability of failure for 40 to 60 years was not calculated, but was assumed to be the same as 25 to 40 years, on the basis that the current inspection regimen will be maintained, or the heads replaced. A Category 3 break was assumed to require multiple nozzle failures, the probability of which was computed via a binomial distribution for the typical number of nozzles in a head. As seen in Table I.1, the probabilities of simultaneous multiple nozzle failures is quite low.

Figure I.4 illustrates similar PFM results (based on the above Weibull model) for the probability of small amounts of leakage from a top head CRDM nozzle. The same inspection regimen was assumed as in the nozzle ejection analysis (no inspections from 0 to 25 years, inspections in accordance with current requirements thereafter). A small leak from a CRDM nozzle was assigned as a Category 0 break [less than 1 gpm (3.8 lpm)] in Table I.1, and the intermediate, Category 1 break size was obtained by logarithmic interpolation between Categories 0 and 2.

Table I.1 Summary of PWR CRDM Nozzle PFM Results

Break Category	Leak Rate >(gpm)	Average LOCA Probabilities During Operating Years:		
		0-25	25-40	40-60
0	1	2.00E-02	5.00E-03	5.00E-03
1	100	1.27E-03	2.75E-04	2.75E-04
2	1,500	2.50E-04	5.00E-05	5.00E-05
3	5,000	4.00E-08	2.00E-09	2.00E-09
4	25,000	-	-	-
5	100,000	-	-	-
6	500,000	-	-	-

I.3 BWR Reactor Vessels

Analyses have been previously submitted and approved [I.3, I.4] that establish reduced inspection requirements for BWR reactor vessels relative to ASME Section XI requirements. Specifically, BWRVIP-05 [I.3] justifies that only axially-oriented welds in the vessel beltline region need be examined on a ten year interval, versus the Section XI requirement to inspect all axial and circumferential welds on this interval. This relief was based on PFM calculations demonstrating, for the BWR fleet, that circumferential weld inspections contribute negligibly to reduction in the already small failure probability of a BWR vessel. The methodology used for the PFM analysis is a computer program (VIPER [I.7]) developed by Structural Integrity Associates for EPRI and the BWRVIP. To address this LOCA frequency contributor, the VIPER software was run for a typical BWR vessel, extending the analysis period from 40 to 60 years. A modification to the software (VIPER-NOZ) was also used to estimate leakage and failure probabilities for BWR Reactor Vessel feedwater nozzles. Feedwater nozzles were selected because they are subject to thermal fatigue cycling, which caused serious nozzle cracking in the 1970s [I.5]. Both analyses take credit for routine ISI programs that are conducted on these components on ten-year inspection intervals. The feedwater nozzle analysis also takes credit for nozzle modifications and thermal sleeve improvements that were installed in all U.S. BWRs to reduce the severity of the thermal fatigue cycling.

I.3.1 BWR Vessel Beltline Region

In the VIPER software, cracks are assumed to exist in BWR vessel welds due to two causes – original manufacturing defects and service-induced cracks which initiate in the stainless steel cladding. These cracks are assumed to grow as a function of operating time due to fatigue crack growth and SCC of the low alloy steel vessel material. Simultaneously, the vessel beltline region is assumed to embrittle due to irradiation. Monte Carlo simulations of these processes are employed in VIPER, which include fracture mechanics crack growth calculations due to fatigue and SCC, and a comparison of predicted crack sizes to the critical crack size due to normal operation as well as possible transient conditions. The governing transient condition was determined to be a LTOP event, since BWRs are not subject to PTS.

The effects of ISI are imposed at appropriate inspection intervals, assuming a POD curve for the inspections. Flaws that are detected during ISI are assumed to be repaired, and thus eliminated from the population, such that they can no longer grow to a leak or vessel failure.

The axial vessel beltline welds are divided into a series of segments, and each segment is analyzed separately to account for axial gradients of irradiation fluence in the welds, which peaks at the core centerline, and decays at elevations above and below that location. The failure frequencies from each segment are weighted by their respective weld volume, and summed to determine failure frequency for the entire vessel.

Modes of failure considered are:
1. Vessel fracture during normal operation ($K_I > K_{Ic}$)
2. Vessel fracture during an assumed LTOP event. The LTOP event considered is pressurization to 1,150 psi (7.93 MPa) at 88°F (31°C), which is assumed to occur at a frequency of 1E-3.
3. Predicted crack growth to 80% of wall thickness before failure modes 1 or 2 occurs (LBB)

The results for a typical BWR are given in the following table:

Table I.2 Summary of BWR RPV Beltline PFM Results

Break Category	Leak Rate >(gpm)	Average LOCA Probabilities During Operating Years:		
		0-25	25-40	40-60
1	100	1.00E-08	2.98E-08	4.57E-08
2	1,500	2.32E-09	4.31E-09	2.84E-08
3	5,000	1.21E-09	1.83E-09	2.30E-08
4	25,000	5.04E-10	5.79E-10	1.73E-08
5	100,000	2.38E-10	2.15E-10	1.36E-08
6	500,000	9.86E-11	6.79E-11	1.02E-08

These provide an estimate of the probability (per vessel year) of breaks of various sizes due to vessel beltline failures. To complete this table, it was assumed that a leak (LBB mode failure) corresponds to a crack of length = 60 inches (1525 mm) that breaks through and begins leaking as a through-wall crack of this length (since the wall thickness is approximately 6 inches (150 mm), and cracks in VIPER are assumed to have a ten to one aspect ratio). Dave Harris ran this case using the PRAISE code leakage rate prediction capability (See Appendix F for description), and computed a leak rate of 193 gpm (733 lpm) for an axial crack of this size in a BWR vessel. Thus, predicted LBB mode failures from the RPV beltline were treated as Category 1 breaks. Predicted vessel fractures, either during normal operation or due to LTOP events were treated as complete RPV ruptures, which were assumed to result in very large, Category 6 breaks. Intermediate break sizes were then determined by log-log interpolation between these two extremes. The sharp increase in large break probability between years 40 and 60 is attributable to the combined effect of two aging mechanisms – crack growth and RPV embrittlement.

I.3.2 BWR Feedwater Nozzles
RPV nozzles constitute another potential non-piping LOCA concern. The example used to address this concern here is BWR feedwater nozzles, which have in the past been subject to thermal fatigue cracking [I.5]. The thermal fatigue problem was caused by mixing of hot reactor water and relatively cold feedwater (see Figure I.5) during reactor startups, shutdowns and other periods of low power operation, when feedwater heating is generally unavailable. Cracking of various depths, up to 1.5 inches (38 mm), was detected in a number of BWRs in the 1970s (see Figure I.6). At that time, the standard feedwater nozzle design incorporated a loose-fitting thermal sleeve/sparger configuration, as shown in Figure I.5. Since then, all U.S. BWRs have installed some type of fix, employing either welded-in spargers or multiple-sleeve designs with shrink fits and piston rings to protect the nozzle from the effects of the cold feedwater. No subsequent cracking has been discovered since the improved thermal sleeves were installed.

In order to perform a base case analysis of this problem, a modification to the software (VIPER-NOZ) was developed to estimate leakage and failure probabilities for BWR Reactor Vessel feedwater nozzles. The substantive changes to the VIPER software in VIPER-NOZ were the addition of thermal fatigue crack initiation and growth algorithms specific to the feedwater nozzle thermal cycling phenomenon, and zeroing out the effects of irradiation embrittlement, since feedwater nozzles are far enough from the reactor core region that neutron fluence effects are small. The VIPER-NOZ software was run for conditions representative of the original nozzle/sparger designs, to confirm that cracking probabilities consistent with early field experience (Figure I.3) are predicted. The boundary conditions were then modified to represent improved nozzle/sparger designs, which reduce the effects of thermal fatigue on the nozzle. The analyses were conducted for a 60 year operating lifetime, and included the effects of periodic ISI, which are performed for these nozzles on ten-year intervals. The results are given in the following table:

Table I.3 Summary of BWR Feedwater Nozzle PFM Results

Break Category	Leak Rate >(gpm)	Average LOCA Probabilities During Operating Years:		
		0-25	25-40	40-60
0	1	<1.00E-06	1.47E-06	1.25E-06
1	100	<1.00E-06	<1.00E-06	<1.00E-06
2	1,500	<1.00E-06	<1.00E-06	<1.00E-06
3	5,000	<1.00E-06	<1.00E-06	<1.00E-06
4	25,000	<1.00E-06	<1.00E-06	<1.00E-06

The predicted leakage cases were treated as Category 0 breaks in this case, and since the nozzle is attached to a 12 inch diameter pipe, the maximum credible break size was assumed to correspond to single ended rupture of a 12 inch pipe, which corresponds to a Category 4 break. A total of 1 million simulations were run, and except for LBB type failures at 40 and 60 years, no other failures were predicted. Thus a failure frequency of less than 1E-6 is given for most entries in the above table.

I.4 Combined LOCA Frequencies due to Reactor Vessel Failures

Tables I.4 and I.5 summarize and combine the above RPV LOCA frequency results for BWRs and PWRs, respectively. For BWRs, three RPV LOCA contributors are addressed: RPV vessel beltline region, large nozzles (6 through 28 inch diameter), and small penetrations (partial penetration welded nozzles, 4 inch in diameter or less, such as CRDs). The individual LOCA probability contributions for each of these are provided in the top three sections of Table I.4, and they are summed in the bottom section of the table. These address all LOCA categories as well as the three time periods under consideration (0-25 yrs, 25-40 yrs, and 40-60 yrs). Note, in Table 4.4 in Section 4 of the main body of this report, only the results for the BWR beltline region and the large feedwater nozzles are presented. The results for the BWR CRDs and other small penetrations listed in Table I.4 of this appendix were not included in Table 4.4 since the estimates for the BWR CRDs and other penetrations are not based on analysis, but instead, were based on engineering judgment, i.e., BWR CRDs and other penetrations LOCA frequencies were simply assumed to be a factor of 10 less than the PWR CRDM LOCA frequencies, which were based on analysis.

The first contributor is the BWR shell region, the failure probabilities for which are dominated by the irradiated reactor vessel beltline region. The upper section of Table I.4 summarizes the results of this analysis from Table I.2, in terms of the probability of leaks of various sizes due to degradation or failure of the RPV beltline region. (Note that these were modified slightly relative to those in Table I.2 to eliminate the negative time factor for Category 5 and 6 LOCAs in the 25 – 40 year period.)

The second section of Table I.4 addresses large nozzle contributions to LOCA probability, which in BWRs are assumed to correspond to the 12 inch diameter Feedwater Nozzles that experienced thermal fatigue cracking in the 1980s [I.5]. LOCA probabilities due to this contributor are given for break Categories 1 through 4, taken from Table I.3 as this nozzle size could not lead to larger break sizes. For the Category 2 through 4 LOCAs for 0-25 year time frame, an assumption was made as to the size factor that each successive LOCA size greater than Category 1 was 5 times less likely to occur than the previous size LOCA. Then, for the 25-40 and 40-60 year time frames, the same time factor as determined for the Category 0 LOCAs in Table I.3 was assumed for the larger size LOCAs. Breaks of the other, larger diameter nozzles, such as recirculation outlet nozzles, are considered to be adequately encompassed by the vessel beltline case.

Finally, the third section of Table I.4 lists LOCA probabilities due to failures of CRDs and other small penetrations in the BWR vessel. These were estimated from the detailed analysis of PWR CRDM penetrations described above (Table I.1) but they assume that the BWR penetrations have about an order of magnitude lower LOCA probability than similar penetrations in a PWR. The order of magnitude reduction is deemed appropriate, because problems in small vessel penetrations in BWRs have occurred at a much lower frequency than the recent PWSCC experience in PWRs, upon which Table I.1 is based. The problems in BWR penetrations have also been attributed to a fairly well-understood phenomenon (IGSCC) and in most cases the nozzles of concern have been mitigated by design and materials changes.

Table I.5 provides a similar summary for PWR RPVs. In this case, LOCA probabilities are reported for only two categories of LOCA contributors, the shell region (RPV beltline) and small penetrations. Again, as was the case for BWRs, Table 4.5 in Section 4 only includes the results for PWR CRDMs. It does not include the results for the PWR beltline region as reported in Table I.5 of this appendix. As was the case for BWR CRDs and other penetrations, the PWR beltline results in Table I.5 are not based on analysis. Again, they are based on engineering judgment, i.e., the PWR beltline LOCA frequencies in Table I.5 of this appendix were simply assumed to be a factor of ten greater than the BWR beltline LOCA frequencies from Table I.4. It was judged that the large nozzles in a PWR RPV do not pose a significant LOCA risk because they are not subject to significant thermal cycles such as the BWR Feedwater nozzles, and except for the safe-ends (which are covered in the piping elicitation), they have not experienced any degradation mechanisms to date. The contributions for the two PWR RPV LOCA contributors are summed in the bottom section of Table I.5.

For the PWR beltline region, results from a prior analysis of a PWR vessel using a third version of the VIPER software (VIPER-PWR) were reviewed. Based on this review, it was estimated that the PWR RPV beltline region presents about an order of magnitude increase in large rupture probability relative to that of a BWR, because PWR beltlines are more highly irradiation embrittled, and because they are potentially subject to PTS transients. Thus, the BWR RPV beltline region LOCA frequency entries in Table I.4 were multiplied by a factor of ten and entered in the upper section of Table I.5.

PWR CRDM penetrations results were entered directly from the above PFM analysis results in Table I.1.

I.5 Summary and Comparison to Piping Base Cases

Figures I.7 and I.8 present plots of these RPV base cases, compared to the piping base cases from Appendices D, E, F, and G. For purposes of this comparison, a single set of piping base case LOCA frequencies were derived that are a composite of the results from the four appendices. Plots are presented for the 0-25 year (Figure I.7) and the 25-40 year (Figure I.8) periods. Since the RPV LOCA frequencies for the 40-60 year period are not significantly different than the 25-40 year results, a separate plot for that case is not included. It is seen from these figures that the RPV base cases are at the low end of the piping

LOCA probabilities for the large break Categories 5 and 6, but are at the high end for small, Category 1 and 2 breaks, due largely to the small penetration (CRDM) contributions discussed above. Note also that the small LOCA probability estimates are substantially lower in the outlying years (25-40 and 40-60) because of inspection programs implemented as a result of these issues. In general, small break LOCA frequency contributors (Categories 1 and 2) from PWR RPVs are seen to be greater than those for BWRs, due to the PWSCC concern in CRDM and other small penetrations. Large break LOCA contributors (Categories 5 and 6) are also estimated to be greater for PWR RPVs due to higher irradiation embrittlement and the potential for PTS transients.

Table I.4 LOCA Frequencies for BWR Reactor Pressure Vessel Base Case
BWR RPV Beltline

Break Cat.	Break Size gpm	Break Size NPS	Average LOCA Probabilities During Operating Years: 0-25 yrs	25-40 yrs	TimeFactor	40-60 yrs	TimeFactor
1	100	0.5	1.00E-08	2.98E-08	2.98	4.57E-08	4.57
2	1,500	1.5	2.32E-09	6.19E-09	2.67	2.84E-08	12.24
3	5,000	3.5	1.21E-09	3.12E-09	2.58	2.30E-08	19.01
4	25,000	7	5.04E-10	1.25E-09	2.47	1.73E-08	34.33
5	100,000	16	2.38E-10	5.65E-10	2.37	1.36E-08	57.14
6	500,000	30	9.86E-11	2.32E-10	2.35	1.02E-08	103.45

BWR FW Nozzles

Break Cat.	Break Size gpm	Break Size NPS	Average LOCA Probabilities during Operating Years: 0-25 yrs	25-40 yrs	TimeFactor	40-60 yrs	TimeFactor
1	100	0.5	1.00E-06	1.47E-06	1.47	1.25E-06	1.25
2	1,500	1.5	2.00E-07	2.94E-07	1.47	2.50E-07	1.25
3	5,000	3.5	4.00E-08	5.88E-08	1.47	5.00E-08	1.25
4	25,000	7	8.00E-09	1.18E-08	1.47	1.00E-08	1.25
5	100,000	16					
6	500,000	30					

BWR CRDs & Other Small Penetrations

Break Cat.	Break Size gpm	Break Size NPS	Average LOCA Probabilities during Operating Years: 0-25 yrs	25-40 yrs	Factor	40-60 yrs	Factor
0			2.00E-03	5.00E-04		5.00E-04	
1	100	0.5	1.27E-04	2.75E-05	0.22	2.75E-05	0.22
2	1,500	1.5	2.50E-05	5.00E-06	0.20	5.00E-06	0.20
3	5,000	3.5	4.00E-09	2.00E-10	0.05	2.00E-10	0.05
4	25,000	7					
5	100,000	16					
6	500,000	30					

BWR Vessel - Totals

Break Cat.	Break Size gpm	Break Size NPS	Average LOCA Probabilities during Operating Years: 0-25 yrs	25-40 yrs	TimeFactor	40-60 yrs	TimeFactor
1	100	0.5	1.28E-04	2.90E-05	0.23	2.88E-05	0.23
2	1,500	1.5	2.52E-05	5.30E-06	0.21	5.28E-06	0.21
3	5,000	3.5	4.52E-08	6.21E-08	1.37	7.32E-08	1.62
4	25,000	7	8.50E-09	1.30E-08	1.53	2.73E-08	3.21
5	100,000	16	2.38E-10	5.65E-10	2.37	1.36E-08	57.14
6	500,000	30	9.86E-11	2.32E-10	2.35	1.02E-08	103.45

Table I.5 LOCA Frequencies for PWR Reactor Pressure Vessel Base Case
PWR RPV Beltline

Break Cat.	Break Size gpm	NPS	Pete Riccardella Estimate 0-25 yrs	25-40 yrs	Factor	40-60 yrs	Factor
1	100	0.5	1.00E-07	2.98E-07	2.98	4.57E-07	4.57
2	1,500	1.5	2.32E-08	6.19E-08	2.67	2.84E-07	12.24
3	5,000	3	1.21E-08	3.12E-08	2.58	2.30E-07	19.01
4	25,000	7	5.04E-09	1.25E-08	2.47	1.73E-07	34.33
5	100,000	14	2.38E-09	5.65E-09	2.37	1.36E-07	57.14
6	500,000	30	9.86E-10	2.32E-09	2.35	1.02E-07	103.45

PWR CRDMs

Break Cat.	Break Size gpm	NPS	Pete Riccardella Estimate 0-25 yrs	25-40 yrs	Factor	40-60 yrs	Factor
0			2.00E-02	5.00E-03		5.00E-03	
1	100	0.5	1.27E-03	2.75E-04	0.22	2.75E-04	0.22
2	1,500	1.5	2.50E-04	5.00E-05	0.20	5.00E-05	0.20
3	5,000	3.5	4.00E-08	2.00E-09	0.05	2.00E-09	0.05
4	25,000	7					
5	100,000	16					
6	500,000	30					

PWR Vessel - Totals

Break Cat.	Break Size gpm	NPS	Pete Riccardella Estimate 0-25 yrs	25-40 yrs	Factor	40-60 yrs	Factor
1	100	0.5	1.27E-03	2.75E-04	0.22	2.76E-04	0.22
2	1,500	1.5	2.50E-04	5.01E-05	0.20	5.03E-05	0.20
3	5,000	3.5	5.21E-08	3.32E-08	0.64	2.32E-07	4.45
4	25,000	7	5.04E-09	1.25E-08	2.47	1.73E-07	34.33
5	100,000	16	2.38E-09	5.65E-09	2.37	1.36E-07	57.14
6	500,000	30	9.86E-10	2.32E-09	2.35	1.02E-07	103.45

I.6 References

I.1 Peter Riccardella, Nathaniel Cofie, Angah Miessi, Stan Tang, Bryan Templeton, "Probabilistic Fracture Mechanics Analysis to Support Inspection Intervals for RPV Top Head Nozzles" U.S. Nuclear Regulatory Commission / Argonne National Laboratory Conference on Vessel Head Penetration Inspection, Cracking, and Repairs, September 29 – October 2, 2003, Gaithersburg, Maryland.

I.2 Materials Reliability Program, MRP-105, "Probabilistic Fracture Mechanics Analysis of PWR Reactor Pressure Vessel Top Head Nozzle Cracking," EPRI Report 1007834 (EPRI Licensed Material), May, 2004.

I.3 EPRI Report, "BWR Reactor Pressure Vessel Shell Weld Inspection Recommendations (BWRVIP-05)," TR-105697, September 1995.

I.4 NRC Report, "Final Safety Evaluation of the BWR Vessel and Internals Project BWRVIP-05 Report (TAC No. M93925)," Division of Engineering, Office of Nuclear Reactor Regulation, May 1998.

I.5 NUREG-0619, "BWR Feedwater Nozzle and CRD Return Line Nozzle Cracking, Resolution of Generic Tech Activity A-10," November 1980.

I.6 U.S. NRC Order EA-03-009, "Interim Inspection Requirements for Reactor Pressure Vessel Heads at Pressurized Water Reactors", issued on February 11, 2003.

I.7 VIPER Version 1.2, Structural Integrity Associates, Report # SIR-95-098 Rev. 1, Feb. 1999.

All inspection data adjusted to 600 °F (Q = 50 kcal/mole)

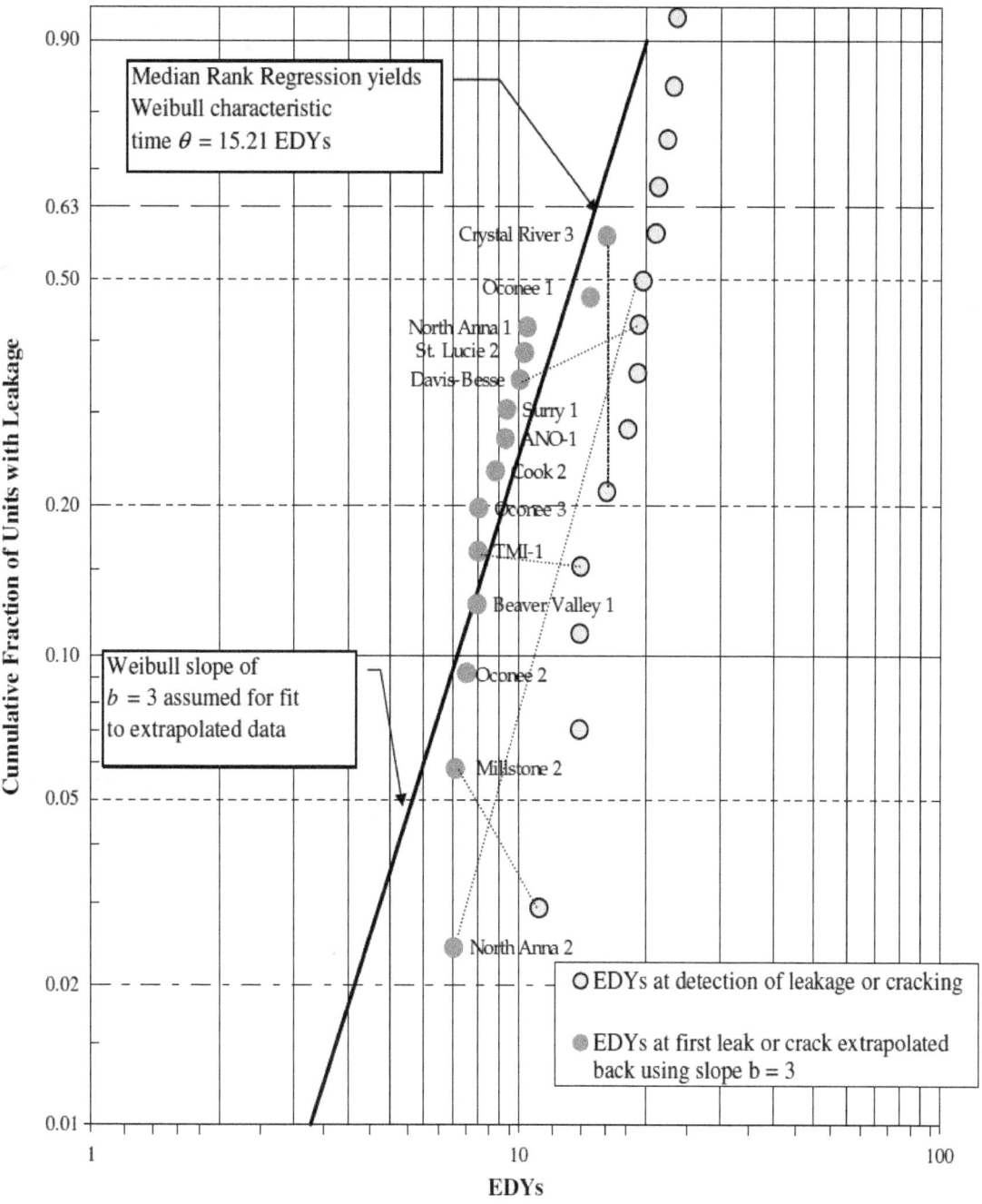

Figure I.1 Weibull Plot of Plant Inspection Data Showing Extrapolation Back to Time of First Leakage or Cracking. Plants that Performed NDE and were Found Clean are Treated as Suspensions

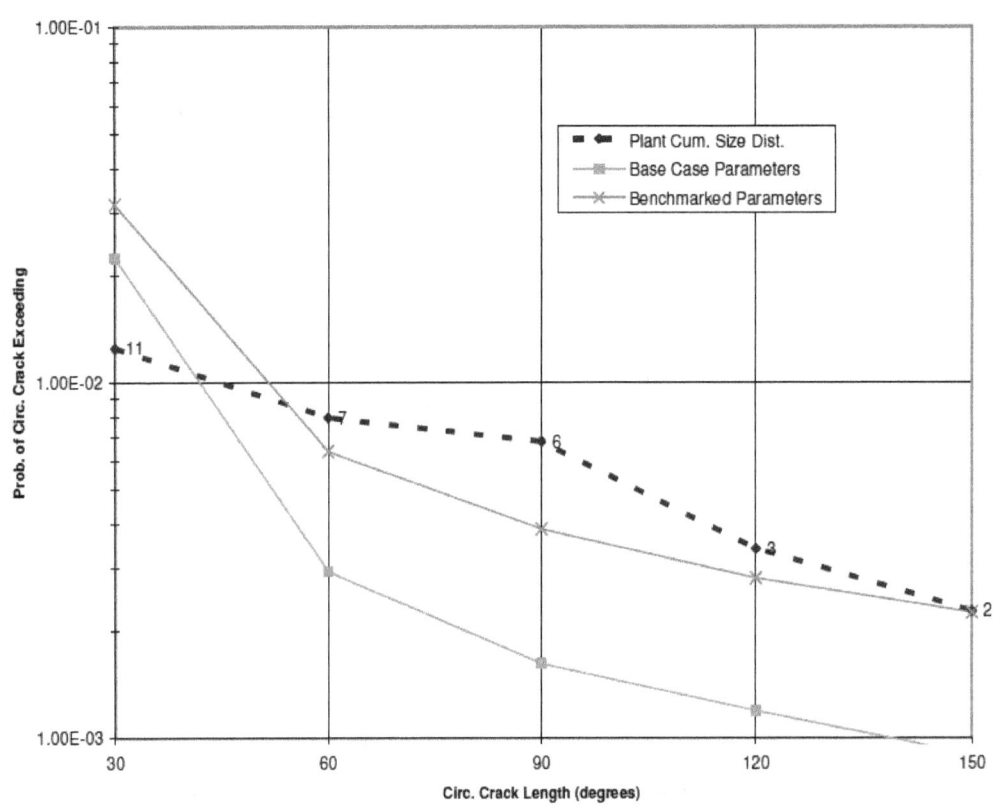

Figure I.2 Benchmarking of PFM Crack Growth Analyses with Respect to Field-Observed Circumferential Cracking of Various Lengths

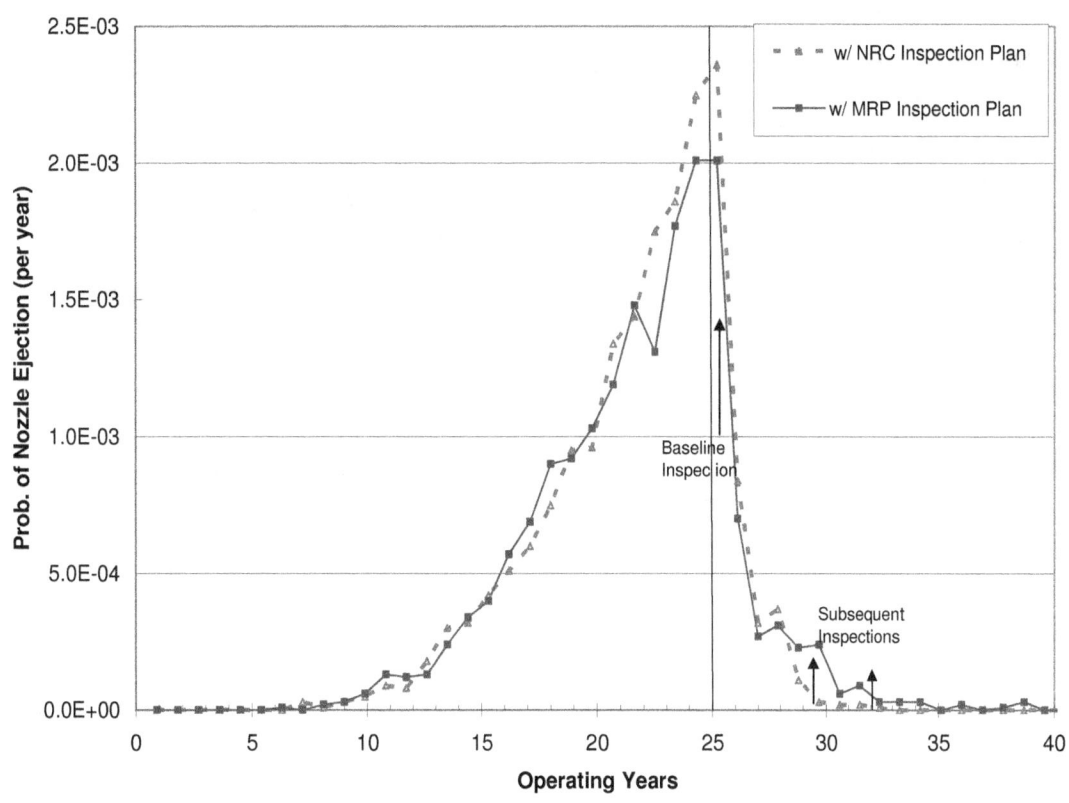

**Figure I.3 RPV Top Head PFM Analysis Results for Plant with 580ºF (304ºC) Head Temperature –
Probability of CRDM Nozzle Failure (i.e. Ejection of Nozzle from Vessel Head)**

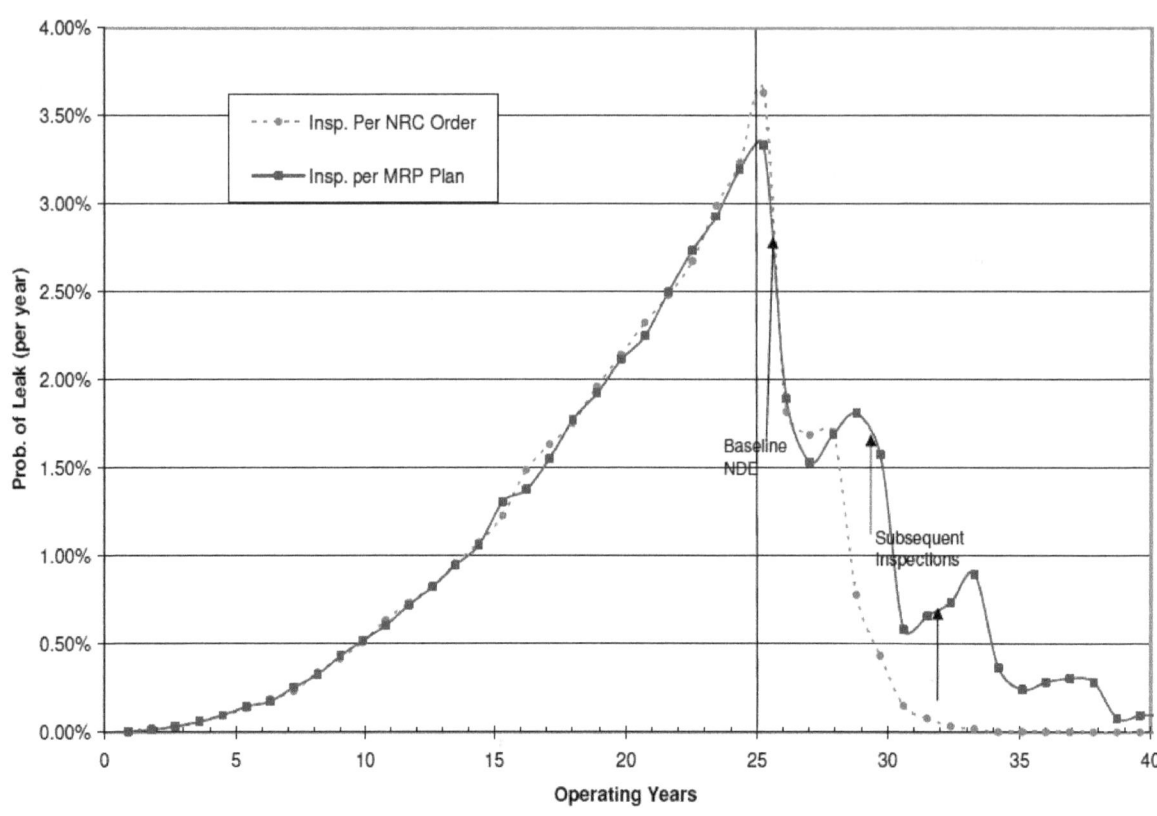

**Figure I.4 RPV Top Head PFM Analysis Results for Plant with 580°F (304°C) Head Temperature –
Probability of Leakage from CRDM Nozzle**

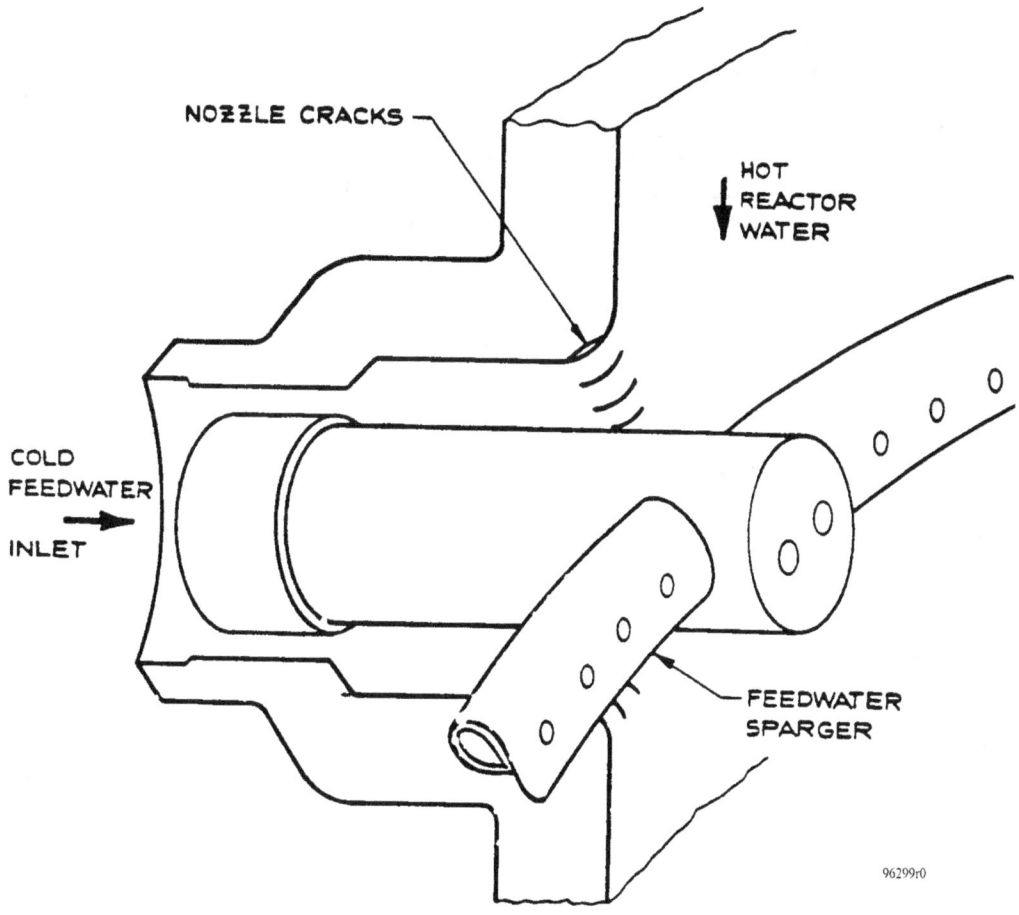

NOZZLE CRACKS

HOT REACTOR WATER

COLD FEEDWATER INLET

FEEDWATER SPARGER

96299r0

Figure I.5 Schematic of Thermal Fatigue Cracking in BWR Feedwater Nozzles

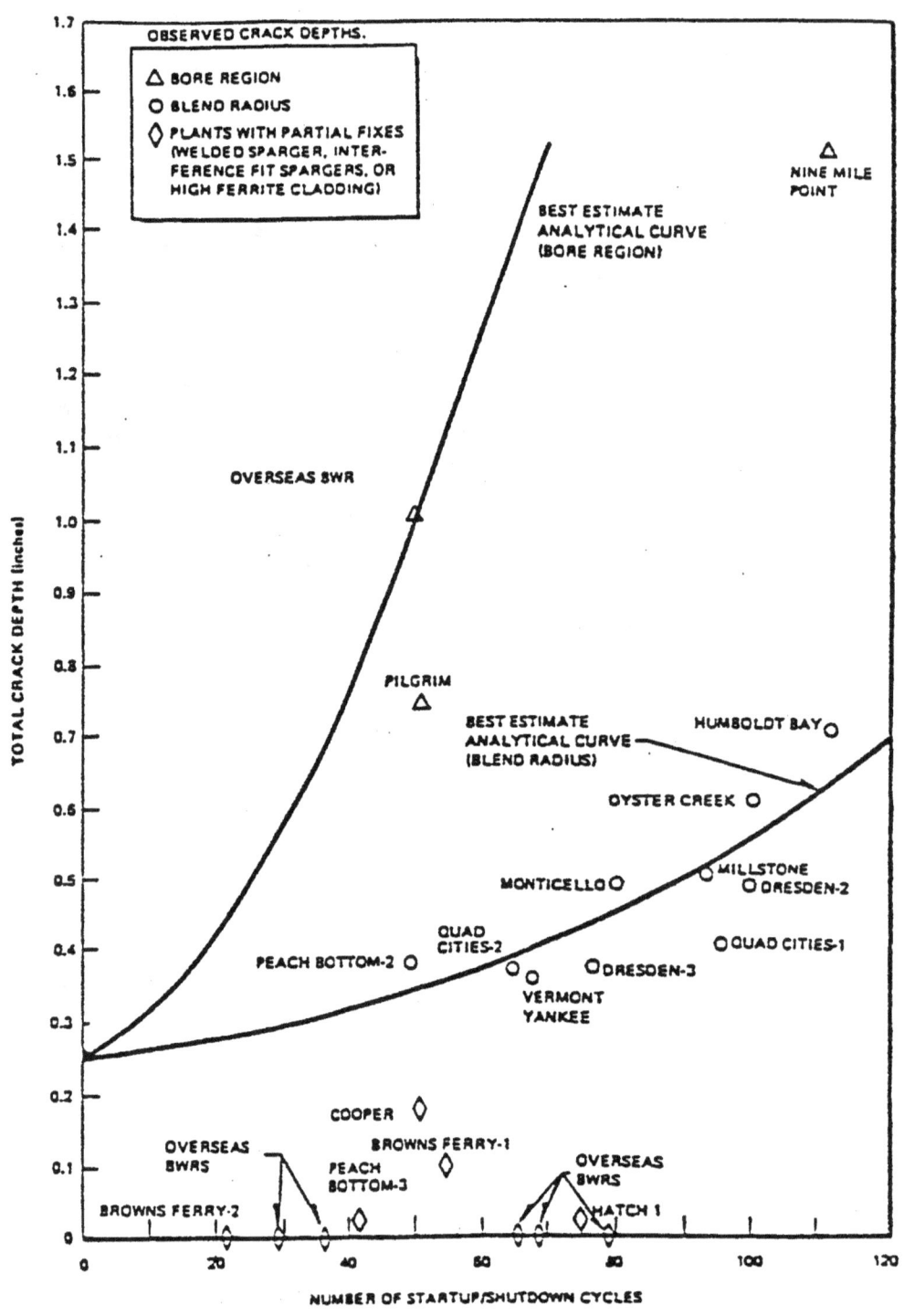

Figure I.6 Historical BWR Feedwater Nozzle Cracking Experience (circa 1980)

Average LOCA Frequencies; 0-25 Years

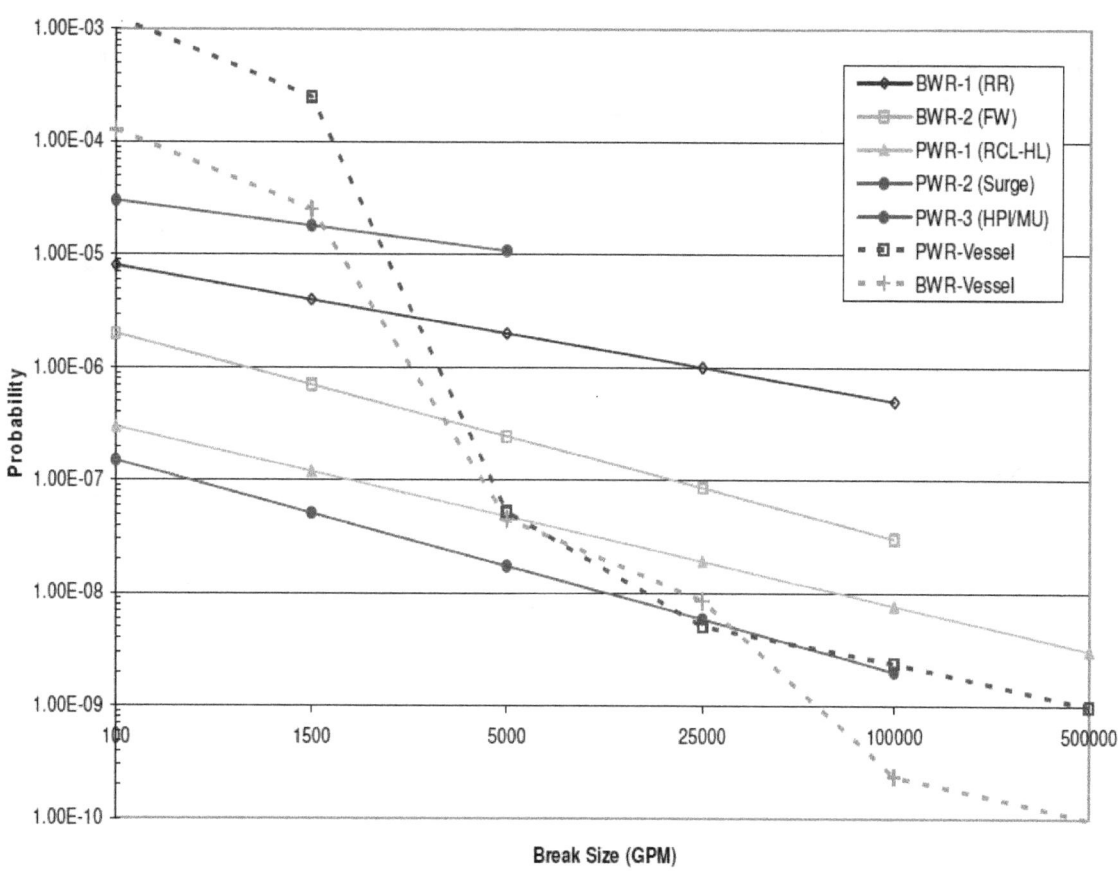

Figure I.7 Comparison of RPV and Piping Base Case LOCA Frequencies Versus Break Size
(0 to 25 Years)

Average LOCA Frequencies; 25-40 Years

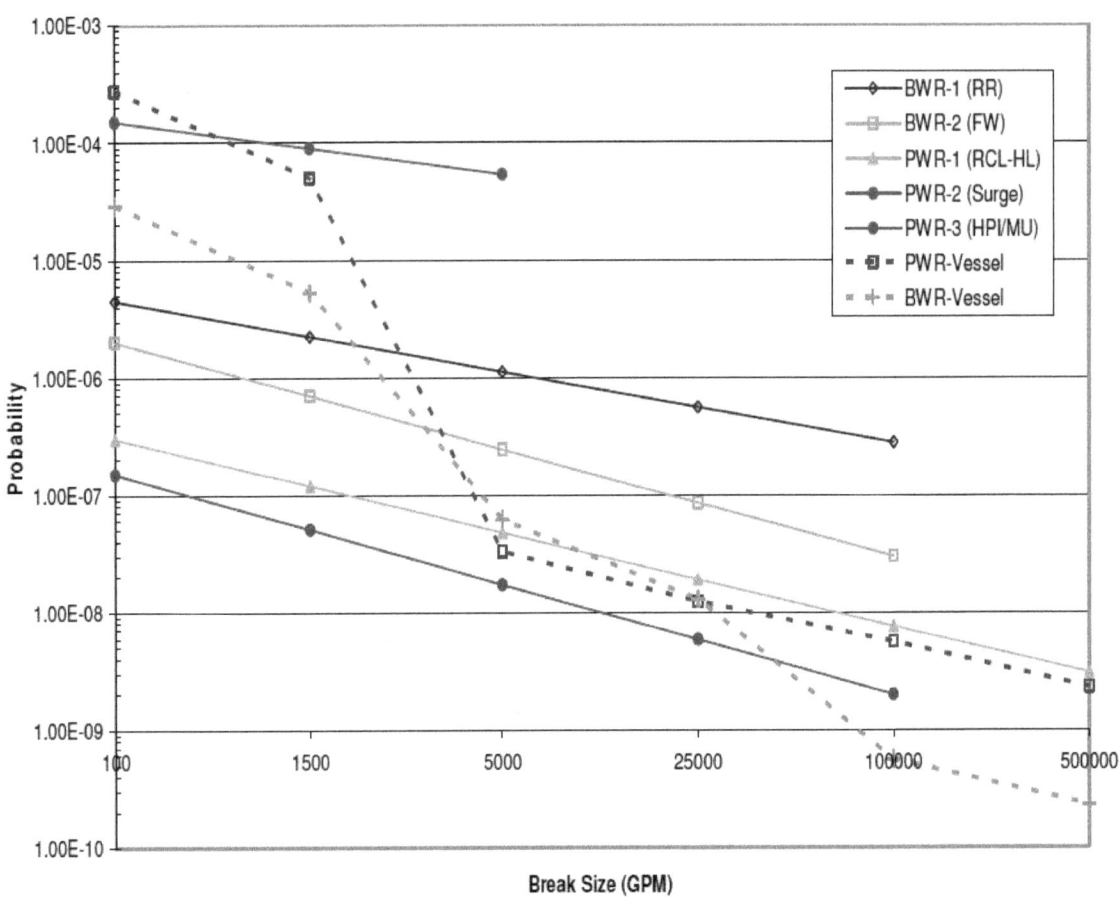

**Figure I.8 Comparison of RPV and Piping Base Case LOCA Frequencies Versus Break Size
(25 to 40 Years)**

APPENDIX J

ELICITATION QUESTIONS

APPENDIX J

ELICITATION QUESTIONS

J.1 Instructions

There are four basic quantities that are the ultimate focus of this exercise: the LOCA frequencies of piping components, the LOCA frequencies of non-piping components, the LOCA probabilities of piping components after emergency faulted loading, and the LOCA probabilities of non-piping components after emergency faulted loading. The elicitation will be structured so that each of these questions can be answered using one of two question sets. The question sets are structured to decompose the underlying issues using different approaches so increase your flexibility.

The bottom-up approaches (3A, 4A, 5A, 6A) could entail significantly more work if every piping and non-piping system is evaluated. It is recommended that people choosing this approach focus on significant contributing issues in only significant piping and non-piping systems to reduce the workload. Similarly the people choosing the top-down approaches (3B, 4B, 5B, 6B) may want to ensure that their significant issues are manifested correctly within relevant systems. These strategies allow you to combine features of each methodology.

Only a few additional examples of these questions are provided in this document. Many examples will be similar to those included in the *Elicitation Question Development* document. Please refer to this document and the *Kick-off Meeting Notes* document as indicated within the notes section for the questions.

J.1.1 Specific Instructions: Minimum Requirements Prior to Your Elicitation
A1.   Provide answers to the questions in the "Base Case Evaluation" area.
A2.   Provide MV estimates for the question set within the "Regulatory and Utility Safety Culture" area.
A3.   Provide MV estimates for at least one question set within the "LOCA frequencies of Piping Components" area.
A4.   Provide MV estimates for at least one question set within the "LOCA frequencies of Non-Piping Components" area.
A5.   Provide MV estimates for at least one question set within the "LOCA Probabilities of Piping Components under an Emergency Faulted Load" area.
A6.   Provide MV estimates for at least one question set within the "LOCA Probabilities of Non-Piping Components under an Emergency Faulted Load" area.
A7.   Categorize the uncertainty ranges (90% coverage intervals) associated with your MV estimates in A2 – A6 as low, medium, or high.

J.1.2 Specific Instructions: Additional Questions During Your Elicitation
We will be asking for your response within the following general areas.
B1.   Provide rationale and discuss those important issues that you identified and quantified in questions A2 – A6.
B2.   Quantify the uncertainty ranges (90% coverage intervals) associated with estimates provided for A2 – A6. This will quantify the initial responses in A7.
B3.   Provide MV estimates for the question sets that you did not initially answer in A2 – A6.
B4.   Quantify uncertainties associated with answers in B3.
B5.   Ensure that the critical issues for LOCA frequencies are captured.

J.2 Elicitation Questions

J.2.1 Elicitation Question 1: Base Case Evaluation

The following questions will be asked to solicit your opinion about the base case evaluation. These questions are necessary to determine how the rest of your responses will be anchored, i.e., if the base case *conditions* will be used for anchoring or if you prefer to anchor using a *specific set of results*. Therefore, you should only consider the *general approach used by each base case team member* and not specific results. You will be given additional opportunity later to provide your assessment of the specific results. Of course, you can also provide additional information that can be used to provide anchoring to the rest of your estimates, including your own set of results for the base cases. All the questions below will be asked for each of the LOCA size categories and the evaluation time periods.

1A.1. Do you think the base case results reflect the same conditions in each of the four team member's calculations?

1A.2. If not, which panelists' results best describe the base case conditions?

1B.1. Do you think that the differences in the four base case team members' results are a reasonable reflection of the range of variability in the true LOCA probability?

1B.2. If not, do the results under or over estimate your opinion of the true uncertainty?

1C. Do you think that any particular base case results are more accurate?

1D. Do you wish to anchor your responses on either the base case conditions, or on a specific team member's results?

J.2.2 Elicitation Question 2: Safety Culture

2A.1. Consider the current utility safety culture that exists after approximately 25 years (current-day) of plant operation and how it influences Category 1 LOCAs. Express the relative change, or ratio, in the utility safety culture's effect on LOCA frequencies after 15 additional years (40 years of operation) compared to its current-day effect. Next, express the ratio of the utility safety culture's effect on LOCA frequencies ratio in 35 years (60 years of plant operation) to its current-day effect. Include the 90% coverage interval for all estimates.

2A.2. Repeat 2A.1 but now considering the effect of the regulatory safety culture on LOCA frequencies.

2A.3. If you believe that safety culture effects are a function of leak rate category, repeat 2A.1 and 2A.2 for Category 2 through Category 6 LOCA frequencies.

2A.4. Do you believe that the utility safety culture and regulatory safety culture are correlated? If so, is the correlation high, medium, or low?

Notes:

a. Some aspects of regulatory and utility safety culture are discussed in the *Kick-off Meeting Notes*. These aspects can be considered independently and then combined or the aspects can be considered in the aggregate in question 2A.

b. Please see EQ 9 in the *Elicitation Question Development* document for additional information and an example for this question.

c. A ratio greater than 1 indicates that the safety culture will result in an proportional increase in the future LOCA frequency compared to the current LOCA frequency. Similarly, ratios less than 1 indicate that the LOCA frequencies will decrease as a function of safety culture.

J.2.3 Elicitation Question 3: LOCA Frequencies of Piping Components

Question Set 3A

3A.1.1. Consider Category 1 LOCAs for the PWR cold leg reference case. Choose a base case to compare with this reference case for this LOCA size at 25, 40, and 60 years of plant operation. Determine the ratio of LOCAs in the cold leg reference case to the chosen base case at each time period. Also, estimate the 90% coverage interval for these ratios.

3A.1.2. Repeat 3A.1.1 for each LOCA size category for the cold leg.

3A.1.3. Repeat 3A.1.1 and 3A1.2 for all other PWR and BWR reference cases.

Notes:
  a. Piping base and reference case conditions are described in the *Kick-off Meeting Notes* document.
  b. Please see EQ 10 in the *Elicitation Question Development* document for additional information and an example for this question.
  c. Any base case can be chosen at any specified period of time (25, 40, and 60 years) for anchoring. Please note the time period of your chosen base case time period is different than the time period being analyzed.

3A.2.1. List the specific combinations of the **variables** (i.e., material, geometry, degradation mechanism, loading, and mitigation/maintenance) which are the most significant contributors to PWR cold leg Category 1 LOCA frequency as a function of plant operating time (25, 40, 60 years). The list should total **at least 80%** of the total contribution to all cold leg Category 1 LOCAs. Estimate the MV contribution of these systems (> 80%). Also, provide the 90% coverage interval for the total contribution estimate of these systems.

3A.2.2. Repeat 3A.2.1 for each LOCA size category for the cold leg.

3A.2.3. Repeat 3A.2.1 and 3A.2.2 for all other PWR and BWR LOCA-susceptible piping systems.

Notes:
  a. The list of possible values for each variable class is provided in the *Kick-off Meeting Notes* document.
  b. Please see EQ 4 in the *Elicitation Question Development* document for additional information and an example for this question.

3A.3.1. Estimate the relative LOCA likelihood, or ratio, of each unique variable combination for Category 1 cold leg LOCA developed in 3A.2 to the cold leg reference case (or another suitable base or reference case) as a function of plant operating time (25, 40, 60 years).

3A.3.2. Repeat 3A.3.1 for each LOCA size category for the cold leg.

3A.3.3. Repeat 3A.3.1 and 3A.3.2 for all PWR and BWR LOCA-susceptible piping.

Notes:
  a. Piping reference case conditions are described in the *Kick-off Meeting Notes* document.
  b. Please see EQ 6 in the *Elicitation Question Development* document for additional information and an example for a similar question.

Question Set 3B

3B.1.1. List the PWR piping systems that provide a minimum of 80% of the total contribution for Category 1 (leak rates > 100 gpm [380 lpm]) LOCAs in US plants after 25, 40, and 60 years of operation. Now estimate the MV contribution of these systems (> 80%). Provide the 90% coverage interval for the total contribution estimate of these systems.

3B.1.2. Repeat 3B.1.1 for Category 2 through 6 LOCAs in PWR piping systems.

3B.1.3. Repeat 3B.1.1 and 3B.1.2 for BWR piping systems.

Notes:
  a. Relevant BWR and PWR piping systems are described in the *Kick-off Meeting Notes* document.

b. Please see EQ 1 in the *Elicitation Question Development* document for additional information and an example for a similar question.

3B.2.1. Estimate the percentage contribution for each PWR piping system in 3B1.1 for Category 1 LOCAs as a function of plant operating time.

3B.2.2. Repeat 3B.2.1 for Category 2 through 6 LOCAs in PWR piping systems.

3B.2.3. Repeat 3B.2.1 and 3B.2.2 for BWR piping systems.

Notes:

a. Please see EQ 2 in the *Elicitation Question Development* document for additional information and an example for a similar question.

3B.3. If a base case piping system(s) is listed within your significant PWR piping systems for Category 1 LOCAs (3B.1.1), go to 3B.5. If not, go to 3B.4.

3B.4.1. Estimate the ratio of the reference case for the Category 1 LOCA contribution of your most important BWR piping system to a suitable base case as a function of plant operating time. Also, provide the 90% coverage range for this ratio.

3B.4.2. Repeat 3B.3 for Category 2 through 6 LOCAs in PWR piping systems.

3B.4.3. Repeat 3B.3 for BWR piping systems.

Notes:

a. Base and reference case conditions for piping systems are defined within the *Kick-off Meeting Notes* document.

b. Please see EQ 10 in the *Elicitation Question Development* document for additional information and an example for a similar question.

3B.5.1. Estimate the ratio of all Category 1 LOCA contributions for this piping system to those contributions represented by the base (or reference) case conditions as a function of plant operating time. Provide the 90% coverage range for this ratio.

3B.5.2. Repeat 3B.3 for Category 2 through 6 LOCAs in PWR piping systems.

3B.5.3. Repeat 3B.3 for BWR piping systems.

Notes:

a. Please see the appendix for an example for this question.

J.2.4 Elicitation Question 4: LOCA Frequencies of Non-Piping Components

Question Set 4A

4A.1.1. Examine the failure scenarios for each of the five PWR non-piping components (pressurizer, valves, pumps, RPV, and steam generator). For each component, list the failure scenarios that provide a minimum of 80% of the total contribution for Category 1 (leak rates > 100 gpm [380 lpm]) LOCAs in US plants after 25, 40, and 60 years of operation. Estimate the MV contribution of these failure scenarios (> 80%). Also, provide the 90% coverage interval for the total contribution estimate of these systems.

4A.1.2. Repeat 4A.1.1 for Category 2 through 6 LOCAs for the non-piping PWR components.

4A.1.3. Repeat 4A.1.1 and 4A.1.2 for BWR non-piping components (valves, pumps, RPV).

Notes:

a. A failure scenario is associated with a specific non-piping component, material, degradation mechanism, etc.

b. Relevant BWR and PWR non-piping failure scenarios and components are discussed in the *kick-off meeting notes* document (called failure modes instead of scenarios in this document). These are also summarized in the elicitation summary tables.

c. Please see EQ 1 in the *Elicitation Question Development* document for additional information and an example for a similar question.

4A.2.1. Choose a piping or non-piping base case which results in the most natural comparison for each of the failure scenarios described in 4A.1.1 for all five PWR non-piping component classes.

Provide a MV estimate of the ratio for the Category 1 LOCA contribution of the chosen non-piping failure scenario to the chosen base case.

4A.2.2. Repeat 4A.2.1 for Category 2 through 6 LOCAs for the non-piping PWR components.

4A.2.3. Repeat 4A.2.1 and 4A.2.2 for BWR non-piping components (valves, pumps, RPV).

Notes:
 a. Please see EQ 6 in the *Elicitation Question Development* document for additional information and an example for a similar question.
 b. Non-piping base cases are currently being quantified to determine the leaking frequencies due to **all degradation mechanisms** for each non-piping component listed in the *kick-off meeting notes* document. There will also be non-piping base cases frequencies for items that have failed such as SGTRs. Additionally, non-piping base cases can still be chosen for making relative comparisons. For instance if a panelist considers valve body failure due to cavitation erosion to be significant for Category 1 PWR LOCAs, then valve body leakage can be chosen as the base case.

Question Set 4B

4B.1.1. List the PWR non-piping failure scenarios that provide a minimum of 80% of the total contribution for Category 1 (leak rates > 100 gpm [380 lpm]) LOCAs in US plants after 25, 40, and 60 years of operation. Now estimate the MV contribution of these failure scenarios (> 80%). Also, provide the 90% coverage interval for the total contribution estimate.

4B.1.2. Repeat 4B.1.1 for Category 2 through 6 LOCAs for the non-piping PWR failure scenarios.

4B.1.3. Repeat 4B.1.1 and 4B.1.2 for BWR non-piping failure scenarios.

Notes:
 a. This question differs from Elicitation Question 4A in that only the significant failure scenarios, regardless of component, need to be considered.
 b. A failure scenario is associated with a specific non-piping component, material, degradation mechanism, etc.
 c. Relevant BWR and PWR non-piping failure scenarios are discussed in the *kick-off meeting notes* document (called failure modes instead of scenarios in this document). These are also summarized in the elicitation summary tables.
 d. Please see EQ 1 in the *Elicitation Question Development* document for additional information and an example for a similar question.

4B.2.1. Estimate the percentage contribution for each PWR non-piping failure scenario in 4B1.1 for Category 1 LOCAs after 25, 40, and 60 years of operation.

4B.2.2. Repeat 4B.2.1 for Category 2 through 6 LOCAs for the non-piping PWR scenarios.

4B.2.3. Repeat 4B.2.1 and 4B.2.2 for BWR non-piping failure scenarios.

Notes:
 a. Please see EQ 2 in the *Elicitation Question Development* document for additional information and an example for a similar question.

4B.3.1 Pick either a piping or a non-piping base case (or a piping reference case) for comparison with one or more of your significant non-piping failure scenarios from 4B1.1 for Category 1 LOCAs. The comparison should be natural, but if possible, should be made with one of the most significant failure scenarios that you listed. Provide a MV estimate of the ratio of the non-piping failure scenario to the chosen base case as a function of operating time (40 and 60 years). Also, provide the 90% coverage range for this ratio.

4B.3.2 Repeat 4B.3.1 for Category 2 through 6 LOCAs for the non-piping PWR failure scenarios.

4B.3.3 Repeat 4B.3.1 and 4B.3.2 for BWR non-piping failure scenarios.

Notes:
 a. Base case conditions for piping systems and are defined within the *kick-off meeting notes* document. Base case conditions for non-piping components are being developed as discussed in the notes to Elicitation Question 4A.2

b. Please see EQ 6 in the *Elicitation Question Development* document for additional information and an example for a similar question.

J.2.5 Elicitation Question 5: LOCA Probabilities of Piping Components under an Emergency Faulted Load

An emergency faulted load represents an initial design consideration for a possible large transient load that was not expected to occur over any particular plant's operating life of 40 years (rare event), or a frequency less than approximately 0.025 yr^{-1}. These loads could be due to seismic loading or any other large pressure transients. Base cases have been developed which examine the conditional failure probability for ASME Service Level B loading. This loading level was estimated for several plants to conservatively approximate a 1*SSE event on a pipe which is flawed up to the allowable technical specification leakage rate for the given piping system and degradation mechanism. An SSE event was initially a design-level earthquake amplitude that was thought to occur once in 40 years; however, operating experience to date suggests that the frequency of an SSE event occurring is less than that.

This question will ask you to list and quantify the effect of the most significant piping systems and degradation mechanisms that contribute to each LOCA category. The quantification will be done for two emergency faulted load sizes (ASME Service Levels B and D) for three assumed damage states. The damage states will consist of tech. spec. leakage rates, the onset of leakage through a slow drip (perceptible leak), and a surface crack with a/t = 0.5. The surface crack length will be assumed by each panelist and may be a function of degradation mechanism and material. A relationship between the failure loads for a circumferential through-wall-crack and surface cracks with a/t = 0.5 and different lengths is provided in the *"Piping Seismic Base Cases"* document. The likelihood of each damage state will also be ascertained by each panelist relative to the operational experience data for the leak-rate frequencies corresponding to each system listed, regardless of degradation mechanism. This assessment will require nine different relative comparisons for each LOCA size category and plant type (BWR or PWR).

The appendix of this document and the *"Piping Seismic Base Cases"* document provide the philosophy behind the seismic piping elicitation questions and detail the seismic piping base case calculations. Both documents should be read prior to answering this elicitation question.

5A.1.1. List the piping systems and degradation mechanisms which most significantly contribute to Category 1 LOCAs given that an assumed emergency faulted load occurs with an equivalent magnitude of an ASME Service Level B (SLB) event for PWRs. This total list should summarize at least the top 80% contributing factors to Category 1 LOCAs under assumed faulted loading conditions. Also, for each system, list the loads which may result in SLB loading and indicate if they are primary (loading-controlled) or secondary (displacement-controlled). Provide the total contribution and also the 90% coverage interval for this estimate.

5A.1.2. Repeat 5A.1.1 for ASME Service Level D (SLD) loading

5A.1.3. Repeat 5A.1.1 and 5A.1.2 for each PWR LOCA size category.

5A.1.4. Repeat 5A.1.1 - 5A.1.3 for BWRs.

   Notes:

   a. Information on relevant piping systems, degradation mechanisms, and piping sizes is contained in the "Elicitation Meeting Notes" from the kick-off meeting.

   b. In this question, pick your piping systems assuming that the pipes will completely fail. Therefore, the LOCA size category will be directly related to the pipe size.

5A.2.1. Pick a representative set of seismic base-case conditions to use for comparison for each of your significant contributors to Category 1 LOCAs in PWRs.

5A.2.2. Repeat 5A.2.1 for each PWR LOCA size category.

5A.2.3. Repeat 5A.2.1 and 5A.2.2 for BWRs.

Notes:

    a. A PWR and BWR base case have been defined in the *"Piping Seismic Base Cases"* document for a specific degradation mechanism, pipe size, and material. Additionally, figures are available which show the effects of changing materials, piping size, and service level loading with respect to the base case definitions.

    b. Comparisons to the selected base cases will be made in subsequent questions.

    c. A relationship between the failure loads for a circumferential through-wall-crack and surface cracks with a/t = 0.5 and different lengths is given at the end of the Piping Seismic Base Case/Background section.

5A.3.1. Consider a single piping system and degradation mechanism combination identified for Category 1 PWR LOCAs in 5.A.1 and the associated seismic base case identified in 5A.2. Determine the ratio of the CFP for this system/degradation mechanism combination (P_{TSL} or $P_{TSL@SLB}$) to the CFP for the chosen seismic piping base case (P_{BC}). Assume that a SLB emergency faulted load occurs and that the piping system is degraded by a through-wall crack that is leaking at the technical specification limit. Also provide the 90% coverage interval of this ratio.

5A.3.2. Consider the same piping system and degradation mechanism combination identified for Category 1 PWR LOCAs in 5.A.3.1. Next, determine the ratio of the CFP for a crack that has just formed a perceptible leak (P_{PL}) to the CFP for a crack leaking at the technical specification limit assuming (P_{TSL}) a SLB load. Also provide the 90% coverage interval of this ratio.

5A.3.3. Again, consider a single piping system and degradation mechanism combination identified for Category 1 PWR LOCAs in 5.A.3.1. Next, determine the ratio of the conditional failure probability for a crack with a maximum depth of 50% of the wall thickness (P_{50}) to the CFP for a crack that has just formed a perceptible leak (P_{PL}) assuming a SLB load. Also provide the 90% coverage interval of this ratio.

5A.3.4. Repeat 5A.3.1 – 5A.3.3 for each significant piping system/degradation mechanism combination listed for PWR Category 1 LOCAs in 5A.1.

5A.3.5. Repeat 5A.3.1 - 5A.3.4 for each PWR LOCA size category.

5A.3.6. Repeat 5A.3.1 - 5A.3.5 for BWRs.

Notes:

    a. The leaking crack size is a function of the degradation mechanism and is the major contributor to the differences with the base-case conditional failure probabilities.

    b. A perceptible leak is a leak which has just initiated.

5A.4.1. Again consider a single piping system and degradation mechanism combination identified for Category 1 PWR LOCAs in 5.A.1. Next, determine the ratio of the CFP for a SLD event ($P_{TSL@SLD}$) to the CFP for a SLB event ($P_{TSL@SLB}$) assuming a crack leaking at the technical specification limit in both cases. Also provide the 90% coverage interval of this ratio.

5A.4.2. Consider the same piping system and degradation mechanism combination identified for Category 1 PWR LOCAs in 5.A.4.1. Next, determine the ratio of the CFP for a crack that has just formed a perceptible leak (P_{PL}) to the CFP for a crack leaking at the technical specification limit (P_{TSL} or $P_{TSL@SLD}$) assuming a SLD load. Also provide the 90% coverage interval of this ratio.

5A.4.3. Again, consider a single piping system and degradation mechanism combination identified for Category 1 PWR LOCAs in 5.A.4.1. Next, determine the ratio of the conditional failure probability for a crack with a maximum depth of 50% of the wall thickness (P_{50}) to the CFP for a crack that has just formed a perceptible leak (P_{PL}) assuming a SLD load. Also provide the 90% coverage interval of this ratio.

5A.4.4. Repeat 5A.4.1 – 5A.4.3 for each significant piping system/degradation mechanism combination listed for PWR Category 1 LOCAs in 5A.1.

5A.4.5. Repeat 5A.6.1 - 5A.6.4 for each PWR LOCA size category.

5A.4.6. Repeat 5A.6.1 - 5A.6.5 for all BWRs.

Notes:
a. If your system and degradation mechanism list in 5A.1.2 for SLD loading is different from that in 5A.1.1 for SLB loading, pick a seismic base for reference in 5A.4.1 instead of referencing with respect to the SLB loading magnitude.

J.2.5.1 Estimation of Piping Damage Likelihood: Now consider the relative likelihood of the occurrence of a particular level of damage (50% through-wall, perceptible leak, tech. spec. leakage) due to the piping system/degradation mechanism combination chosen in 5A.1. All answers will be ultimately referenced to a piping base-case damage probability as in earlier questions. However, there are no numbers assigned to the base-case damage probabilities at this time, so the comparisons should be made with respect to a piping base-case damage *condition*. A separate piping base-case condition is defined for each piping system and LOCA size category identified in 5A.1, as the operational experience frequency of *all* leaks *regardless* of the degradation mechanism. This frequency will be determined from operational experience data.

5A.5.1   Consider a single piping system and degradation mechanism combination identified for Category 1 PWR LOCAs in 5.A.1. Next, determine the ratio of the likelihood of a pipe having a perceptible leak due to that degradation mechanism in that piping system (L_{PL}) after 25 years of operation (L_{PL}) to the base case (L_{BC}), which is the likelihood of a leak due to any degradation mechanism. Also provide the 90% coverage interval for this estimate.

5A.5.2   Consider the same single piping system and degradation mechanism as in 5A.5.1. Next, determine the ratio of the likelihood of a technical specification leak (L_{TSL}) to a perceptible leak (L_{PL}) due to that degradation mechanism after 25 years of operation. Also provide the 90% coverage interval for this estimate.

5A.5.3   Consider the same single piping system and degradation mechanism as in 5A.5.1. Next, determine the ratio of the likelihood of a 50% through-wall leak (L_{50}) to a perceptible leak (L_{PL}) due to that degradation mechanism after 25 years of operation (current-day). Also provide the 90% coverage interval for this estimate.

5A.5.4   Now determine if you believe the relative likelihood ratios in 5A.5.1 – 5A.5.3 will increase, decrease, or remain constant with continued operating time. First consider all three likelihood estimates (L_{PL}/L_{BC}, L_{TSL}/L_{PL}, and L_{50}/L_{PL}) at 40 years and then 60 years of continued operation. Determine the ratio of these estimates at 40 years of operation to the current-day estimates. Next, determine the ratio these estimates at 60 years of operation to current-day estimates.

5A.5.5   Repeat 5A.5.1- 5A.5.4 for each significant piping system/degradation mechanism combination listed for PWR Category 1 LOCAs in 5A.1.

5A.5.6   Repeat 5A.5.1 - 5A.5.5 for each PWR LOCA size category.

5A.5.7   Repeat 5A.5.1 - 5A.5.6 for all BWRs.

J.2.6 Elicitation Question 6: LOCA Probabilities of Non-Piping Components under an Emergency Faulted Load

An emergency faulted load represents an initial design consideration for a large transient load that was not expected to occur over any particular plant's operating life of 40 years (rare event). These loads could be due to seismic loading or any other large pressure transients. Similar to the piping evaluation, base cases will be used for anchoring on the conditional failure probability. However, the actual base cases will not be developed until after the panelists' identify the non-piping components which provide significant LOCA contributions. In the interim, each panelist should use a particular set of base case conditions for anchoring. More information on this selection will follow in Elicitation Question 6A.2.

This question will ask you to list and quantify the effect of the most significant non-piping systems and degradation mechanisms that contribute to each LOCA category. The quantification will be done for two emergency faulted load sizes (SLB and SLD) for three assumed damage states. The damage states will consist of tech. spec. leakage rates, the onset of leakage through a slow drip (perceptible leak), and a surface crack with a/t = 0.5. The surface crack length will be assumed by each panelist and may be a function of degradation mechanism and material. The likelihood of each damage state will also be ascertained by each panelist relative to the operational experience data for the leak-rate frequencies corresponding to each non-piping component listed, regardless of degradation mechanism. This assessment will require nine different relative comparisons for each LOCA size category and plant type (BWR or PWR).

The structure of this question is almost identical to Elicitation Question 5. The appendix contains information on the philosophy behind these two questions.

6A.1.1. List the non-piping components and degradation mechanisms (or failure scenarios) which most significantly contribute to Category 1 LOCAs given that an assumed emergency faulted load occurs with an equivalent SLB magnitude for PWRs. This total list should summarize at least the top 80% contributing factors to Category 1 LOCAs under assumed faulted loading conditions. Also, for each component, list the loads which may result in SLB loading and indicate if these loads are primary (load-controlled) or secondary (displacement-controlled). Provide the total contribution and also the 90% coverage interval for this estimate.

6A.1.2. Repeat 6A.1.1 for Service Level D (SLD) loading

6A.1.3. Repeat 6A.1.1 and 6A.1.2 for each PWR LOCA size category.

6A.1.4. Repeat 6A.1.1 - 6A.1.3 for BWR non-piping components.

Notes:

a. Information on relevant non-piping components and degradation mechanisms, are contained in the "Elicitation Meeting Notes" from the kick-off meeting and subsequent revisions to Tables B.1.13 – B.1.17 in this document.

b. In this question, pick your non-piping component assuming that it will completely fail. Therefore, the LOCA size category will be directly related to the component size.

6A.2.1. Pick a representative set of seismic base-case conditions to use for comparison for each of your significant contributors to Category 1 LOCAs in PWRs.

6A.2.2. Repeat 6A.2.1 for each PWR LOCA size category.

6A.2.3. Repeat 6A.2.1 and 6A.2.2 for BWR non-piping components.

Notes:

a. The base case conditions should correspond to at least one (or several, or all) of the significant non-piping LOCA contributors identified in 6A.1. Assume that the component is damaged with a fatigue flaw which results in technical specification leakage. Assume that the base case loading magnitude is an SLB load. Assume that absolute size of this flaw and the actual conditional failure probability to a SLB load magnitude will be quantified at a later date.

b. Comparisons to the selected base cases will be made in subsequent questions.

6A.3.1. Consider a single non-piping component and degradation mechanism combination identified for Category 1 PWR LOCAs in 6.A.1 and the associated seismic base case identified in 6A.2. Determine the ratio of the CFP for this system/degradation mechanism combination (P_{TSL} or $P_{TSL@SLB}$) to the CFP for the chosen seismic non-piping base case assuming (P_{BC}) that an SLB emergency faulted load occurs and that the non-piping component contains a through-wall crack that is leaking at the technical specification limit. Also provide the 90% coverage interval of this ratio.

6A.3.2. Consider the same non-piping component and degradation mechanism combination identified for Category 1 PWR LOCAs in 6.A.3.1. Next, determine the ratio of the CFP for a crack that has just formed a perceptible leak (P_{PL}) to the CFP for a crack leaking at the technical specification limit (P_{TSL}) assuming a SLB load magnitude. Also provide the 90% coverage interval of this ratio.

6A.3.3. Again, consider the single non-piping component and degradation mechanism combination identified for Category 1 PWR LOCAs in 6.A.3.1. Next, determine the ratio of the CFP for a crack with a maximum depth of 50% of the wall thickness (P_{50}) to the CFP for a crack that has just formed a perceptible leak (P_{PL}) assuming a SLB load. Also provide the 90% coverage interval of this ratio.

6A.3.4. Repeat 6A.3.1 – 6A.3.3 for each significant non-piping component/degradation mechanism combination listed for PWR Category 1 LOCAs in 6A.1.

6A.3.5. Repeat 6A.3.1 - 6A.3.4 for each PWR LOCA size category.

6A.3.6. Repeat 6A.3.1 - 6A.3.5 for BWR non-piping components.

Notes:
    a. The leaking crack size is a function of the degradation mechanism and is the major contributor to the differences with the base-case conditional failure probabilities.

6A.4.1. Again consider a single non-piping component and degradation mechanism combination as identified for Category 1 PWR LOCAs in 6.A.1. Next, determine the ratio of the CFP for a SLD event ($P_{TSL@SLD}$) to the CFP for a SLB event ($P_{TSL@SLB}$). Assume that a crack exists which is leaking at the technical specification limit in both cases. Also provide the 90% coverage interval of this ratio.

6A.4.2. Consider the same non-piping component and degradation mechanism combination identified for Category 1 PWR LOCAs in 6.A.4.1. Next, determine the ratio of the CFP for a crack that has just formed a perceptible leak (P_{PL}) to the CFP for a crack leaking at the technical specification limit (P_{TSL} or $P_{TSL@SLD}$). Assume a SLD loading magnitude in each case. Also provide the 90% coverage interval of this ratio.

6A.4.3. Again, consider the same non-piping component and degradation mechanism combination identified for Category 1 PWR LOCAs in 6.A.4.1. Next, determine the ratio of the CFP for a crack with a maximum depth of 50% of the wall thickness (P_{50}) to the CFP for a crack that has just formed a perceptible leak (P_{PL}). Assume a SLD loading magnitude in each case. Also provide the 90% coverage interval of this ratio.

6A.4.4. Repeat 6A.4.1 – 6A.4.3 for each significant non-piping component/degradation mechanism combination listed for PWR Category 1 LOCAs in 5A.1.

6A.4.5. Repeat 6A.6.1 - 6A.6.4 for each PWR LOCA size category.

6A.4.6. Repeat 6A.6.1 - 6A.6.5 for all BWR non-piping components.

Notes:
    a. If your system and degradation mechanism list in 6A.1.2 for SLD loading is different from that in 6A.1.1 for SLB loading, pick a seismic base for reference in 6A.4.1 instead of referencing with respect to the SLB loading magnitude.

J.2.6.1 Estimation of Piping Damage Likelihood: Now consider the relative likelihood of the occurrence of a particular level of damage (50% through-wall, perceptible leak, tech. spec. leakage) due to the non-piping component/degradation mechanism combination chosen in 6A.1. All answers will be ultimately referenced to a non-piping base-case damage probability. However, there are no numbers assigned to the non-base-case damage probabilities at this time. Comparisons should therefore be made with respect to a non-piping base-case damage *condition*. A separate non-piping base-case condition is defined for each non-piping component identified in 6A.1, as the operational experience frequency of all component leaks regardless of the degradation mechanism.

6A.5.1 Consider a single non-piping component and degradation mechanism combination identified for Category 1 PWR LOCAs in 6.A.1. Next, determine the ratio of the likelihood of the non-piping component having a perceptible leak after 25 years of operation (L_{PL}) due to that degradation mechanism to the base case (L_{BC}), which is the likelihood of a leak due to any degradation mechanism. Also provide the 90% coverage interval for this estimate.

6A.5.2 Consider the same non-piping component and degradation mechanism as in 6A.5.1. Next, determine the ratio of the likelihood of a technical specification leak (L_{TSL}) to a perceptible leak

(L_{PL}) due to that degradation mechanism after 25 years of operation. Also provide the 90% coverage interval for this estimate.

6A.5.3   Consider the same single non-piping component and degradation mechanism as in 6A.5.1. Next, determine the ratio of the likelihood of a 50% through-wall leak (L_{50}) to a perceptible leak (L_{PL}) due to that degradation mechanism. Also provide the 90% coverage interval for this estimate.

6A.5.4   Now determine if you believe the relative likelihood ratios in 6A.5.1 – 6A.5.3 will increase, decrease, or remain constant with continued operating time. First consider all three likelihood estimates (L_PL/L_{BC}, L_{TSL}/L_{PL}, and L_{50}/L_{PL}) at 40 years and then 60 years of continued operation. Determine the ratio of these estimates at 40 years of operation to the current-day estimates. Next, determine the ratio these estimates at 60 years of operation to current-day estimates.

6A.5.5   Repeat 6A.5.1- 6A.5.4 for each significant non-piping component/degradation mechanism combination listed for PWR Category 1 LOCAs in 6A.1.

6A.5.6   Repeat 6A.5.1 - 6A.5.5 for each PWR LOCA size category.

6A.5.7   Repeat 6A.5.1 - 6A.5.6 for all BWR non-piping components.

ATTACHMENT A TO APPENDIX J: ADDITIONAL EXAMPLE FOR ELICITATION QUESTION 3B.5.1

3B.5.1. Estimate the ratio of all Category 1 LOCA contributions for this piping system to those contributions represented by the base (or reference) case conditions as a function of plant operating time. Provide the 90% coverage range for this ratio.

For question set 3B, Panelist A has previously listed the following important PWR piping systems and their individual contributions to Category 1 LOCAs after 25 years of plant operation (see Table J.A.1). For this example, this panelist does not expect the relative contributions to change after either 40 or 60 years of plant operation.

Table J.A.1 PWR Piping System Contributions for Category 1 LOCAs

Case	Piping System Lines	Percentage Contribution
1	Instrumentation	50
2	Drain Lines	10
3	Reactor Coolant Pressure Hot Leg	10
4	Chemical Volume Control System	10
5	Safety Injection System Accumulator	10

The reactor coolant pressure hot leg has an associated base case. The base case geometry is a 30 inch diameter pipe, manufactured from Type 304 stainless steel with an Alloy 600 safe end. The safe end to pipe weld is a nickel-based bimetallic weld. The base case degradation mechanisms are thermal fatigue and PWSCC. The loading is pressure, thermal, residual stress, dead weight, with a pressure pulse transient. Panelist A next needs to estimate the ratio between all Category 1 LOCAs in the hot leg compared to those represented solely by the base case conditions. His results are summarized in Table J.A.2.

Table J.A.2 Panelist A's Ratio Between Entire Piping System and Base Case Contributions for Category 1 PWR Hot Leg LOCAs

Base/ref case	25 Years			40 Years			60 Years		
	5% LB	MV	5% UB	5% LB	MV	5% UB	5% LB	MV	5% UB
SL BC	1	2	3	1	3	4	1	4	5

Panelist A believes that the total hot leg Category 1 LOCAs are twice the number represented by those captured by the base case conditions after 25 years of service (present day). However, this ratio increases with time until after 60 years, the total hot leg Category 1 LOCA probability is 4 times what is predicted by the base case conditions. This initial difference is due to the fact that Panelist A believes that an equal number of Category 1 LOCAs will be due to thermal fatigue of other piping materials than are represented by the base case (specifically cast stainless steel and stainless steel clad carbon). This panelist also believes that these mechanisms will become more apparent with time than either thermal fatigue or PWSCC of stainless steel material. Future increases are also included for unanticipated mechanisms.

ATTACHMENT B TO APPENDIX J: PHILOSOPHY BEHIND ELICITATION QUESTION 5

The conditional LOCA probability for a given piping system, degradation mechanism, and emergency faulted load can be determined by multiplying the likelihood curve (L, red in Figure J.B.1) by the conditional piping failure probability (P, black in Figure J.B.1) and then integrating over all the possible damage states. This conditional LOCA probability will likely be a function of LOCA size (pipe size), piping system, applied emergency faulted load, and degradation mechanism. Figure J.B.1 below provides a schematic for a fixed ASME Service Level B (SLB) load, piping system (instrument lines), LOCA size (Category 1), and degradation mechanism (PWSCC). The curve shapes/trends in Figure J.B.1 are an illustration and do not represent any panelist opinion. A separate set of likelihood and conditional failure probability curves exists for each unique combination of these four variables.

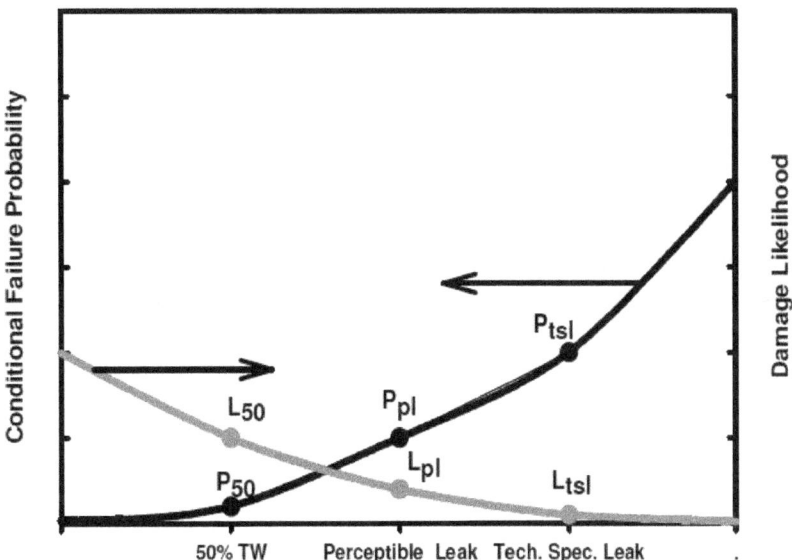

Figure J.B.1 Conditional Failure Probability for Service Level B Loading for Category 1 LOCAs in PWR Instrument Lines due to PWSCC

The elicitation question first asks each panelist to identify only the significant piping system and degradation mechanisms to conditional LOCA for each LOCA size category (EQ 5A.1). Then, the panelist will pick seismic base case conditions for either the PWR hot leg or BWR feedwater seismic base cases described in the *"Seismic Base Case"* document (EQ 5A.2).

Next, the panelist will provide ratios of the relative likelihood and conditional failure probabilities (Figure J.B.1) among three different damage states (a 50% through-wall crack, a perceptible leak, and a technical specification leak) at a fixed load for each LOCA size category. All the estimates are initially for a

Service Level B load. Here are the ratios that will be provided: P_{tsl}/P_{bc} (EQ 5.A.3.1), P_{pl}/P_{tsl} (EQ 5A.3.2), and P_{50}/P_{pl} (EQ 5A.3.3). The P_{bc} value is defined in the *"Seismic Base Case"* document while all other variables are defined in Figure J.B.1. In this way, all answers are linked back to the case quantification of the base-case conditions. The next elicitation question (5A.4.1) asks for the relationship between the Service Level D and Service Level B event for the given degradation mechanism and system which is the ratio of $P_{tsl\ at\ SLD}/P_{tsl\ at\ SLB}$. The other questions ask for the ratio of P_{pl}/P_{tsl} (EQ 5A.4.2), and P_{50}/P_{pl} (EQ 5A.4.3) for solely the SLD loading event.

Finally, the likelihood of the 3 damage states is estimated. The likelihood base case to be used for each piping system and degradation mechanism listed by each panelist will be the leaking frequency for that system over the piping size range of interest for all degradation mechanisms (L_{bc}). Obviously, this frequency will be more than the frequency for any single degradation mechanism, but dominant degradation mechanisms may provide a ratio close to 1. This frequency has not been quantified, but it will be after the elicitation once a complete listing of important systems and degradation mechanisms has been provided by the panelists. However, leaking rate propensity has been provided as part of the piping base case analysis for those systems (i.e., see Bill Galyean and Bengt Lydell's results). The elicitation questions ask each panelist to provide the following three likelihood ratios: L_{pl}/L_{bc} (5A.5.1), L_{tsl}/L_{pl} (5A.5.2), and L_{50}/L_{pl} (5A.5.3). These variables are defined in Figure J.B.1.

The curve is developed for the 3 damage states and 2 loads in an attempt to capture the most significant contributions to the conditional LOCA probability. These events will be interpolated and extrapolated as necessary to develop continuous relationships as a function of damage state and loading magnitude. This information can then be combined with a plant's seismic hazard curve as well as knowledge about the relationship between the hazard curve and actual piping stresses to determine actual LOCA frequencies due to a seismic event. They could also be used to determine the LOCA contribution of other large transients knowing both the transient frequency and the relationship between the transient applied loading magnitude and the ASME service level loading magnitudes.

APPENDIX K

**GENERAL APPROACH AND PHILOSOPHY
OF EACH PANEL MEMBER**

APPENDIX K

GENERAL APPROACH AND PHILOSOPHY OF EACH PANEL MEMBER

In this appendix the general approach and philosophy that each panelist followed as part of this elicitation exercise is presented.

BRUCE BISHOP

For PWR piping frequencies, the median probability of a 5,000 gpm (19.000 lpm) leak after 40 years of operation comes from the average point estimate for 7 plants that used the PFM methodology for the WOG Piping RI-ISI (WCAP-14257, Rev. 1-NP-A, Supplement 1). These seven plants were selected to provide a representative sampling of all plants with a Westinghouse NSSS design. Characteristics considered in the sampling included number of primary loops, old and new design vintage and foreign and domestic utility operators. The variability in 40-year probability with leak-rate comes from a WOG supported sensitivity study that reflected both the decrease in probability with increasing leak rate of one pipe size and the number of pipes of a given size that could contribute to a given leak rate. All piping leak probabilities consider the effects of LBB with a minimum detectable leak of 1 gpm (3.8 lpm) per typical plant tech-spec requirements. The increase in failure probability in going from 40 years to 60 years of operation is based upon another WOG sensitivity study. This study and its results are described in a paper presented at the 1999 Pressure Vessel and Piping Conference of ASME and included in PVP-Volume 383.

Non-Piping Frequencies are based upon the degradation mechanism of PWSCC initiation and through-wall growth, which is currently the primary cause of unexpected leaks in non-piping components in the primary system. Most other degradation mechanisms are being effectively mitigated. The relative frequencies by component type are based upon a proprietary best-estimate of PWSCC susceptibility by Westinghouse experts for unmitigated Alloy 600/182 base/weld metal. The uncertainties are based upon the variability between the best-estimate susceptibility for PWSCC and observed leak experience.

VIC CHAPMAN

In order to derive a basic set of failure probabilities, the values generated by the 'Base Cases' analysis were initially considered. However, in the end, a decision was made to use the results from some previous work that involved a 'Risk-Informed ISI' application. That work considered a full plant assessment using fatigue as the basic degradation mechanism. Initially, the results from this full plant assessment were compared with the appropriate base cases in order to ascertain whether they were in general agreement with each other. Once it was decided that the two sets of results were in agreement, it was decided to proceed with using the full plant assessment results. These results provided a set of pipe weld failures over a full range of pipe weld sizes that could be considered as a form of global values for each weld size. Factoring would then be from this base set.

Since the leak rate, given a failure, is independent of the failure probability, this can be evaluated separately to obtain a conditional probability. The basic method developed for the base case was expanded to include lower and upper estimates at each step. These basic steps are as follows:
1   Use expert judgment to estimate the COD, up to full rupture, as a function of defect size.
2   Evaluate the defect cross-sectional area for a given defect size using its associated COD.
3   Evaluate the leak rate from a given defect size using some data supplied by the USNRC.
4   Use expert judgment to assess the distribution of the defect length at failure.
5   Combine Steps 3 and 4 to obtain the conditional probability of a leak rate greater than the prescribed leak rate.

The final probabilities were obtained by combining the conditional probability above with the basic fatigue failure probability.

The effect of leak detection was introduced via a factor that was a function of the leak rate. This reduction factor varied from about 5 for a Category 1 leak, up to about 50 for a Category 6 leak.

For non-weld areas, such as the pump bowls and nozzle crotch corners, the basic probabilities were first factored. Next, the basic steps to derive the conditional leak rates as discussed above were followed to adjust the distributions as appropriate.

The effect of PWSCC was introduced as a multiplying factor on the basic fatigue failure rates. It was assumed that for small pipes, 2 inch diameter, that they would still have a significant contribution from fatigue, but that for the largest pipes, the full three orders of magnitude implied by the PWR-1 Base Case (i.e., hot leg base case) should be applied.

Finally, the failure rates for each system were derived by simply summing over all the elements within a given system.

WILLIAM (BILL) GALYEAN

The approach taken by Bill Galyean is based on the total operating experience of U.S. commercial nuclear power plants. This experience consists of approximately 2,650 LWR-years with zero category-1 (> 100 gpm [380 lpm]) loss of coolant accidents. The average age of these plants is approximately 25 years, with a number of plants being 30-plus years old. During this time, a number of RCS degradation issues have arisen and been addressed, for example, IGSCC in BWRs and thermal fatigue in PWRs. The operating experience therefore indicates that degradation will occur, but it will likely be detected and corrected before it can lead to a catastrophic failure. Consequently, this data is the basis for estimating an average LOCA frequency using a Bayesian update of a non-informative prior distribution. Since both PWRs and BWRs have zero LOCAs, the reasonable assumption is that the two designs share a similar LOCA frequency. The operating experience for the two designs is therefore pooled (i.e., use zero failures and the total 2,650 LWR-years of experience). Assuming the LOCA frequency has been (and will be) relatively constant over time (again, this seems reasonable given the history of degradation mechanisms being detected and subsequently mitigated), the resulting LOCA frequency of 1.9E-4/LWR-year produces a probability of one or more LOCA events in the 2,650 LWR years of experience of 39% (again not unreasonable, given there have been zero LOCA events). By contrast, separating the PWR and BWR experience and analyzing them separately produces LOCA frequencies of 2.8E-4/PWR-yr and 5.6E-4/BWR-yr, and a probability of seeing one or more LOCA events (either PWR or BWR) in 2,650 LWR-years experience of 63%. Again, given that there have been no LOCA events, the first (pooled) estimate seems to be the more realistic.

This assumed LOCA frequency (1.9E-4/LWR-yr) was used for the category 1 LOCA (> 100 gpm [380 lpm]). Note that as defined in the elicitation effort, category 1 LOCA includes all larger size categories. So the approach followed by this panel member was to assume a ½ order of magnitude reduction in frequency for each next larger size category. This general approach (if not the precise value of the reduction) has been followed by virtually every LOCA frequency estimate ever made, and is supported by studies on precursor events documented in NUREG/CR-5750, Appendix J.

The time-independent assumption for the LOCA frequency is also based upon the historical experience, if only qualitatively. There seems to be no doubt that the LOCA frequency fluctuates over the age of the plant, but there is reason to believe it will both increase and decrease over time. The IGSCC experience seems to support the assertion that times of increasing frequency will be followed by times of decreasing frequency as degradation mechanisms are identified, understood, and mitigated. Indeed, even the recent RPV-head corrosion event at Davis Besse supports this model of a LOCA frequency increase as degradation occurs undetected, then a decrease as mitigation programs are implemented (e.g., in the case of Davis Besse, replacing RPV head).

The last issue to be addressed is the allocation of the total LOCA frequency among the systems and components that compose the RCS. This aspect again relies upon operating experience data, this time in the form of the relative frequency of crack and leak events (i.e., precursor events). Basically, these precursor data were collected from LER and foreign reactor experience, and then sorted by degradation mechanism and RCS subsystem/component. In many cases the information provided on the precursor event was somewhat unclear or incomplete. Also, there is little assurance that all precursor events have been captured. However, assuming there is no bias in the reporting of the events such that the data samples for each subsystem/component are equally representative of the all events for that subsystem/component, then the data can be used to support estimates of the relative contribution from each subsystem/component. That is, the precursor events do not have to be completely reported, just consistently reported. Further, the RCS subsystem/component boundaries have not been clearly defined. Hence, the relative contributions to the overall LOCA frequency would likely change somewhat if the precursor data were reviewed and categorized by a different analyst. Nevertheless, this aspect of the analysis was performed simply to allocate the total LOCA frequency (described above) to the general subsystems/components that make-up the RCS.

In summary, the entire U.S. LWR operating experience is used to estimate an average industry-wide total LOCA frequency. This frequency is used for both BWRs and PWRs, not because they are believed to be the same, but on the basis that the operating experience does not support different frequencies. Time-independence is assumed using the rationale that variation (both increases and decreases) in the frequency will occur as degradation mechanisms manifest themselves and are subsequently addressed by the industry. This total LOCA frequency is allocated by LOCA size categories using the argument that as pipe-size increases, the LOCA frequency decreases. This argument is supported by a number of studies on precursor data and if nothing else has been reflected in all LOCA frequency estimates since WASH-1400 (1975). The total LOCA frequency is also allocated by RCS subsystem/component using data collected on primary system crack and leak events (although the details of this allocation are view as somewhat subjective with respect to the boundaries of the different subsystems/components).

KAREN GOTT

The approach taken by Karen Gott to the elicitation was to first consider how her experience from the Swedish nuclear fleet was applicable to the US fleet of nuclear power stations. In this respect she took into account the known histories of the various degradation mechanisms which have troubled the two fleets as well as the mitigation methods which have been developed. This led her to amongst other things to the conclusion that the likelihood of an unexpected mechanism leading to failure is probably larger than the likelihood of a known mechanism resulting in failure in a region which is inspected on a regular basis. In general a new area of concern with regard to component degradation has arisen on a seven to ten year cycle over the lifetime of commercial nuclear plants.

To produce the numbers she used her database of failures and degradation in mechanical components in Swedish plants. The degradation mechanisms are the same, but the numerical figures are different because of differences in design and construction. She based her elicitation figures on the number of leaks in proportion to the number of reported cases (many were detected early) and took these to be the current figures for a good safety culture situation. The database includes other mechanical components than pipes, but does not cover steam generator tubes, so she was able to generate figures for pump and valve housings, for example. She then considered the differences in the philosophies concerning qualification and application of inspection programmes between the two countries. This she incorporated into her thinking about the safety culture aspects, both for the current time and for the extrapolation to 40 and 60 years.

DAVE HARRIS

For each plant type and for piping and non-piping Dr. Harris selected a reference system and attempted to scale other systems relative to that reference system. He tried to use estimates based on operating experience to the maximum extent, and then scaled the relative frequencies for the LOCA categories using results from the PFM analyses. In many instances, operating experience was not applicable, so he then relied more on the PFM results. If he felt that a given system was not a significant contributor to the leak frequency for a given LOCA category, then he was less concerned about the accuracy of the frequency estimate for that system.

PWR Non-seismic LOCA: For the PWR case, he used the hot leg as the reference system for large leak flow rates. Operating experience is not readily applicable. The PFM analyses for the hot leg showed a very wide range of results depending on the assumptions and input to the analyses. Therefore, he scaled the hot leg results by use of the RPV reference case results. Results presented at the wrap-up meeting in February 2004 provided an estimate of the RPV > 500,000 gpm (1,900,000 lpm) as 10^{-10} (per plant-year) for the first 25 years. He doubled this value to account for 2 hot legs. He assumed that the leak frequency for 100 gpm (380 lpm) LOCA is 3 ½ orders of magnitude higher, and then interpolated on a log-log scale. This fixes the frequency-leak rate for the hot leg at 25 years. He assumed that the frequency for > 500,000 gpm (1,900,000 lpm) in the time increment 25-40 years is twice that for 0-25 years, and four times as large in the increment 40-60 years. The leak frequency for 100 gpm (380 lpm) is assumed to be independent of time.

The cold leg is then assumed to have frequencies 1/3 those of the hot leg, because the cold leg operates at a somewhat lower temperature. The surge line is assumed to have leak frequencies 100 times as large as the hot leg, because the surge line sees a lot more cycles than the hot leg, and is just as hot. These estimates then define the very large leak frequencies for the entire plant.

At the low end of the leak rate scale, he assumed the plant results to be bounded by the past operating experience for steam generator tubes. An estimate from the wrap-up meeting for the steam generator LOCA frequencies is 3.5×10^{-3} per plant-year.

He used the HPI make-up nozzle as a surrogate for all 2 to 6 inch diameter lines. He used the reference case results from the PFM results that was presented at the wrap-up meeting, but reduced the leak frequencies by an order of magnitude at 5,000 gpm (19,000 lpm). He assumed that the SIS accumulator and RHR systems have about an order of magnitude less contribution than the surge line, so they have a small contribution to the overall plant.

This procedure provides his best estimate. The 5% and 95% estimates are scaled up and down from the best estimate. He estimated the 5% to be 1 ½ orders of magnitude below the best estimate (multiply by 0.03), independent of time and leak rate. He varied the multiplier for the 95% estimate, making it larger for the larger leak categories. The multiplier varied from 30 to 1000. He believes that we have a better handle on the smaller leak rates, because they are bounded by steam generator tubing experience, which is plentiful.

BWR Non-seismic LOCA: He selected the recirculation system for the reference for BWRs. For intermediate leak rates 100 to 25,000 gpm (380 to 95,000 lpm), the 12 inch diameter portion of the recirculation system dominates. He used the base case results from the wrap-up meeting, but reduced them by an order of magnitude, because his PFM analysis underestimated the benefit of the post-remedial action residual stress.

The feedwater system is also important, because it has lots of welds, and is prone to FAC (which is not related to welds). Since this is a dominant system, he assumed it to be comparable to the surge line in a PWR (which is a dominant system for that type of plant). The steam line is about the same size and same material as the feedwater, but is not prone to FAC, therefore he assumed the steam line to have leak frequencies that are two orders of magnitude below the feedwater system. He assumed the RHR line to be the same as the PWR surge line, which is about the same size. NUREG/CR-6674 shows very low probabilities of through-wall cracks in the HPCS/LPCS system, so the contribution of this system was assumed to be negligible. The recirculation, feedwater, steam line and RHR are assumed to be the dominant systems, and no estimates were made for other systems.

The estimated uncertainty bands are generally tighter than the PWR estimates, because they are based more on experience for the dominant system (recirculation). They are independent of time, but do vary somewhat with leak category.

Non-piping: Dr. Harris felt less confident making estimates for non-piping components, because most of his experience is related to piping. He relied heavily on results provided by others in the wrap-up meeting, and used CRDM nozzle PWSCC, rector pressure vessel and steam generator tubing data for reference purposes. He also scaled relative to piping in many instances. He did not estimate time dependencies. For instance, for pumps and valves, he figured that they are less failure prone than the piping system in which they are located (passive failures). He estimated probabilities, and then calculated relative contributions of failure scenarios.

The approach is documented in "Base Case Report No. 2." For systems other than the five Base Case Systems, the base case results established anchor distributions for BWR and PWR Code Class 1 reference piping systems. As an example, the base case results for PWR hot legs were applied to PWR cold legs but adjusted to account for insights about the service conditions and degradation susceptibility specific to cold legs. For the other BWR and PWR piping systems not covered by the base case study, the base case results were adjusted downwards or upwards as appropriate by accounting for unique piping design features (e.g., size, material, and weld population), service conditions and field experience. For non-piping passive components the base case report again was used as the main reference (or source of calibration parameters) in combination with reviews of relevant operating experience. In summary, this Panel Member's response to the elicitation questions is based on insights from degradation mechanism analyses in combination with reviews and statistical evaluations of operating experience.

SAM RANGANATH

One of the challenges in the LOCA frequency estimation is trying to predict the probability of an event that has never happened before, but which has enormous consequences if it did. It is important to maintain a sense of balance in this effort and aim for a realistic approach that is based strictly on technical considerations. As in any probabilistic analysis, the success depends on how realistic the inputs are and how the approach reflects actual field experience. Having worked in the BWR industry for almost 30 years, Dr. Ranganath felt that his most important contribution to the elicitation process was to make sure that that frequency estimates reflect BWR field experience. For example, use of probabilistic defect distribution data is acceptable as long as the prediction is consistent with actual field behavior. Dr. Ranganath's philosophy was to start from actual field data and to predict future behavior based on his understanding of failure mechanisms, mitigation measures and BWR systems design. Since his knowledge is mainly on BWR systems, he focused his attention on BWRs rather than PWRs. He did not want to speculate in areas where he did not necessarily have the expertise. There are other people who are more knowledgeable about PWRs and they can do a better job on the estimates. He felt that the diversity of the elicitation panel and their expertise and the open mindedness of the NRC team helped in coming up with the best estimates.

PETE RICCARDELLA

The first step in the expert panel elicitation was to develop an amalgamated set of base case LOCA frequencies upon which the elicitation responses are anchored. The generic base case LOCA frequencies developed for the panel represented the work of four teams: two teams used an empirical approach based on operating plant experience with leakage and other precursor events, while two other teams used theoretical, PFM analyses. Each of these approaches has different strengths and weaknesses, such that a better estimate of base case LOCA probabilities can be achieved by selectively combining the results in a manner that optimizes the strengths of both. The method and rationale for combining the base case results of the individual teams were documented, ultimately producing a revised set of LOCA frequencies for the five piping base cases.

LOCA frequencies for each of the LOCA sensitive piping systems identified for PWRs and BWRs were then estimated. This was done by picking the base case which is most representative of the specific LOCA sensitive system, considering plant type, material of construction, operating conditions and relevant degradation mechanisms, and then scaling the base case frequencies for each LOCA category based on judgment of any substantive differences between the base case and the system under evaluation. One of the main factors accounted for in this scaling process was differences in the size of the systems, in terms of number of welds of various pipe sizes (based on a system-by-system weld census that was provided to the panel). Scaling was also used to account for system specific factors, such as whether repairs or mitigation have been applied to address degradation mechanisms considered in the base cases, and the timing of such actions. An estimate of the probability of breaks in small diameter socket welded piping (instrument, vent and drain lines) due to vibration fatigue was made, which was not included in any of the base cases. This estimate was based on prior experience with this relatively common failure mechanism.

It was felt that there is a relatively large uncertainty band in all of the above probability estimates; plus or minus an order of magnitude. Included in this uncertainty band is the potential development of new, as yet unseen degradation mechanisms in the future, which obviously weren't considered in the base cases.

A set of base cases for non-piping LOCAs was developed, the methodology for which is documented in Appendix I to this NUREG report. These base cases included potential breaks due to small vessel penetrations such as CRD nozzles, medium size breaks due to larger diameter nozzles (excluding safe-end ruptures which are included in the piping base cases), and very large breaks due to pressure vessel ruptures (specifically addressing irradiation embrittlement of the RPV). The resulting base cases were then used to estimate contributions to LOCA frequency from non-piping LOCAs.

The detailed rationale used in developing the elicitation response for each system was documented in a report, to permit the reconstruction of the logic in the future if it becomes necessary.

HELMUT SCHULZ

The general approach and philosophy used follows the approached taken by GRS to estimate frequency of LOCA initiating events at passive systems for German PSAs.

The major steps and assumptions of this approach are as follows:

-   In principle, wall penetration of pipes which would result into a leak follows in their geometries either

-   • a slit type penetration originating from cracks caused by fatigue or corrosion or
    • a bulging type penetration caused by wall thinning.

Beyond critical dimensions wall penetrating stable defects turn into a full break. This means in practice that for each pipe size there are two or more leak sizes which are of a distinct different probability of occurrence governed by the likelihood of the respective active failure mechanism and the reliability to detect leaks and to take actions to avoid aggravation of the situation e. g. isolation of the leak, stop operation.

The maximum leak size related to a wall penetrating stable defect (undercritical crack, bulge, pit) depends on the actual load specifically the relationship between membrane and bending stresses. The majority of systems being considered in the safety analysis of NPP's fall either into the category of high pressure or low pressure systems. For reasons of simplicity UB values can be taken to describe maximum leak sizes connected to wall penetrating stable defects for each pipe size. Based on experimental evidence as well as fracture mechanics calculations the maximum leak size resulting from an undercritical crack is rather limited, expressed in terms of fractions of the pipe cross section it is only a few percent. This approach uses 2% of the cross section as a rule of the thumb for high pressure systems. Through wall corrosion pits are generally very small. Bulge-type wall penetrations caused by wall thinning have a potential for stable leaks of a considerable size.

-   The frequency of leaks is estimated based on the operating experience of the national population of nuclear power plants and in addition the worldwide experience is considered as far as applicable and available. In general, the operating experience give indications of leaks in a sense of precursors for most classes of piping or give indications of zero failures statistics only.

-   To estimate the probability of a break (which is connected by the diameter of the piping to the flow rate) a relationship is used to describe the frequency of breaks in relation to the frequency of leaks as the function of the diameter of piping systems being designed to the same design parameters. For the small bore piping (less than 2 inch) the relationship between leak and break is arrived from operating experience. For the largest pipe (main primary pipe) the relationship between leak and break is based on a number of technical arguments and PFM analyses. For the pipe sizes in between a linear relationship is used between the UB and LB as described before.

-   For reasons of simplicity and in accordance with technical experience it is assumed that within the piping systems only so called leak relevant elements are contributing to the frequencies. These leak relevant elements are essentially welds which are adjacent to changes in geometry (nozzles, branches, reducers etc.) which in itself would introduce enhanced loads and to some extend represent more difficult areas for manufacturing and inspection.

-   The whole population of pipes, nozzles and penetrations connected to the main components are divided into subpopulations taking pipe diameters as orientation values, using e. g. 5 or 6 subpopulations to represent the difference pipe sizes, materials and operating conditions. For each subpopulation the frequency of leaks is based according to the procedure described before (operating experience, zero failure statistics), the frequency of leaks is adjusted to the size of the relevant population each time and the frequency of breaks within the subpopulation is estimated using the described relationship.

-    The frequency of the different subpopulation which could contribute in a different way (leak or break) to the specified LOCA classes is then summed up. In view of the limitation regarding the verification of very low values of estimated frequencies a cut-off value is used.

FRED SIMONEN

Operating experience was applied as the best method for estimating frequencies for more common failure events such as small detectable leakage and of ruptures for small diameter piping. Operating experience has the advantage of reflecting contributions from all degradation mechanisms and is not limited to a particular mechanism that can be addressed by a PFM model. For lower frequency events, for which there are little or no data from operating experience, the data were therefore supplemented by trends from PFM models. The fracture mechanics models were taken to provide <u>relative</u> frequencies such as for (for a given pipe size) the ratios of frequencies of for different categories of failures (in terms of leak-rates). Similarly, models can indicate ratios for one failure category of leak for differing pipe sizes. Reports of small detectable leakage (from data bases) were taken to be precursor events that can be used to calibrate estimates of frequencies for categories of larger leakage events. Another consideration was that only the larger pipe sizes contribute to the frequencies for larger leak rates. It was implicitly assumed that contributions from smaller pipe sizes dominate for the smaller categories of leak rate frequencies.

Non-nuclear experience was also used to support estimates of failure frequencies for nuclear components. Component designs, materials, construction codes, operating conditions etc. are much the same for nuclear applications and as for many non-nuclear applications. Non-nuclear experience however provides a much larger number of years of plant operation (by orders of magnitudes) than available from nuclear experience. Non-nuclear experience therefore provides additional justification for very low failure frequencies for components such as pump bodies, tube sheets, manways, etc. that imply large extrapolations from the limited years of nuclear plant operation.

GERY WILKOWSKI

The piping non-seismic LOCA evaluations were conducted for PWR and BWR piping separately from a bottom-up approach using reference cases for certain pipe systems in each type of plant. The reference cases were determined from a combination of the base case results supplied to the elicitation panel. The base cases supplied consisted of two independent PFM analyses, and two independent operating-experience evaluations for certain pipe systems. The probabilistic pipe fracture mechanics analyses (PRAISE or PRODIGAL) base cases were not chosen since Dr. Wilkowski did not believe that those computer codes properly determined the probability of a long surface crack occurring, which is the actual way that a LOCA would occur, i.e., a through-wall leaking crack will be readily discovered by leakage before failing at normal operating conditions. Consequently, the two historical base cases were averaged, but only up to 25 years of operation (current time period). Dr. Wilkowski did not believe that the historical based approaches would be that good for extending the LOCA reference cases to 40 or 60 years. Consequently, his reference cases were only for 25-year time period (present), and the 40- and 60-year evaluations were adjusted depending if he thought the particular pipe system would be susceptible to some near term or long-term degradation mechanism (e.g., PWSCC), and if that mechanism could produce a large surface crack. These evaluations were done for 12 different PWR pipe systems and 13 different BWR pipe systems, with six different LOCA flow-rate categories. The uncertainties (5 and 95 percent bounds) in the predictions generally increased as the amount of time increased, i.e., the uncertainty for 25 years was less than 40 years, and the uncertainty in the 40-year predictions was less than for the 60-year predictions.

For non-piping, Dr Wilkowski felt he did not know enough about failure modes of all the different categories of non-piping components (with the exception of a few categories like CRDM nozzle ejection). He therefore chose the steam generator tube historical failure frequencies for small LOCA as a controlling PWR case, but all the other cases were governed by the piping failure probabilities.

APPENDIX L

DETAILED RESULTS

APPENDIX L

DETAILED RESULTS

The detailed results from the elicitation procedure are presented in this appendix. Each panelist's quantitative elicitation responses can be found through the "Electronic Reading Room" link on the NRC's public website (http://www.nrc.gov/) using ADAMS. The documents are found in ADAMS using the following accession number: ML080560005. Both the quantitative (i.e., numerical LOCA frequency estimates) and qualitative results (i.e., rationale) are presented in this appendix. The quantitative results are often presented in the form of *box and whisker plots*. Box and whisker plots (often referred to simply as *box plots*) are a graphical representation of a data set. A box plot is typically based on five data points: the minimum and maximum of the data set plus the median, lower quartile (LQ), and upper quartile (UQ) associated with the data set. The LQ and the UQ are the 25th and 75th percentiles, respectively, of the data set.

An example box plot is shown in Figure L.1. The region defined by the LQ and UQ is drawn as a shaded box with a vertical line through the median. In this appendix, the 10th and 90th percentiles are also indicated in the box plots with short vertical lines. Note, for the plotting program used to generate these box plots, the vertical lines (representative of the 10th and 90th percentiles) and the associated horizontal connecting them only appear when there was at minimum of nine points in the data set being plotted. Finally, the box plots are also overlaid with a horizontal scatter plot of the data set, with the left-most point being the minimum value and the right-most point being the maximum value in the data set. Note that the 10th and 90th percentiles are not necessarily points in the data set. The range in the data set encompassed by the shaded box (i.e., the range between the 25th and 75th percentiles) is referred to as the interquartile range (IQR) of the box plot. In the box plots in this appendix, letter designators are often included for the minimum and maximum values (e.g., the letters G and D in the example shown in Figure L.1). The letters designate the code for the panelist whose estimated value was either the lowest or the highest value of all the panelists who provided responses for the quantities in the data set.

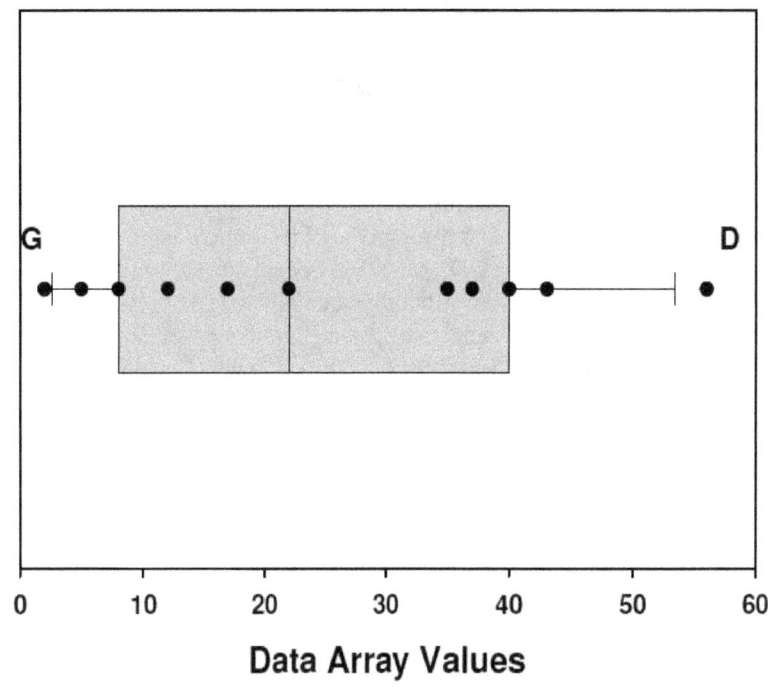

Figure L.1. Example "Box and Whisker" Plot

Generally, the source of the qualitative responses (i.e., the rationale) came from the individual elicitations although there were some opinions expressed during the various plenary panel meetings that were also included. For each of the individual elicitations, minutes were taken. Minutes were also taken at each of the plenary panel meetings (see Appendix B). In addition, the participants often provided a handout to lead the discussion at their individual elicitations. After each elicitation, most of the participants also provided formal written responses to the elicitation questions. It was from these minutes, handouts, and written responses that the rationale provided below was extracted. Finally, each of the elicitations was audio taped and each meeting was video taped to provide a permanent record of the proceedings.

Most of the panelists believe that precursor events (e.g., cracks and leaks) are a good barometer of LOCA susceptibility. This is reflected in the fact that almost all the panelists anchored their responses against some form of the available operational experience data. A distinct advantage of the operational experience data is its inclusion of all degradation mechanisms which have emerged to date, while the PFM approaches only address selected degradation mechanisms. The advantage of the PFM approaches is that they are best suited for addressing LOCA size and operating time effects. A number of panelists used the PFM results as a basis for adjusting the operational experience data in this manner.

A major assumption made in the elicitation procedure is that all components, both piping and non-piping, were fabricated in accordance with applicable code standards, e.g., there were no counterfeit bolts used.

L.1 Safety Culture

Figures L.2 and L.3 show the effect of the industry and regulatory safety culture, respectively, on the LOCA Ratio (i.e., the ratio of the LOCA frequency in the future to the LOCA frequency at 25 years) for Category 1 LOCAs. Figures L.4 and L.5 show the effect of industry and regulatory safety culture on the

LOCA Ratio for Category 4 LOCAs. Ratios less than 1.0 are indicative of a perceived reduction in the LOCA frequency as a result of improvements in the safety culture mindset. As can be seen in these figures, the panel members overwhelming expected the safety culture to either improve or remain constant over the next ten to fifteen years and beyond. The panel felt that the industry as whole was acting in a consistent manner. However, a few plants with a less diligent safety culture mindset would provide the greatest challenge from a LOCA perspective. It was thought that these outlier plants may not affect the mean trends, but could strongly influence the bounds. The Davis-Besse experience was frequently cited as an example of this effect. The panel also expressed the opinion that the industry and regulatory safety culture are highly positive correlated. Therefore, regulatory and industry changes are expected to occur virtually simultaneous.

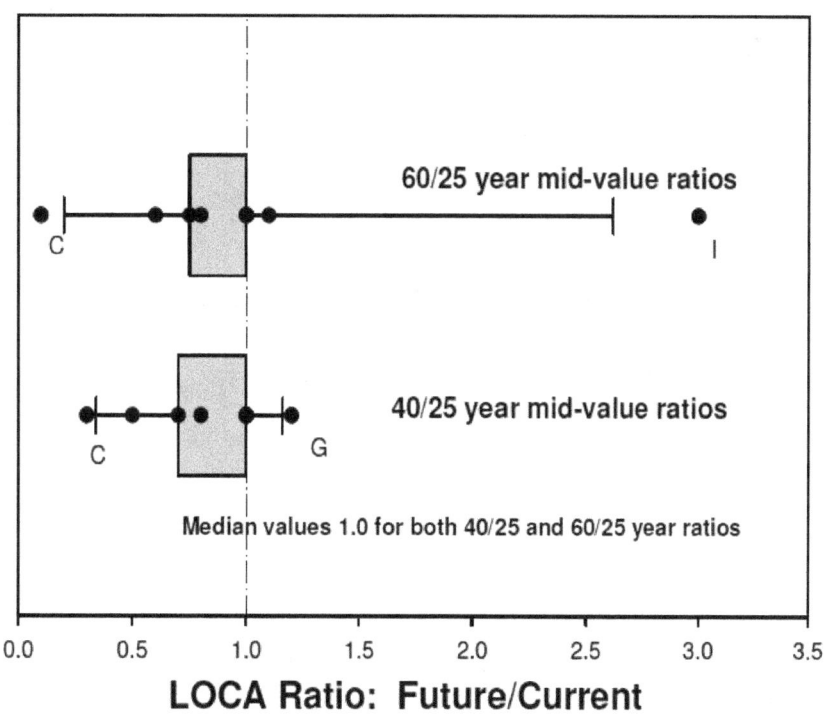

Figure L.2 Effect of Utility Safety Culture on Category 1 LOCAs

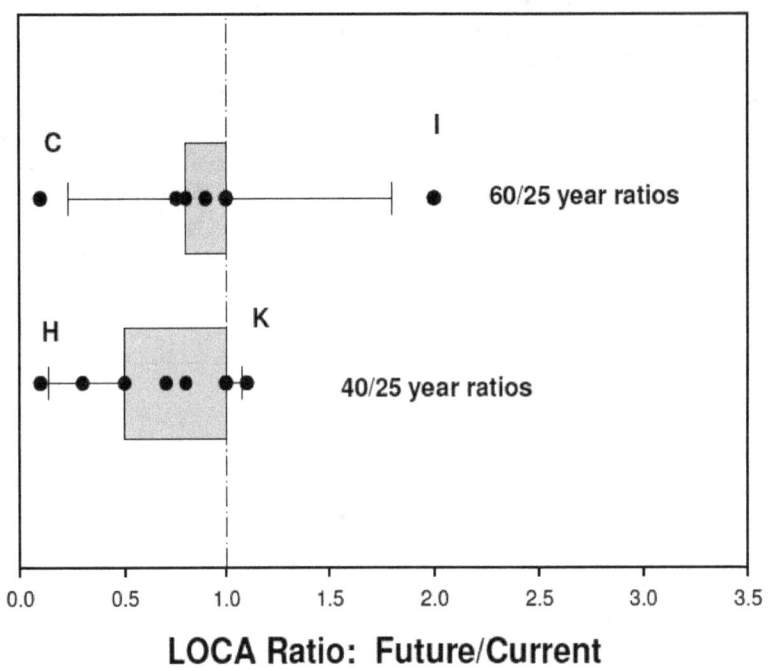

LOCA Ratio: Future/Current

Figure L.3 Effect of Regulatory Safety Culture on Category 1 LOCAs

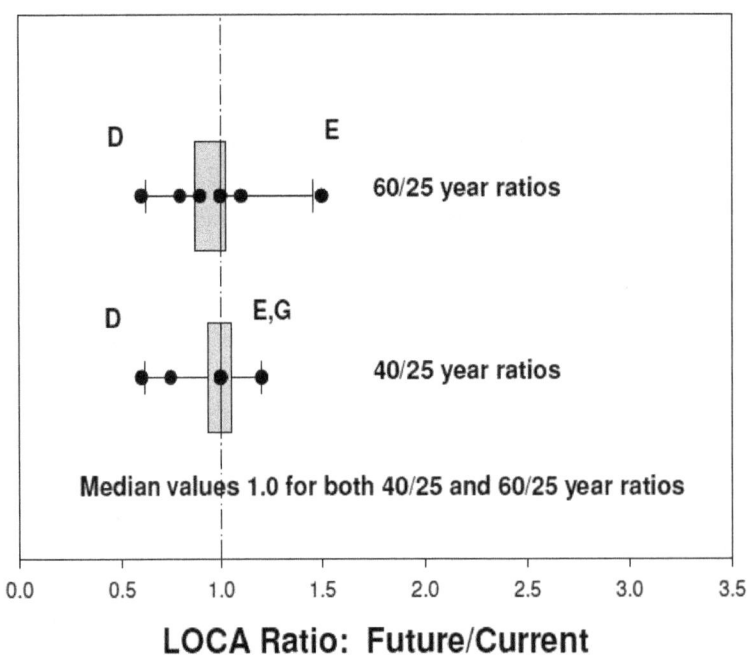

LOCA Ratio: Future/Current

Figure L.4 Effect of Utility Safety Culture on Category 4 LOCAs

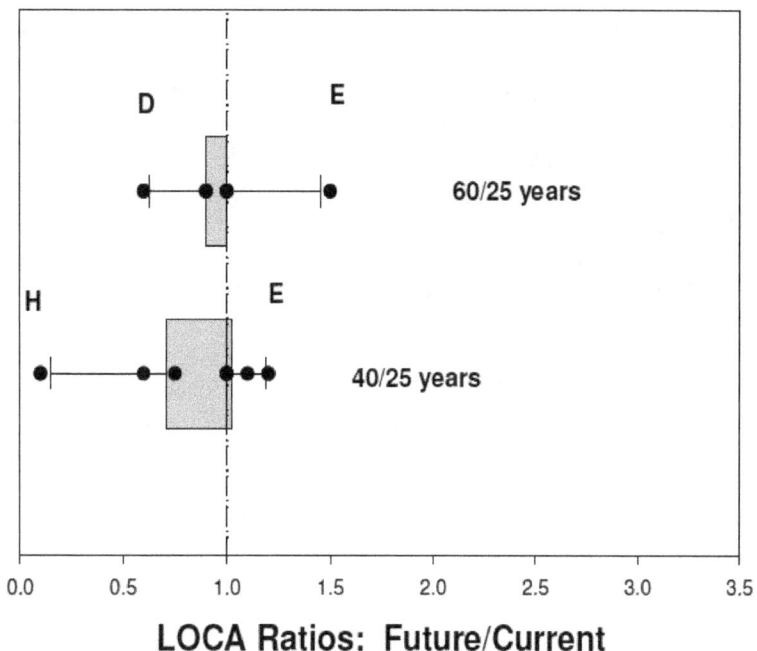

Figure L.5 Effect of Regulatory Safety Culture on Category 4 LOCAs

As can be seen in comparing Figures L.2 with L.4 and L.3 with L.5, the panel members felt that any improvements in safety culture would be more beneficial for the smaller LOCA categories than their larger counterparts because the smaller LOCA categories constitute the bulk of the experience base. The frequency of the larger LOCA categories due to safety culture effects is expected to remain relatively constant over time.

The bottom line from this discussion is that because the panel members felt that the effect of safety culture was relatively minor, the LOCA frequencies developed during this exercise were not modified to account for this effect. The main caveat to this general conclusion is the previously mentioned concern that the LOCA frequencies developed through the elicitation process could be significantly degraded by a safety-deficient plant operating philosophy. The other concern frequently expressed was that the industry safety culture mindset may deteriorate near the end of a plant's license as management tries to "squeeze out" the final few years of operations without investing in the necessary maintenance activities. Also, near the end of the plant's license there was a concern expressed that the morale of the plant's operating staff may begin to erode as they foresee a potential loss of employment. These concerns are manifested in the higher LOCA Ratios for the 60/25 year results when compared with the 40/25 year results in Figures L.2 through L.5.

L.2 BWR Piping

The participants generally thought that the important degradation mechanisms for BWR piping were thermal fatigue, FAC, and IGSCC. It was argued that BWR plants are more prone to thermal fatigue problems than the PWRs because they experience a greater temperature fluctuation during the normal operating cycle. In BWRs, thermal fatigue is a concern for the feedwater lines, the main steam lines, and the RHR system. From a LOCA perspective, thermal fatigue is an important aging mechanism because it does not manifest itself as a single crack, but as a family of cracks over a wide area. As such, it can lead to a large LOCA. Thermal fatigue cracks also tend to propagate rapidly, and since it is not material sensitive (i.e., it can attack a number of materials), it is difficult to prioritize critical areas for inspections.

Only the feedwater piping system is highly susceptible to FAC. The main steam line is the other major carbon-steel piping system which experiences constant fluid flow. However, it is not as susceptible to FAC because the erosion rates associated with two-phase flow are less severe. While FAC caused a serious accident in the secondary side piping at Surry 15 years ago, the panel members generally thought that the industry had inspection programs in place today to prevent the reoccurrence of such an event, especially for the primary side piping systems. However, a number of panel members expressed the concern that the water chemistry improvements which mitigate IGSCC could lead to unexpected FAC problems.

The panel consensus is that the susceptibility to IGSCC is greatly reduced compared to the past. Measures such as improved HWC, weld overlay repairs, and pipe replacement with more crack resistant materials had essentially reduced the likelihood of IGSCC. However, there is still residual concern about the failure likelihood of the large recirculation piping material that has not been replaced. Furthermore, even for the pipe which has been replaced, the question was raised as to whether the new replaced pipe was immune to this type of degradation, or is the problem simply been move out into the future. The German experience with Type 347 stainless steel was raised in this regard. There was also concern expressed about the effects of increased sulfate levels in the future due to efforts focused at extending the life of some of the filters in the plants.

Another aging mechanism of concern is mechanical fatigue. This is primarily a problem in smaller diameter piping, especially those with socket welds, and is caused by an adjacent vibration source. From a LOCA perspective, it was noted that locations susceptible to mechanical fatigue damage were not always obvious. It is impossible to eliminate all plant vibrations, and furthermore, changing the configuration of the plant can result in newly susceptible areas.

As part of this elicitation exercise a total of 14 LOCA-susceptible piping systems were considered for the BWR plants. Of these, however, most of the participants focused on a few common systems as being the important LOCA contributors. Figure L.6 shows the Category 1 LOCA frequencies for each of these piping systems at 25 years of plant operation (present day). Note, the results for the HPCS and LPCS systems are combined as a single entry in Figure L.6 (HPCS/LPCS). For these smaller category LOCAs, the main concern is with the smaller diameter lines, such as the instrument and drain lines. Most of the participants believe that it is more likely to have a complete break of a smaller diameter line than a comparable size opening in a larger diameter pipe. One reason for this is that for a given crack size, the crack is a larger percentage of the pipe circumference in the smaller diameter pipes, and it was thought that a small diameter pipe was just as likely to have a crack of a certain length as a larger diameter pipe. Furthermore, smaller diameter lines are often fabricated from socket welded pipe which has a history of mechanical fatigue damage from plant vibrations. These lines may also be susceptible to external failure mechanisms arising from human error (e.g., damaging with equipment, such as fork trucks). Finally, these smaller diameter lines are often subject to fabrication flaws and they are typically more difficult to inspect, if they are inspected at all. In-service inspection is not routinely performed on these lines. Conversely, the larger diameter lines are inspected more rigorously and routinely.

Besides the instrument and drain lines, the recirculation and, to a slightly lesser extent, the CRD and RHR lines are also of concern, primarily as a result of SCC susceptibility.

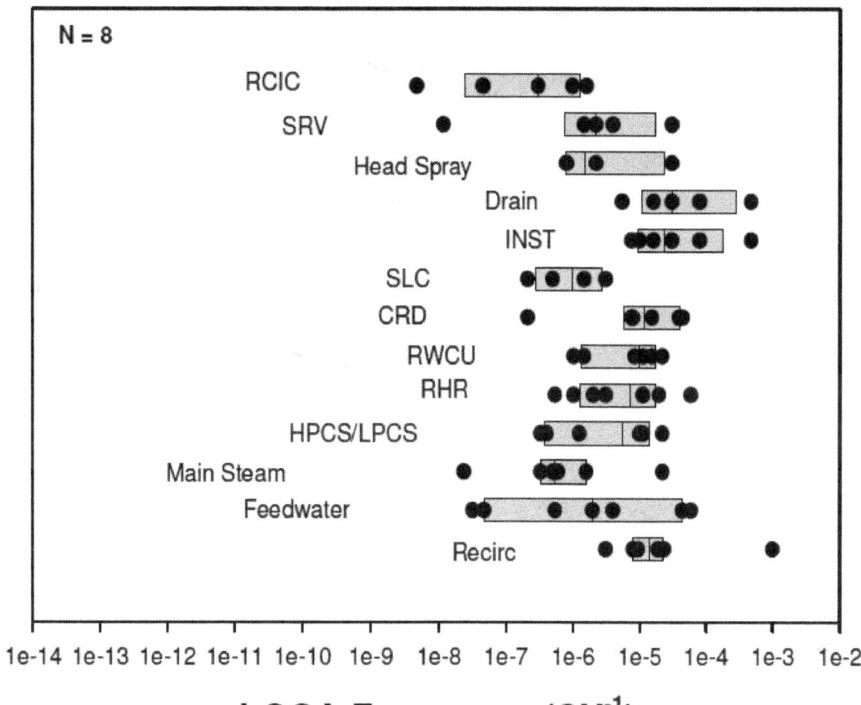

Figure L.6 Category 1 LOCA Frequencies for BWR Piping Systems at 25 Years of Plant Operation

For larger Category 3 LOCAs, the recirculation system was the largest contributor to the overall LOCA frequencies, see Figure L.7. (Note in this figure that the instrument and drain lines, as well as the CRD lines, are no longer shown in that these smaller diameter lines cannot support a Category 3 LOCA.) The fact that the recirculation system is the largest contributor is a slight departure from the PWR estimates where the smallest diameter piping system that can support a particular LOCA category consistently had the highest LOCA frequencies. The main concern with the recirculation system piping continues to be SCC, even when considering the effective mitigation programs in place today. Of secondary importance were the feedwater, RHR, RWCU, core spray, and SRV systems. There was wide variability expressed for the feedwater system. Several participants thought that its susceptibility was similar to that of the recirculation system while others thought that it would make an inconsequential contribution. This latter group generally thought that the mitigation programs in place for the feedwater system were overall effective. The RHR system was deemed important by some panel members due to the relatively larger number of precursor events reported and the relatively high number of welds. A number of the participants used the weld census data provided to differentiate the relative contributions between systems for those systems that have similar operating experience. The SRV lines were judged to be potentially problematic by four of the eight respondents who addressed the question of BWR piping. They pointed out that the SRV lines are subject to high dynamic loads during the relatively common SRV discharge events, however, only a short section of these lines are actually susceptible to a LOCA event. Overall, in comparing Figure L.6 with Figure L.7, one can see approximately a one order of magnitude reduction in the LOCA frequency between the Category 1 and 3 LOCAs for most of the BWR piping systems considered.

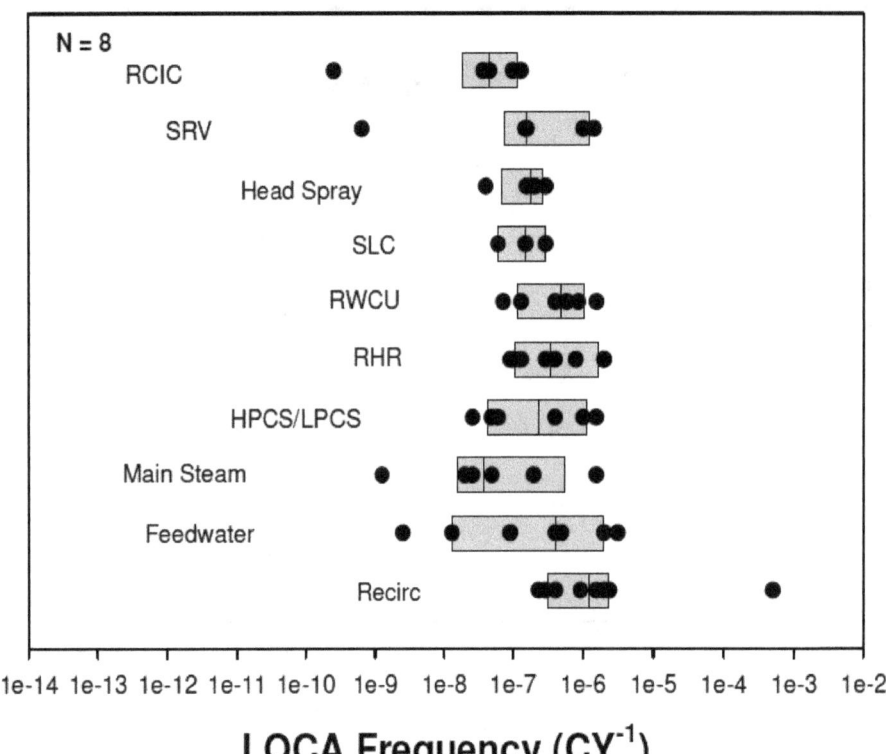

Figure L.7 Category 3 LOCA Frequencies for BWR Piping Systems at 25 Years of Plant Operation

For the largest category BWR piping LOCAs (Category 5), the recirculation system remains the main contributor to the overall LOCA frequencies, see Figure L.8. The RWCU system had about the same median value, however, there was a question expressed as to whether the RWCU system could actually sustain such a high flow rate LOCA. One of the participants thought that the maximum diameter for this system was only 6-inches, not 24-inches as specified in the development of the elicitation questions. Besides the recirculation, and RWCU systems, the next two largest contributors to the BWR Category 5 LOCA frequencies were the feedwater and RHR systems. As for the Category 3 LOCAs, the RHR system was deemed important due to the large number of precursor events reported and the large number of potentially susceptible welds. Several of the participants indicated that these lines are susceptible to SCC.

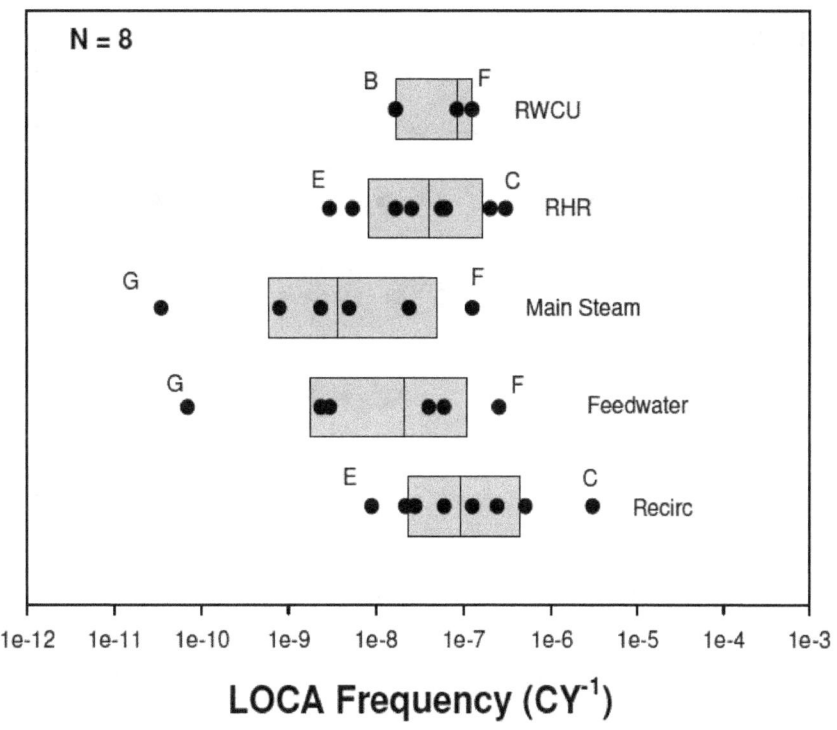

Figure L.8 Category 5 LOCA Frequencies for BWR Piping Systems at 25 Years of Plant Operation

Figure L.9 is a plot of the cumulative BWR piping LOCA frequencies (including contributions from all of the piping systems) for Category 1 through 5 LOCAs. The BWR piping LOCA frequency decrease with LOCA size is relatively shallow, i.e., approximately ½ order of magnitude per LOCA category. The results tend to be governed by the results from the recirculation system. It was noted that for the recirculation system that the mitigation programs in place for controlling IGSCC promote a more uniform residual stress field which can in turn promote longer cracks which are more likely to cause a LOCA. This effect will potentially offset the overall reduction in crack growth due to the mitigation program. It is also of note from Figure L.9 that the variability in the results as expressed by the interquartile range and the difference between the minimum and maximum values does not vary much with LOCA size. It is also of note that the expert ranking is relatively consistent with LOCA size, i.e., Participant C always predicted the highest LOCA frequencies and Participants E and G consistently predicted the lowest LOCA frequencies.

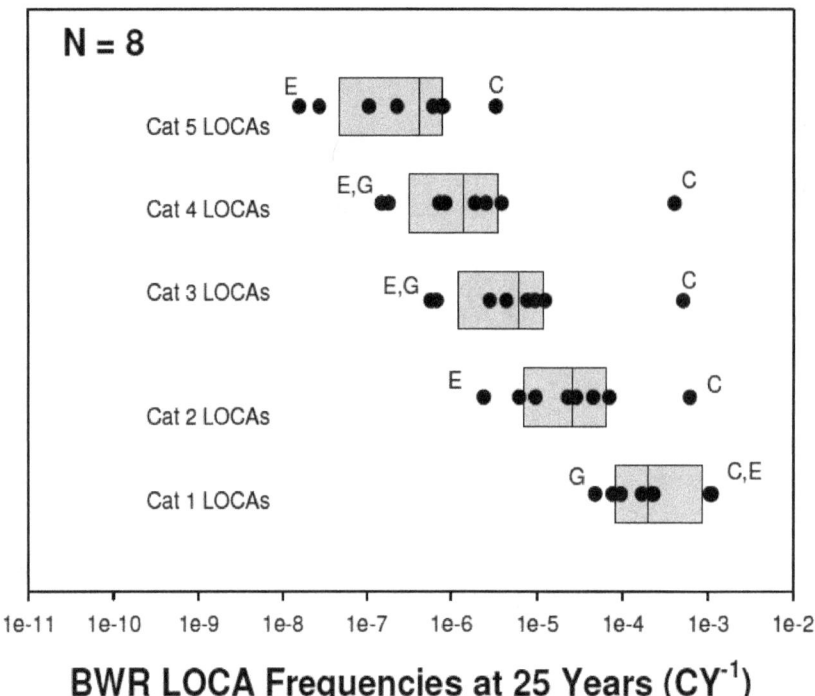

Figure L.9 Cumulative BWR LOCA Frequencies at 25 Years of Plant Operations

Figure L.10 shows the effect of operating time on the cumulative Category 1 LOCA frequencies for BWR piping systems. As can be seen in Figure L.10, there is not much of an effect of operating time on the cumulative Category 1 frequency. Similar findings were evident for the larger Category 3 and 5 LOCAs. Obviously, any unabated degradation mechanism would cause an increase in the overall LOCA frequencies. However, it was generally assumed by the panel members that any new degradation mechanism that came on the scene would be aggressively met by the industry and NRC, just like the IGSCC problem in BWRs was met in the past and the PWSCC problem in PWRs is being met today. The minimal changes in LOCA frequencies with time evident in Figure L.10 were the result of a number of compensating factors considered by the panel members. From the perspective of potential decreases in the LOCA frequencies, the recirculation lines should see a decrease in the LOCA frequencies with respect to the current-day estimates that are based on an analysis of operational experience data due to improved mitigation strategies that have been put in place. The panelists generally felt that the IGSCC issue for BWRs had been effectively mitigated for the foreseeable future. In addition, the core spray systems may see a decrease in the LOCA frequencies with time as the segments of stainless steel piping potentially susceptible to IGSCC are replaced with carbon steel piping. Finally, future inspection and mitigation programs are expected to lead to additional decreases in the predicted LOCA frequencies. In this regard, having the industry focus its inspection resources on the more important systems through risk-informed ISI should help reduce the propensity for LOCAs. Counteracting these potential decreases are potential increases due to bigger thermal fatigue and FAC concerns in the future. Concern was expressed about the high usage factors that will exist near the end-of-plant license. Also, there is the concern with new, previously unknown degradation mechanisms that may arise in the future. In this regard, the inspection methods of today may not be reliable for these new mechanisms. Furthermore, these new mechanisms may not manifest themselves in the same locations of concern today. Finally, while timely and proper maintenance programs are always beneficial, there are instances in which they may prove counterproductive. The frequent opening and closing of systems for inspections increases the likelihood for human error such as having tools and other debris left behind or bolts not being torqued properly. Also, improper service of active components (e.g., valves) can lead to passive system failures.

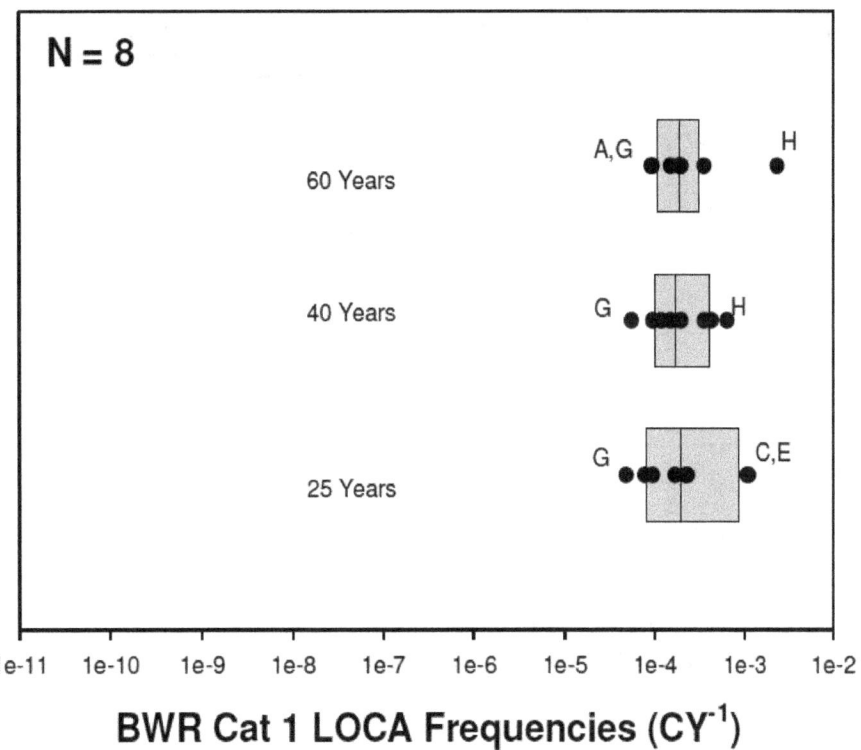

Figure L.10 Effect of Operating Time on the Cumulative Category 1 LOCA Frequencies for BWR Piping Systems

Figures L.11 and L.12 show the cumulative MV estimates, along with the 5% and 95% bound values for the various participants for the Category 1 and 3 LOCAs, respectively. The uncertainty range (difference between 5% LB and 95% UB values) for the Category 3 LOCAs are comparable (or slightly greater than) for the Category 1 LOCAs. Only participants A, E, and F expressed considerably more uncertainty for the Category 3 LOCAs than they did for the Category 1 LOCAs. Similar findings were found when comparing the Category 5 results with the Category 3 results. Overall, the panelists appeared more confident about their BWR estimates than they did for the corresponding PWR estimates. They had less uncertainty about future and bigger size LOCA frequencies compared with their PWR predictions. There was also less uncertainty among the panelists about the magnitude of the dominant contributing factors. In addition, the panel members used more consistent approaches and more consistent base case estimates for the BWR estimates than they did when making their estimates for PWRs.

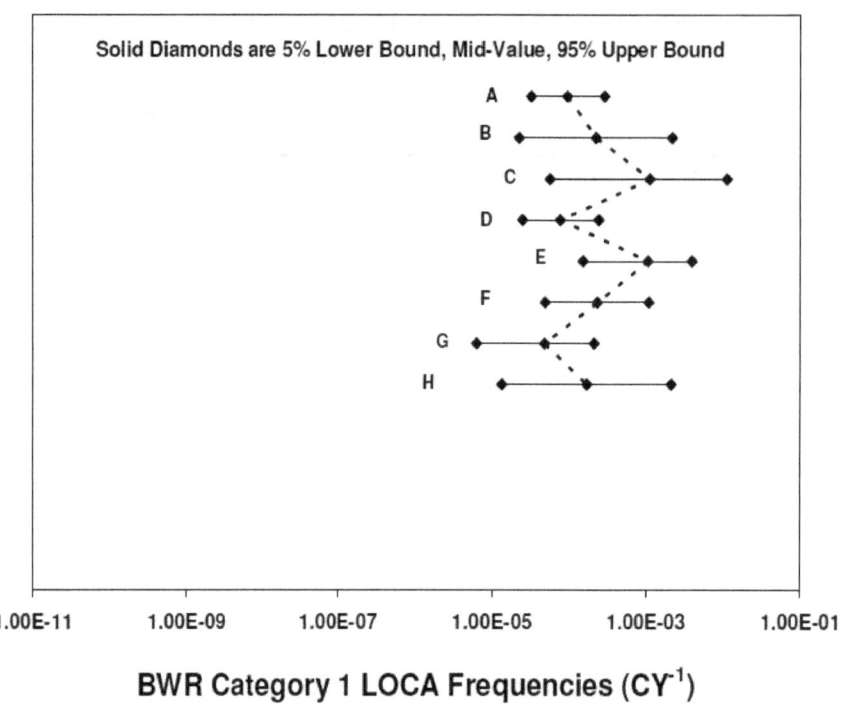

Figure L.11 BWR Category 1 LOCA Frequencies Showing MVs, 5% LB, and 95% UB Values for All Participants Who Responded to the BWR Piping Questions

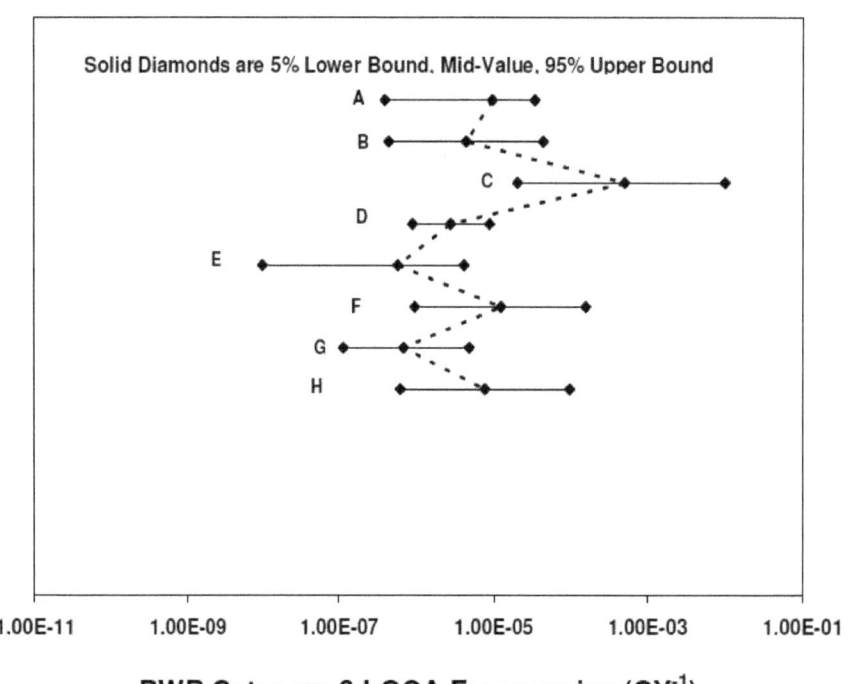

Figure L.12 BWR Category 3 LOCA Frequencies Showing MVs, 5% LB, and 95% UB Values for All Participants Who Responded to the BWR Piping Questions

L.3 PWR Piping

The primary aging mechanisms identified by the participants for PWR plants are thermal fatigue, PWSCC, and mechanical fatigue. The concerns associated with thermal and mechanical fatigue in PWR plants are similar to those in BWR plants. In PWRs, the surge line is thought to be susceptible to thermal fatigue due to cyclic thermal stratification stresses. Also, the DVI and CVCS lines were thought to be susceptible due to periodic testing which imposes additional thermal cycles. In addition, the DVI and Accumulator SIS lines have experienced some thermal fatigue cracking issues due to leakage of cold water past the check valves in these systems.

PWSCC is a relatively new mechanism that has manifested itself in this country over the last 5 years. It has many similar characteristics to the IGSCC problem experienced in BWR reactors in the past. It is a temperature dependent mechanism that attacks Alloy 600 type materials such as bimetallic Inconel 82/182 welds. Many panel members believe that PWSCC problems will be resolved (i.e., mitigated) over the next 15 years. Therefore, its contribution to the overall LOCA frequencies may peak between the 25 and 40 year time period, but then decrease in the future. Today, instances of PWSCC have been observed in surge lines at the surge line to pressurizer weld in the United States at Three Mile Island, in Belgium and Japan, and in hot legs at the hot leg to RPV weld in the United States at V.C. Summer and Sweden. Other lines in which PWSCC may become an issue in the future are the cold leg and the pressurizer spray lines. However, since the cold leg operates at lower temperatures than the hot legs and surge lines, any problems that may materialize in the cold leg will be delayed until later in the operating life.

As part of this elicitation a total of 12 LOCA sensitive piping systems were considered for the PWR plants. However, as was the case for BWR piping, most of the participants focused on a few common systems as being important LOCA contributors. Figure L.13 shows the Category 1 LOCA frequencies at 25 years for the 12 PWR LOCA sensitive piping systems. As was the case for the BWR piping systems, the Category 1 LOCA frequencies are dominated by the small diameter instrument and drain lines. The estimated Category 1 LOCA frequencies for these lines are two orders of magnitude higher than the hot leg and surge lines. This again reflects the belief that a complete break of a smaller pipe is more likely than a partial break of comparable size of a larger pipe. The reasoning provided by the panel members for why these small diameter lines are so susceptible to these smaller category LOCAs are the same as provided for the case of BWRs. These 1- and 2-inch socket welded lines are susceptible to vibration fatigue concerns. Also, they are more difficult to inspect, if they are inspected at all, than their larger diameter counterparts. Generally speaking, the benefits of leak detection and ISI decrease with decreasing pipe size. Finally, they are more susceptible to other, non-aging related failure mechanisms, e.g., an externally applied overload.

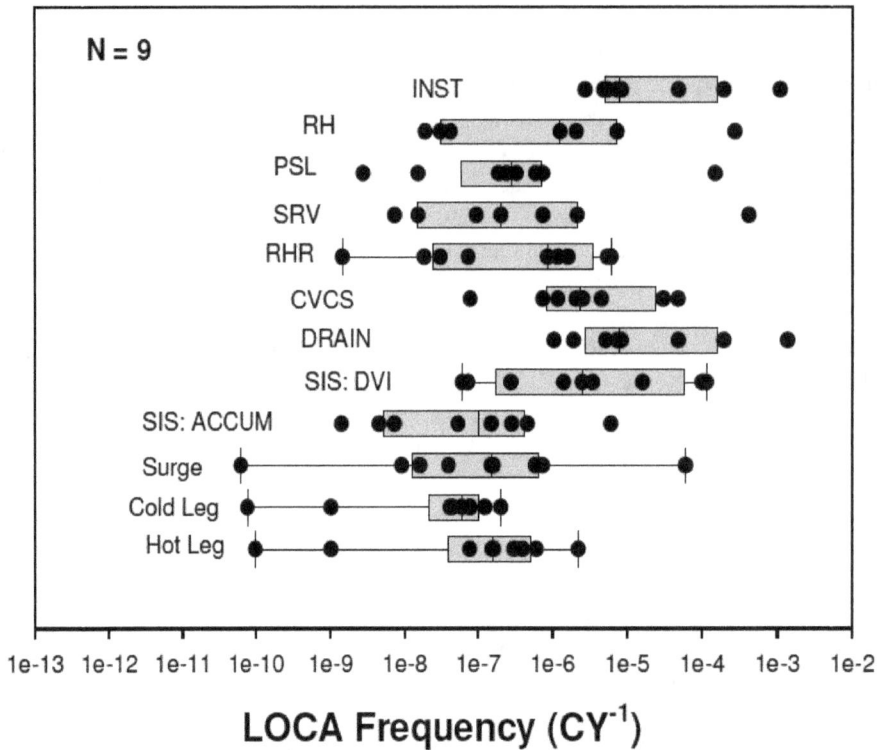

N = 9

INST
RH
PSL
SRV
RHR
CVCS
DRAIN
SIS: DVI
SIS: ACCUM
Surge
Cold Leg
Hot Leg

1e-13 1e-12 1e-11 1e-10 1e-9 1e-8 1e-7 1e-6 1e-5 1e-4 1e-3 1e-2

LOCA Frequency (CY⁻¹)

Figure L.13 Category 1 LOCA Frequencies for PWR Piping Systems at 25 Years of Plant Operation

Figure L.14 shows the breakdown of PWR Category 3 LOCA frequencies by piping system at 25 years of plant operations (present day). The small diameter instrument and drain lines, as well as the RH lines, do not appear on this figure in that they are of such size that they could not sustain a Category 3 LOCA. Again, as was the case for the PWR Category 1 LOCAs, the smallest diameter lines that can sustain this size (i.e., category) of LOCA are the dominant contributors. These include the CVCS, SIS-DVI, RHR, surge, and PSL. This is different than what was observed for the BWR Category 3 LOCAs where the larger recirculation system was the dominant contributor, primarily due to its susceptibility to IGSCC. The two most listed systems as being major contributors to this category of LOCA for PWR piping were the CVCS and SIS-DVI lines. For both, the primary concern was fatigue. One participant commented that the CVCS line was one of the most fatigue sensitive locations in the entire plant. Another commented that they were concerned with environmentally-assisted fatigue for this system. With regard to the SIS-DVI (and the SIS-Accumulator lines for that matter), several participants indicated that both lines had experienced thermal fatigue cracking in the past due to cold water leaking past the check valves. Another line that a number of participants thought would be a major contributor to this category of LOCA was the RHR lines. The concern with these lines was with environmental attack due to the stagnant nature of the flow in these lines. The pressurizer spray lines were of a concern due to the chance for PWSCC at one of the bimetal welds.

L-14

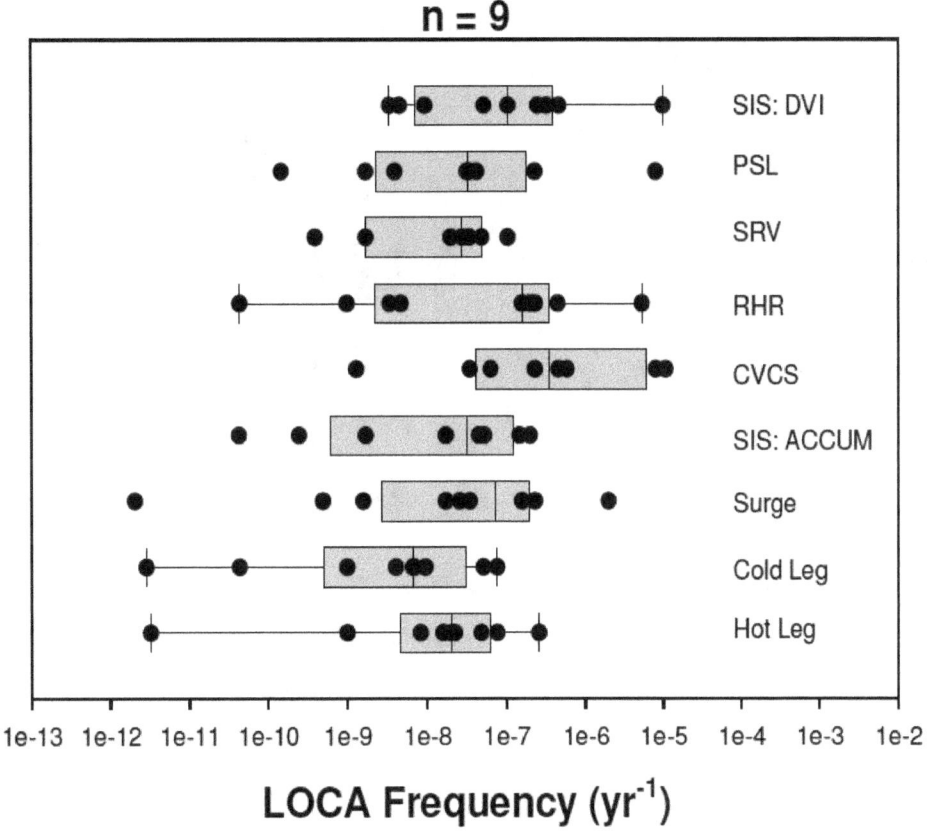

Figure L.14 Category 3 LOCA Frequencies for PWR Piping Systems at 25 Years of Plant Operation

For the largest categories of PWR piping LOCAs (Categories 5 and 6), the hot leg, cold leg, surge line, and RHR lines all contribute to the overall LOCA frequencies, see Figure L.15 for the Category 5 LOCAs. Of these, the median value of the LOCA frequency for the cold leg is about a half order of magnitude less than the median values for the other three piping systems. This slight reduction is primarily due to the fact that the cold leg is less susceptible to PWSCC than either the hot leg or surge line at this time (25 years of plant operations) due to the fact that it operates at a slightly lower temperature. Somewhat surprisingly in examining Figure L.15, a number of the participants felt that the hot leg would have a greater propensity for a Category 5 LOCA than the surge line. Both lines are susceptible to PWSCC due to the presence of bimetallic welds and the high operating temperatures, but the surge line was also judged to be susceptible to thermal fatigue due to thermal stratification and thermal striping stresses. Also, the surge line is smaller diameter, which based on the thought that smaller diameter lines are more prone to LOCAs than their larger counterparts, would imply that the Category 5 LOCA frequencies for the surge line should be higher. Finally, at least one participant argued that the surge line to pressurizer bimetallic weld was one of their biggest concerns in the entire plant due to its susceptibility to PWSCC and the fact that it is a very difficult weld to inspect. Counteracting these arguments, however, is the fact raised by a number of the participants that there are more hot leg to RPV bimetal welds (2 to 4 depending on the number of loops) in a plant than there are surge line to pressurizer bimetal welds (one).

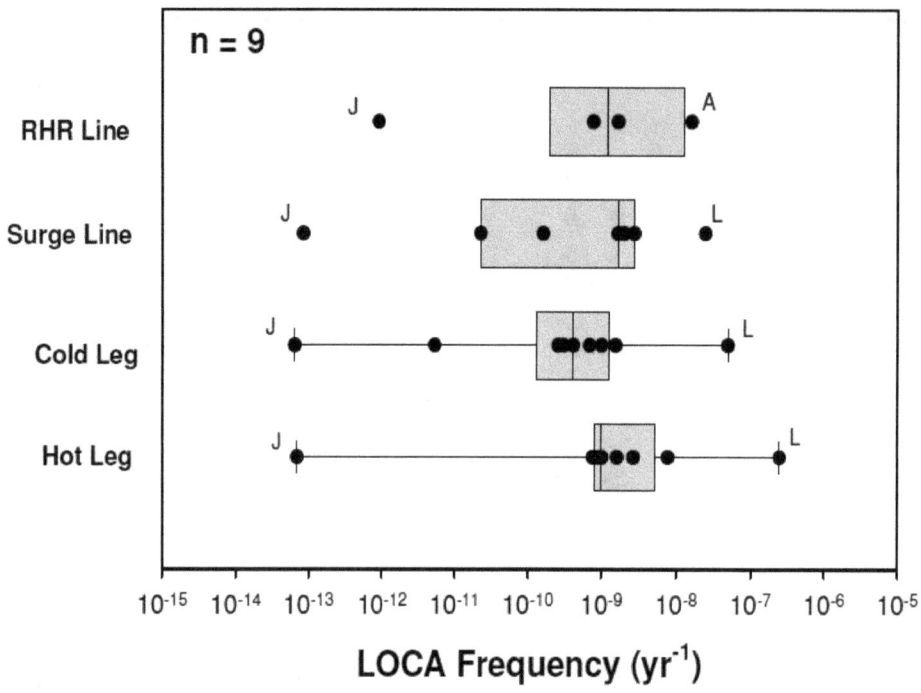

Figure L.15 Category 5 LOCA Frequencies for PWR Piping Systems at 25 Years of Plant Operation

Figure L.16 is a plot of the cumulative PWR LOCA frequencies at 25 years of plant operation. Cumulative frequencies are shown for Category 1, 3, and 6 LOCAs. Based on a review of Figure L.16 there appears to be approximately a one order of magnitude reduction in LOCA frequency between each successive LOCA category.

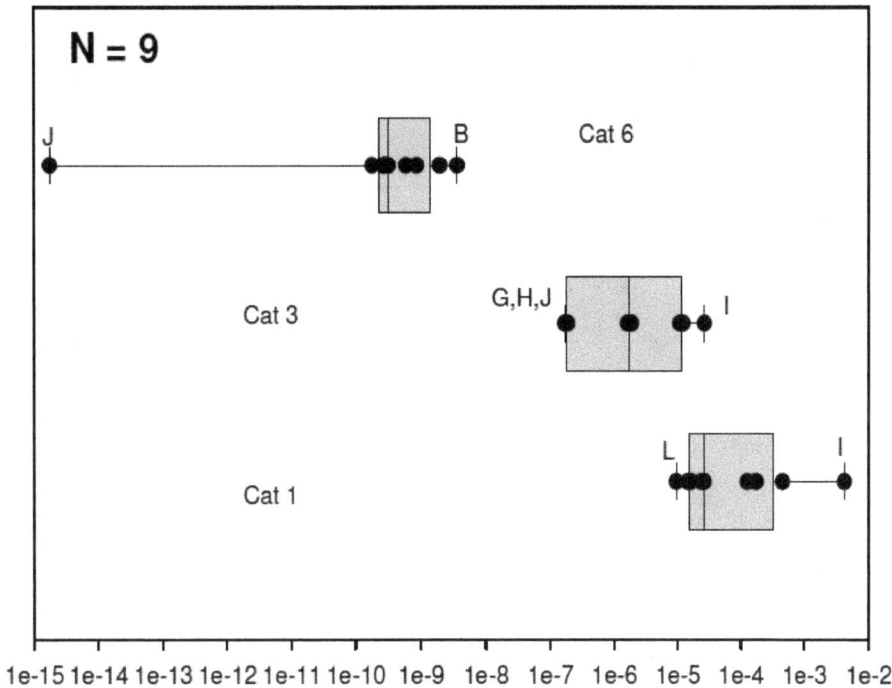

PWR LOCA Frequencies at 25 Years (CY⁻¹)

Figure L.16 Cumulative PWR LOCA Frequencies at 25 Years of Plant Operations

Figure L.17 shows the effect of operating time on the cumulative Category 1 LOCA frequencies for PWR piping systems. Several participants felt that the operational experience is sufficient to expect the frequencies to remain relatively constant out to 60 years of life. Degradation and aging will naturally continue to occur. However, the inspection and mitigation strategies will effectively identify and temper the frequency increases caused by this aging. Some panelists expected a short term frequency increase due to PWSCC before effective mitigation is developed. This trend is consistent with the historical response to evidence of emerging degradation by the industry. Also, at least one participant expressed a concern about the high usage factors that will exist at 60 years at many locations. All of these concerns are reflected in the results showing the effects of operating time and aging in Figures L.17 and L.18 for Category 1 and 3 LOCAs, respectively. As can be seen in Figures L.17 and L.18, there is a slight increase in the cumulative Category 1 and 3 LOCA frequencies between 25 and 40 years, but not much of an effect between 40 and 60 years. The median LOCA frequencies for the Category 1 and 3 LOCAs at 40 years are an order of magnitude higher than the median LOCA frequencies for the Category 1 and 3 LOCAs at 25 years. Similar findings were evident for the larger Category 6 LOCAs. The rationale behind this is that this size of LOCA (and associated pipe size) is most affected by aging. These pipes are not as easily inspected, or as leak sensitive, as their larger counterparts and these pipes have not experienced the infant mortality as their smaller counterparts.

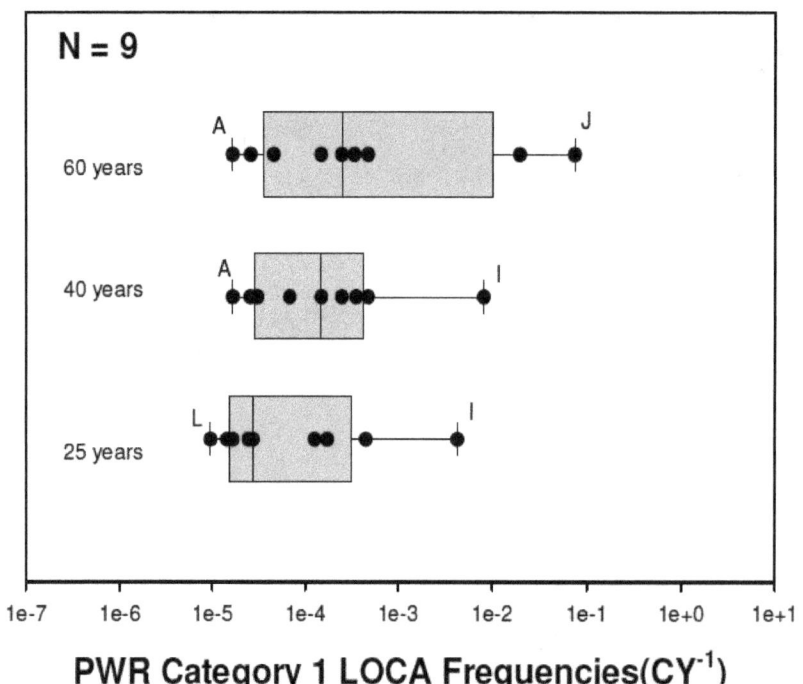

Figure L.17 Effect of Operating Time on the Cumulative Category 1 LOCA Frequencies for PWR Piping Systems

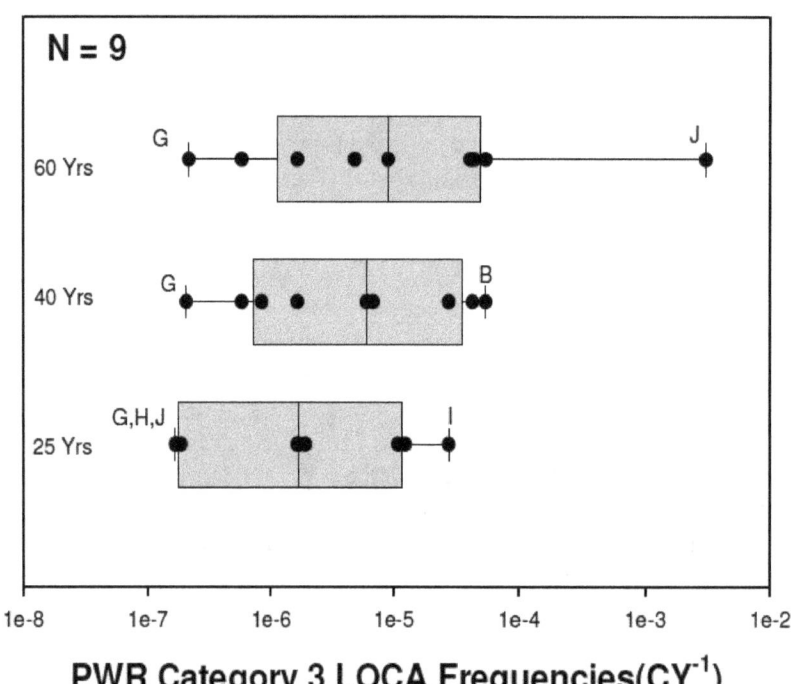

Figure L.18 Effect of Operating Time on the Cumulative Category 3 LOCA Frequencies for PWR Piping Systems

Figures L.19 and L.20 show the cumulative MV estimates, along with the 5% and 95% bound values for the various participants for the Category 1 and 3 LOCAs, respectively. The uncertainty range (difference between 5% LB and 95% UB values) for the Category 3 LOCAs are typically greater than for the Category 1 LOCAs for most of the participants. In a similar vein, the level of uncertainty for the Category 6 estimates were much greater than for the Category 1 or 3 estimates, see Figure L.21. All of the panelists had at least two orders of magnitude difference between the LB and UB values for their Category 6 estimates, and some of the panelists (C, E, and J) had greater than four orders of magnitude difference.

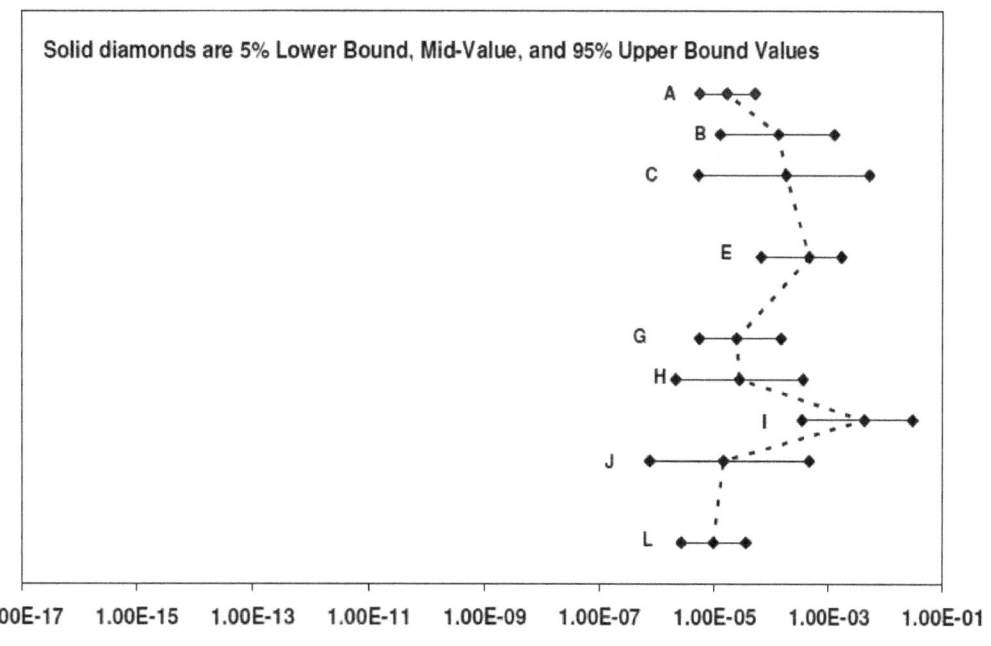

PWR Category 1 LOCA Frequencies (CY^{-1})

Figure L.19 PWR Category 1 LOCA Frequencies Showing MVs, 5% LB, and 95% UB Values for All Participants Who Responded to the PWR Piping Questions

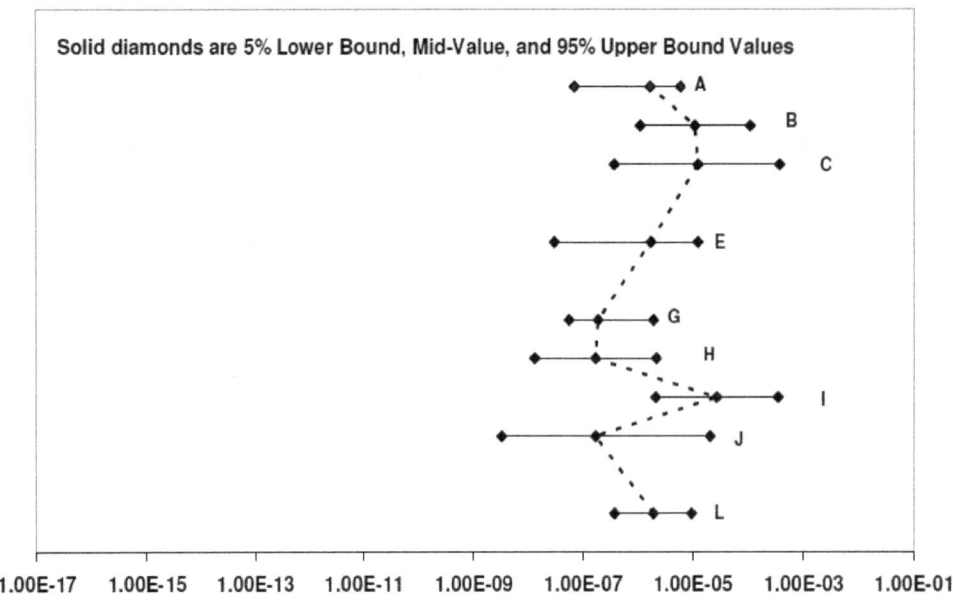

PWR Category 3 LOCA Frequencies (CY^{-1})

Figure L.20 PWR Category 3 LOCA Frequencies Showing MVs, 5% LB, and 95% UB Values for All Participants Who Responded to the PWR Piping Questions

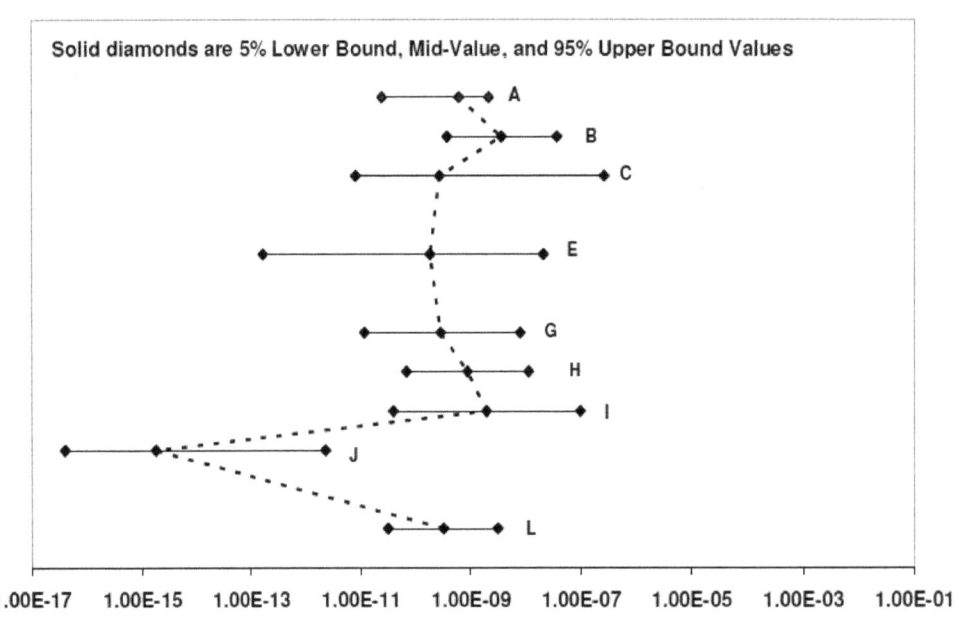

PWR Category 6 LOCA Frequencies (CY^{-1})

Figure L.21 PWR Category 6 LOCA Frequencies Showing MVs, 5% LB, and 95% UB Values for All Participants Who Responded to the PWR Piping Questions

In general, the results for PWR piping appear consistent. The quantitative results and the qualitative opinions and rationale were for the most part consistent. The variability between participants stems from the different approaches used and the basis for their estimates. Several different approaches with different anchoring points were used by the different panelists. The variability between the participants seems reasonable given the frequency magnitudes being computed.

L.4 BWR Non-Piping

Generally speaking making estimates of LOCA frequencies for non-piping components is more challenging than making estimates for piping systems. There are multiple components to consider, each having different operating requirements, design margins, materials, and inspectability. There are also widely varying failure modes and scales to consider. For PWRs for the smaller category LOCAs, one must consider SGTRs and small penetration failures. For the larger category LOCAs, common cause bolting failures and component shell failures need to be considered. For the larger components, the bigger design margins (compared to those for piping) are somewhat offset by the decreased inspection quantity and quality. Compounding all of this is the fact that there is generally not as much precursor information available for the non-piping components as there is for piping.

For the BWR plants, the three major non-piping components that were considered were the RPV, the pumps, and the valves. In general, many of the same degradation mechanisms that are of concern for BWR piping are also a concern for BWR non-piping components. Stress corrosion cracking (specifically PWSCC) is a concern for many of the smaller Alloy 600 components, such as the CRDMs and other penetrations. As with piping, multiple cracks and fast propagation rates could lead to LOCAs. While the mechanism (PWSCC) is more severe at higher temperatures associated with PWRs, this mechanism could become a more significant issue later in the life of the BWRs. Thermal fatigue is another degradation mechanism associated with BWR non-piping components that is common with BWR piping. Thermal fatigue is especially of concern at inlet nozzles and other locations that experience thermal stratification, especially at the feedwater nozzles. For the same reasons as highlighted above for BWR piping, thermal fatigue can possible lead to larger leaks or LOCAs.

Other mechanisms for non-piping components that were not of concern for BWR piping are radiation embrittlement, common cause bolting failures, and thermal aging of cast stainless steel components, such as pump and valve casings. Radiation embrittlement reduces the base metal toughness of the RPV. Fortunately, for BWRs, it is not as of much concern as it is for PWRs due to the increased shielding available with the BWRs. Common cause bolting failures are important for manways and bolted valves. The common cause mechanisms may possibly include: improper installation or maintenance of the bolts, i.e., improper torque, external corrosion of multiple bolts, and steam cutting of bolts. One participant thought that these common cause failures will cause the greatest risk. Thermal aging of cast stainless steels can cause a significant reduction in the fracture toughness of these materials, however, fortunately to date no cracking mechanisms have emerged for these materials.

Figure L.22 shows the Category 1 LOCA frequencies for the RPV, pumps, and valves at 25 years. The RPV shows the biggest expected Category 1 LOCA frequencies. The Category 1 RPV LOCA frequencies are driven by the CRDM penetration failures. However, the severity of the CRDM failures for BWRs was reduced by about one order of magnitude with respect to PWR CRDM failures due to the BWR heads operating at a lower temperature. Other than these head penetrations, nozzle and body cracking were mentioned as possible sources of failures. A number of precursor cracking incidents have been seen in service. The valves and pumps contribute to a lesser extent. Most of the panelists generally treated these components the same. At least one panelist (F) had some experience with manufacturing defects in valve bodies which led to some increased concern with valves. Other issues with valves, and pumps, included the potential for bolt failures for the reasons outlined above, and the fact that the material they are made of (cast stainless steel) is notoriously difficult to inspect and is subject to toughness degradation due to thermal aging.

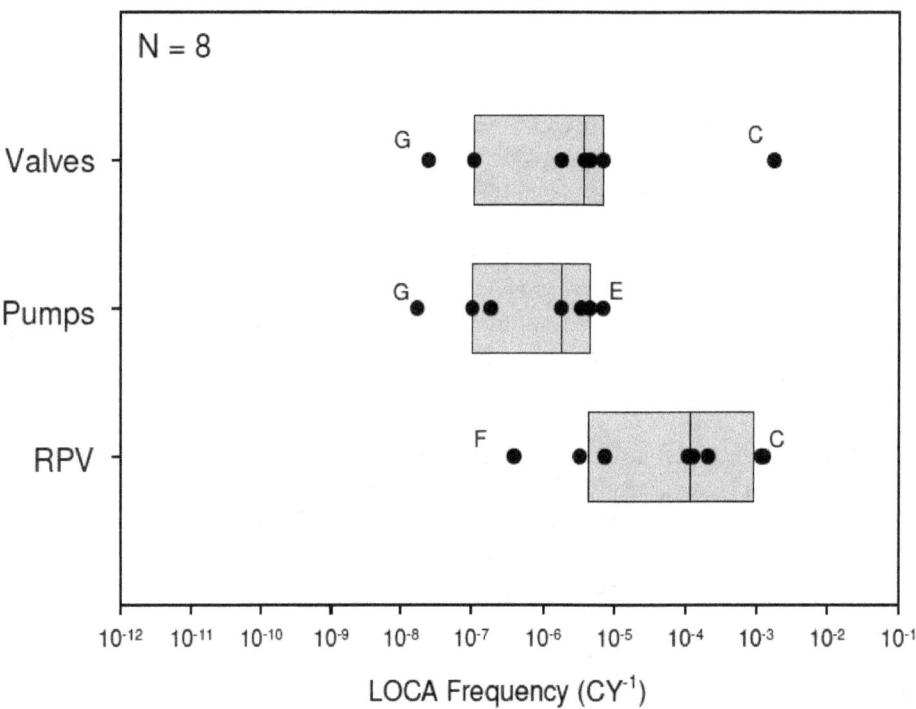

Figure L.22 BWR Category 1 Non-Piping LOCA Frequencies by Major Component at 25 Years of Plant Operations

Figure L.23 shows the Category 3 non-piping LOCA frequencies at 25 years of plant operations. The most noticeable difference between the Category 3 and Category 1 LOCA frequencies is the three orders of magnitude reduction in the median value of the estimated LOCA frequencies for the RPV as the CRDM concerns disappear. A single CRDM failure cannot support a Category 3 LOCA. Only about half of the participants were concerned about RPV nozzle failures, but those that were assigned comparatively high frequencies to them. For the pumps and valves, the corresponding decrease in LOCA frequency is only about one order of magnitude. Consequently, the pumps, valves, and vessel now contribute about equally to the overall LOCA frequency.

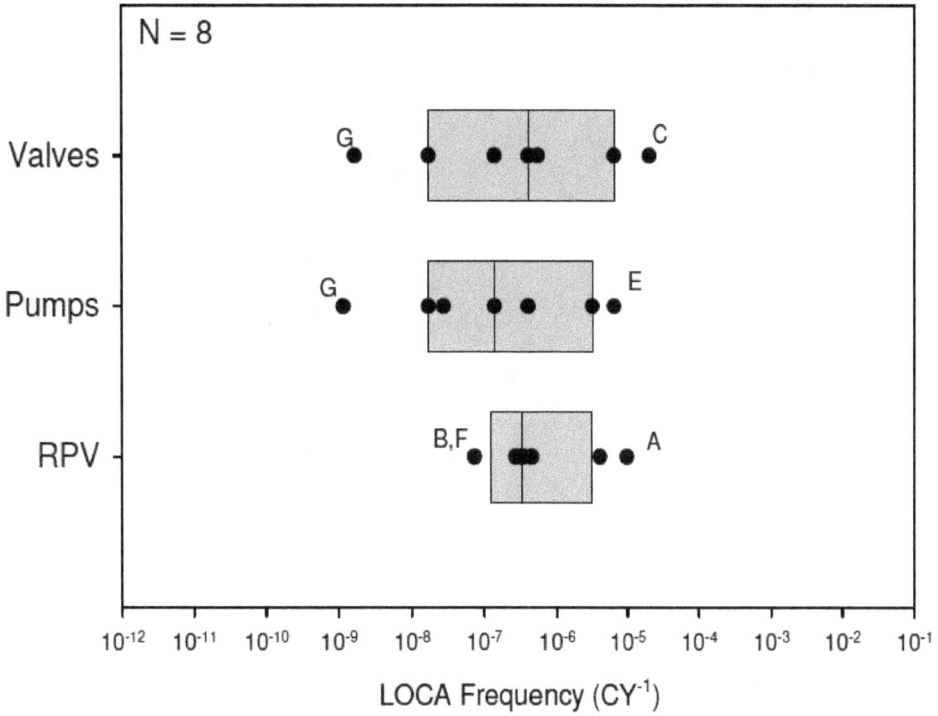

Figure L.23 BWR Category 3 Non-Piping LOCA Frequencies by Major Component at 25 Years of Plant Operations

Figure L.24 shows the Category 5 non-piping LOCA frequencies at 25 years for the BWR non-piping components. As was the case for the Category 3 LOCAs, the pumps, valves, and vessel are all now of about equal importance. For these large LOCAs, the panelists felt that the valve, pump, and vessel bodies were the most likely subcomponents to fail. For the vessel body, the concern was with LTOP while for the valve and pump bodies, the concern was with fatigue and SCC.

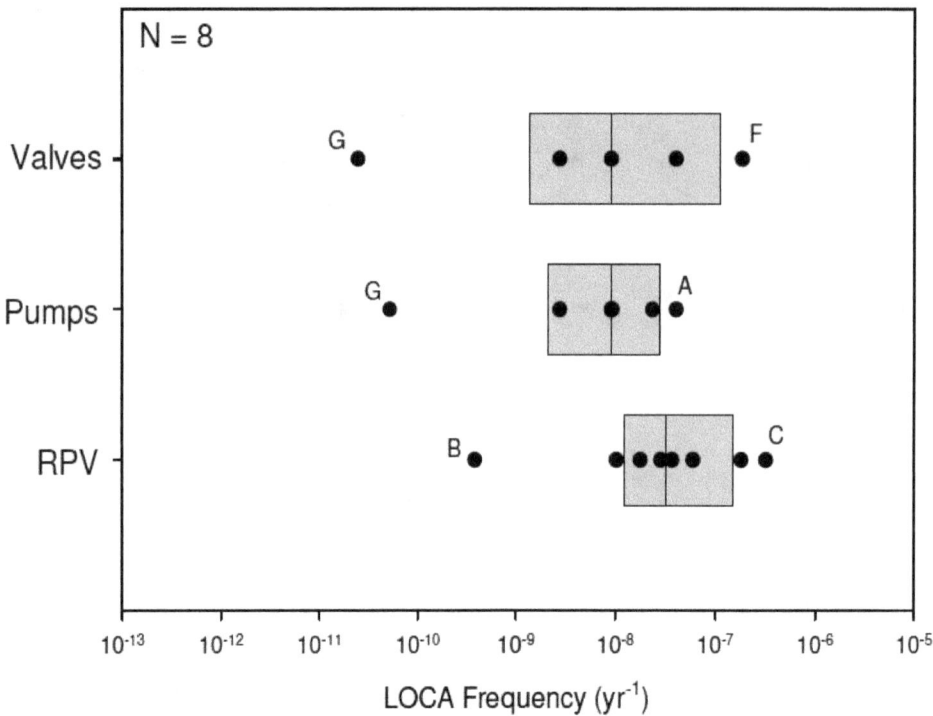

Figure L.24 BWR Category 5 Non-Piping LOCA Frequencies by Major Component at 25 Years of Plant Operations

Figure L.25 shows the cumulative LOCA frequencies for the BWR non-piping components at 25 years of plant operations. On average there is about a one order of magnitude shift in the cumulative LOCA frequency between each successive LOCA category. The median value for the estimate of the Category 1 LOCA frequency is approximately 10^{-4} while the median value for the Category 6 LOCA frequency is about 10^{-9}.

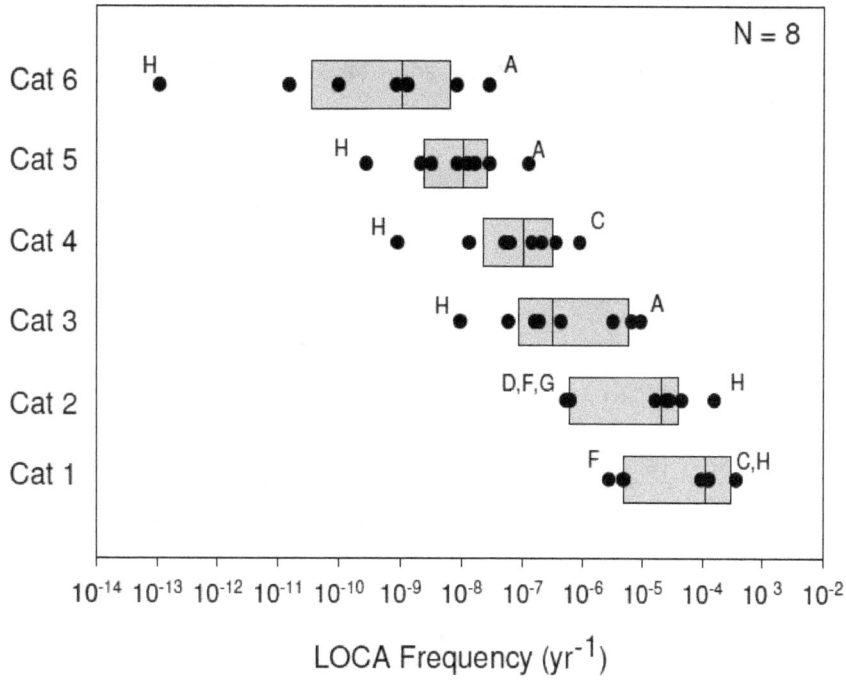

Figure L.25 Cumulative BWR Non-Piping LOCA Frequencies at 25 Years of Plant Operations

Figures L.26 and L.27 show the effect of time on the Category 1 and 3 cumulative LOCA frequencies, respectively, for the BWR non-piping components. For all intents and purposes there is almost no effect of time on the predicted LOCA frequencies. The median values do not change nor does the variability, i.e., the interquartile ranges remain the same. Non-piping components are affected by similar partially compensating factors as the piping components. In addition, a number of participants expressed the belief that the maintenance and mitigation issues raised for piping also apply for non-piping components. The only thing that changes is the minimum value predicted by Participant H for LOCA Category 3. Participant H foresees the non-piping LOCA frequencies increasing at both the 40 and 60 year time interval. Figure L.28 shows the effect of time on the Category 5 frequencies. In this case the median values do not vary with time, nor does the maximum values, however, a number of participants started to see the LOCA frequencies increasing near the end-of-plant license renewal (60 years) such that the lower end of the IQR (i.e., the 25th percentile) increased an order of magnitude at 60 years over what it was at 40 and 25 years. This increase in the Category 5 predictions was driven by aging concerns of a few of the panelists at 60 years.

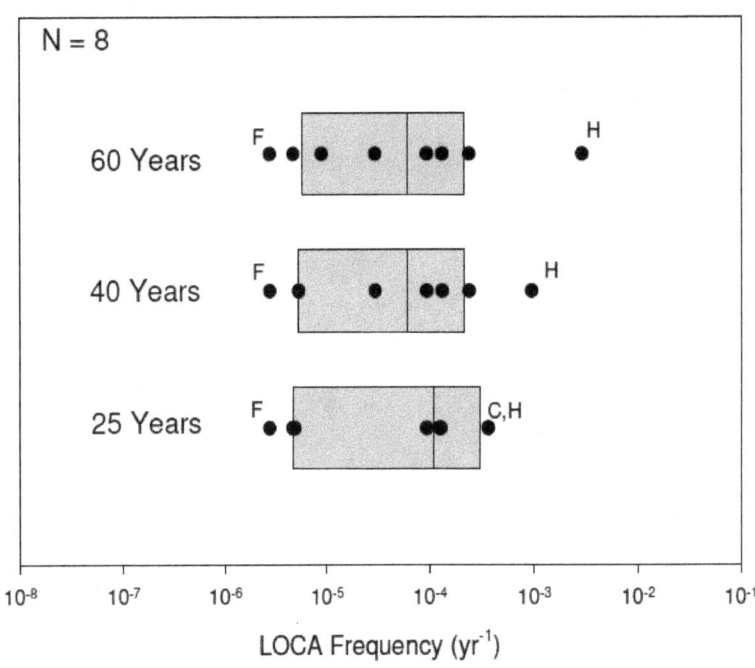

Figure L.26 Effect of Operating Time on the Cumulative Category 1 LOCA Frequencies for BWR Non-Piping Components

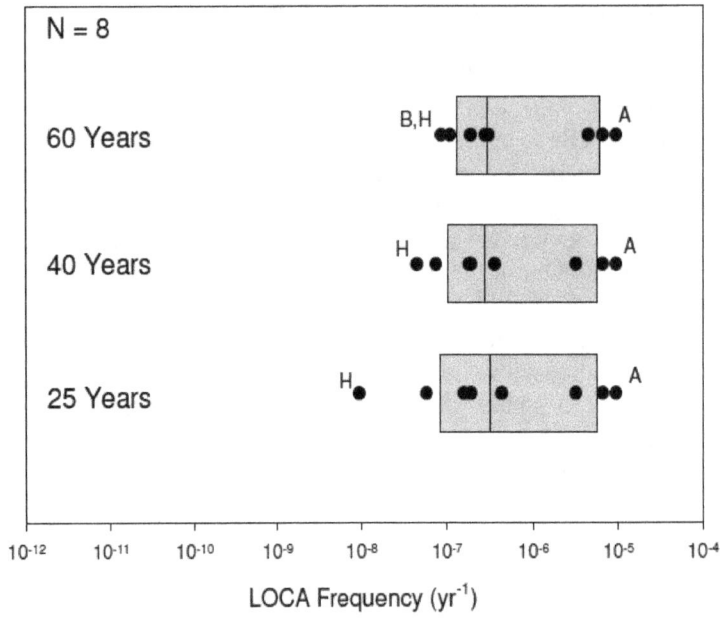

Figure L.27 Effect of Operating Time on the Cumulative Category 3 LOCA Frequencies for BWR Non-Piping Components

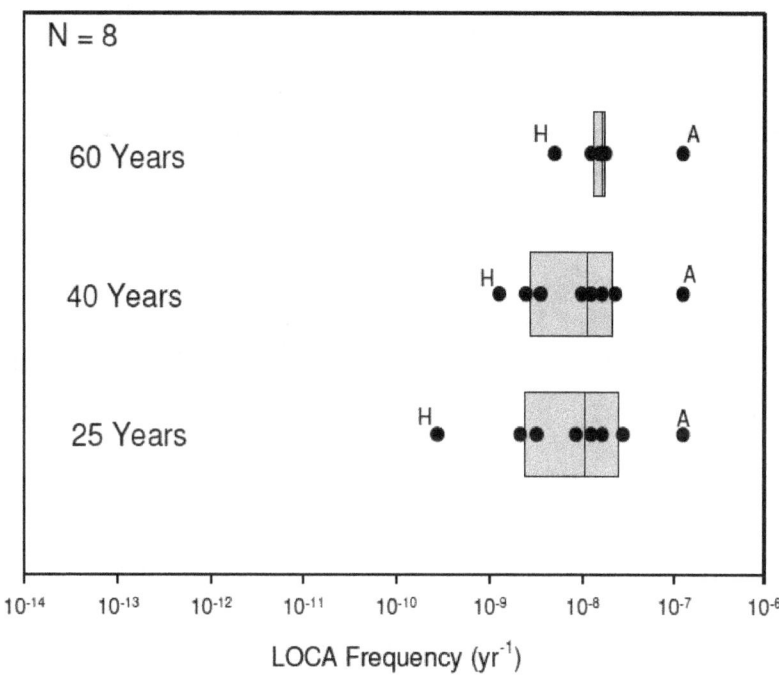

Figure L.28 Effect of Operating Time on the Cumulative Category 5 LOCA Frequencies for BWR Non-Piping Components

Figures L.29 and L.30 show the cumulative MV estimates, along with the 5% and 95% bound values for the various participants for the BWR Category 1 and 3 non-piping LOCA frequency estimates, respectively, at 25 years of plant operating time. Of note from these figures is the fact that a number of the participants (A, E, F, and H) predicted greater uncertainty for the Category 3 LOCAs than they did for the Category 1 LOCAs. This is not unusual in that one would expect the uncertainty to increase for lower frequency events, such as larger LOCAs. It is probably somewhat more surprising that the other four participants predicted comparable uncertainty for the Category 1 and 3 LOCAs. Overall the predictions for BWR non-piping were more consistent than for PWR non-piping discussed next. For the BWRs, there are less components and failure modes to consider and the approaches used to estimate the frequencies were more closely related.

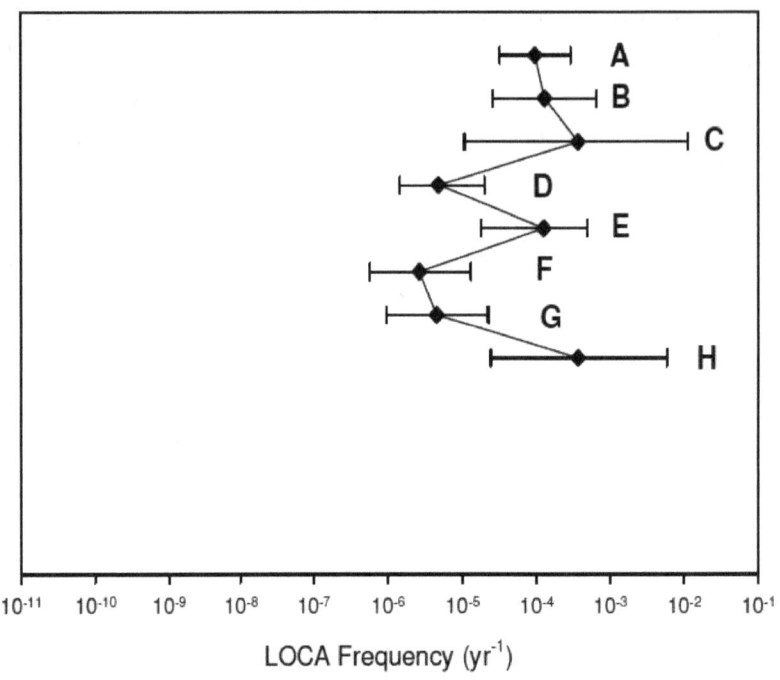

Figure L.29 BWR Non-Piping Category 1 LOCA Frequencies Showing MVs, 5% LB, and 95% UB Values for All Participants Who Responded to the BWR Non-Piping Questions

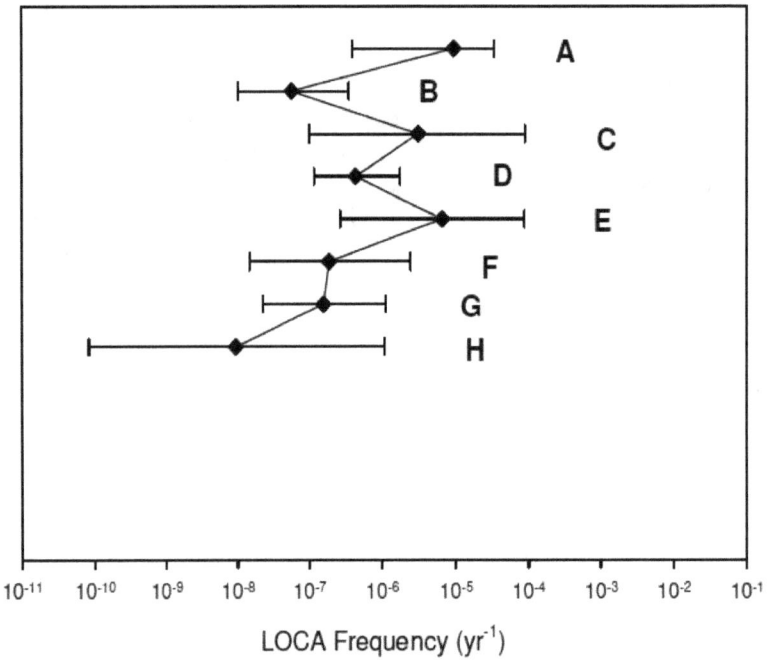

Figure L.30 BWR Non-Piping Category 3 LOCA Frequencies Showing MVs, 5% LB, and 95% UB Values for All Participants Who Responded to the BWR Non-Piping Questions

L.5 PWR Non-piping

The same three major non-piping components (RPV, valves, and pumps) as considered for BWRs are considered for PWRs, plus the steam generator and pressurizer are added. One of the operational experience databases showed an order of magnitude higher incident rate for PWR non-piping than BWR non-piping. This was partially attributed to the fact that there are more PWRs than BWRs. However, this comparison is also biased by the large number of steam generator tube failures reported in the databases. Steam generator tubes are subjected to a host of degradation mechanisms: fatigue, denting, external SCC, PWSCC, and overload failures. It was almost universally accepted that SGTRs would be the dominant contributor to the PWR Category 1 non-piping LOCA frequency. In fact the PWR steam generator tube failure frequency is the dominant contributor to the overall PWR small-break LOCA (Category 1 LOCAs) frequency when considering both the piping and non-piping contributions. Even so, it is the expectation of a number of the panel members that the steam generator tube contribution to the small-break LOCA frequency will decrease with time due to steam generator tube replacement programs and improvements made to the secondary side water chemistry.

In general, many of the same degradation mechanisms that are important for PWR piping are important for the non-piping components as well. PWSCC is an important degradation mechanism for many of the smaller Alloy 600 components such as the CRDMs, heater sleeves, steam generator tubes, and other penetrations. As was the case for piping, the likelihood of multiple cracks forming, and possibly coalescing, and the relatively fast propagation rates associated with this type of cracking makes this mechanism a major concern from a LOCA perspective. Also, since this mechanism is more severe at the higher temperatures associated with PWRs, it is considered to be a bigger threat for the PWRs than the BWRs, at least in the short term. As was the case for PWR piping, thermal fatigue is also a concern for PWR non-piping components. It is especially of concern at nozzle inlets and other locations where thermal stratification may exist. Furthermore, for all the reasons highlighted for piping, thermal fatigue is a mechanism that can lead to large leaks, i.e., fast propagation rates, attacks a wide area, and difficult to prioritize inspection protocols due to the fact that it can attack a variety of materials. Mechanical fatigue is another common degradation mechanism to both PWR piping and non-piping components. Mechanical fatigue is most important for smaller components, such as heater sleeves and small penetrations that are subjected to vibratory stresses due to equipment operation.

Another mechanism of special concern to non-piping components is common cause bolting failures. This is especially relevant to manways and bolted valves. The common cause mechanism could be improper installation or maintenance of bolts, e.g., improper torque, external corrosion of multiple bolts, or possibly steam cutting of multiple bolts.

Also, boric acid corrosion of carbon steel components such as RPV and steam generators can be aggressive under certain conditions.

Figure L.31 shows the Category 1 LOCA frequencies for the major PWR non-piping components at 25 years. As expected, the expected failure frequencies are highest for the steam generator. The higher LOCA frequency for the steam generator is driven by the SGTR data.

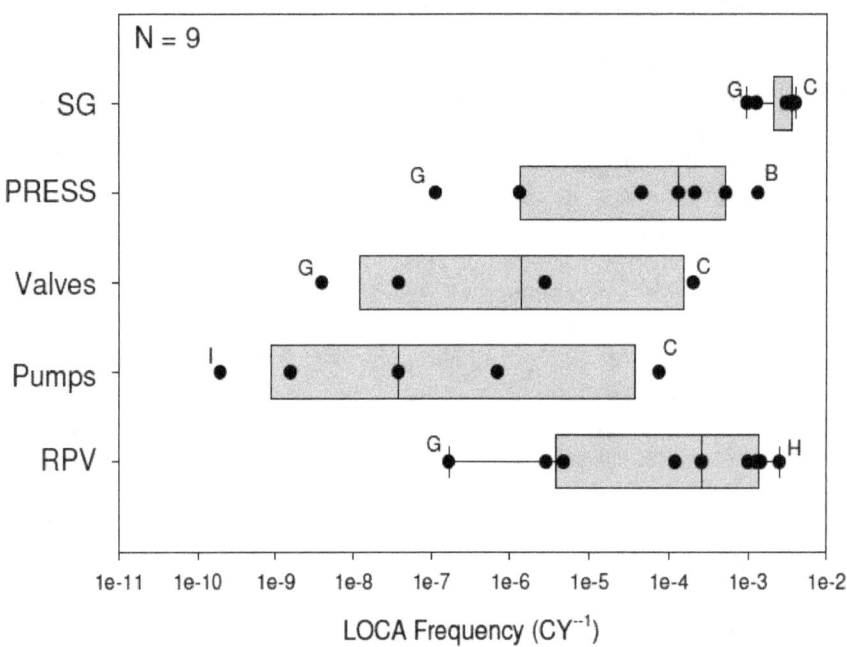

Figure L.31 PWR Category 1 Non-Piping LOCA Frequencies by Major Component at 25 Years of Plant Operations

The other major contributors to the Category 1 LOCA frequencies for PWR non-piping were the RPV and the pressurizer. The main subcomponent contributing to the RPV frequency is the CRDMs while the main subcomponent contribution to the pressurizer frequency is the heater sleeves. For Category 2 LOCAs, a single SGTR cannot sustain such a leak. Thus, for the Category 2 LOCAs, the CRDM and pressurizer heater sleeves became the main contributors.

For Category 3 and 4 LOCAs there was no consistent agreement among the panelists as to the major contributors. As one can see in Figure L.32, all five major components contribute fairly equally to the Category 4 LOCA frequencies. As such, there is tremendous variability about the frequency associated with each component. This variability was also apparent for the Category 6 LOCAs. This variability reflects the inconsistent opinions and approaches followed by the panelists, as well as the difficulty of this type of assessment. As is to be expected, there was a wide array of possible failure modes for dissimilar components to be considered, and the panelists tended to gravitate towards a few of the failures that they personally thought were most credible. Given all of this, the level of variability was thought to be reasonable given the event frequencies. This was one reason for adopting the elicitation approach in the first place. The highest LOCA frequencies were for the pressurizer nozzle. In addition, many of the panelists considered manway or shell failures important, irrespective of the component type. Thus, they anticipated similar distributions for both the steam generator and pressurizer. There were also major differences of opinion among the panelists as to the most important failure modes.

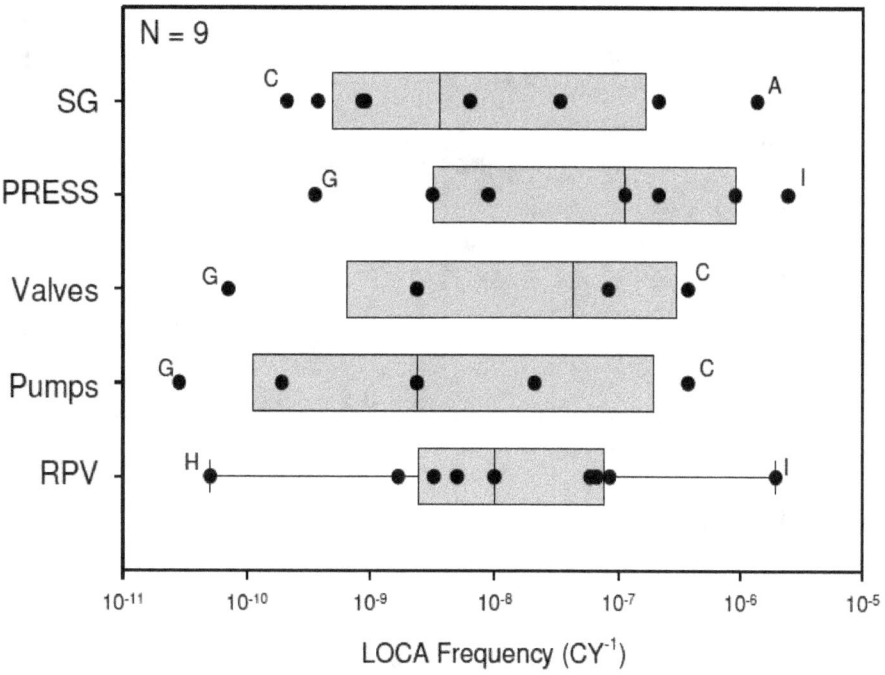

Figure L.32 PWR Category 4 Non-Piping LOCA Frequencies by Major Component at 25 Years of Plant Operations

Figure L.33 shows the cumulative LOCA frequencies for the PWR non-piping components at 25 years of plant operations. The Category 1 LOCA frequencies for PWR non-piping are the highest frequencies estimated by the elicitation panel for piping or non-piping, BWR or PWR. The median frequency is almost 5×10^{-3}. The variability among the panelists was very small. The difference between the minimum and maximum predictions was less than an order of magnitude. These high frequencies and low variability were driven by the SGTR data for which ample data exist in the operational experience databases; thus explaining both the high frequencies and excellent agreement between participants. For the Category 2 LOCAs, the agreement, at least on a minimum and maximum basis, is not nearly as good. However, the agreement on the basis of the spread in the IQR is nearly as good as it is for the Category 1 LOCAs. Again, for the Category 2 LOCAs, the major contributors are the CRDMs and the pressurizer heater sleeves. The much wider variability for the Category 3 through 6 LOCAs reflects differences in opinion among the panelists as to the important failure modes and their associated frequencies.

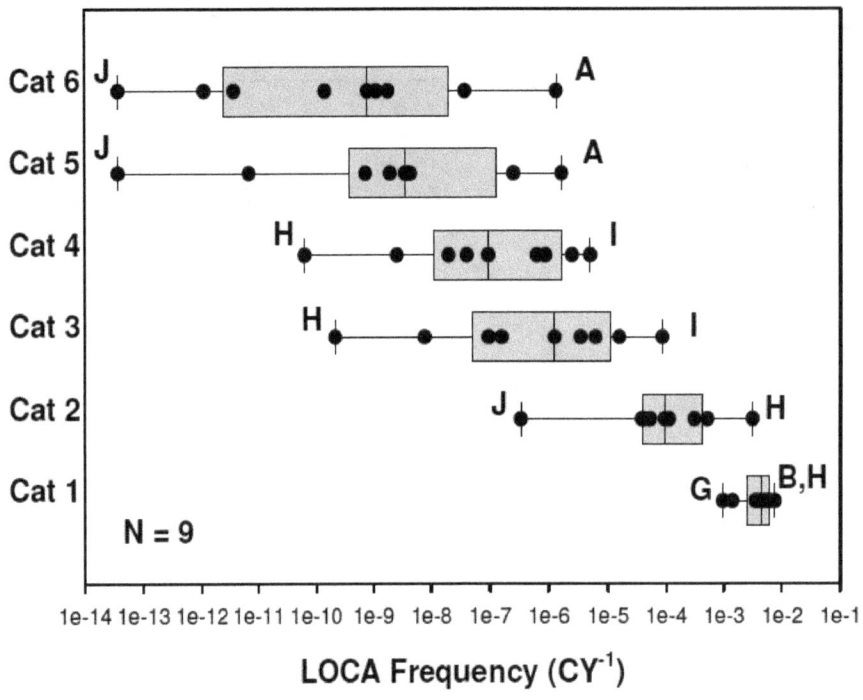

Figure L.33 Cumulative PWR Non-Piping LOCA Frequencies at 25 Years of Plant Operations

Figure L.34 shows the effect of time on the PWR non-piping Category 1 LOCA frequencies. There is a very slight decrease in the frequency between 25 years (present day) and 40 years (end of plant license) due mostly to steam generator replacement programs and improved inspection and mitigation programs, e.g., improved eddy current inspection programs and improved secondary side water chemistry. There was also an expected decrease in the LOCA frequencies associated with CRDMs due to on-going head replacement programs and improved CRDM inspection programs that may go into effect over the next few years. However, there was some concern expressed that the maintenance and inspection programs for the larger component bodies (pressurizer, steam generator, RPV) may not be as rigorous as for the piping systems. Figure L.35 shows the effect of operating time on the PWR non-piping Category 6 LOCA frequencies. As can be seen in Figure L.35, the median values remain constant with time and the variability among the participants (at least on the basis of the IQR) also remains fairly constant. This tends to indicate that the participants did not foresee any significant aging effects to occur.

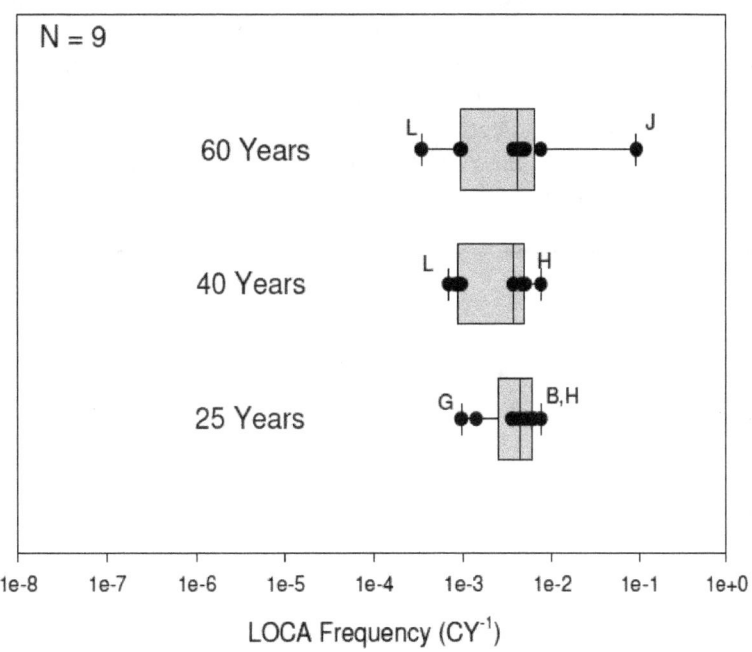

Figure L.34 Effect of Operating Time on the Cumulative Category 1 LOCA Frequencies for PWR Non-Piping Components

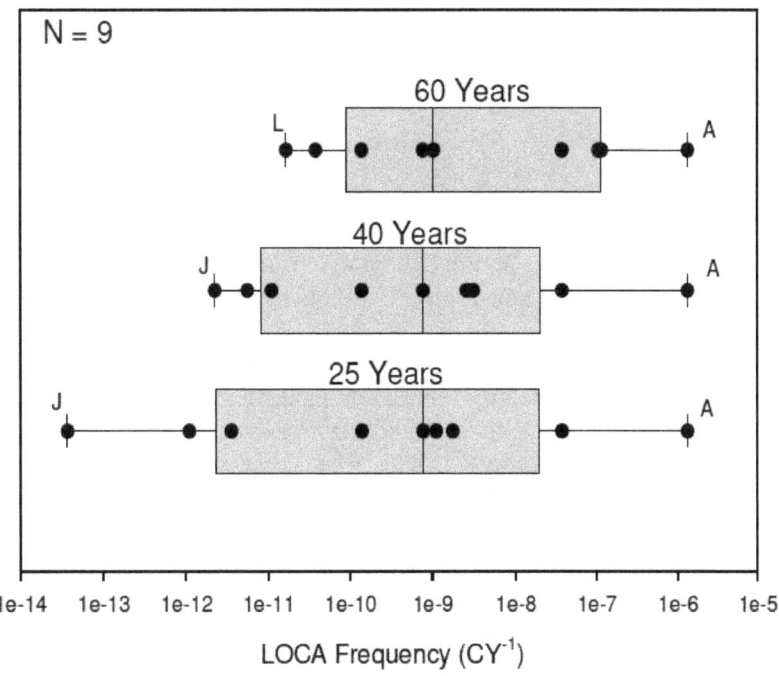

Figure L.35 Effect of Operating Time on the Cumulative Category 6 LOCA Frequencies for PWR Non-Piping Components

Contrary to what was observed for the Category 1 and 6 LOCA frequencies, the median value of the Category 4 LOCA frequencies increases an order of magnitude between 25 and 40 years and then remains constant after that, see Figure L.36. As was the case for PWR piping, aging was thought to have the largest impact on LOCA Categories 3 and 4. It was thought by some that aging could accelerate near the end of the plant license faster than the effects of mitigation and inspection could become effective, especially if the plant operators do not see a return on their investment for such inspection and mitigation programs near the end of the plant's license.

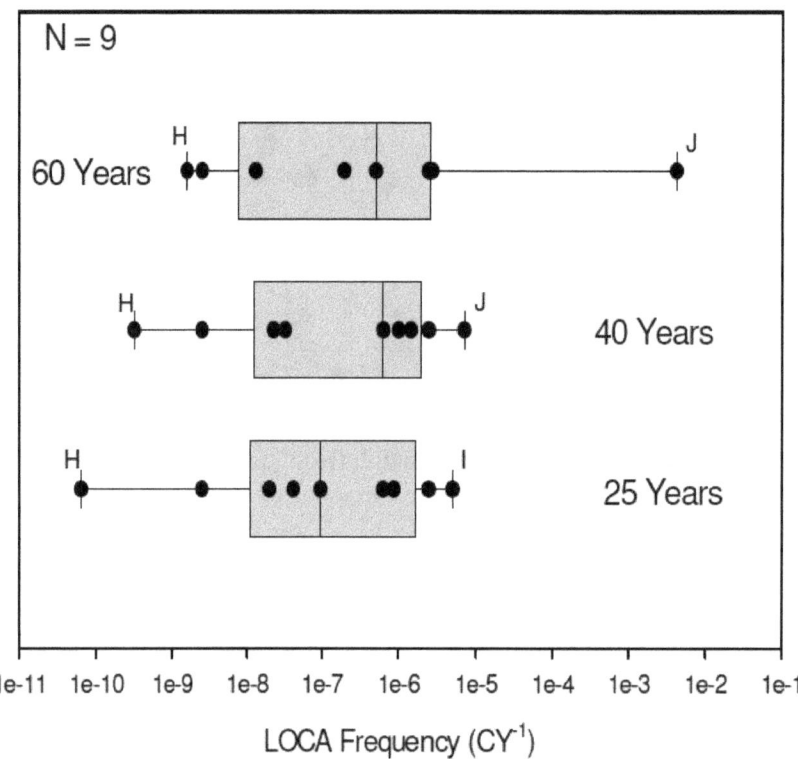

Figure L.36 Effect of Operating Time on the Cumulative Category 4 LOCA Frequencies for PWR Non-Piping Components

Figures L.37 and L.38 show the cumulative MV estimates, along with the 5% and 95% bound values for the various participants for the PWR Category 1 and 4 non-piping LOCA frequency estimates, respectively, at 25 years of plant operating time. Of note from these figures is higher uncertainty among almost all of the participants for the Category 4 LOCAs when compared with the Category 1 LOCAs. A number of the panelists showed 3 to 4 orders of magnitude of uncertainty for the Category 4 LOCAs while all of the panelists had less than approximately 2 orders of magnitude of uncertainty in their Category 1 results. In addition, the variability in the panelist's MV estimates was within 1 order of magnitude for their Category 1 results while the variability among their results spread over a range of almost five orders of magnitude for their Category 4 results. The fact that the agreement among the panelists was so good for the Category 1 predictions plus the low uncertainty of their individual predictions is a reflection that there was near consensus agreement that the single overwhelming dominant contributor to this class of LOCAs was SGTRs, for which ample field experience is available in the operational experience databases.

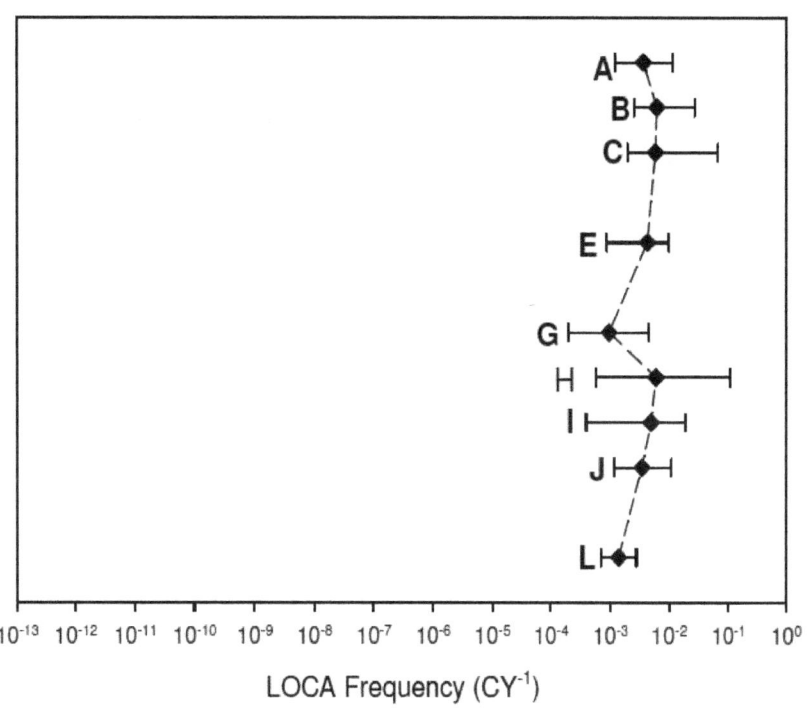

Figure L.37 PWR Non-Piping Category 1 LOCA Frequencies Showing MVs, 5% LB, and 95% UB Values for All Participants Who Responded to the PWR Non-Piping Questions

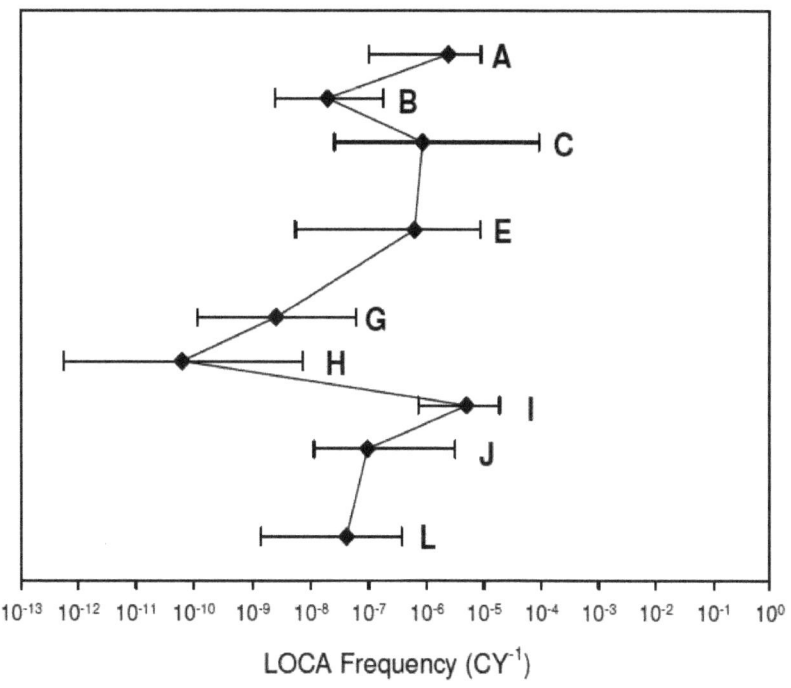

Figure L.38 PWR Non-Piping Category 4 LOCA Frequencies Showing MVs, 5% LB, and 95% UB Values for All Participants Who Responded to the PWR Non-Piping Questions

APPENDIX M

PUBLIC COMMENT RESPONSES

APPENDIX M – PUBLIC COMMENT RESPONSES

General Comments

Comment Number: GC1
Submitted by: Bill Galyean – Idaho National Laboratory
Comment: [Note: The footnote indications below do not appear in the submitted comment; they were added as reference points for the response.] Aside from the fact that I was a contributing panel member in the elicitation process, I want to express my compliments on the effort made by the NRC management and staff to produce realistic and useful results. I believe with the significant research performed in recent years coupled with the accumulated operating experience, reasonable estimates of LOCA frequencies can be made. NRC has recognized these facts and acted accordingly and appropriately. That said, I also wish to express my opinion on the some of the details of the elicitations process and portions of the subsequent analyses about which I disagree, but with the acknowledgement that had they been done differently, the results would not change significantly (i.e., the reported results would probably be reduced by less than an order of magnitude). The first issue relates to the interpretation of the LOCA frequencies and associated uncertainties. The instructions given to the panel members stated that we were to make a best-estimate[1] of the "single 'true' value"[2] for the industry-wide (or more accurately BWR-wide and PWR-wide) LOCA frequency. This is an issue because I do not believe a "single true value" exists for the LOCA frequency. Specifically, I believe each plant has differences in design, construction, operations, age, and maintenance such that in reality the plant-to-plant variation in LOCA frequencies will be quite large. These two interpretations can be made consistent however, if the "single true value" is viewed as an average or mean value of the population of frequencies. While on the surface this might seem to be a question of semantics and appears to have little significance, the implication of the interpretation on the uncertainty characterization is significant. If the "single true value" interpretation is employed, the question becomes, what does the uncertainty associated with this value represent. Since the implicit assumption is that plant-to-plant variability does not exist (otherwise how can there be a single frequency appropriate for the entire industry?), then there is no stochastic or aleatory uncertainty associated with the estimate. That is, the presence of outliers (or event plant-to-plant variability) in the population of nuclear power plants (NPP) is ignored. The uncertainty therefore represents the level of confidence of each panel member's estimate of this single true frequency.[3] Given that the uncertainty represents an individual's confidence in their own estimate, what is the basis for automatically assuming the probability distribution associated with this confidence uncertainty is not symmetrical?[4] Or to ask more specifically, with this interpretation of the point estimate value and associated interpretation of the uncertainty, why do the authors assume that the uncertainty surrounding each panel member's estimate should be a lognormal distribution that is weighted toward higher (conservative) values? While I agree that the uncertainty associated with LOCA frequencies should be asymmetrically weighted toward higher values; this is based on the observation that not all plants are identical, and that if a LOCA occurs at a plant, that plant will likely be shown to be a poor performer, not representative of the fleet as a whole.[5] This is not the same as assuming the confidence uncertainty surrounding the estimates provided by the panel members, of a "single true value" should be represented with a lognormal (which was done in the elicitation).[6] [Note: the use of the lognormal distribution became entrenched in PRA with its use in WASH-1400 (1975). However, the model employed then was motivated by the sparse data available from the commercial nuclear power industry at that time. Data used in that study were collected from many different industries and sources. These data were used to develop a probability distribution of the population of possible values, effectively capturing the random component of the uncertainty. Specifically, the authors of WASH-1400 did not know which value in the population of values collected, was the most appropriate value to use. Therefore the entire population of values was characterized in the probability distribution. The data were not combined, or

M-1

averaged to estimate a single value, and the uncertainty modeled with the lognormal distribution was not meant to describe a statistical confidence on a single true value, but used to describe the variability in possible values. Quoting directly from WASH-1400: "Because of the large spread, the failure rate data were treated as random variables, incorporating both the physical variability and the uncertainty associated with the rates. Moreover, since the study's results were to apply to a population of approximately 100 nuclear plants, it was important to show the possible variability and uncertainty in this population." The consequences of employing the lognormal distribution to characterize each individual panel member's uncertainty, manifests itself in how the estimates were interpreted and processed (by the authors) and in the effects of the overconfidence adjustment made to the base case results. Specifically, the best-estimates solicited by the authors of this "single true value" were interpreted as median values of the lognormal distribution, allowing for the derivation (by the authors) of a higher mean value. [Note that for any probability distribution skewed toward higher values, the lognormal being one example, the mean will always be greater than the median.] This calculated mean value was then used to represent the panel member's input to the aggregation process used for generating the LOCA frequency results. Additionally, this assumption of a lognormal for the uncertainty on each panel member's estimate, has the additional impact of calculating an even higher (compared to the non-adjusted calculated mean) mean value after the widening of the uncertainty in the overconfidence adjustment. My concerns are twofold. First, the opinions of the panel members were solicited, and then modified (increased) by the authors.[7] Irrespective of the instructions and discussions during the elicitation process, this point was viewed with dismay by more than one panel member. Second, the processes and analyses employed have introduced a conservative bias into the final base case results, with additional conservative bias[8] inserted in the various sensitivity studies.[9] This "creeping conservatism" is not necessarily undesirable given the significant uncertainties and the uses to which the results will be employed; however, it should be explicitly acknowledged and clearly stated rather than obscured in the details of the analyses.[10]

Response: The authors have identified ten separate issues in this comment, as indicated by the inserted footnotes. The responses to these issues are provided below.

Issue 1: The instructions given to the panel members stated that we were to make a best-estimate[1] of the "single 'true' value"[2] for the industry-wide (or more accurately BWR-wide and PWR-wide) LOCA frequency.

1. Issue 1 Response: The panel members were not asked to provide a "best-estimate" of any quantity. Rather, they were asked to provide a MV and LB and UB values for each question. The MV was defined such that, in the panel member's opinion, the unknown true value for that particular question has a 50% chance of falling above or below the MV, with similar definitions for the LB and UB values (Section 3.8.5).

Issue 2: The instructions given to the panel members stated that we were to make a best-estimate[1] of the "single 'true' value"[2] for the industry-wide (or more accurately BWR-wide and PWR-wide) LOCA frequency.

2. Issue 2 Response: Additionally, the panel members were not asked to estimate a "single 'true' value" of any quantity. Rather, the elicitation focused on estimating generic, or average, LOCA frequencies for the commercial fleet by combining the contributions from individual component failures. As stated in Section 2 of NUREG-1829, the generic BWR and PWR estimates were determined by first estimating the separate LOCA frequency contributions associated with specific BWR piping, BWR non-piping, PWR piping, and PWR non-piping failures for each panelist. These individual piping and non-piping component failure frequencies were then combined to estimate parameters of the total passive system LOCA frequency distributions for BWR and PWR plants at each distinct LOCA category and time period. Panelists were specifically instructed to consider broad plant differences in estimating these component failure

frequencies and their uncertainties (Section 3). More information related to the instructions given to the panel can be found in Sections 2 and 3 of NUREG-1829.

Issue 3: If the "single true value" interpretation is employed, the question becomes, what does the uncertainty associated with this value represent. Since the implicit assumption is that plant-to-plant variability does not exist (otherwise how can there be a single frequency appropriate for the entire industry?), then there is no stochastic or aleatory uncertainty associated with the estimate. That is, the presence of outliers (or event plant-to-plant variability) in the population of nuclear power plants (NPP) is ignored. The uncertainty therefore represents the level of confidence of each panel member's estimate of this single true frequency.[3]

3. Issue 3 Response: More precisely, as stated in Section 3.8.5, the elicited quantities for each question (MV, LB and UB) are percentiles of each panel member's subjective distribution. As stated above, this subjective distribution should consider broad plant difference related to plant design that can affect the LOCA frequency. Also, except for the base case frequencies used to anchor the results, panel members were never asked about absolute frequencies, only about relative frequencies of specific components, subsystems or systems. Their responses were then combined to form estimates of LOCA frequencies.

Issue 4: Given that the uncertainty represents an individual's confidence in their own estimate, what is the basis for automatically assuming the probability distribution associated with this confidence uncertainty is not symmetrical?[4]

4. Issue 4 Response: The distribution associated with the response to any question was assumed to be asymmetrical only if the stated LBs and UBs were not symmetrical about the MV.

Issue 5: While I agree that the uncertainty associated with LOCA frequencies should be asymmetrically weighted toward higher values; this is based on the observation that not all plants are identical, and that if a LOCA occurs at a plant, that plant will likely be shown to be a poor performer, not representative of the fleet as a whole.[5]

5. Issue 5 Response: As noted above and in NUREG-1829, the estimated LOCA frequencies are industry-wide or "generic, or average, estimates for the commercial fleet" (Section 2). Consequently, the uncertainties associated with the estimates pertain to these generic estimates and not to any individual plants whose LOCA frequencies may differ from the generic estimates. As stated in Section 2 of the NUREG, the panelists were instructed to account for broad plant-specific factors which influence the generic LOCA frequencies in providing uncertainty bounds, but not consider factors specific to any individual plants. Thus, the uncertainty bounds should include both contributions related to the uncertainty of the generic estimates as well as uncertainty due to broad plant-specific fleet differences.

Issue 6: This is not the same as assuming the confidence uncertainty surrounding the estimates provided by the panel members, of a "single true value" should be represented with a lognormal (which was done in the elicitation).[6]

6. Issue 6 Response: The lognormal, or split lognormal when the responses are asymmetrical, is a reasonable distribution for representing the responses to the various questions so that the responses can be combined to estimate LOCA frequencies. Based on the sensitivity analyses conducted in Section 7, the authors expect that the log-normal distribution assumption has an inconsequential impact on the bottom line parameter estimates, especially in light of the large uncertainties observed in the final results. Note that the report does not estimate the distributions of LOCA frequencies, but only the four bottom-line parameters (mean, median, 5th and 95th percentiles) of these distributions.

Issue 7: First, the opinions of the panel members were solicited, and then modified (increased) by the authors.[7]

    7. Issue 7 Response: It is misleading to state that "…the opinions of the panel members [were] modified (increased) by the authors". Rather, an overconfidence adjustment was applied to increase only the uncertainties for those panelists whose responses indicated more confidence than the group average. The panelists' median estimates were not modified. The justification for the overconfidence adjustment is provided in Section 5.6.2. Responses were not modified in any other manner.

Issue 8: Second, the processes and analyses employed have introduced a conservative bias into the final base case results, with additional conservative bias[8] inserted in the various sensitivity studies.[9]

    8. Issue 8 Response: The only "conservative bias" that may have been introduced in the summary estimates was through the use of the overconfidence adjustment. However, as discussed in Section 5.6.2 and by reviewing the wide range of uncertainty estimates provided by the panelists, there is good reason to believe that at least some of the panelists may have been overconfident. Only those panelists that were more confident than the group median were adjusted. Furthermore, sensitivity studies indicated that the error factor adjustment used was a reasonable adjustment scheme. This method was not the most conservative adjustment scheme which could have been used.

Issue 9: Second, the processes and analyses employed have introduced a conservative bias into the final base case results, with additional conservative bias[8] inserted in the various sensitivity studies.[9]

    9. Issue 9 Response: Sensitivity analyses were conducted to examine the impact of the assumptions used to process the results. The assumptions and methods used to calculate the "baseline" LOCA estimates are clearly stated in Sections 5 and 7. Additionally, the methods used to conduct each sensitivity analysis are clearly explained in Sections 5 and 7 and comparisons are made to the baseline results so that the effects on the LOCA frequency estimates are readily apparent. Those sensitivity analyses resulting in the largest differences from the baseline estimates are clearly discussed in the report.

Issue 10: This "creeping conservatism" is not necessarily undesirable given the significant uncertainties and the uses to which the results will be employed; however, it should be explicitly acknowledged and clearly stated rather than obscured in the details of the analyses.[10]

    10. Issue 10 Response: The NUREG report systematically documents the assumptions and analysis used to calculate the LOCA frequency estimates. The NUREG also discusses the effects of alternative assumptions and analysis methods using sensitivity studies.   The analysis procedures utilized in the elicitation process were fully discussed with the panelists at several times throughout the process: during the base case review, prior to conducting the individual elicitations, during the presentation of the preliminary results, and during the preparation of the draft NUREG. Each panelist was also provided numerous opportunities to modify their results, including changes if they believed the initial instructions were unclear, or if the processing of their results did not actually reflect their solicited opinion. None of the panelists elected to provide wholesale changes to their estimates based on these issues. General feedback from the panelists about the elicitation process in general, and the processing of the results in particular, was favorable.

Comment Number: GC2
Submitted by Joseph Conen of the BWR Owners Group
Comment: Another issue is the inclusion of thermal fatigue as a degradation mechanism for the BWR feedwater line. We are not aware of any thermal fatigue issue other than the feedwater nozzles; that issue was resolved in the early 1980s through several mitigation measures including the installation of GE-

designed triple thermal sleeve. A rigorous inspection program per NUREG-0619 is currently in place. Not a single one of hundreds of these inspections have turned up any evidence of cracking. Thus, thermal fatigue is not an issue in the NSSS portion of the BWR feedwater line. In addition, please refer to the letter from T. Essig (U.S.NRC) to T. J. Rausch (BWROG), dated June 5, 1998, subject: BWROG-Safety Evaluation of Proposed Alternative to BWR Feedwater Nozzle Inspections (TAC M94090). This letter provides evidence that the thermal fatigue of the FW nozzles is effectively managed at BWRs and provides the basis for revising the examination frequencies.

Response: The NUREG report acknowledges the improvements that have been implemented in BWRs as a result of the NUREG-0619 inspection requirements. As stated in Section 6.3.2, "There was a rash of feedwater nozzle cracks reported in the 1970 to early 1980 time period in BWRs. Plant and system modifications were implemented after a detailed study of the problem and augmented inspections are being conducted based on NUREG-0619 requirements. These mitigation measures have proven effective as no new thermal fatigue cracks have been discovered in these BWR feedwater nozzles over the last 20 years." However, as stated in Section L.2 when discussing the relative contributions for various BWR piping systems, "There was wide variability expressed for the feedwater system. Several participants thought that its susceptibility was similar to that of the recirculation system while others thought that it would make an inconsequential contribution." This latter group generally thought that the mitigation programs in place for the feedwater system were generally effective and additional thermal fatigue locations within the feedwater system were not significant LOCA contributors.

However, rationale provided by some of the panelists who believe that thermal fatigue is a significant contributor to the LOCA frequency estimates is summarized in Section L.2 and Section 6.3.2. As stated in Section 6.3.2, "The BWR plants are expected to be more prone to thermal fatigue problems compared with the primary side of PWR plants because they experience greater temperature fluctuations during the normal operating cycle. In BWR plants, thermal fatigue remains an important contributor for the feedwater lines and the RHR system." Additionally, "...thermal fatigue is an aging mechanism that could lead to a large LOCA because it does not manifest itself as a single crack, but as a family of cracks over a wide area. Thermal fatigue cracks also tend to propagate rapidly, and since it is not material sensitive (i.e., it can attack a number of materials), it is difficult to prioritize critical areas for inspections." These reasons explain why thermal fatigue is still regarded as an important LOCA contributor by many panelists.

The general variability in opinion expressed by the elicitation panelists for feedwater, main steam, and RHR systems is summarized in the box and whisker plots in Figures L.6 through L.8. Some of the differences associated with the illustrated variability in these figures results from the diversity in opinion about the significance for thermal fatigue in BWR plants.

Comment Number: GC3
Submitted by Nuclear Energy Institute (NEI)
Comment: As the NUREG-1829 report may be considered to be the "most recent applicable data" upon finalization, it is important that the final report provide an alternative to continue using operational experience data for the determination of small break LOCA frequencies. Most PRAs currently reference NUREG-5750, which used such a basis (at the time there were 1,250 reactor years of operating experience) to estimate small break LOCA frequencies. Since issuance of NUREG-5750, over one thousand additional reactor years of operational experience have confirmed the conclusions of NUREG-5750 relative to small break LOCAs. Draft NUREG-1829 notes that, when steam generator tube rupture data are excluded, there is general correlation on small break frequencies with NUREG-5750. However, our review of the report indicated that draft NUREG-1829 estimates these frequencies over one order of magnitude higher than the estimate of NUREG-5750. Using the NUREG-1829 small break LOCA frequency estimation, the US reactor fleet should be experiencing one small break LOCA on average

every 4 years. However, no such LOCAs have occurred in the operating history of the US plants. Obviously, the incorporation of this frequency estimate into existing PRAs would lead to unwarranted impacts that are out of context with reality.

Response: Comparing results from NUREG-1829 and NUREG/CR-5750 must be done with some care due to differences in how the LOCA frequencies were calculated. The SB LOCA frequencies typically reported in NUREG-1829 for PWRs include contributions from SGTRs while the SB LOCA frequencies reported in NUREG/CR-5750 exclude SGTRs. Furthermore, the LOCA frequencies in NUREG-1829 are based on threshold leak rates (Category 1 LOCAs include all leaks greater than 100 gpm [380 lpm], Category 2 LOCAs include all leaks greater than 1,500 gpm [5,700 lpm], etc.). Conversely, the SB LOCA results in NUREG/CR-5750 include only those events with leak rates between 100 and 1,500 gpm (380 and 5,700 lpm).

These distinctions were considered in making the comparisons provided in Section 7.9 that are summarized in Table 7.20. The reported SB mean LOCA frequency estimates for NUREG/CR-5750 are 4.0E-04 per calendar year and 7.4E-03 per calendar year for BWR and PWR plants, respectively. Note that these PWR SB LOCA estimates include the steam generator rupture frequencies calculated in NUREG/CR-5750. The revised NUREG-1829 SB LOCA estimates are 5.2E-04 and 6.6E-03 for BWR and PWR plants, respectively (geometric mean with overconfidence adjustment). The ratio of the NUREG/CR-5750 to the NUREG-1829 results is 0.76 for BWRs and 1.12 for PWRs. These SB LOCA estimates are therefore similar.

It is also interesting to compare the PWR SB LOCA frequency estimates after excluding SGTR frequencies from the Category 1 LOCA estimates (Table 7.19). As reported in Section 7.8 of NUREG 1829, the SGTR rupture frequencies predicted by the elicitation, NUREG/CR-5750, and operational experience are consistent. The cumulative SB LOCA Category 1 and 2 estimates without SGTR contributions are 1.9E-03 and 4.2E-04 per calendar year (Table 7.19). Therefore, the interval value that corresponds to the historical SB LOCA definition (i.e., breaks between 100 and 1,500 gpm [380 and 5,700 lpm]) is 1.48E-03 per calendar year. This value is approximately 3.7 times higher than the mean SB LOCA frequency from NUREG/CR-5750 of 4E-04 per calendar year.

The increase in the elicitation result reflects the panelists' opinion that current PWR SB LOCA frequencies for components other than steam generator tubes are higher than historical averages. This increase primarily stems from current PWSCC concerns (Section 6.3.2). The practical implications from these differences however are not striking. The NUREG/CR-5750 estimate translates to an expected PWR SB LOCA every 36 years for the current fleet of 69 operating PWRs. The elicitation frequency corresponds to an expected PWR SB LOCA every 10 years. Additional comparisons between the NUREG-1829 and NUREG/CR-5750 estimates are contained in Section 7.9.

Because of this and similar comments, Section 7.10 has been added to NUREG-1829 to compare the elicitation LOCA frequency estimates with estimates derived from operating experience. This comparison shows that the differences between the elicitation and operating experience-based estimates of the non-SGTR, PWR Category 1 LOCA frequencies are not statistically significant. However, even though these differences are not statistically significant, this does not imply that the two approaches are estimating the same frequency. Because an operating experience-based estimate is an historical average based on many years of operation, a difference will exist if the panelists believe that the current failure frequency differs from the historical average. In fact, the increased elicitation estimate is supported by the panelists' qualitative and quantitative responses. As noted above, the panelists indicated that medium and, to a lesser extent, small LOCAs in PWRs are most dramatically impacted by PWSCC in relatively small diameter passive system component (e.g., CRDMs, instrument nozzles, etc.) (Section 6.3.2).

Additional details on this comparison are found in Section 7.10. Related information is also found in the responses to GC4, GC5, GC6, GC7, and 7-8.

Comment Number: GC4
Submitted by Nuclear Energy Institute (NEI)
Comment: The Nuclear Energy Institute offers the following comments on the subject Federal Register notice, which solicited public comments on draft NUREG-1829. This NUREG was intended to provide technical support for the proposed rulemaking to 10 CFR 50.46 which would establish the option to revise the design basis LOCA break size. Thus, the emphasis of the expert elicitation was on estimating frequencies for large break LOCAs. Our comments are limited to the report's estimation of small break LOCA frequencies, which, unlike large break LOCAs, are important contributors to PRA risk profiles.

PRA standards have been developed by consensus bodies and endorsed by NRC, with the expectation that plants will be expected to conform to these standards to support regulatory applications. ASME PRA standard RA-S-2002 (endorsed through NRG Regulatory Guide 1.200) contains the following requirement relative to initiating event frequency estimation:

> IE-C1: Calculate the initiating event frequency from plant-specific data, if sufficient data are available. Otherwise, use generic data. Use the most recent applicable data to quantify the initiating event frequencies.....

Response: Contrary to the commenter's assertion, the objective of the expert elicitation (Section 2) was to calculate LOCA frequencies for all break sizes, not just large break LOCAs. Also, in general, the objectives and results of the expert elicitation as summarized in NUREG-1829 are consistent with the above statement in ASME PRA standard RA-S-2002. The NUREG-1829 small break LOCA estimates represent generic values. Calculations based on plant-specific data, specifically for PRA, are acceptable for use provided that a sufficient technical basis supporting these alternative estimates has been established. There is also an implication by this comment (and subsequent Comment GC5) that the NUREG-1829 estimates do not compare favorably with operator experience. As documented in the newly-added Section 7.10, the differences between the NUREG-1829 estimates and operating experience are not statistically significant. Those differences that do exist are also supported by the panelists' qualitative and quantitative responses. More details on the comparison between operating experience and the NUREG-1829 LOCA frequency estimates are provided in Section 7.10 which was added to address this and similar comments. See also the responses to GC3, GC5, GC6, GC7, and 7-8 for similar discussions.

Comment Number: GC5
Submitted by Nuclear Energy Institute (NEI)
Comment: Draft NUREG-1829 used plant experiences to estimate the steam generator tube rupture (SGTR) frequency which amounts to greater-than 50% of the total small LOCA frequency. Estimate of the remaining 50% of Category 1 LOCA was entirely based on expert elicitation. The resulting Category 1 frequency estimates from the panel showed a significant divergence of opinions. It is recommend that Category 1 LOCA frequency estimate should continue to be related to the large number of years of plant experiences similar to the method used in NUREG-5750. The current lengths of those experiences amount to thousands of reactor-years, and are statistically significant for use in estimating the annual frequency of events at the 1E-2, 1E-3, and 1E-4 levels. Similar estimates are used in PRA models for numerous other important PRA parameters (such as SGTR).

Response: The commenter incorrectly states that plant experiences were used to estimate the SGTR frequency while the remaining contributions were based on expert elicitation. All contributions to the Category 1 LOCA frequency estimates were determined by the expert elicitation process. For the reasons documented in Section 1 of NUREG-1829, the authors believe that the expert elicitation results are more accurate than results calculated simply from operational experience.

However, the SGTR contributions provided by the panelists are nearly identical to estimates determined simply from operational experience. This consistency, however, simply reflects the panelists' combined opinion that the operational experience data is relevant for calculating the current day SGTR contributions. The contributions of non-SGTR failures to the PWR LOCA Category 1 frequencies and the BWR LOCA Category 1 estimates are also not inconsistent with estimates based solely on operational experience (See Section 7.10). However, the panelists do expect some increase in the PWR LOCA Category 1 frequencies compared to operational experience-based estimates as a result of current PWSCC issues (Sections 6.3.2 and 7). See Section 7.10, which was added to address this and similar comments, for additional details on the comparison between the elicitation estimates and operational experience data. Additional relevant information is also provided in the responses to GC3, GC4, GC6, GC7, and 7-8.

Comment Number: GC6
Submitted by Nuclear Energy Institute (NEI)

Comment: The report should provide a discussion on statistical validation of small LOCA frequency. By using the method of Jeffrey's non-informative prior (over the past 2,500 reactor years with zero events excluding steam generator tube ruptures), the expected small LOCA frequency is at or below the 1E-04 level. This frequency is over one order of magnitude lower than the frequency reported in the draft NUREG. Plant operational experience of over 2,500 reactor-years should be considered as a valid predictor of small LOCAs. That consideration is further strengthened by improved methods and increased requirements for in-service-inspections and leak detections.

Response: The use of Jeffrey's non-informative prior to estimate a frequency if no events have been observed leads to a significant underestimate of the mean frequency. Using this method, NEI states that the expected frequency is no more than 1E-04, based on a denominator of 2,500 reactor years. With zero events, a reasonable estimate of the mean is based on using a 50% confidence bound or a value of 0.7 in the numerator. This yields $0.7/2,500 = 2.8E-04$ which is a factor of 3 larger than the estimate using Jeffrey's non-informative prior. Furthermore, the use of 2,500 reactor years as the denominator clearly combines both PWR and BWR operating experience. More rigorous evaluation of operational experience data (See Section 7.10) has demonstrated that the operational experience-based and elicitation-based SB LOCA estimates are not inconsistent. Furthermore, differences are supported by the panelists' responses and rationale. Section 7.10, which was added to address this and similar comments, provides additional details on the comparison between the elicitation estimates and operational experience data. Also, see the responses to GC3, GC4, GC5, GC7, and 7-8 for related information.

Comment Number: GC7
Submitted by Nuclear Energy Institute (NEI)
Comment: The report should provide a discussion on probabilistic validation of the small LOCA frequency. Using a Poisson distribution with failure rate of 2.9E-03 (NUREG-1829 Category 1 LOCA frequency, excluding steam generator tube rupture events), and considering the approximately 2,500 reactor-years of operation experience, the probability of no small LOCA events (actual industry performance) is around 1 percent. This result shows an excessive conservatism in the category 1 LOCA frequency estimation of NUREG-1829.

Response: A new section (Section 7.10) has been added to the revised NUREG which compares the small break LOCA frequency estimates from the elicitation with operating experience. As discussed in this section, the elicitation estimates for BWR and PWR small break LOCA frequencies are generally consistent with the historical data. See the response to Comment 7-8. Related insights are also provided in the responses to Comments GC3, GC4, GC5, GC6, and 7-8.

Two additional points should be noted. First, the 2,500 reactor-years of operating experience referred to in the comment clearly combines BWR and PWR plants. The NUREG provides separate estimates for BWRs and PWRs and treats their operating experience separately. The number of reactor-years used in Section 7.10 is 1,023 for BWRs and 1,986 for PWRs, all as of December 2006. Second, as discussed in Section 7.10, there has been one pipe break in a PWR plant that exceeded the LOCA Category 1 flow rate threshold defined in this study. Additionally, a second PWR pipe break resulted in flows that were close to the Category 1 threshold. The comment claims that there have been no non-SGTR Category 1 events.

Comment Number: GC8
Submitted by Professor Larry Hochreiter of Penn State University
Comment: We have revised the data for pipe breaks and leaks following the suggestions from the NRC staff and have made comparisons to the NRC expert solicitation panel's break spectrum failure frequency plots that were developed in support of the revisions to 10CFR50.46. We had discussion with C. Gary Hammer and other members of the NRC staff on our original comparisons and they suggested that we limit our comparisons to just the class 1 piping for the reactor since that was the class of piping systems that the NRC break spectrum frequency study were originally developed for. Therefore, we eliminated all piping failure, leaks, and cracking data from the mixed data base that we had originally developed such that only the class 1 piping information remained. A CD with the revised data is included with this letter. We also reviewed the class 1 systems with the NRC to make sure that we isolated only those systems which should be compared to the estimated break frequency curves. When comparing the remaining data to the NRC break frequency curves, we used approximately the same break size bins that were used in the NRC study. We also normalized the individual bins of data using the approach recommended by the NRC in which the number of breaks or breaks and leaks were normalized on the numbers of effective full power days.

We have separated out "Breaks" from the total data base as separate plots for both PWRs and BWRs as shown in Figures 1 and 2. As these figures indicate, there are no large breaks in the class 1 piping. However, for the smaller breaks, the data clearly lies above the estimated break frequencies estimated in the NRC solicitation study. The data is above the 95% percentile as compared to the NRC break frequency plots indicating that even smaller breaks are more frequent as compared to the estimated frequencies from the solicitation study.

We also have combined "Breaks" and "Leaks" together on separate plots for PWRs and BWRs as seen in Figures 3 and 4. The rational for this is that leaks are really breaks since the pipe has failed and is leaking. We used the size of the pipe for when combining the leaks with the breaks. This may not be the best method of comparing the data to the break frequency distribution and it may appear to bias the results since there are only leaks for the larger pipes and not breaks. However, it is not clear that, given a set of conditions, a leak in a large pipe could grow to a full pipe size break. Say for example, under a safe shutdown earthquake condition. Also, this grouping could be viewed as being conservative since the original premise is that the pipes should not leak in the first place.

All the data, breaks and leaks, when combined in this form indicate that there is a significant difference between the existing data and the break spectrum failure frequencies from the NRC expert solicitation study. This indicates to me that we should not be revising 10CFR50.46 by introducing a "transitional break size" and reducing the mitigation capabilities of the plant's ECC systems and defense in depth for the larger break sizes. I believe that we need to maintain the full functional capability of the robust ECC systems that we have in the current operating plants for all possible break sizes up to and including the full double-ended break of the reactor coolant piping. The break and leak data for the class 1 piping indicates to me that we need the full functional ECC systems and their supporting systems and they should not be compromised or sacrificed by revising 10CFR50.46.

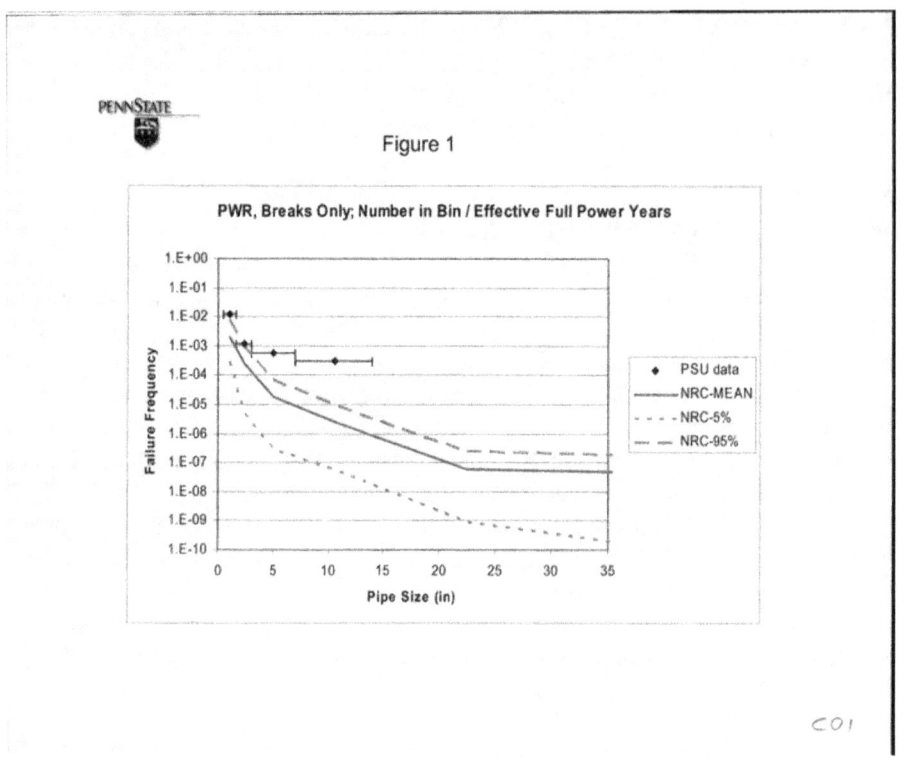

Figure 1

PWR, Breaks Only; Number in Bin / Effective Full Power Years

Figure 2

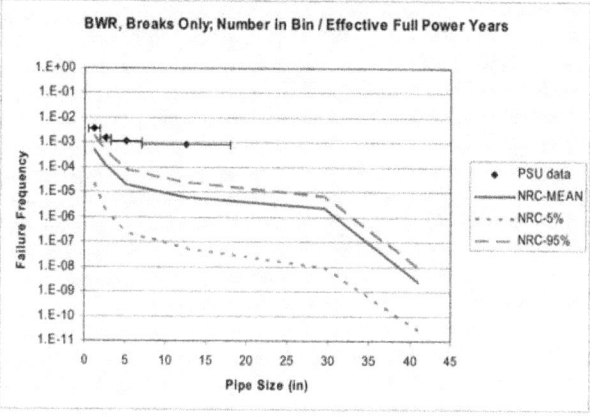

Figure 3

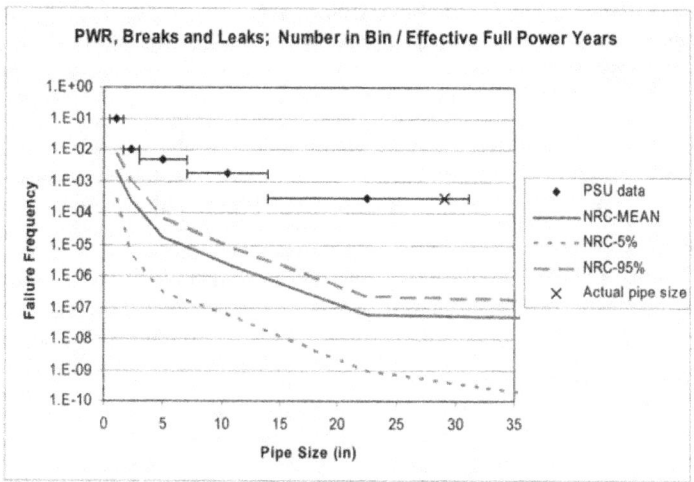

M-11

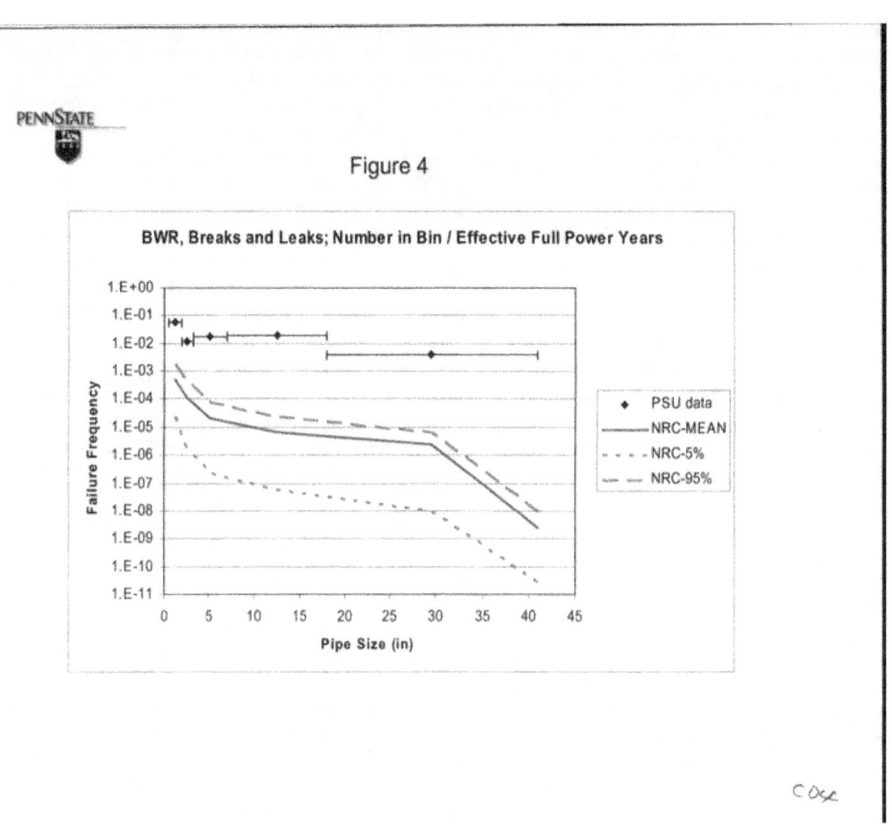

Figure 4

Response: There are several issues with the analysis performed by the commenter with respect to both the pipe "break" frequencies plotted in Figures 1 and 2 and the pipe "break and leak" frequencies in Figures 3 and 4. Figures 1 and 2 above were created by identifying "breaks" from a database of piping failures to construct separate PWR and BWR break frequencies. As depicted in Figures 1 and 2 above, the resulting plot clearly lies above the estimated break frequencies estimated in the expert elicitation documented in NUREG-1829.

The biggest issue pertaining to the commenter's results is with the integrity of the database used in the analysis which was supplied as part of the commenter's public comment. The database appears to be similar or identical to the database developed in Bush, et. al, "Piping Failures in US Nuclear Plants: 1961-1995," SKI 96.20. This can be surmised from the fact that there are only two events as recent as early 1996 in the database, and the database contains the similar erroneous information to the Bush et. al. report for selected audited events. The Bush et. al. database was never validated and it was also the foundation of EPRI TR-111880. An independent review of this database[1] found that a large percentage of the database records involved non-piping failures. The EPRI report was subsequently withdrawn due to requests by EPRI members.

The authors reviewed the larger pipe events that could be considered "breaks" and used in constructing Figures 1 and 2. The authors examined events classified as a breakage, rupture, failure, or severance in the database for 2 inch diameter or greater pipes in both BWR and PWR plants. This approach was necessary because the commenter neither definitively identified the data used to construct Figures 1 and 2, nor provided details supporting the calculations. Events were checked against the OPDE Database Rev. 0.e, dated 24 March 2004 which has been verified. In several cases, the original failure references were obtained for verification.

---

[1] Lydell, B., "Independent Review of SKI 96-20 Database," Technical Note 1996-01, SKI Ref. No. 14.2-940477, February 1996.

Nineteen events were reviewed in all and this data verified many of the reported problems with the Bush et. al. database. Almost all of the database records contain some error or inconsistency. Many of the listed events cannot be referenced to a known event. Common problems include incorrect event dates, references, pipe sizes, or break sizes. The failure classification type (i.e., leak, rupture, severance, etc.) was also found to be both inconsistent and inaccurate in several cases. In spite of these problems, the authors searched the OPDE Database for events that likely correspond to the break events identified in the commenter's database. Approximately 15 events were identified which either could be definitively matched to an event in the commenter's database or at least represented by a pipe break at the listed plant in a similar system. None of break events occurred in an unisolable segment of the reactor coolant pressure boundary piping. This is the fundamental definition of a LOCA, and the objective of the elicitation was to estimate LOCA frequencies. The only basis for comparison with the elicitation results is an estimation of LOCA frequencies. No other basis, including a comparison of piping failure frequencies, is consistent with the elicitation results. Hence, the Figure 1 and 2 comparisons with the elicitation results are invalid without further consideration.

However, there are other inconsistencies and errors in the commenter's analysis. The largest "failed" PWR piping in the database is for a 4 inch diameter pipe. Leak data is reported in the commenter's database for PWR piping up to 13 inch diameter. However, Figure 1 above depicts breaks up to 12 to 15-inch diameter. The basis for reporting PWR break data greater than 4 inch diameter is not supported by the supplied database. There are other, less significant, but still consequential problems with the commenter's analysis. Contrary to the commenter's statement that all the breaks and leaks represent "Class 1 piping" systems, most of the events occurred in lower grade piping. This is not a trivial distinction because Class 1 piping is subject to more rigorous design, testing, and inspection requirements than other piping systems within the plants. In addition, the commenter's database contains nearly twice the number of carbon steel failures in PWR plants than in BWR plants. The finding is heavily biased by the consideration of non-ASME Code, FAC-susceptible piping in the commenter's database. Also, several of the rupture sizes appear to overestimate either the actual pipe size or the rupture size which occurred. The distinction is important because the size of the rupture, not the piping size, determines the rate of loss of system fluid. A partial failure of a large pipe can lead to substantially lower fluid loss rates than expected if the pipe completely burst.

The leak data comparisons presented in Figures 3 and 4 are similarly misleading. The most substantive issue is the commenter's contention that "leaks are really breaks since the pipe has failed and is leaking." This is simply untrue. Pipe breaks occur once a flaw reaches a critical size such that it grows unstably at the applied load level. The piping can then not support required operational loading, a readily-apparent hole or breach forms, and internal fluid is released at a rate beyond the reactor water make-up system of the plant. Conversely, piping having leaking joints or through-wall cracks typically remains intact and continues to support internal and external stress without imminent failure. These cracks may continue to grow until imminent failure occurs, but the difference between the leaking crack size and the rupture crack size increases with pipe size. Hence, larger pipes provide more margin against failure after a leak appears. The NRC's LBB approach for large pressure boundary piping is based on the premise that a leaking through-wall flaw can be detected at normal operating loads via the plant's normal leakage detection systems before the flaw reaches a size that could grow unstably at emergency or faulted load conditions. The commenter's database records support this premise as the larger diameter, Class 1 piping events are leaks that were found and repaired prior to a catastrophic failure. The LBB approach has used to ensure compliance of GDC-4 requirements for over 20 years.

Additionally, the commenter's analysis of the leak data using the supplied database suffers from similar issues as the break data. A majority of these leak events actually occurred in secondary-side plant systems and the pipe leak did not result in loss of pressure boundary coolant. Therefore, longer term failures of these pipes (presuming that they remained undetected and the associated flaw continued to grow to failure) would not have resulted in a primary LOCA event.

In contrast to the commenter's analysis, a review of some of the more well-established and well-recognized piping failure databases (PIPEex, OPDE, and SLAP) has found 1 event that could be characterized as a SB LOCA in PWR plants and no events that could be characterized as SB LOCAs in BWR plants (See Section 7.10 of NUREG-1829). No MB or LB LOCA events have occurred. As previously discussed (See responses to GC3, GC4, GC5, GC6, and GC7) the LOCA elicitation results for the present day estimates are consistent with this operating experience. The comparison of the elicitation results and operating experience is also provided in Section 7.10 of the revised NUREG report.

Comments Related to Executive Summary of NUREG-1829

Comment Number: ES1
Submitted by Westinghouse Owners Group
Comment: Method of Aggregating Individual Frequency Responses into a Group Response. For this study the geometric mean aggregation method was used instead of the arithmetic mean method or mixture distribution methods, which would give higher mean values of LOCA frequency. The reviewers concur that the geometric mean is most representative of the consensus of the group (expert panel). As an example, consider possible individual responses the different group means for the distribution of factors on a baseline frequency in the following table (see WOG response for table):

Number or Responses	Value of Factor
1	0.01
2	0.1
3	1.0
2	10
1	100

For this example, the arithmetic mean value for the 9 responses is 13.69 while the geometric mean value (average of the logarithms) is 1.0, which seems to be much more representative of the group's opinions. Part of the reason for this is that the probabilities and frequencies of failures of structural components, such as piping, are normally expressed as orders of magnitudes much less than one. Uncertainties on these values are also expressed as factors instead of differences because the physical contributions to structural failures (leaks and breaks), such as flaw sizes and crack growth rates, are also known to be log-normally distributed. Use of logarithmic distributions and geometric means is also consistent with NRC Guidance on Risk-Informed In-service Inspection for Piping (Draft Report NUREG-1661, January 1991). Figures 3.3 and 3.4 of this guidance show the range of frequency estimates from expert elicitation, plotted on a logarithmic scale, for failure of auxiliary feedwater system components and the reactor pressure vessel, respectively. Figure 4.6 of this same report shows the uncertainty in the best estimate (median value) of piping failure probability, calculated using probabilistic fracture mechanics methods, to also be logarithmically distributed. Scanned copies of these figures are provided in Appendix A[2] of the WOG's public comment document.

Response: The authors agree that the geometric mean aggregation supports the elicitation objective of developing a consensus group estimate by reducing the effect of either wide differences in the individual

---

[2] Letter from Frederick P. Schiffley, II, Chairman to USNRC, "Westinghouse Owners Group Comments on Draft NUREG-1829, 'Estimating Loss-of-Coolant Accident (LOCA) Frequencies Through the Expert Elicitation Process' (MUHP-3062)", WOG-05-517, dated November 28, 2005, ADAMS Accession # ML0503340274.

estimates, or the effect of a single estimate which is significantly higher than the others. As documented in Section 7, geometric-mean aggregation, in this study, produces group estimates which approximate the median of the individual estimates. In the example cited by this commenter, the arithmetic mean estimate of 13.7 is larger than all but one of the responses and does not approximate the median of the individual estimates.

However, as indicated in the Executive Summary, because the alternative aggregation methods can lead to significantly different results, a particular set of LOCA frequency estimates is not generically recommended for all risk-informed applications. The purposes and context of the application must be considered when determining the appropriateness of any set of elicitation results. This position is consistent with the recommendation of the NRC's Advisory Committee on Reactor Safeguards (Letter from W.J. Shack to D.E. Klein dated December, 20, 2007, Subject: Draft Final NUREG-1829, "Estimating Loss-of-Coolant Accident (LOCA) Frequencies Through the Elicitation Process," and Draft NUREG-XXXX, "Seismic Considerations for the Transition Break Size", ADAMS Accession Number ML073440143).

Comment Number: ES2
Submitted by Joseph Conen of the BWR Owners Group
Comment: It is apparent that the panel has not given appropriate credit to the IGSCC mitigation measures for the NSSS stainless steel piping that the BWR owners have implemented since the early 1980s. For example, the second paragraph from bottom on page xvii states, in part: "...the biggest frequency contributors for each LOCA size tend to be systems having the smallest pipes, or component, which can lead to that size LOCA. The exception to this general rule is the BWR recirculation system, which is important at all LOCA sizes due to lingering IGSCC concerns." Since the largest pipe size in the recirculation piping system can be up to 28-inches, the preceding statement essentially implies that LB LOCA redefinition is not applicable to BWRs. The panel did not seem to give adequate credit for several effective mitigation measures in terms of better material (e.g., use of nuclear grade stainless steel in replacement lines), stress improvement (e.g., induction heating stress improvement [IHSI], last pass heat sink welding [LPHSW], and mechanical stress improvement process [MSIP]) and water environment (e.g., hydrogen water chemistry [HWC]) and repair measures such as the weld overlays and elimination of creviced geometries. The panel apparently did not consider the report GE-NEA41-00110-00-1, Rev. 0, *A Review of NUREG/CR-5750 IGSCC Improvement Factor and Probability of Rupture Given a Through-Wall Crack*, provided to NRC by the BWROG on April 25, 2002 (ADAMS Accession NO. ML021210417). In addition, the panel did not recognize the contribution of BWRVIP-75, which provides evidence that IGSCC is effectively managed at BWRs and provides the basis for revising examination frequencies. We consider these significant oversights, given their relevance to the panel's conclusions. On the other hand, the panel did accept the future effectiveness of mitigating measures for PWSCC issue for the PWR small diameter piping (p. 6-5) in reducing failure rates for this piping. The NUREG should provide similar credit for the BWR IGSCC mitigation measures noted above with regard to break frequencies.

Response: The authors disagree with the contention that the panel has not given appropriate credit to the IGSCC mitigation measures for the NSSS stainless steel piping that the BWR plants have implemented since the early 1980s. The report referenced above (GE-NEA41-00110-00-1) was provided as background documentation to the peer review panel. Additionally, several of the panelists had extensive experience with the assessment of the IGSCC issue and the development of appropriate mitigation strategies. One of the BWR base cases specifically investigated the failure probability of primary recirculation piping due to IGSCC. The base case model plant was assumed to follow the Generic Letter 88-01 inspection technique, used weld overlay to reinforce the flawed piping, and utilized normal water chemistry.

This base case definition was chosen for convenience to improve the expected accuracy of the PFM analysis to be conducted by only considering one easily modeled mitigation strategy, i.e., weld overlay. It was well recognized and stressed during meetings that this base case was generally not representative of BWR plants. It was noted that most BWR plants employ hydrogenated water chemistry and may employ mitigation strategies other than weld overlays. Additional sensitivity analyses of the base case results were conducted to evaluate the effectiveness of BWR mitigation strategies for IGSCC. These sensitivity studies are summarized in Sections 4.3.3 and 4.3.4 of NUREG-1829, while more detail is provided in Appendices D – F. These results were presented to the panelists during the base case review meeting and copies were available to support the preparation of their individual elicitations. This information, the background information provided, and the experience of individual panelists was sufficient to ensure that the panelists were sufficiently informed about the effectiveness of IGSCC mitigation so that it was properly credited during the elicitation.

However, although the mitigation has been effective in reducing the associated LOCA frequencies, there is still risk associated with failure of BWR systems containing pre-existing flaws. As summarized in Section 6.3.2 of NUREG-1829, "The panel consensus is that the susceptibility of BWR piping systems to IGSCC is greatly reduced compared to what it was in the past. Measures such as improved HWC, weld overlay repairs, stress relief, and pipe replacement with more crack resistant materials have led to this reduction. Inspection quality has also improved such that the probability of crack detection is much better than in the past. However, as indicated earlier, there remains concern about the failure likelihood of the large recirculation piping and the RHR lines that have not been replaced. The original piping materials are much more susceptible to IGSCC and many lines retain preexisting cracks that initiated and grew before HWC was adopted."

Additionally, at least one panelist was also concerned that the more IGSCC-resistant replacement piping materials may still crack under service conditions. This panelist cited the German plant experience with cracking in Type 347 stainless steel. Another panelist raised the possibility that cold work (e.g. due to grinding) could increase the IGSCC susceptibility of the low carbon (L grade) stainless steel that has been used as a replacement material in many U.S. plants. However, the U.S. BWR experience with L grade stainless steel piping was widely recognized by the panel as being very good thus far. For these reasons, many panelists believe that continued vigilance is required through the augmented inspection requirements in Generic Letter 88-01 and NUREG-0313.

Key elements of this response have been used to modify the Executive Summary and Sections 3.5, 4.3, and 6.3 of the revised NUREG.

Comment Number: ES3
Submitted by Joseph Conen of the BWR Owners Group
Comment: Table 1 on page xxi and Table 3.8 mention effective break size of 42 inches for BWRs. It is likely to be an artifact of an assumed LOCA size of 500,000 gpm and not representative of the BWR NSSS geometries.

Response: The LOCA categories were defined in terms of "threshold" flow rates, see Table 3.2. A Category 6 LOCA was defined as a LOCA which resulted in a flow rate greater than 500,000 gpm (1,900,000 lpm). For BWR thermal/hydraulic conditions for a liquid (not steam), this corresponded to an effective break size of 41 inches (1040 mm) assuming a circular opening (see Table 3.8). Similarly a Category 5 LOCA (>100,000 gpm [380,000 lpm]) would result in a 18 1/2 inch (470 mm) circular opening for BWR liquid conditions. The panel members recognized the fact that when considering BWR piping systems, the largest piping system in the plant (~28 inch diameter recirculation system) would not support a Category 6 (>500,000 gpm [1,900,000 lpm]) LOCA. However, a significant RPV rupture (classified as a non-piping failure) could result in a Category 6 LOCA. Table 1 combines both piping and non-piping LOCA contributions which is why this size LOCA is possible for a BWR plant.

The sources of BWR Category 6 LOCA contributions have been clarified in Sections 3.4.1 and 3.7 of the revised NUREG.

Comments Related to Section 1 of NUREG-1829

None

Comments Related to Section 2 of NUREG-1829

None

Comments Related to Section 3 of NUREG-1829

Comment Number: 3-1
Submitted by Joseph Conen of the BWR Owners Group
Comment: Tables 3.7 and B.1.9 depict the assumed conditions for the base cases. For the BWR recirculation line, the assumed plant water chemistry condition is NWC. This is not representative of the current US BWR fleet where most of the plants are operating on HWC. It is not clear to this reviewer if the panel has factored in the improvements in reactor water conductivity, irrespective of whether the plant is on NWC or HWC, which most BWR plant owners have put in place in the last decade.

Response: The authors agree that most BWRs no longer operate with normal water chemistry and that there have been improvements in reactor water conductivity. However, the purpose of the base case analyses was to provide anchoring points as a basis for panel members' elicitation responses. The panel members accounted for improvements in water chemistry through their adjustment ratios which were used to modify these base case frequencies.

Additionally, the operating-experience base case estimates did directly consider the effect of IGSCC mitigation on the cracking rates. Bill Galyean's estimates were developed by only considering post-1985 operating experience so that IGSCC cracking frequencies were not overestimated. The year of 1985 was used in the base case analyses to demark the period where wholesale mitigation methods were adopted by the industry. See Appendix E for more details. Bengt Lydell's estimates also considered the effects of all mitigation techniques (e.g., HWC, weld overlays, etc.) on post-1985 through wall IGSCC cracking frequencies by using Bayesian analysis to determine the effect of mitigation techniques. See Appendix D for more details. As a result of their similar consideration of operating experience, these BWR recirculation system LOCA Category 1 base case results are within approximately a half order of magnitude of each other. See Section 4.3.4 for a more detailed discussion on IGSCC mitigation for these base case analyses.

Additional base case sensitivity analyses also examined the effect of a weld overlay (another common IGSCC mitigation technique) on one of the PFM base case analyses. This effect is documented in Section 4.3.3. Additionally, the general effectiveness of all IGSCC mitigation measures was provided to the panelists by analyzing the operating experience both prior to and after 1985. The pre and post-1985 cracking frequencies are discussed in Section 4.3.4.

Key elements of this response have been incorporated into Sections 3.5.1 and 4.3.4 of the revised NUREG.

Comment Number: 3-2
Submitted by Joseph Conen of the BWR Owners Group
Comment: For the BWR feedwater line base case, Tables 3.7 and B.1.9 mention assumed water chemistry condition as NWC and include flow-assisted corrosion (FAC) as an aging mechanism. This reviewer believes that the FAC is a potential issue in BWR feedwater lines only when the oxygen level is very low (e.g., few ppb) – a possibility with HWC. During the NWC condition assumed in the base case, the oxygen level is high enough that FAC is not likely to be an issue. Also, most BWR plants with HWC have implemented controls to maintain a certain minimum oxygen level in the incoming feedwater to mitigate likelihood of FAC.

Response: Several elicitation panelists largely echoed the sentiment in this comment. As summarized in Section 6.3.2 of NUREG-1829, most panel members believe that the industry has inspection programs in place today to prevent FAC, especially in the primary side piping systems. However, one panel member expressed the concern that the water chemistry improvements (HWC) that mitigate IGSCC could lead to FAC in unanticipated locations that are not monitored as part of these inspection programs. The elicitation responses generally reflect the group expectation that FAC is not as significant a contributor to the LOCA frequency contribution as is IGSCC and other mechanisms. Typically panel members estimated that the FAC LOCA frequency contribution is between ½ to 1 order of magnitude less than the contribution from the recirculation piping. Accordingly, the contribution of FAC to the feedwater system does not significantly contribute to the BWR LOCA frequency estimates.

Comment Number: 3-3
Submitted by Nuclear Energy Institute (NEI)
Comment: The equivalent break diameters used in various PRA models that form boundaries between various LOCA categories do not necessarily match those used in the draft NUREG. For example, a small LOCA range may extend up to 0.03 square feet equivalent break area (derived through existing capability of high pressure safety injection). The draft NUREG should give a clear guideline on interpolations between the various LOCA frequency values, including advice on arithmetic or geometric preference for interpolation.

Response: Because LOCA frequencies were estimated only for the six specified LOCA categories, interpolation between them may be problematic. In the figures, the plotted points were connected with straight lines for visual clarity, but this should not be taken to mean that linear interpolation is recommended. Interpolation of frequencies between category sizes can be done at the user's discretion depending on the conservatism required by the application. Some common interpolation schemes are linear, multi-point nonlinear and cubic spline. A step-wise or stair-step interpolation between two categories where the frequency for the lower category size is used for all flow rates or corresponding break sizes between the two categories provides the most conservative interpolation scheme. Note that any interpolation scheme does not reflect the uncertainty in the interpolated frequencies.

These guidelines were added to the Executive Summary and to the introduction to Section 7 of the revised NUREG.

Comments Related to Section 4 of NUREG-1829

Comment Number: 4-1
Submitted by Joseph Conen of the BWR Owners Group
Comment: Table 4.1 shows a six-order of magnitude difference between the PFM and the field history estimates of through-wall cracking frequencies for the BWR-2 base case. Although the report suggests

that service history data could be analyzed to resolve this difference, it is not clear if this was actually done. This also points out to the need for a rigorous examination of the statistical methods used to translate field leak data or service experience data into pipe break frequencies.

Response: The complexity of translating field leak data or operating-experience data into pipe break frequencies was part of the rationale for conducting the elicitation (Section 1). It was also a major consideration for performing multiple base case analyses so that results from different approaches could be compared. The small break LOCAs and through-wall cracking frequencies represent the categories where operating experience and PFM comparisons are most meaningful because the least extrapolation is required. For the small break LOCAs (Category 1), the results are typically within 2 orders of magnitude for most of the base case team members. However, the total spread in both the through-wall cracking leak frequencies (Figure 4.1) and SB LOCA frequencies (Figure 4.2) can be greater than three orders of magnitude. As the commenter indicates, the BWR-2 Category 0 frequencies (Figure 4.1) vary significantly (4-5 orders of magnitude in final NUREG)

Furthermore, while the PFM and the operating-experience predictions of the base case analyses did not always readily agree, this simply reflects the current uncertainty in calculating these estimates. Differences among these estimates reflect the different assumptions and approaches used in the various analyses. Specifically, the BWR-2 operating-experience frequency estimate was based on 20 reported incidents in BWR feedwater systems. However, none of these events resulted in through-wall leakage. Therefore, the operating-experience estimate is undoubtedly conservative.

The elicitation reported in NUREG-1829 did not attempt to resolve these discrepancies; rather, the elicitation's goal was to reflect the scientific uncertainty in the technical community. Therefore, the purpose of the base case estimates was not to obtain convergence among the predictions. Rather, it was to understand how the assumptions and approaches associated with particular analyses contributed to the disparity in the results. These analyses and their differences were clearly presented to the elicitation panelists so that they could judge their effect as part of their elicitation. As a result, most panel members chose to anchor their current-day LOCA Category 1 responses using operating experience and used PFM, if at all, to inform their assessments for LOCA sizes greater than Category 1 or for future LOCA frequency estimates. These are the domains where relevant operational experience data does not exist.

Key elements of this response were incorporated into Section 4.2 of the revised NUREG.

Comments Related to Section 5 of NUREG-1829

None

Comments Related to Section 6 of NUREG-1829

None

Comments Related to Section 7 of NUREG-1829

Comment Number: 7-1
Submitted by Zouhair Elawar – Palo Verde Nuclear Generating Station
Comment: Draft NUREG-1829 used plant experiences to estimate the steam generator tube rupture (SGTR) frequency which amounts to greater than 50% of the total small LOCA frequency. Estimate of the remaining 50% of Category 1 LOCA was entirely based on expert elicitation. The resulting Category 1 frequency estimates from the panel showed a significant divergence of opinions. I strongly recommend

that Category 1 LOCA frequency estimates should continue to be related to the large number of years of plant experiences similar to the method used in NUREG 5750. The current lengths of those experiences amount to thousands of reactor-years. They are statistically significant to be used to estimate the annual frequency of events at the 1E-2, 1E-3, and 1E-4 levels. Similar estimates are used in PRA models for numerous other important PRA parameters (such as SGTR).

Response: Operating experience becomes more relevant as the LOCA size decreases, especially for events such as SGTR, because the database is populated with actual events. However, there is still scant data on SB LOCAs due to limited passive piping or non-piping component failures. The panelists were made aware of this operating experience data, and SGTR was specifically used as a base case estimate, along with frequencies for other piping and non-piping precursor events (e.g., cracking). However, one of the cautions with using operating experience solely to calculate Category 1 LOCA frequency estimates is that past experience is not necessarily indicative of current or future performance. Common cause material degradation can result in systematic increases in generic frequencies as a function of time compared to frequencies based on operating experience. Conversely, wholesale mitigation programs (as with IGSCC) can result in systematic decreases in LOCA frequencies over time.

These common cause considerations are one reason why elicitation is preferable to simple operating experience estimates for even small break LOCAs. In fact, the elicitation results are most justified for this LOCA category because of operating experience data. The operating experience data was fully considered by the panelists to estimate past LOCA frequencies, and they then used their best judgment to modify these estimates, as appropriate, based on their current knowledge. It is preferable to use current knowledge in this way rather than wait for several years for operating experience to adequately reflect the new conditions. It is worth noting, however, that the elicitation results and operating experience data are generally consistent for LOCA Category 1. The only differences, which are not statistically different, are for the frequencies associated with non-SGTR, Category 1 PWR failures. The panelists justified the higher NUREG-1829 estimates as a result of the effects of PWSCC on small diameter piping failures.

Key elements of this response have been incorporated into Section 7.10 of the revised NUREG. This comment is also very similar to GC5. Please see the response to this comment for additional information.

Comment Number: 7-2
Submitted by Zouhair Elawar – Palo Verde Nuclear Generating Station
Comment: The draft NUREG combined a variety of LOCA sources into each LOCA category. Piping LOCAs and several non-piping LOCAs were pooled together to form each of the LOCA categories. It would be useful for each of the 6 LOCA categories to add a table of LOCA sources and frequency contributions. This breakdown is particularly important for the small and medium LOCA categories. Some contributors to the small and medium LOCAs are modeled separately in most PRA models (SGTR, RCP seals, inter-system LOCAs and others). If the end user does not subtract the separately-modeled LOCA contributors, then the contribution to CDF (core damage frequency) from those contributors would be conservatively and redundantly modeled.

Response: First, it is important to recall that this study was only concerned with estimating passive system failure frequencies of structural system components (SSCs) within the primary system (Section 2). Therefore, LOCAs associated with RCP seals, interfacing system LOCAs, and active system LOCAs are not included in the NUREG-1829 estimates. There are previous estimates for these types of events (e.g., NUREG/CR-5750). The revised NUREG (Section 7.8) includes a table of SGTR frequencies as requested by this comment because all the panelists provided sufficient information to determine individual LOCA frequency estimates for this system. Section 7.8 of the revised NUREG also contains estimates for all non-SGTR passive-system SSC failures.

In general, however, it was not possible to develop information for individual system failures from the elicitation responses. For each LOCA category there was a potential contribution from either 12 or 13 different piping systems (12 for the PWRs and 13 for the BWRs) plus a contribution from either 3 or 5 non-piping components (3 for the BWRs and 5 for the PWRs). In addition, for each of the non-piping components, there were up to 8 (for the RPVs) subcomponents (e.g., vessel body, CDRMs, nozzle, head bolts, etc.) that were potential contributors. In order to make the responses tractable, the panelists were requested to only provide information for systems that they expected to make up at least 80 percent of the failure frequency contribution. Thus, not every panelist provided responses for every piping system or non-piping component or subcomponent. Furthermore, it was not possible to impute this data accurately because the relative frequency contributions to each panelist's estimates are small. Consequently, it was not possible to calculate group estimates of individual system failure frequencies because this would have required each panelist to provide estimates for every piping system and non-piping component..

Key elements of this response have been incorporated into Section 2 of the revised NUREG. Additionally, Section 7.8 has been added to the revised NUREG to provide the SGTR and non-SGTR estimates for PWR plants.

Comment Number: 7-3
Submitted by Zouhair Elawar – Palo Verde Nuclear Generating Station
Comment: The steam generator tube rupture frequency (merged with LOCA category 1) was reported as mean value = 3.5E-03 based on number of industry events averaged over years of reactor operations. This mean value frequency should be separated from the main small LOCA category and estimated as a "range" consisting of upper and lower bounds. Plants with aging steam generators are encouraged to use the upper bound of the range. And, plants with new steam generators may use the lower bound of the range. Of course, numerous plants would use the overall mean value of 3.5E-03.

Response: Separate SGTR frequencies have been calculated from the elicitation responses and are provided in the revised NUREG (Section 7.8). The current-day median STGR frequency is 2.6E-3 and the mean frequency is 3.7E-3 (Table 7.18). These values compare favorably with the operating-experience Category 1 frequency of 3.5E-3 for SGTRs as reported in Section 4.4.1. The 5th and 95th percentiles of the group estimates are 5.0E-4 and 1.0E-2, respectively. Additional discussion concerning the SGTR estimates is also contained in Section 7.10 of the revised NUREG.

Comment Number: 7-4
Submitted by Zouhair Elawar – Palo Verde Nuclear Generating Station
Comment: The various LOCA frequencies are reported in the several tables as cumulative values. In order to isolate the frequency of each LOCA category, one has to subtract the frequency of the next higher ranking category. This reporting format may lead to human errors. Some users may not become aware of the cumulative table format since that description is briefly stated at the later sections of a very large report. Please add a footnote under each LOCA frequency table explain how to obtain the frequency of each LOCA category.

Response: The six LOCA categories were defined in terms of cumulative thresholds because the panelists felt more comfortable with providing their responses as cumulative thresholds rather than to intervals defined by consecutive thresholds. (Appendix B) Because the differences between the results for consecutive LOCA categories are typically much smaller than their uncertainties (see Table 1, Figure 1, Figure 7.36 and Figure 7.37), there is no statistical difference between using the cumulative frequencies or interval frequencies determined by subtraction. If interval-defined LOCA frequencies are required, simply use the NUREG cumulative threshold estimates in Section 7.7. However, Section 7.9 does use interval-defined frequencies for consistency in comparing the NUREG-1829 estimates with other prior study estimates. This point has been clarified in the revised NUREG in the Executive Summary, Section 3.4.1, and Section 7.9.

Comment Number: 7-5
Submitted by Zouhair Elawar – Palo Verde Nuclear Generating Station
Comment: The equivalent break diameters used in various PRA models that form boundaries between various LOCA categories are not necessarily matching those used in the draft NUREG. For example, a small LOCA range may extend up to 0.03 square feet equivalent break area (derived through existing capability of high pressure safety injection). The draft NUREG should give a clear guideline on interpolations between the various LOCA frequency values, including advice on arithmetic or geometric preference for interpolation.

Response: This comment duplicates Comment 3-3. See the response to Comment 3-3 for guidance on interpolation schemes.

Comment Number: 7-6
Submitted by Zouhair Elawar – Palo Verde Nuclear Generating Station
Comment: The various LOCA frequency tables provided results for the past 25-year operating time period and for the 40-year average plant life. The more suitable value to select for use in PRA models may well be the expected LOCA frequencies applicable to the next 15 years. That particular selection allows for frequency penalties at aging plants as well as credits at plants with new steam generators and/or improved methods of inspections & leak detections. In order to extract the next 15-year LOCA frequencies from the existing tables, the user has to perform simple arithmetic calculations which may lead to human errors and inconsistent applications across the industry. Please provide frequency estimates for the next 15-year time period.

Response: As stated in the Executive Summary, the frequency estimates are not expected to change dramatically over the next fifteen years for any size LOCA, or even the next thirty-five years for LOCA Category 4 and smaller. Because of the predicted stability in these estimates over the near-term, it is recommended that the 25 year (i.e., current-day) results be used to estimate the average LOCA frequencies over the next 15 years of fleet operation. This last point was incorporated in both the Executive Summary and Section 7.4 of the revised NUREG.

Comment Number: 7-7
Submitted by Zouhair Elawar – Palo Verde Nuclear Generating Station
Comment: There is insufficient description of small LOCA frequency' "comparison results" relating to those in NUREG 5750 (which is most typically used in current PRA models). If one excludes the contribution of steam generator tube rupture frequency, the draft NUREG-1829 small LOCA value is ONE order of magnitude higher. Please add justifications for that large difference.

Response: This comment is very similar to Comment GC3. See the response to Comment GC3 for the comparison of NUREG-1829 with both NUREG/CR-5750 and operating experience estimates. More detailed information on these comparisons is available in Sections 7.9 and 7.10.

Comment Number: 7-8
Submitted by Zouhair Elawar – Palo Verde Nuclear Generating Station
Comment: Provide a section on statistical validation of small LOCA frequency. By using the method of Jeffrey's non-informative prior (over the past 1,500 reactor years with ZERO events excluding steam generator tube ruptures), the expected small LOCA frequency is at the 1E-04 level. This frequency is one order of magnitude lower than the frequency reported in the draft NUREG. Plant experiences of >1,500 reactor-years of operating history should be considered as valid predictor of small LOCAs. That

consideration is further strengthened by improved methods and increased requirements for in-service-inspections & leak detections.

Response: A new section has been added to the NUREG (Section 7.10) to compare SB LOCA frequency estimates from the elicitation with operating experience. As discussed in this section, the elicitation estimates for BWR and PWR small break LOCA frequencies are generally consistent with operational experience estimates. The BWR and total PWR elicitation mean frequency estimates are only 20 percent and 100 percent higher, respectively, than operational experience estimates. Further the PWR SGTR frequencies are virtually identical to operating experience predictions. The biggest difference between the elicitation results and operating experience occurs for SB LOCA estimates that are determined without SGTR contributions. The elicitation mean frequency estimate is approximately 5 times higher than the operating experience estimate which accounts for nearly all of the 100% increase in the total PWR SB LOCA frequencies indicated above. Although the five-fold increase in the elicitation non-SGTR SB LOCA frequencies is not inconsistent with operating experience (Section 7.10), this difference is physically supported by the panelists' qualitative and quantitative responses. The panelists indicated that medium and, to a lesser extent, small LOCAs in PWRs are most dramatically impacted by current PWSCC concerns (Section 6.3.2). This increase reflects this concern.

More detail is found in Section 7.10 of the revised NUREG. Related insights are also provided in the responses to Comments GC3, GC4, GC5, GC6, and GC7.

Comment Number: 7-9
Submitted by Zouhair Elawar – Palo Verde Nuclear Generating Station
Comment: Provide a section on probabilistic validation of the small LOCA frequency. Using a Poisson distribution with failure rate of 2.9E-03 (NUREG-1829 category 1 LOCA frequency excluding steam generator tube rupture events). And, approximately 1500 reactor-years, the probability of ZERO small LOCA events (actual industry performance) is ONLY 0.013 (i.e.1.3% chance). This result shows an excessive conservatism in the category 1 LOCA frequency. Are we to expect an average of ONE small LOCA event (with 74% chance) per 345 reactor-years (equivalent to ~4 calendar years)? I think not.

Response: This comment is similar to both GC7 and 7-8. See the responses to these comments for comparisons between the NUREG-1829 Category 1 LOCA estimates and operating experience. Section 7.10 has also been added to provide an in-depth comparison with operating experience.

Comment Number: 7-10
Submitted by Westinghouse Owners Group
Comment: It would be useful to the PRA community, and help facilitate plant to plant consistency if, for the smaller break sizes, more information or guidance were provided to help separate out the frequencies of SGTR from small break LOCA, and CRDM nozzle breaks from medium LOCAs, as well as any other contributors other than primary system piping. Although there is no current intention to use the results of the expert elicitation to update the various LOCA frequencies assumed in individual plant PRAs, as plants go forward with peer reviews of PRAs, it is likely that LOCA frequencies for small, medium and large LOCAs will be compared with NUREG-1829 results. The NUREG-1829 frequencies for the smaller LOCA sizes are several times higher than the values presented in NUREG-5750. The NUREG-1829 values listed for the small break frequency (> 100 gpm) are so high that one would expect to have seen an event in the US every three or four years, whereas to date we have seen none. It would be helpful if the conservatism in the estimates for the smaller beak sizes was discussed and some caveat provided so that plant PRAs don't end up with excessive conservatism in their small LOCA risk estimates.

Response: This comment is similar to Comment 7-2. See this response to this comment for a discussion about separating failure of individual components like steam generator tubes and CRDMs from the total

LOCA frequency estimates. Section 7.8 in the revised NUEREG provides separate SGTR estimates. Also, the responses to Comments GC3, GC4, GC5, GC7, 7-1, 7-3, 7-8, and 7-11 summarize comparisons between NUREG-1829 and either NUREG/CR-5750 or operating experience estimates. See also Sections 7.9 and 7.10 of the revised NUREG for more detailed comparisons.

Comment Number: 7-11
Submitted by Nuclear Energy Institute
Comment: The draft NUREG combined a variety of LOCA sources into each LOCA category. Piping LOCAs and several non-piping LOCAs were pooled together to form each of the LOCA categories. It would be useful for each of the 6 LOCA categories to add a table of LOCA sources and frequency contributions. This breakdown is particularly important for the small and medium LOCA categories. Some contributors to the small and medium LOCAs are modeled separately in most PRA models (SGTR, RCP seals, inter-system LOCAs and others). If the end user does not subtract the separately-modeled LOCA contributors, then the contribution to CDF (core damage frequency) from those contributors would be conservatively and redundantly modeled.

Response: This comment is similar to Comment 7-2. See the response to this comment for a discussion about separating failure of individual components like steam generator tubes and CRDMs from the total LOCA frequency estimates. Section 7.8 of the revised NUREG provides separate SGTR estimates.

Comment Number: 7-12
Submitted by Nuclear Energy Institute
Comment: The various LOCA frequencies are reported in the several tables as cumulative values. In order to isolate the frequency of each LOCA category, one has to subtract the frequency of the next higher ranking category. This reporting format may lead to human errors. Some users may not become aware of the cumulative table format since that description is briefly stated at the later sections of a very large report. Please add a footnote under each LOCA frequency table explain how to obtain the frequency of each LOCA category.

Response: This comment is identical to Comment 7-4. See the response to this comment for guidance on obtaining interval values from the cumulative results reported in NUREG-1829. This point has been clarified in the revised NUREG in the Executive Summary, Section 3.4.1, and Section 7.9.

Comments Related to Section 8 of NUREG-1829

Comment Number: 8-1
Submitted by Westinghouse Owners Group
Comment: In Chapter 8, Ongoing Work, it is noted that the LOCA elicitation results were for normal operating conditions only. The effects of Service Level D transients, of which seismic was found by NRC to be the most prominent, were not considered in the elicitation efforts. The reason seismic loading was not explicitly considered was that most of the expert panel did not believe that it would significantly change the LOCA frequencies for normal operation. Experience from probabilistic fracture mechanics calculations indicates that severe seismic loading, such as that from a design-basis safe shutdown earthquake, could increase the conditional probability of failure in flawed piping by one to two orders of magnitude. However, the probability of having the severe seismic loading during the worst time in life, such as the 40th or 60th year of operation, would be a maximum of 0.001 and would likely be much less. Thus, the maximum effect of this severe seismic loading would be to increase the LOCA frequency during normal operating conditions by 10 percent. This increase was deemed to be insignificant relative to the other uncertainties that were considered by the expert panel in the elicitation process for LOCA frequencies.

Response: The results from a separately-sponsored NRC-led seismic LOCA study (Reference: Chokshi, N.C., Shaukat, S.K., Hiser A.L., DeGrassi, G., Wilkowski, G., Olson, R., and Johnson, J.J., "Seismic Considerations For the Transition Break Size," NUREG-1903, U.S. Nuclear Regulatory Commission, February 2008) tend to support this comment. In this study, both unflawed and flawed piping analyses were conducted in order to ascertain the magnitude of any potential adjustments to the baseline TBS for the proposed rule change to 50.46a due to failures associated with seismic loading.

The principal findings from this study are that the critical flaws associated with the stresses induced by seismic events of 10^{-5} and 10^{-6} annual probability of exceedance are large. When considering the effects of mitigation strategies to preclude large flaws in service, the probabilities of pipe breaks larger than the TBS are likely to be less than 10^{-5} per year. Similarly, for the cases studied, the probabilities of indirect failures of large RCS piping systems are less than 10^{-5} per year.

These findings tend to support the contention of the commenter that seismic loading would not significantly change the LOCA frequencies under normal operation. As a result of this and related comments, the NRC report on *Seismic Considerations for the Transition Break Size* is now referenced in Section 2 of NUREG-1829. In addition, a summary of the seismic LOCA analysis and results is provided in the Executive Summary and in Section 7.2 of the report. Additionally, Section 2 clearly identifies that the elicitation LOCA frequency estimates do not consider rare event loading from seismic, severe water hammer, and other similar sources.

Comments Related to Section 9 of NUREG-1829

None

Comments Related to Appendix A of NUREG-1829

None

Comments Related to Appendix B of NUREG-1829

None

Comments Related to Appendix C of NUREG-1829

None

Comments Related to Appendix D of NUREG-1829

Comment Number: D-1
Submitted by Joseph Conen of the BWR Owners Group
Comment: Figure D.7 in Appendix D shows two through-wall IGSCC cases for 22 inch and 28 inch stainless steel pipe field history data. This reviewer is not aware of any through-wall IGSCC cracks in large diameter (>20-inch) BWR stainless steel pipes. A primary reason for this is the presence of mid-wall compressive weld residual stresses in such pipe that tend to retard deep cracks.

Response: Figure D.7 in the draft NUREG only shows *selected* IGSCC data points (only weld flaws for which detailed sizing data are available). In actuality there are have been other leaks in 22-inch and 28-

inch diameter recirculation lines in BWRs. According to the expanded OPDE database used as the basis of this query (currently 1,215 records on IGSCC), there have been 10 instances of circumferential through-wall cracking in large diameter (D=22 inch to 28 inch recirculation system piping, 8 of which were leaks. Three of the leaks were in 22-inch diameter piping: a cap-to-manifold leak at Hatch-1 (LER 82-089, November 1982) and two welds at Monticello (LER 82-013, October 1982). The other five leaks were associated with the 28-inch diameter recirculation line at Brunswick-1 (LER 85-026, July 1985): Weld 1B32-RR-28-A-4, Weld 1B32-RR-28-A-14, Weld 1B32-RR-28-B-4, Weld 1B32-RR-28-B-8, and Weld 1B32-RR-28-A-15.

Comments Related to Appendix E of NUREG-1829

None

Comments Related to Appendix F of NUREG-1829

None

Comments Related to Appendix G of NUREG-1829

None

Comments Related to Appendix H of NUREG-1829

None

Comments Related to Appendix I of NUREG-1829

None

Comments Related to Appendix J of NUREG-1829

None

Comments Related to Appendix K of NUREG-1829

None

Comments Related to Appendix L of NUREG-1829

None

NRC FORM 335 (9-2004) NRCMD 3.7	U.S. NUCLEAR REGULATORY COMMISSION	1. REPORT NUMBER (Assigned by NRC, Add Vol., Supp., Rev., and Addendum Numbers, if any.)
	BIBLIOGRAPHIC DATA SHEET *(See instructions on the reverse)*	NUREG-1829, Vol. 2

2. TITLE AND SUBTITLE		3. DATE REPORT PUBLISHED	
Estimating Loss-of-Coolant Accident (LOCA) Frequencies Through the Elicitation Process Appendices A through M		MONTH	YEAR
		April	2008
		4. FIN OR GRANT NUMBER	
		N6360	

5. AUTHOR(S)	6. TYPE OF REPORT
Robert Tregoning and Lee Abramson, US NRC Paul Scott, Battelle	Technical
	7. PERIOD COVERED *(Inclusive Dates)*
	N/A

8. PERFORMING ORGANIZATION - NAME AND ADDRESS *(If NRC, provide Division, Office or Region, U.S. Nuclear Regulatory Commission, and mailing address; if contractor, provide name and mailing address.)*

Division of Engineering
Office of Regulatory Research
U.S. Nuclear Regulatory Commission
Washington, DC 20555-0001

Battelle-Columbus
505 King Avenue
Columbus, OH 43201

9. SPONSORING ORGANIZATION - NAME AND ADDRESS *(If NRC, type "Same as above"; if contractor, provide NRC Division, Office or Region, U.S. Nuclear Regulatory Commission, and mailing address.)*

Same as above

10. SUPPLEMENTARY NOTES
A. Csontos, NRC Project Manager

11. ABSTRACT *(200 words or less)*

The NRC is developing a risk-informed revision of the design-basis pipe break size requirements in 10 CFR 50.46, Appendix K to Part 50, and GDC 35 which requires estimates of loss-of-coolant-accident (LOCA) frequencies as a function of break size. Separate BWR and PWR piping and non-piping passive system LOCA frequency estimates were developed as a function of effective break size and operating time through the end of license extension. The estimates were based on an expert elicitation process which consolidated service history data and insights from probabilistic fracture mechanics studies with knowledge of plant design, operation, and material performance.

The elicitation required each member of an expert panel to qualitatively and quantitatively assess important LOCA contributing factors and quantify their uncertainty. The quantitative responses were combined to develop BWR and PWR total LOCA frequency estimates for each contributing panelist. The individual estimates were then aggregated to obtain group estimates, along with measures of panel diversity. Sensitivity studies were conducted to examine the effects of distribution shape, correlation structure, panelist overconfidence, measures of panel diversity, and aggregation method. The group estimates are most sensitive to the method used to aggregate the individual estimates.

12. KEY WORDS/DESCRIPTORS *(List words or phrases that will assist researchers in locating the report.)*	13. AVAILABILITY STATEMENT
piping risk-informed emergency core cooling system (ECCS) loss-of-coolant accident (LOCA) break frequencies design-basis break size LOCA frequency estimates expert elicitation aging	unlimited
	14. SECURITY CLASSIFICATION
	(This Page) unclassified
	(This Report) unclassified
	15. NUMBER OF PAGES
	16. PRICE

PRINTED ON RECYCLED PAPER